Studies in Mechanobiology, Tissue Engineering and Biomaterials 1

Series Editor: Amit Gefen

Studies in Mechanobiology, Tissue Engineering and Biomaterials

Series Editor

Prof. Amit Gefen
Tel Aviv University
Fac. Engineering
Dept. Biomedical Engineering
69978 Ramat Aviv
Israel
E-mail: gefen@eng.tau.ac.il

Further volumes of this series can be found on our homepage: springer.com

A. Gefen (Ed.)
Bioengineering Research of Chronic Wounds, 2009
ISBN 978-3-642-00533-6

Amit Gefen (Ed.)

Bioengineering Research of Chronic Wounds

A Multidisciplinary Study Approach

Springer

Editor

Prof. Amit Gefen
Tel Aviv University
Fac. Engineering
Dept. Biomedical Engineering
69978 Ramat Aviv
Israel
E-mail: gefen@eng.tau.ac.il

ISBN 978-3-642-00533-6 e-ISBN 978-3-642-00534-3

DOI 10.1007/978-3-642-00534-3 ISSN 1868-2006

Studies in Mechanobiology, Tissue Engineering and Biomaterials

Library of Congress Control Number: 2009934348

Typeset & Cover Design: Scientific Publishing Services Pvt. Ltd., Chennai, India.

Printed in acid-free paper

5 4 3 2 1 0

springer.com

Preface

I am delighted to launch the Book Series *Studies in Mechanobiology, Tissue Engineering and Biomaterials* with this first thematic volume on *Bioengineering Research of Chronic Wounds*. Chronic wounds such as pressure ulcers, diabetic foot ulcers and venous leg ulcers are one of the most significant health problems nowadays, as highlighted in many of the chapters in this book. Taken together, these wound types has a socioeconomic impact which is comparable to that of cardiovascular diseases, osteoporosis and cancer.

Populations at risk are multitude, and include for example elderly with neuromuscular impairments, cognitive disorders and peripheral vascular diseases. However, chronic wounds are not a problem confined to elderly and to patients at the end of their life, as young individuals, with spinal cord injury or lesions (e.g. multiple sclerosis), or diabetic patients are also susceptible. With an ever-growing elderly population (bedfast or chairfast elderly in particular) on the one hand, and the epidemic nature of diabetes in Western countries on the other hand, the impact of chronic wounds on human society is expected to escalate quickly over the next few years.

Unfortunately, the attention that chronic wounds received from the bioengineering community over the last 20 years or so was negligible compared to the attention focused on other major health problems such as cardiovascular, respiratory and orthopaedic disorders. Accordingly, in terms of quantity, bioengineering research in chronic wounds is very much lagging behind. In terms of quality however, research in this field has started to bloom for the last few years, adopting state-of-the-art approaches from the more mature fields, including e.g. sophisticated computer modeling, cellular and tissue engineering for basic research as well as therapy, biomolecular markers and proteomics for early diagnosis, biomaterials for tissue repair and more. This book has captured this flourishing research work, and is reflecting it in full. The 19 chapters in this book, written by well-known experts in the various fields of bioengineering research of chronic wounds who conduct their work in 7 different courtiers, present the frontier of knowledge in the field, and are an excellent guide to the kind of research that should lead us forward to understanding why and how these wounds happen, how they can be treated, and even better, how they can be avoided.

Amit Gefen, Ph.D.
Editor, *Bioengineering Research of Chronic Wounds*
Series Editor, *Studies in Mechanobiology, Tissue Engineering and Biomaterials*

Contents

VI Innovative Technologies and Devices in the Diagnosis and Treatment of Chronic Wounds

Fundamentals of Pressure, Shear and Friction and Their Effects on the Human Body at Supported Postures

R.H.M. Goossens

Delft University of Technology, Faculty of Industrial Design Engineering,
Landbergstraat 15, 2628 CE, Delft, The Netherlands

Abstract. The human body is supported throughout most of life (while we sleep, travel, work and in our spare time). It is estimated that when one is 72 years of age, 62 years have been spent in some form of body support, such as an office chair and a bed. (Goossens 1994).

Body-supporting surfaces are often related to complaints in the neck, back, buttocks, legs and soft tissue (Grieco 1986). Numerous solutions for cushioning the contact area between product and user have been and implemented during the last centuries, but only in the last decades systematic research has been conducted on the underlying mechanical principles and the results of mechanical load imposed on the human skin and muscular skeletal system.

The skin is the interface between body and body support. Within the skin, all kinds of mechanoreceptors continuously monitor the deformation and warn us of damage (pain). The stress-strain relationship shows that there can be a high degree of deformation of the skin (up to 50%) before yield (mechanical damage) occurs. Physiological response however starts at lower deformation of the skin and is more important when evaluating the effect of body support.

Friction turned out to be a significant component factor of mechanical load on the skin (Reichel 1958, Goossens 1994). Friction on the skin consists of three components: the deformation, ploughing and adhesion components. The adhesion component (μ_a) between skin and support surface is the most important component for body support interfaces and it can be influenced by the choice of material and texture.

In the human body, at places where the mechanical load can become too high on bone, muscle tissue or other tissue, bursae are found. Bursae are found in regions where muscles, tendons, or bones rub against other muscles, tendons, or bones. The bursae function in two ways, lubricating points of friction, and dissipating force by distributing it through a fluid medium. Normally, the bursae produce just enough synovial fluid to reduce friction by reducing the tissue bonding. From an engineer's point of view, the advantage this provides is that a bursa separates moving tissue, thereby eliminating tissue bonding (μ_a the adhesion component is relatively small).

Different studies on comfort and discomfort have shown that mechanical load (pressure as well as shear) influences discomfort. It was also concluded that equal pressure and low shear increase the amount of time that passes before discomfort is noticed. A test (Garcia Lechner and Goossens 2003) showed that, on a hard seat surface without sharp edges that produce high pressure, at least 30 minutes will pass before the comfort limit is crossed i.e. there is a feeling of discomfort. This would explain the fact that people do not value the effect of good cushioning (i.e. low shear and equal pressure) during a short test period.

A. Gefen (Ed.): Bioengineering Research of Chronic Wounds, SMTEB 1, pp. 1–30.
springerlink.com

1 Introduction

The human body is supported throughout most of life (while we sleep, travel, work and in our spare time). It is estimated that when one is 72 years of age, 62 years have been spent in some form of body support, such as an office chair and a bed (Goossens 1994).

Body-supporting surfaces are often related to complaints in the neck, back, buttocks, legs and soft tissue (Grieco 1986). Numerous solutions for cushioning the contact area between product and user have been made and implemented during the past centuries, but only in recent decades has systematic research been conducted on the underlying mechanical principles and the consequences of mechanical load imposed on the human skin and muscular skeletal system.

In an attempt to identify the factors of comfort and discomfort in sitting, Zhang (Zhang et al., 1996) concludes that, amongst other factors, an excessively high mechanical load was one of the factors causing discomfort. In some studies, this relationship between pressure and discomfort was demonstrated (Diebschlag and Hormann 1987; Grindley and Acres 1996; Ballard 1997; Buckle and Fernandes 1998). Most of the cushioning innovations, therefore, focus only on pressure reduction at the contact interface. However, since Reichel (1958) pointed out that shear stress is also a significant factor of the mechanical load and its effect on tissue, attention should also be given to this aspect.

The aim of this paper is to present the fundamentals of mechanical load (pressure, shear and friction) in relation to body support from an engineering point of view. To begin, terms and definitions in this area of research are presented. Since all external forces on the human body enter through the skin, first a short introduction to the human skin and sensory system is presented. Then, the effect of mechanical load on the skin and deeper tissue is discussed in relation to human anatomy. Thereafter, the theory of body support principles and the research on comfort and discomfort follows.

2 Terms and Definitions

Throughout this document, terms are used that have their origin in both technical and medical fields. In order to make this paper readable to people from both disciplines, the most important terms are defined in this section.

2.1 Bursae

Encyclopædia Britannica (2003) defines a bursa as follows:

Bursa (plural BURSAS, OR BURSAE) *any small pouch or sac within the mammalian body between tendons, muscles, or skin and bony prominences, at points of friction or stress. The bursas are classified by type as adventitious, subcutaneous, or synovial.*

Adventitious, or accidental, bursas *arise in soft tissues as a result of repeated subjection to unusual shearing stresses, particularly over bony prominences.*

Subcutaneous bursas *ordinarily are ill-defined clefts at the junction of subcutaneous tissue and deep fasciae (sheets of fibrous tissue); these bursas acquire a distinct wall only when they become abnormal, and they are classified as adventitious by some authorities.*

Synovial bursas *are thin-walled sacs that are interposed between tissues such as tendons, muscles, and bones and are lined with synovial membrane, so called because it exudes synovia, a lubricating fluid. In the human body a majority of synovial bursas are located near the large joints of the arms and legs.*

2.1.1 Synovial Bursae

As an example, Figure 1 shows some synovial bursae found in and around the knee joint. At the juncture bones meet, and synovial fluids and membranes reduce the friction. In the body, at places where the mechanical load can become too high on bone, muscle tissue or other tissue, bursae are found. Bursae are found in regions where muscles, tendons or bones rub against other muscles, tendons or bones. The bursae function in two ways: lubricating points of friction and dissipating force by distributing it through a fluid medium. Normally, the bursae produce just enough synovial fluid to reduce friction by reducing the tissue bonding (See 'Friction and the skin'). This reduction of friction, for example, can also be experienced when a car hydroplanes due to a thin layer of water between the tire and the road. However, constant irritation may lead to over-secretion and consequent enlargement and inflammation of the bursa, a condition known as bursitis.

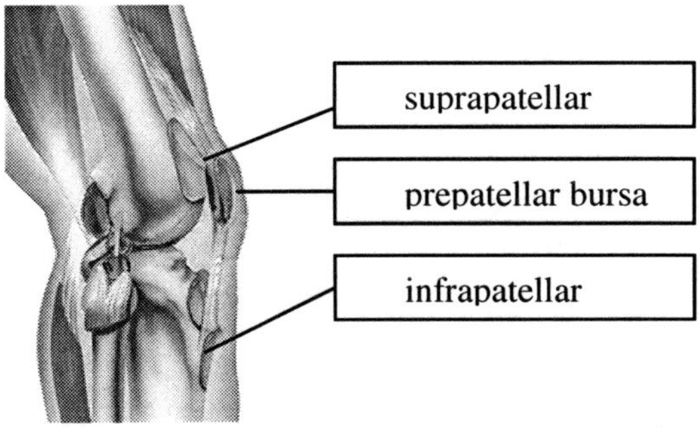

Fig. 1. The knee with some bursae. The suprapatellar bursa is located between the deep surface of the quadriceps muscle and the distal part of the femur. The prepatellar bursa is located between the superficial surface of the patella and the skin. An infrapatellar bursa is located between the patellar ligament and the skin. Other bursae decrease friction at the attachment points of the various muscles (Jenkins, 1991).

2.2 Mechanical Load

Throughout this document, the effect of mechanical load on the skin and deeper tissue is studied. Mechanical load can be resolved in pressure and shear or friction. In the case of shear and friction, there may be some uncertainty about the definitions. So in this section these terms are defined.

2.2.1 Shear/Friction

Shear (Gere and Timoshenko, 1990)

In the view of mechanical engineers, a shear force (V) is a force that acts parallel or tangential to the surface. The average shear **stress** equals the force V divided by the area over which it acts, see figure. Shear stresses are customarily denoted by the Greek letter τ (tau).

The surface can be any surface, including a surface within the tissue. It can be shown that any load on the skin surface (pressure, shear, pressure and shear) always results in a shear stress in certain cross-sections in the skin.

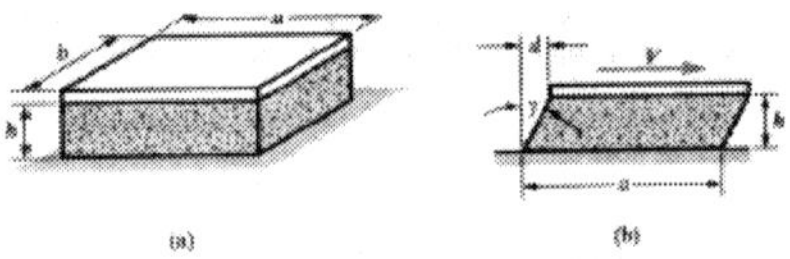

$$\tau = \frac{V}{a \cdot b}$$

Friction (Fishbane et al., 1996)

Static friction

Friction is a contact force that impedes sliding, and works parallel to the contact area. Suppose that you want to slide a crate from one place to another. You push on it with a small horizontal force, but nothing happens. This is because static friction is acting between the floor and the crate in the absence of motion in such a way as to prevent motion. This force must be variable because it balances each of your own different pushes.

Kinetic friction

Suppose that you finally get the crate moving. The force overcame the static friction because static friction has a maximum magnitude. Once the crate is moving, it is easier to keep it moving at a constant speed. There is still friction opposing your push, but it is now kinetic (or sliding) friction; that is friction associated with motion. The magnitude of kinetic friction is smaller than the maximum of static friction. The entire sequence of getting the crate started and keeping it moving can be seen in the figure below.

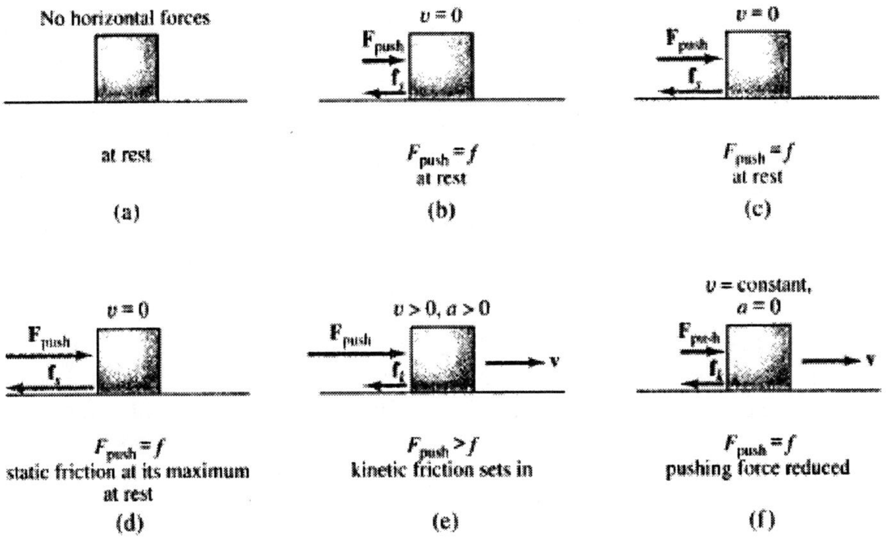

Static and kinetic friction. Source: Fishbane et al. 1996

Coefficient of friction

The proportionality constant that relates the friction force and the normal force is the coefficient of friction, μ. The (unit less) constant μ is determined experimentally. The maximum value of static friction is generally not equal to the force of kinetic friction, so we distinguish two coefficients: μ_s for static friction and μ_k for kinetic friction. If we write the force of static friction as fs and that of kinetic friction as fk, their magnitudes are given by

$$\text{Static friction: } 0 \leq fs \leq \mu_s F_N$$
$$\text{Kinetic friction: } fk = \mu_k F_N$$

However, it has been found that for soft surfaces like the skin the maximum static friction can be described roughly as proportional to the applied pressure (::F_N). Several experiment results (Yamaguchi 1990) indicate that the coefficient of friction is generally due to various combined effects of asperity deformation (μ_d), ploughing in the surface by wear particles (μ_p) and molecule adhesion between surfaces (μ_a). (See also ‘The skin’).

More accurately, it can be described as a non-linear function of the normal force fs = μ $(F_N)^q$. Mossel (1998), using that as a basis, came up with the following logarithmic model for friction between the forefinger and stainless steel.

$$0 \leq f_s \leq K \cdot c_p \cdot (E \cdot A_r)^{1-q} \cdot F_N{}^q$$

In which:

K	is a dimensionless factor
c_p	is a pressure distribution factor
E	is the modulus of elasticity of the skin (the amount of stress (force per unit area) required to produce a given amount of strain (stretching))
A_t	is the contact area
q	is a dimensionless exponent smaller than 1
F_N	is normal force

2.2.2 Pressure

Pressure is a force acting perpendicular to the surface, and is defined as the perpendicular force per unit area. In the 1950s, the first pressure measuring devices were developed. Today different systems are on the market to measure the interface pressure between the skin and the body support surface.

2.2.3 Summarized

Pressure is a force acting perpendicular to the surface. Shear and friction are caused by a force acting parallel to the surface. Shear and friction are, in one sense, overlapping definitions. Shear is used for the static situation (no movement) and can act on the contact surface (outside the tissue) but also on a cross-section inside the tissue. Friction is used for both static and dynamic situations and always acts on the contact surface (outside the tissue) (unpublished data: Goossens and Bain in ISO/DIS 16480-1 pre-Draft).

The definitions overlap in the static situation, when acting on the contact surface (outside the tissue); in that case friction and shear are the same force.

2.2.4 Comfort versus Discomfort

Helander and Zhang (1997) performed a study in which 143 emotions or feelings about office chairs were identified and then reduced with cluster analyses to a checklist with 14 items. They concluded that the use of a unidimensional scale – from extreme discomfort to extreme comfort – would be inappropriate, since it is possible that users gave simultaneous ratings of average comfort and average discomfort. The authors therefore presented a conceptual model for sitting comfort and discomfort (See ‘Comfort, discomfort and pain’) in which there is a link between comfort and discomfort. High values of comfort can be attained only if discomfort is low. They postulated an operational definition of comfort and discomfort, based on empirical evidence, and that will be used throughout the document.

Comfort

Comfort is based on aesthetics and the plushness of (chair) design and a sense of relaxation and relief (Helander 2003).

Discomfort
Discomfort is based on poor biomechanics and fatigue (Helander, 2003).

Pain
An unpleasant sensation occurring in varying degrees of severity as a consequence of injury, disease or emotional disorder (Webster, 2003).

Ischemia
Decreased flow of oxygenated blood to tissue.

Necrosis
The localized death of living cells associated with infection or the interruption of blood supply.

3 The Human Body

The entire human body is involved when the body is supported. A load on the body is generally transfered to the skeletal system via the skin and soft tissue. In this section a closer look is taken at the aspects of the human body that are of importance during body support.

3.1 The Skeletal System

A load on the body is generally transferred to the skeletal system. Soft tissue, like skin, underlying fat and muscle act thereby as a cushioning interface for the transfer of the load to the bones (Hobson 1988). The transmitted forces have a potential of damaging soft tissue. The critical factor is not simply the force, but the ratio of the force to the surface area over which it acts. At some parts of the human body, the bony prominences are covered by a relatively thin layer of tissue like the sacrum, hip, heel and ankle. At these locations, the so-called 'force intensity' or stress and deformation is high when using a body support surface. Therefore it is mainly at these bodily places where tissue damage and feelings of discomfort and pain caused by mechanical load can be found (Petersen, 1976).

3.2 The Skin

Because the skin is the contact area for all external forces that act on the body, the skin and the receptors in the skin for mechanical load are discussed here in more detail.

With a total surface area of 1.7 m^2 (2 square yards), the skin is the largest organ of the human body. The skin consists of three parts, namely the epidermis, the dermis, and subcutaneous tissue (Figure 2).

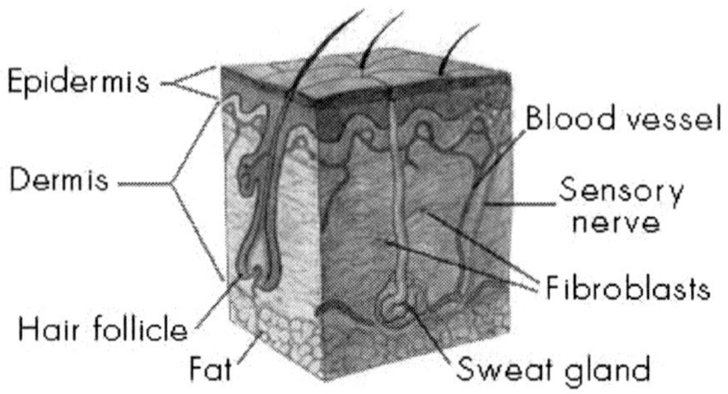

Fig. 2. Different layers of the skin. The epidermis, dermis and subcutaneous tissue.

The epidermis is the outer layer of the skin and contains living cells in the deeper parts, and dead cells on the surface. Underneath the epidermis, in the dermis, receptor cells can be found for touch, pain, pressure and temperature (Figure 3). The receptor cells have different functions:

- Mechanoreceptors: for touch, vibration, pressure (these are divided in fast and slow adapting receptors);
- Thermo receptors: for temperature;
- Nociceptors: for pain.

The receptors detect specific stimuli and convert them into electrical signals that are conducted through the nervous system to the brain. The mechanoreceptors are sensitive to deformations induced by mechanical load (shear and pressure). The mechanoreceptors measure deformation because an important task of the skin is to resist mechanical trauma. Although the skin is well adapted to many types of

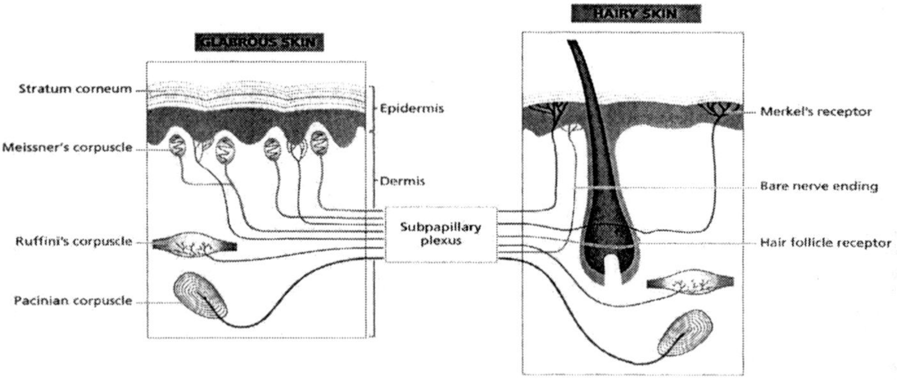

Fig. 3. Receptors for hairy and glabrous skin. [Source: Textbook of Dermatology, 1998, Rook, Wilkinson, Ebling].

trauma, excessive mechanical load (shear and pressure) can result in the formation of various dermatoses (blisters) (Kanerva 1990) and in pressure ulcers. Blisters occur in this deeper part of the epidermis and between the epidermis and the dermis. Pressure ulcers occur at the surface of the skin as well as underneath the skin in muscles and bone.

3.3 Mechanical Properties of the Skin

The ability of skin to undergo extreme elongation results in its being deformed around objects. This deformation increases the area of contact, decreases stress, increases friction and, at low force levels, decreases trauma (Armstrong, 1985).

3.3.1 Stress-Strain Relationship

The stress-strain relationship of the skin, can be divided into three distinctive phases that correspond to straightening and alignment, extension and finally fracture of the fibrous components of the skin (Bader and Bowker, 1983).

Many investigators have made stress-strain measurements of the skin to calculate the modus of elasticity; the amount of stress (force per unit area) required to produce a given amount of strain (stretching).

Kenedi (Kenedi, Gibson et al., 1975) applied variable force to excised tissue and measured the strain. The resulting stress-strain curve shows the skin to be non-linear with a large initial deformation occurring at the application of force. They also examined the time required by skin to recover from a periodically maintained stress and found that the skin will not return to its original condition unless it experiences the stress and strains to which the skin normally is subjected. In other words, skin remains deformed when put under higher stress than it experiences in normal situations.

Kenedi et al. (1967) reported that the modulus of elasticity increases from near zero at strains of less than 20% to approximately 138 MPa (mega Pascal) at strains between 30 % and 50 % (Figure 4). Also Ridge and Write (1967) reported such a curve linear force-displacement relationship for skin under tension. They showed that initial stretching of the skin is associated with straightening of the collagen bundles, which act like weak springs. The second stage of strain was associated with the stretching of the uncoiled fibers, which act like stiff springs. The final stage corresponds to an ultimate elongation, before yielding, of 30%-50%.

The amount of stress needed to produce mechanical yielding is called tensile stress. Tensile strengths are based strictly on mechanical responses, not on biological responses. Physiological changes such as occlusion of blood flow occur at stress levels much less than those required to produce mechanical yielding. Kenedi et al. (1967) define the biological usefulness of skin as that amount of strain which produces blanching or occlusion of the cutaneous blood flow. Blanching limits of skin based on in situ measurements at various locations of the forearm can be seen in Figure 5. Therefore, physiological responses to stress may be more important in the design of body supports than are passive mechanical responses (Armstrong 1985).

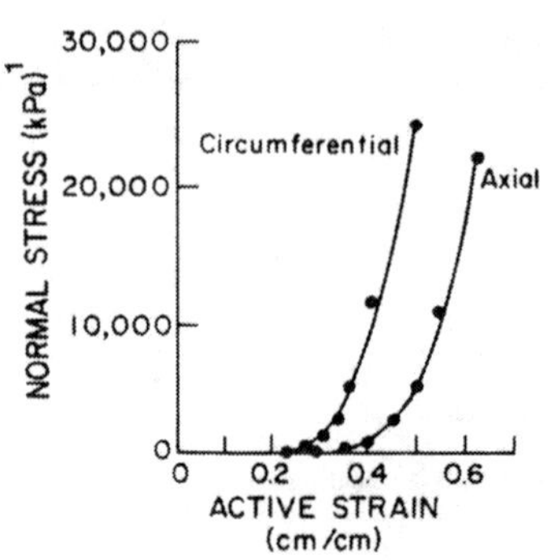

Fig. 4. Elasticity of the skin [Source: Kenedi et al., 1967]

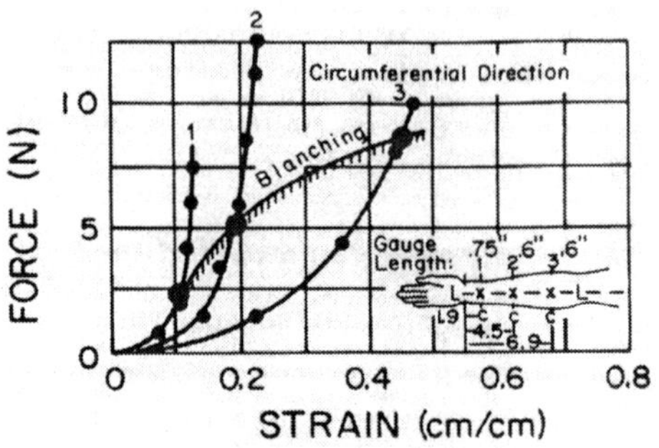

Fig. 5. Blanching of the skin [Source: Kenedi et al., 1967]

3.3.2 Friction and the Skin

Skin and polymers appear to behave similarly with respect to friction and do not follow the basic laws of friction that exist for solid materials (Bobjer et al., 1993). Several experimental results (Yamaguchi 1990) indicate that the coefficient of friction is generally due to various combined effects of asperity deformation (i.e. deformation caused by microscopic irregularities on the surface) (μ_d), ploughing in the surface by wear particles (μ_p) and molecule adhesion between surfaces (μ_a). The surface of the skin contributes to these different factors according to the following mechanisms (Bobjer et al., 1993):

- μ_d The friction due to asperity deformation. The deformation of the dermal ridges of the skin contributes to this factor during static and sliding conditions.
- μ_p The ploughing component is low when the soft surface of the skin slides against a hard surface.
- μ_a The adhesion component depends on lubricants at the interface. Oil and lard have a reducing effect on this component of friction, there is also an unknown effect of proteins and an increasing effect of sweat on textured surfaces.

3.3.3 Techniques to Measure Properties of the Skin

Diridollou et al. (Diridollou et al., 1998) provide an overview of several techniques that have been used to study in vivo the mechanical properties of human skin, with those most frequently encountered in the literature being based on:

- Tonometric (pressure) measurements that evaluate the ability of the skin to withstand vertical forces of extension (pressure);
- Traction, which applies linear displacement in the horizontal displacement of the skin;

- Indentation, which uses the application of a negative trust through a disc or point glued to the skin;
- Torsion, which applies a torque to the skin;
- Suction, which applies a negative pressure to the skin;
- Durometer ratings of the skin. The durometer rating of the skin can be addressed when dealing with shear. Materials with a low durometer tend to bond with solid surfaces of contact. This bond must be broken before movement can occur. The skin will tear if the force required to break the bond is greater than the tensile strength of our skin at the points of contact. Excessive wear will occur if the bond is at or near 50% of the tensile strength of our skin. An example of this condition occurs when one places the tongue against a cold metal object.

3.4 Conclusion

The skin is the interface between body and body support. Within the skin all kinds of mechanoreceptors continuously monitor the deformation and warn for damage (pain).

The stress strain relationship shows that there can be large deformations of the skin (up to 50%) before yield (mechanical damage) occurs. Physiological response, however, starts at lower levels of deformation of the skin and is more important when evaluating the effect of body support.

Friction of the skin consists of three components: the deformation, ploughing and adhesion components. The adhesion component between skin and support surface is the most important component for body support interfaces and can be influenced by the choice of material and texture.

4 The Effects of Mechanical Load on the Body

The prior sections identified friction, shear and pressure as the elements of mechanical load. The properties of the skin were also discussed. This section addresses the effect of mechanical load on the body in relation to the body's contact points and the anatomy of the contact points. Different reactions of the body can be seen, varying from severe damage, such as pressure ulcers and blisters, to discomfort and pain.

4.1 Contact with the Body

4.1.1 Deformation at the Bony Prominences

Figure 6 shows a model of a seated subject in contact with a body support surface (Staarink, 1995). It can be seen that the load L of the body is transmitted to the support surface as support forces F. In his study, Staarink implemented 14 pressure sensors (dots) in a dummy to record the pressure at different places underneath the skin. He showed that the highest pressures were measured at the bony prominences, in this example 22.5 kPa at the ischial tuberosities, and that

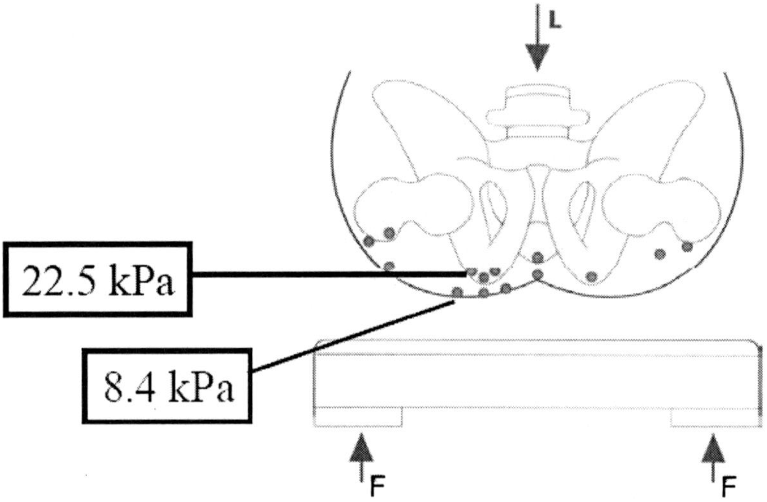

Fig. 6. Pressure sensors in a dummy to record pressure underneath the skin [Source: Staarink 1995)

much lower pressures were measured at the contact surface with the cushion (8.5 kPa) in the example. These high differences in pressure cause high levels of deformation at the tuberosities. High levels of deformation of the skin will trigger the mechanoreceptors (and in some cases the nociceptors) in the dermis of the skin and disturb blood circulation. Because soft tissue is compressed most near the bony prominences, these areas of the body receive special attention when evaluating body support surfaces.

4.1.2 Different Layers at the Interface

Figure 7 again shows a subject seated on a cushioning material, but now in greater detail, showing the different layers that can be found between skin and the outside world.

A shear force transverses all of these layers and it is the coefficient of friction (μ) between each layer (in this case from μ_1 to μ_6) that defines the maximum shear force that can be guided through. Because the shear force has to go through all the layers, it is the lowest coefficient of friction that determines the maximum shear force that can act on the skin ($fs = \mu\ (F_N)^q$). The pressure (equivalent to F_N) is the same through all surfaces. The other defining factor in the maximum shear stress is the distribution of pressure over the surface. This is because highest shear stress occurs at regions of highest pressure, which are the bony prominences (Goossens 1997).

In daily life, situations with the lowest friction coefficients between different layers can be recognized easily. For example, when sitting on a couch with a cushion that is not fixed to the seat surface (the frame). In this case, μ_5 determines

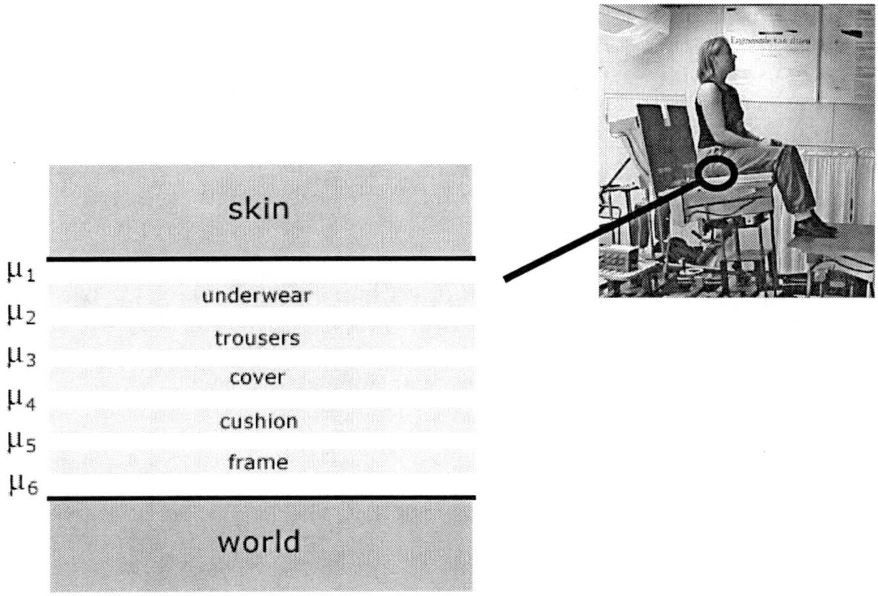

Fig. 7. Different layers at the interface. From skin to world contains 6 surfaces that make contact. Each two surfaces with their own coefficient of friction.

the maximum allowable shear force for static equilibrium and when the shear force becomes too high, the sitter and the seat cushion slide out of the couch.

Another example is when sitting in a hospital bed wearing a pair of pajamas. In this case, μ_3 (between trousers and sheet) determines the maximum allowable shear force, and the subject slides over the sheet out of the bed when his position requires too high a shear force for equilibrium (Goossens 1994).

4.2 Reactions of the Body to Mechanical Load

Mechanical contact forces on the body lead to a reaction of the body that can be categorized from severe injuries to discomfort. In the following paragraphs these reactions are discussed.

4.2.1 Pressure Ulcers

A pressure ulcer is localized injury to the skin and/or underlying tissue usually over a bony prominence, as a result of pressure, or pressure in combination with shear and/or friction. A number of contributing or confounding factors are also associated with pressure ulcers; the significance of these factors is yet to be elucidated. [National Pressure Ulcer Advisory Panel, 2007]. Pressure ulcers are classified in the following grades, based on the severity of the lesions. [citation from www.npuap.org website].

Pressure Ulcer Stages

Suspected Deep Tissue Injury
Purple or maroon localized area of discolored intact skin or blood-filled blister due to damage of underlying soft tissue from pressure and/or shear. The area may be preceded by tissue that is painful, firm, mushy, boggy, warmer or cooler as compared to adjacent tissue.

Further description
Deep tissue injury may be difficult to detect in individuals with dark skin tones. Evolution may include a thin blister over a dark wound bed. The wound may further evolve and become covered by thin eschar. Evolution may be rapid exposing additional layers of tissue even with optimal treatment.

Stage I
Intact skin with non-blanchable redness of a localized area usually over a bony prominence. Darkly pigmented skin may not have visible blanching; its color may differ from the surrounding area.

Further description
The area may be painful, firm, soft, warmer or cooler as compared to adjacent tissue. Stage I may be difficult to detect in individuals with dark skin tones. May indicate "at risk" persons (a heralding sign of risk).

Stage II
Partial thickness loss of dermis presenting as a shallow open ulcer with a red pink wound bed, without slough. May also present as an intact or open/ruptured serum-filled blister.

Further description
Presents as a shiny or dry shallow ulcer without slough or bruising.* This stage should not be used to describe skin tears, tape burns, perineal dermatitis, maceration or excoriation.

*Bruising indicates suspected deep tissue injury.

Stage III
Full thickness tissue loss. Subcutaneous fat may be visible but bone, tendon or muscle are not exposed. Slough may be present but does not obscure the depth of tissue loss. May include undermining and tunneling.

Further description
The depth of a stage III pressure ulcer varies by anatomical location. The bridge of the nose, ear, occiput and malleolus do not have subcutaneous tissue and stage III ulcers can be shallow. In contrast, areas of significant adiposity can develop extremely deep stage III pressure ulcers. Bone/tendon is not visible or directly palpable.

Stage IV
Full thickness tissue loss with exposed bone, tendon or muscle. Slough or eschar may be present on some parts of the wound bed. Often include undermining and tunneling.

Further description
The depth of a stage IV pressure ulcer varies by anatomical location. The bridge of the nose, ear, occiput and malleolus do not have subcutaneous tissue and these ulcers can be shallow. Stage IV ulcers can extend into muscle and/or supporting structures (e.g., fascia, tendon or joint capsule) making osteomyelitis possible. Exposed bone/tendon is visible or directly palpable.

Unstageable
Full thickness tissue loss in which the base of the ulcer is covered by slough (yellow, tan, gray, green or brown) and/or eschar (tan, brown or black) in the wound bed.

Further description
Until enough slough and/or eschar is removed to expose the base of the wound, the true depth, and therefore stage, cannot be determined. Stable (dry, adherent, intact without erythema or fluctuance) eschar on the heels serves as "the body's natural (biological) cover" and should not be removed.

Most pressure ulcers develop near bony prominences such as the sacrum, ischium, hip, heel and ankle (Figure 8). At these spots, the skin and underlying tissue are relatively thin, and the load distribution under the skin is restricted to a relatively small area. This results in high load concentrations and relatively large deformations (because of concentrated pressure and shear) at these bony prominences when the weight of the body is supported by the body support surfaces.

Pressure Ulcer Factors

Pressure ulcers are caused by factors that are classified generally as intrinsic and extrinsic. The intrinsic factors are related to the patient's clinical condition and both the nature of the illness and its severity are relevant. The extrinsic factors, which can be influenced directly, are concerned with pressure, shear, temperature and humidity. All authors in scientific literature agree that the leading cause of pressure ulcers is the mechanical load (pressure and shear) on the skin.

Although most authors agree that pressure ulcers are due to prolonged tissue ischemia caused by the mechanical load, through which the capillaries are closed and diffusion of oxygen and metabolites to the cells is also hindered, other extra mechanisms are reported in literature.

Reddy et al. (1981) studied the effects of external pressure on interstitial fluid dynamics using a simple mathematical model, concluding that the squeezing of interstitial fluid may also play a role in ulcer formation. Meijer (1991) states that it is most likely that local blood circulation under influence of the mechanical load is also controlled by regulatory mechanisms, which partly can be nervous.

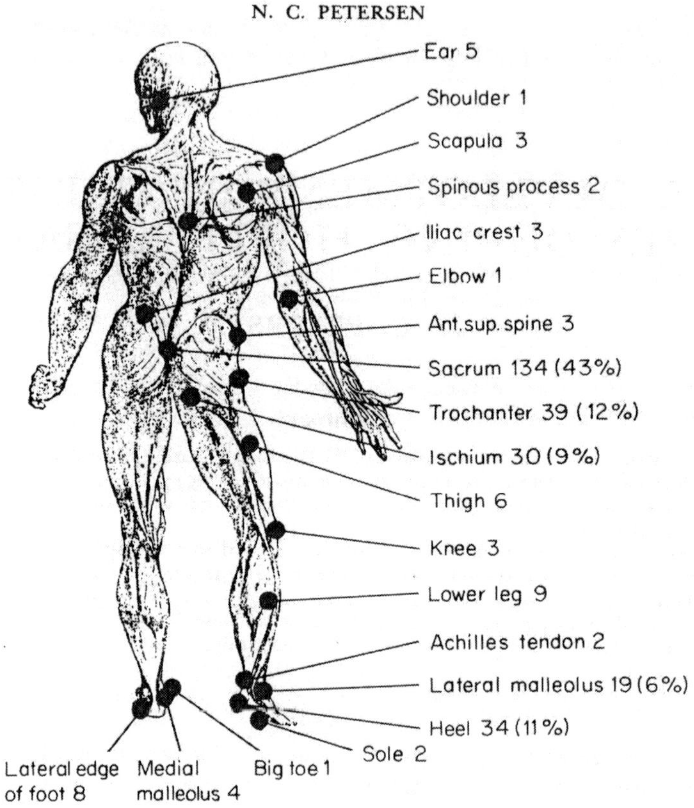

Fig. 8. Most pressure ulcers can be found near the bony prominences [Source: Petersen 1976]

When studying the mechanical properties of the skin, it was found that forces acting on the body surface will lead to stresses and strain (deformation) in the underlying tissue, and in a final stage to wounds known as decubitus or pressure ulcers. Various studies have been reported in literature that focused on the effect of exerted force and effect on the skin.

Kosiak (Kosiak 1959) applied different levels of pressure for different periods of time on dogs' hind limbs. He found that ischemic ulcers in dogs were produced by both high pressure applied for short periods and low pressure applied for long periods. This was also found in humans by Reswick and Rogers (1976) (see Figure 12).

More recent insights in the cause of (deep) pressure ulcers differentiate between the individual roles of deformation and ischemia in the onset of skeletal muscle damage (Stekelenburg et al., 2008). Based on a review of the results of as well in vitro as in vivo studies the authors propose a sequence of events leading to cell necrosis in which both deformation and ischemia can play a role (Stekelburg et al., 2008).

Various other studies confirm that the role of deformation is substantial to the problem of pressure ulcers. For example, an in vitro study on engineered muscle tissue by Gawlitta et al. (2006) showed that hypoxia did not lead to significant cell death over a 22 h period. By contrast, compression led to immediate cell death that increased with time. Another study confirmed the relation of the levels of compressive strain and the necrotic cell death of bio-artificial muscles (Gefen et al., 2008).

The role of tissue deformation on tissue injury was also confirmed in an animal study by Ceelen et al. (2008), in which the authors concluded that the relationship between maximum shear strain and damage seems to reflect an intrinsic muscle property, as it applied across a considerable number of the experiments. In another study (Stekelenburg et al., 2006) magnetic resonance (MR) in an animal model was used to study early damage after compressive loading of 2 hours by means of an indenter. The authors conclude that the MR images showed localized areas with increased signal intensity immediately after unloading, Histological examination at 1 and 4 hour after unloading showed large necrotic regions with complete disorganization of the internal structure of the muscle fibers.

Shear

It was Reichel (1958) who started to focus attention on shear force as an important component of the mechanical load and its effect on tissue. Since the publication of his article, various authors have confirmed the importance of shear stress as a factor in pressure ulcers.

Dinsdale (Dinsdale 1974) studied the effect of repeated pressure with and without friction in normal and paraplegic swine. He found that in those animals that received both pressure and friction, ulceration occurred at lower pressure levels than in those animals that received only pressure.

In relation to this last study, cut-off pressure can be defined as the level of external pressure on the skin at which ischemia of the skin can be expected. Goossens (Goossens, Zegers et al., 1994) found that cut-off pressure is significantly lower when a combination of pressure and shear are applied to the skin, compared to a situation in which only pressure is applied.

4.2.2 Blisters

Blisters (Falcon-Braun, 1969; Stoughton 1964) result from the accumulation of fluid between the cells of the epidermis of the skin or between the cells of the epidermis and dermis of the skin. Some form of cell damage is necessary to cause the fluid to enter the spaces of the previously damaged sites (Moschella, Phillsbury et al., 1975). This blister resulting from the cell damage ends up with some benefit of protection. And although blisters are produced in a wide variety of ways (eczema, infection, burns), the type of blisters that this document focuses on are called friction blisters in literature, but ‘mechanically induced blisters’ would be a better term.

Mechanically Induced Blisters

In order to understand friction blister formation in greater detail, it is necessary to have some idea of the anatomy of the dermo-epidermal junction (the junction between the epidermis and the dermis), as revealed by electron microscopy (Burton 1990). The dermo-epidermal junction includes the basal cell layer, the basal lamina (which is electron dense) and the upper papillary dermis (Figure 9).

The narrow space that separates the membrane of the basal cell from the basal lamina is called the lamina lucida. This space contains a few anchoring filaments that cross from the basal cells to the basal lamina. Friction blisters are intra-epidermal blisters; the following mechanisms occur for these types of blisters (Figure 10):

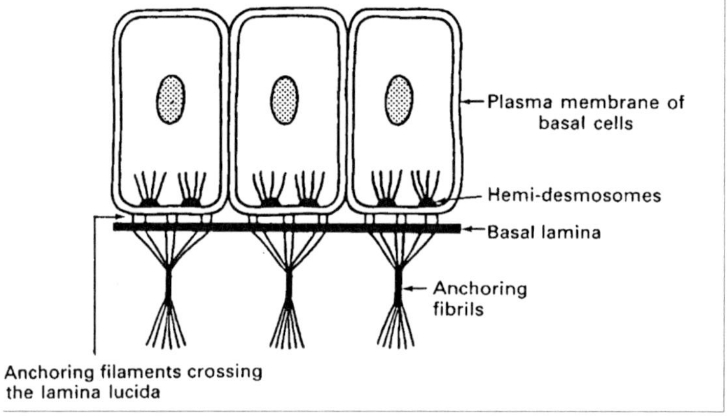

Fig. 9. The dermo-epidermal junction. [Source: Essentials of Dermatology, 1990, Burton].

Spongiosis	The cells become separated by accumulation of oedema fluid.
Epidermal cell necrosis	The cells become swollen and vacuolated to produce an appearance called 'balloon degeneration'
Damage to intercellular 'cement'	Cells that form intercellular cement lose their cohesion and drift apart

The factor responsible for friction blisters is reportedly a mechanical fatigue phenomenon rather than that induced by wear, heat, enzymes, pressure, stretching of ischemia (Comaish, 1973). These blisters are more easily produced in the old than in the young, in women than in men, and in hot than in cold skin (Paechy, 1971).

Shear

The shear force parallel to the skin surface is the leading factor in the formation of these blisters (Samitsz 1985 in Kanerva 1990). This force results in epidermal cell necrosis (cell death) and this causes an intra-epidermal slit which fills with fluid. The necrosis is due to repeated cellular distortion, which denatures the cellular protein (Moschella, Phillsbury et al., 1975).

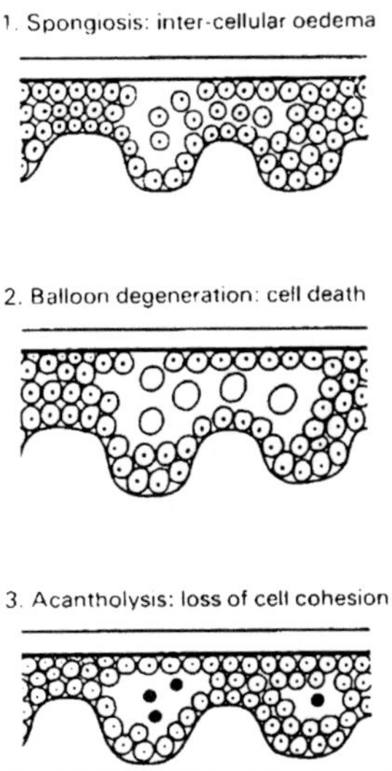

Fig. 10. Mechanisms for intra-epidermal blister formation [Source: Essentials of Dermatology, 1990, Burton]

Pressure
Pressure bullae, can be a sign of a variety of neurological diseases (Arndt et al., 1973). This phenomenon occurs in patients who have been immobilized for prolonged periods on pressure sites. The cutaneous lesions are due to pressure-induced ischemia (no blood flow through capillaries).

4.3 Comfort, Discomfort and Pain

In this paragraph the effect of shear and pressure on comfort and discomfort are discussed.

In a review of literature, Lueder (Lueder, 1983) gave a general overview of approaches to the assessment of the comfort-relevant design of office furniture. The author concluded that although substantial research exists, little insight is available into the meaning of comfort. More recently Zhang (Zhang et al., 1996) concluded that comfort and discomfort are two different and complementary entities in ergonomic investigations. In an attempt to identify the factors of comfort and discomfort in sitting, the authors conclude that amongst other factors, poor biomechanics (meaning an excessive mechanical load) was one of the factors that caused discomfort. In some studies, this relationship between pressure and discomfort was demonstrated (Diebschlag and Hormann 1987; Grindley and Acres 1996; Ballard 1997; Buckle and Fernandes 1998).

Comfort is an extensive concept that is mainly related to subjective influences. Therefore, comfort is not easy to quantify (Annett, 2002). Despite this fact, several authors, such as Shackel et al. (1969), Corlett and Bishop (1976), Helander and Zhang (1997) and Straker et al. (1997), developed 'comfort-scales' with which the amount of comfort and discomfort experienced can be assessed. However, it must be stressed that this assessment remains a subjective process. Therefore, several authors have tried to find objective parameters to which comfort could be related. The pressure distribution, i.e. the variation of pressure on the seat surface, was pointed out by several authors to correlate well with seat comfort (Kamijo et al., 1982, Yun et al., 1992, Thakurta et al., 1995, Park and Kim 1997, Milivojevich et al., 2000, Tewari and Prasad 2000, Ebe and Griffin 2001, Demontis and Giacoletto 2002).

Although comfort and discomfort are studied by various authors, discomfort is apparently not exactly the opposite of comfort (Helander and Zhang 1997).

Helander and Zhang (1997) performed a study in which 143 emotions or feelings about office chairs were identified, and reduced them with cluster analyses to a checklist with 14 items. They concluded that the use of a unidimensional scale from extreme discomfort to extreme comfort would be inappropriate, since it is possible that users gave simultaneous ratings of average comfort and average discomfort. The authors also presented a conceptual model for sitting comfort and discomfort (Figure 11) in which there is a link between comfort and discomfort. High values of comfort can be attained only if discomfort is low. They postulated an operational definition of comfort and discomfort, based on empirical evidence.

- Discomfort is based on poor biomechanics and fatigue.
- Comfort is based on aesthetics and the plushness of chair design and a sense of relaxation and relief.

Seat discomfort may be experienced when sitting over a long period of time, or when a wrong sitting position is adopted. When seat discomfort is experienced for a long time, circulatory deficit and even a preliminary stage of decubitus lesions may be developed. This deficiency, which is called ischemia, caused by an increase of tissue pressure, results in un-physiological changes at the capillary level (e.g. deformation of vessels, decrease of lumen and oedema). As a result, tissue destruction can occur. Resistance to these conditions of diminished blood supply varies from tissue to tissue. Nerve and muscle tissue is very prone to being harmed by the lack of blood supply, even for short periods of time. Whereas bones and tendons will still be vital after several hours of total blood occlusion.

Reswick and Rogers (1976) published their famous ‘pressure-time tolerance curve’ (PTTC) as shown in Figure 12. This curve served as a guideline for the time a person could spend with a certain amount of pressure before exceeding a limit that was taken as acceptable. After that limit, an unacceptable zone is entered in which there is a high risk of developing decubitus lesions. The variables that describe this limit are the amount of pressure and time.

In the etiology of decubitus injuries, three factors are of main importance: 1) the level of pressure exerted on the person, 2) the level of shear force and 3) the period during which pressure and shear force are exerted. There is a presumption that, as in the Reswick and Rogers curve, there might also be a ‘pressure-time tolerance curve’ (PTTC) for discomfort. In contrast to decubitus lesions, there is no such curve reported in literature. In an unpublished study by Garcia Lechner and Goossens (2003), such a curve was defined, with the aim of getting designers

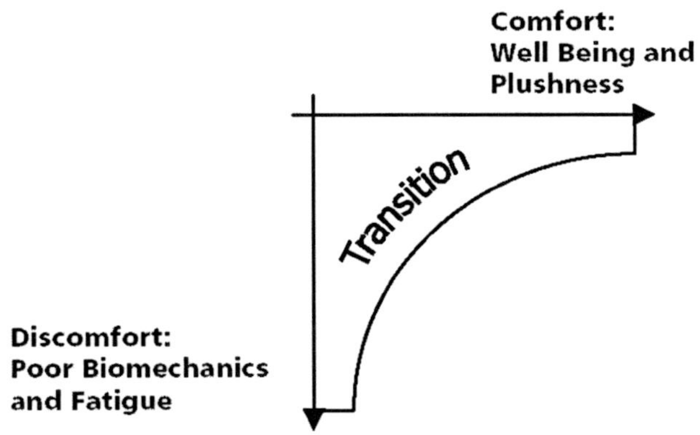

Fig. 11. A link between comfort and discomfort. High values of comfort can be attained only if discomfort is low [Source: Helander 2003].

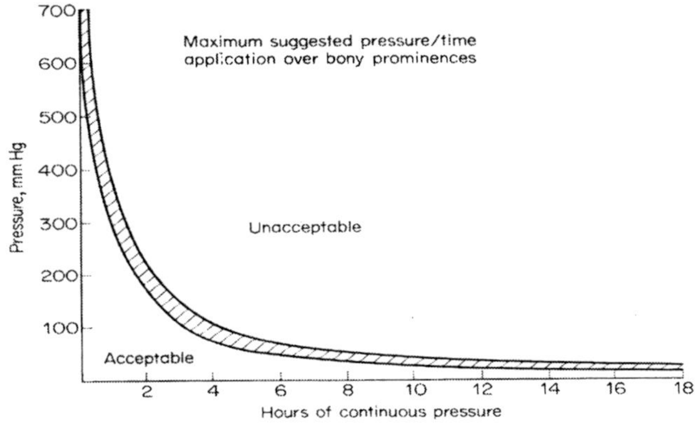

Fig. 12. Pressure time tolerance curve for pressure ulcers [Source: Reswick and Rogers 1976]

of chairs and other body supports to take the pressure-time tolerance for discomfort into account while designing these products. When a designer has to develop a chair for sitting during short periods, as for example in a subway, he can keep in mind that people will sit on that seat for no longer than 15-20 minutes. This would mean that users could be liable to higher pressures (deformations because of mechanical load) before the discomfort limit is exceeded. Conversely, if the designer has to develop a chair for longer sitting periods, as for example in a flight of a couple of hours, the pressure (better it is to say: the deformations because of mechanical load caused by pressure and shear) exerted should be kept lower so that comfort is experienced for a longer time. The results of the study of Goossens and Lechner show that on a hard as well as on a soft surface subjects found themselves comfortable for some time (about 30 minutes), but on the hard surface without sharp edges with high pressure the discomfort limit is reached earlier than on the soft surface with lower pressure.

In another study Goossens (Goossens 2000) showed that different combinations of pressure and shear (for example high shear and low pressure, and high pressure and low shear) when applied to the outside of the skin still have the same effect inside the skin. In this way it was demonstrated that not only pressure relates to discomfort but also shear stress. For both aspects of the mechanical load (pressure and shear) it can be concluded that a reduction leads to less discomfort.

4.3.1 Conclusion

It can be concluded that different studies have shown that there is an influence of mechanical load (pressure as well as shear) on discomfort. It can also be concluded that equal pressure and low shear increase the amount of time that passes before discomfort will be noticed. A test showed that on a hard seat surface without sharp edges that produces high pressure, at least 30 minutes will pass before the comfort limit is crossed, i.e. there was a feeling of discomfort. This would explain the fact that people do not value the effect of good cushioning (i.e. low shear and equal pressure) during a short test period.

4.4 Pain

Referred pain is a phenomenon that has been known almost since the practice of medicine began. This section is mainly focused on cutaneous pain, although trauma, disease process or vascular problems may also result in nervous signals.

Pain, pressure, warmth, cold and location can all be discerned by humans whose skin is appropriately stimulated. The exact nature and degree of the specificity became apparent with the discovery of isolated peaks of sensitivity to touch, warmth, cold, or pain over most of the body surface (Boring, 1942 in Boff et al., 1986). Pain was identified as a separate identity with the finding of a separate receptor group for the stimulus, the nociceptors. This group of nerve endings carries the message 'pain' to the brain. In studies it was shown that the nociceptors show responses to firm pressure, to heat at painful levels and to pricking (Torebjörk, 1979).

In addition to the variation in the physiological response of the nociceptors, many studies are conducted into the psychophysical response of humans to painful stimuli. Depending on the kind of stimulus employed to produce pain, the categories of response may range from 'something felt' through 'something strong and unpleasant but not painful' to something strong and clearly painful' (Boff et al., 1986). To determine the pain threshold, it was concluded that although one may know the stimulus level, the threshold for pain is a highly labile quantity. It therefore can be concluded that the psychophysical response to a range of stimuli of the nociceptors may lead to a range of responses varying from day-to-day and person-to-person.

4.4.1 Behavior

Although the psychophysical responses to the stimuli from the nociceptors lie in a large range, they are partly responsible for (sitting) behavior (Branton 1969). In naturalistic studies of sitting (Branton and Grayson 1967) it was observed that spontaneous behavior regularly produces a variety of postures with highly significant differences in frequency and duration. Branton (1969) concludes that the variety of postures that represented an 'urge to move' is caused by ischemia of the tissue. The author concludes that the high pressure on the skin and tissue under the ischial tuberosities creates bodily states which make change of position desirable, and thus is a necessary condition for the 'urge to move' to become manifest. This finding is also confirmed in a study of Bar (1991) in which he studied the pressure-time relationship on the ischial tuberosities using dynamic pressure measurements. Bar (1991) concluded after 2 hours of dynamic pressure measurement on the ischial tuberosities of each subject that postural changes or repositioning were characterized by significant changes in pressure lasting longer than a few minutes (macro-movements), and that variations in trunk position and limb movements lead to changes in pressure of a few seconds duration (micro-movements). He observed that a change of posture could occur, represented by a large shift in pressure distribution, followed by the presence of small pressure oscillations indicating further small trunk or limb movements.

Bhatnager et al. (1985) observed the postural changes of four seated individuals. They found that the frequency of postural changes or fidgeting increased by more than 50% over the three hours of observation. The total frequency of posture changing was an indicator of postural stress and discomfort. In a study of fourteen subjects, Pustinger et al. (1985) found fewer episodes of motion and fewer health complaints at an adjustable workplace compared to a fixed workplace. Increase in movement was also found to have an adverse effect on productivity. Mark et al. (1985) also found considerably less movement at properly adjusted computer workstations.

4.4.2 Movement and the Skin

Bader (1990) studied the recovery characteristics of soft tissue subjected to externally applied cyclic loads using transcutaneous gas tension measurements. The author found that the response of all subjects to cyclic loading at the sacrum and ischial tuberosities could be characterized under one of two distinctive forms, both having physiological implications. The first type of response, in terms of loading and recovery, was a normal physiological reaction, as typified by reactive hyperemia. The author states that this normal response may be a direct result of the mechanical stresses that can induce the release of biochemicals, which are vasodilators. The alternative response suggests an impaired control mechanism. The time permitted for tissue recovery was inadequate and would inevitably lead to diminished oxygen levels on repeated loading and, eventually, tissue ischemia.

Because most patients with spinal injury do not shift their weight (change posture) because a lack of sensation in the legs and buttocks, Cumming et al. (1986) designed a microprocessor-based device in order to reduce the risk of tissue breakdown for paraplegic patients. The device measures the time interval between weight shifts of the patient with sensors in the seat, and uses audible alarms to tell the patient when to perform a pressure relieving liftoff.

Bar (1991) found that relatively small changes in position (micro-movements) could produce large changes in pressure on the ischial tuberosities when sitting on foam cushions. He also found that on gel cushions the micro-motions could be absorbed to some extent by the ‘flow’ of the gel leading to lower changes in pressure. He also hypothesized that if the gel was allowed to move freely it would distribute pressure evenly throughout the supporting medium, and absorb micro-motions better.

4.5 Conclusions

The body reacts to mechanical load in different ways. As a result of prolonged pressure and shear, localized damage to the skin and underlying tissue can be found. The resultant wounds are known as pressure ulcers.

In some cases the more superficial epidermal layer is damaged and cells in the epidermis become separated, damaged or die (necroses) because of an external pressure or shear. When the damaged area fills with fluid, these wounds are called blisters. This blister resulting from the cell damage ends up with some protective benefits.

The natural response of the body to areas where the mechanical load on bone and muscle tissue becomes too high is the formation of a sac with a smooth membrane called a bursa. An early warning to high mechanical load on the body is the feeling of discomfort.

When the nociceptors are stimulated, pain will be felt, but the psychophysical response varies from day-to-day and person-to-person. However, all subjects will show an ‘urge to move’ when ischemia of tissue occurs.

Redistribution of pressure and neutralization of shear leads to less discomfort and can only be obtained though the cushion material of the body support surfaces.

5 Body Support Surfaces

In order to optimize the mechanical load that acts on the skin during body support, different cushions have been developed. Especially for the prevention of pressure ulcers, many foam cushions, cushions incorporating gel or water, and even air fluidized cushioning can be found. All cushioning materials, except foam, need a cover to keep the medium (gas, liquid, gel) together. To obtain an understanding of the principle by which the different cushioning materials work, this paragraph discusses them from a biomechanic perspective.

When considering the biomechanical behavior of different body support surfaces, including consideration of the cushion and the cover (top layer), the following models can be seen (Holscher, Goossens et al., 1994). Figure 13.

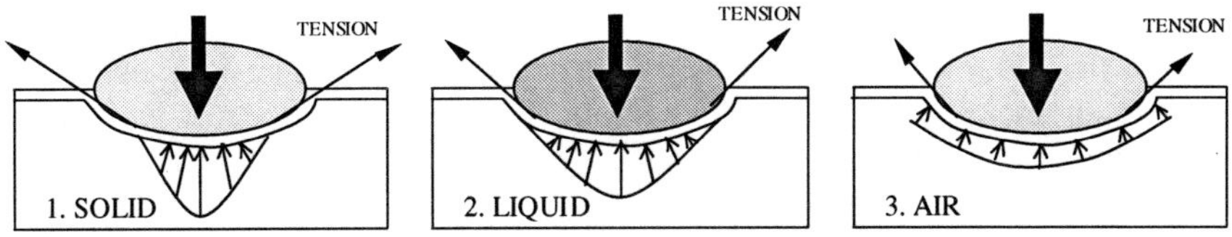

Fig. 13. Biomechanical behavior of three different cushions, with a solid, liquid and air medium between body and cushion

The medium of the cushions may be divided into three groups.

5.1 Solid

This group contains solid cushions (foam, high-density gel) in gradations from normal polyurethane foams and latex, to special foams such as the so-called ‘memory foam’ and gel with high density. In order to establish a large contact area between body and cushion, the material must be soft. The force equilibrium between body and cushion is achieved by three factors:

a) The increase of the force with the increase of the imprint of the body (depth of the body) into the cushion [cushion behaves like a spring];
b) The enlargement of the contact area;
c) The tension in the cover.

Because of factor a) and c) there is no uniform pressure distribution over the cross-section but rather an un-equalized pressure build-up with the maximum in the middle.

And because of factor c), the tension in the cover, there is a shear force between body and cushion. In this situation the cover of the cushion will be under tension and helps carry the load. This is the so-called ‘hammock effect’ (Nicol and Min 1997), resulting in an extra pressure and shear force at the contact surface.

5.2 Liquid

This group contains fluid media such as water or gel that have to be separated from the body by some kind of cover. The purpose of the cushions is to keep the body ‘floating’ and therefore the density of the media must be greater than the density of the body (approximately 1000 kg/m3). In order to separate the body from the medium, an enveloping cover is used. The equilibrium between body and cushion is achieved by three factors:

a) The increase of the force with the increase of the imprint of the body (depth of the body) into the cushion;
b) The enlargement of the contact area;
c) The tension in the cover.

There is no uniform pressure distribution because of factors a) and c). Compared to situation 1 (solid), the influence of factor a) on pressure build-up is less, because the pressure build-up of liquids is smaller than of solid media. In this situation the cover that keeps the fluid together will be under tension, this is the so called 'hammock effect' (Nicol and Min 1997), resulting in extra pressure and shear force at the contact surface.

5.3 Air

This group contains all gaseous media (mostly air) that are put in a closed cushion system. Also this medium has to be separated from the body by some kind of cover. The equilibrium between the body and the cushion is now achieved by two factors:

a) The enlargement of the contact area;
b) The tension in the cover.

There would be a uniform pressure over the cross section of the medium, because there is no increase in force in gas with the increase of the imprint of the body. But because of the tension in the cover (factor b) there is no uniform pressure distribution on the body. Also in this situation, a shear force in introduced because of the tension in the cover.

5.4 Conclusion

The theory shows that the support principles can be divided in three main groups, namely solid, liquid and air.

6 Conclusions

The body reacts to mechanical load (pressure and shear) in different ways.

In the normal mechanical load situation, the mechanoreceptors in the skin are sensitive for deformations induced by mechanical load (shear and pressure). The mechanoreceptors measure deformation because an important task of the skin is to resist mechanical trauma. In cases when mechanical load is too high, different responses from the skin and underlying tissue can be seen. In some cases the more superficial epidermal layer is damaged -mostly by means of friction - and cells at the epidermis become separated, damaged or die (necroses) because of an external pressure or shear. When the damaged area fills with fluid, these wounds are called blisters. As a result of prolonged pressure and shear, localized damage to the skin and underlying tissue can be found and the resultant wounds are known as pressure ulcers. The natural response of the body to areas where the mechanical load on bone and muscle tissue becomes too high is the formation of a sac with a smooth membrane called a bursa.

Different studies on comfort and discomfort have shown that mechanical load (pressure as well as shear) influences discomfort. It was also concluded that equal pressure and low shear increase the amount of time that passes before discomfort is noticed. A test showed that, on a hard seat surface without sharp edges that produced high pressure, at least 30 minutes passes before the comfort limit is crossed, i.e. there is a feeling of discomfort. This would explain the fact that people do not value the effect of good cushioning (i.e. low shear and equal pressure) during a short test period.

References

Annett, J.: Subjective Rating Scales: science or art? Ergonomics 45, 966–987 (2002)

Armstrong, T.J.: Mechanical Considerations of Skin in Work. Am. J. Ind. Med. 8, 463–472 (1985)

Arndt, K.A., Mihm Jr., M.C., Parrish, J.A.: Bullae: A Cutaneous Sign of a Variety of Neurologic Diseases. J. Invest. Dermatol. 60(5), 312–320 (1973)

Ballard, K.: Pressure-relief Mattresses and Patient Comfort. Prof. Nurse 13(1), 27–32 (1997)

Bader, D.L., Bowker, P.: Mechanical Characteristics of Skin and Underlying Tissues in vivo. Biomaterials 4, 305–308 (1983)

Bader, D.L.: The Recovery Characteristics of Soft Tissue following Repeated Loading. J. of Rehab. Res. and Dev. 27, 141–150 (1990)

Bar, C.A.: Evaluation of Cushions using Dynamic Pressure Measurement. Prosthetics and Orthotics International 15, 232–240 (1991)

Bhatnager, V., Drury, C.G., Schiro, S.G.: Posture, Postural Discomfort, and Performance. Human Factors 27, 189–199 (1985)

Bennett, L., Kavner, D., et al.: Shear vs Pressure as Causative Factors in Skin Blood Flow Occlusion. Arch. Phys. Med. Rehabil. 60(7), 309–314 (1979)

Bennett, L., Kavner, D., et al.: Skin Blood Flow in Seated Geriatric Patients. Arch. Phys. Med. Rehabil. 62(8), 392–398 (1981)

Bennett, L., Kavner, D., et al.: Skin Stress and Blood Flow in seated Paraplegic Patients. Arch. Phys. Med. Rehabil. 65(4), 186–190 (1984)

Bobjer, O., Johansson, S., Piguet, S.: Friction between Hand and Handle. Effects of oil and lard on textured and non-textured surfaces; perception of discomfort. Applied Ergonomics 24(3), 190–202 (1993)

Boff, K.R., Kaufman, L., et al.: Handbook of Perception and Human Performance. John Wiley & Sons, Chichester (1932)

Branton, P.: Behaviour, Body Mechanics and Discomfort. Ergonomics 12, 316–327 (1969)

Branton, P., Grayson, G.: An Evaluation of Train Seats through Observation of Sitting Behaviour. Ergonomics 10, 35–51 (1967)

Buckle, P., Fernandes, A.: Mattress Evaluation–assessment of contact pressure, comfort and discomfort. Appl. Ergon. 29(1), 35–39 (1998)

Bursa. Encyclopædia Britannica. Encyclopædia Britannica Premium Service (retrieved, September 2003)

Burton, J.L.: Essentials of Dermatology. Edinburgh, Churchill Livingstone (1990)

Ceelen, K.K., Stekelenburg, A., Loerakker, S., Strijkers, G.J., Bader, D.L., Nicolay, K., Baaijens, F.P., Oomens, C.W.: Compression-induced damage and internal tissue strains are related. J. Biomech. 41(16), 3399–3404 (2008)

Corlett, E.N., Bishop, R.P.: A Technique for assessing Postural Discomfort. Ergonomics 19, 175–182 (1976)

Cumming, W.T., Tompkins, W.J., Jones, R.M., Margolis, S.A.: Microprocessor-based Weight Shift Monitors for Paraplegic Patients. Arch. Phys. Med. Rehabil. 67, 172–174 (1986)

Demontis, S., Giacoletto, M.: Prediction of Car Seat Comfort from Human-seat Interface Pressure Distribution. In: SAE Conference 2002, SAE no 2002-01-0781 (2002)

Diebschlag, W., Hormann, M.: Improving sitting comfort in wheelchairs for the prevention of pressure sores. Rehabilitation (Stuttg) 26(4), 153–183 (1987)

Diridollou, S., Berson, M., Vabre, V., et al.: An in-vivo Method for Measuring the Mechanical Properties of the Skin using Ultrasound. Ultrasound in Med. Biol. 24(2), 215–224 (1998)

Dinsdale, S.M.: Decubitus Ulcers: role of pressure and friction in causation. Arch. Phys. Med. Rehab. 55, 147–152 (1974)

Ebe, K., Griffin, M.J.: Factors Affecting Static Seat Cushion Comfort. Ergonomics 44, 901–921 (2001)

National Pressure Ulcer Advisory Panel, `http://www.npuap.org`

Fishbane, P.M., Gasiorowicz, S., Thornton, S.T.: Physics for Scientists and Engineers, 2nd Extended edn. Pretince Hall International, Inc., Englewood Cliffs (1996)

Garcia Lechner, E., Goossens, R.H.M.: Seat Comfort; a research on discomfort in time. MSc report Delft University (2003)

Gawlitta, D., Li, W., Oomens, C.W., Baaijens, F.P., Bader, D.L., Bouten, C.V.: The relative contributions of compression and hypoxia to development of muscle tissue damage: an in vitro study. Ann. Biomed. Eng. 35(2), 273–284 (2007)

Gefen, A., van Nierop, B., Bader, D.L., Oomens, C.W.: Strain-time cell-death threshold for skeletal muscle in a tissue-engineered model system for deep tissue injury. J. Biomech. 41(9), 2003–2012 (2008)

Gere, J.M., Timoshenko, S.P.: Mechanics of Materials, 2nd SI edn. PWS Engineering, Boston (1990)

Goossens, R.H.M., Zegers, R., et al.: Influence of Shear on Skin Oxygen Tension. Clin. Physiol. 14(1), 111–118 (1994)

Goossens, R.H.M.: Fundamentals of body support. The bursa-like interface. Whitepaper (January 2004)

Goossens, R.H.M., Teeuw, R.: Sensitivity for Pressure Differences at the Ischial Tuberosities. Ergonomics 48(7), 895–902 (2005)

Goossens, R.H.M., Snijders, C.J.: Lipoatrophia Semicircularis: A Biomechanical Approach. World Biomechanics, Calgary (2002)

Goossens, R.H., Snijders, C.J., et al.: Shear Stress measured on Beds and Wheelchairs. Scand. J. Rehabil. Med. 29(3), 131–136 (1997)

Goossens, R.H.M.P.M., Teeuw, R., Snijders, C.J.: Decubitus Risk: is shear more important than pressure? IEA 2000/ HFES 2000, Congress, San Diego (2000)

Grindley, A., Acres, J.: Alternating Pressure Mattresses: comfort and quality of sleep. Br. J. Nurs. 5(21), 1303–1310 (1996)

Helander, M.G., Zhang, L.: Field Studies of Comfort and Discomfort in Sitting. Ergonomics 40(9), 895–915 (1997)

Helander, M.G.: Forget about Ergonomics in Chair Design? Focus on Aesthetics and Comfort! Ergonomics 46(13/14), 1306–1319 (2003)

Holscher, T.G., Goossens, R.H.M., et al.: A New Low-cost Anti-decubitus Mattress for Home Care: requirements and development. J. Rehab. Sc. 7(2), 53–58 (1994)

ISO/DIS 16840-1 pre-Draft Title: Wheelchair seating – Part 1: Measurement of Postural Support Surfaces and Body Segments – Vocabulary ISO TC 173 SC1

Jenkins, D.B.: Hollinshead's Functional Anatomy of the Limbs and Back. Harcourt Brace & Company, Philadelphia (1991)

Liao, M.H., Drury, C.G.: Posture, Discomfort and Performance in a VDT task. Ergonomics 43(3), 345–359 (2000)

Lueder, R.K.: Seat Comfort: a review of the construct in the office environment. Hum. Factors 25(6), 701–711 (1983)

Kamijo, K., Tsujimura, H., Obara, H., Katsumata, M.: Evaluation of Seating Comfort, SAE Technical paper No. 820761 (1982)

Kanerva, L.: Physical Causes of Occupational Skin Disease. Occupational Skin Disease. R. M. Adams, W.B. Saunders Company (1990)

Kenedi, R., Gibson, T., Daly, C.: Bioengineering Studies of the Human Skin. In: Kenedi, R. (ed.) Biomechanics and related topics. Pergamon Press, Oxford (1967)

Kenedi, R.M., Gibson, T., et al.: Tissue Mechanics. Phys. Med. Biol. 20(3), 699–717 (1975)

Kosiak, M.: Etiology and Pathology of Ischemic Ulcers. Arch. Phys. Med. Rehab. 40, 62–69 (1959)

Mark, L.S., Vogele, D.C., Dainhoff, M.J., Cone, S., Lassen, K.: Measuring Movement at Ergonomic Workstations. In: Trends in Ergonomics/Human Factors, pp. 431–438. Elsevier, Amsterdam (1985)

Milivojevich, A., Stanciu, R., Russ, A., Blair, G.R., Van Heumen, J.D.: Investigating Psychometric and Body Pressure Distribution Responses to Automotive Seating Comfort. In: SAE Conference 2000, SAE no 2000-01-0626 (2000)

Moschella, S.L., Phillsbury, D.M., et al.: Dermatology. Saunders Company, Philedephia (1975)

Mossel, W.P.: Modelling Skin Friction. In: Scott, P.A., Bridger, R.S., Charteris, J. (eds.) Global Ergonomics, pp. 429–435. Elsevier, Amsterdam (1998)

Nicol, K., Min, L.: Coupling Force Distribution and Finite Element Model for Calculating the Consequences of Distributed Force Input. Clin. Biomech. (Bristol, Avon) 12(3), S13 (1997)

Black, J., Baharestani, M., Cuddigan, J., Dorner, B., Edsberg, L., Langemo, D., Posthauer, M.E., Ratliff, C., Taler, G.: NPUAP: National Pressure Ulcer Advisory Panel's updated pressure ulcer staging system. Dermatol. Nurs. 19(4), 343–349 (2007)

Park, S.J., Kim, C.B.: The Evaluation of Seating Comfort by Objective Measures. In: SAE Conference 1997, SAE no 970595 (1997)

Petersen, N.C.: The Development of Pressure Sores during Hospitalization. In: Kenedi, R.M., Cowden, J.M. (eds.) Bed sore Biomechanics, pp. 219–224. The Macmillan Press, Basingstoke (1976)

Price, D.D., Bush, F.M., Long, S., Harkins, S.W.: A Comparison of Pain Measurement Characteristics of Mechanical Visual Analogue and Simple Numerical Rating Scales. Pain 56, 217–226 (1994)

Pustinger, C., Dainoff, M.J., Smith, M.: VDT Workstation Adjustability: effects on worker posture, productivity, and health complaints. In: Trends in Ergonomics/Human Factors, pp. 445–451. Elsevier, Amsterdam (1985)

Reichel, S.M.: Shearing Force as a Factor in Decubitus Ulcers in Paraplegics. JAMA 166, 762–763 (1958)

Reswick, J.B., Rogers, J.E.: Experience at Rancho Los Amigos Hospital with Devices and Techniques to Prevent Pressure Sores. In: Kenedi, R.M., Cowden, J.M. (eds.) Bed sore Biomechanics, pp. 301–310. The Macmillan Press, Basingstoke (1976)

Ridge, M., Wright, V.: A Rheological Study of the Skin. In: Kenedi, R. (ed.) Biomechanics and Related Topics. Pergamon Press, Oxford (1967)

Rook, A., Wilkinson, D., Ebling, J.: Textbook of Dermatology, 6th edn., vol. I. Blackwell Science, Malden (1998)
Shackel, B., Chidsey, K.D., And Shipley, P.: The Assessment of Chair Comfort. Ergonomics 12, 269–306 (1969)
Staarink, H.A.M.: [In Dutch] Sitting Posture, Comfort and Pressure. The quality of wheelchair cushions, Delft University of Technology (1995)
Stekelenburg, A., Oomens, C.W., Strijkers, G.J., Nicolay, K., Bader, D.L.: Compression-induced deep tissue injury examined with magnetic resonance imaging and histology. J. Appl. Physiol. 100(6), 1946–1954 (2006)
Stekelenburg, A., Gawlitta, D., Bader, D.L., Oomens, C.W.: Deep tissue injury: how deep is our understanding? Arch. Phys. Med. Rehabil. 89(7), 1410–1413 (2008)
Straker, L.M.: Body Discomfort Assessment Tools. The Occupational Ergonomics Handbook, 1239–1252 (1999)
Susten, A.S.: The Chronic Effects of Mechanical Trauma to the Skin: a review of the literature. Am. J. Ind. Med. 8(4-5), 281–288 (1985)
Tewari, V.K., Prasad, N.: Optimum Seat Pan and Back-rest Parameters for a Comfortable Tractor Seat. Ergonomics 43, 167–186 (2000)
Thakurta, K., Koester, D., Bush, N., Bachle, S.: Evaluating Short and Long-term Seat Comfort. In: Human Factors in Vehicle Design: lighting, seating and advanced electronics, Publication No SP-1088, pp. 33–37 (1995)
Torebjörk, H.E.: Activity in C Nociceptors and Sensation. In: Kenshalo, D.R. (ed.) Sensory Functions of the Skin of Humans. Plenum, New York (1979)
Yamaguchi, Y.: Tribology of Plastic Materials. Elsevier, Amsterdam (1990)
Yu, B., Westreich, A., Cahalan, T., Schwen, E., An, K.: Assessment of Seat Pressure during Bicycling. Mayo Clinic/Mayo Foundation, Biomechanics Laboratory and the Sports Medicine Center (1995)
Yun, M.H., Donges, L., Freivalds, A.: Using Force Sensitive Resistors to Evaluate the Driver's Seating Comfort. In: Advances in Industrial Ergonomics and safety IV, pp. 403–410 (1992)
Zhang, I., Helander, M.G., Drury, C.G.: Identifying Factors of Comfort and Discomfort in Sitting. Human Factors 38, 377–389 (1996)
Zhang, M., Roberts, V.C.: The Effect of Shear Forces Externally Applied to Skin Surface on Underlying Tissues. J. Biomed. Eng. 15(6), 451–456 (1993)

Mechanobiology of Cutaneous Wound Healing and Scarring

Rei Ogawa[1,2] and Dennis P. Orgill[1]

[1] Division of Plastic Surgery, Department of Surgery, Brigham and Women's Hospital, Harvard Medical School, Boston, USA
[2] Department of Plastic, Reconstructive and Aesthetic Surgery, Nippon Medical School, Tokyo, Japan

Abstract. Mechanical forces are related to cutaneous wound dealing. A better understanding of the mechanobiological environment of skin will be helpful in designing new skin expansion and wound healing therapies. Mechanical forces are also related to cutaneous scarring. Scar formation follows the normal injury repair process in skin. Control and regulation of extrinsic / intrinsic mechanical forces are important to reduce abnormal scarring (hypertrophic scar or keloid).

1 Introduction

Human skin is a remarkable organ that protects us from the external environment providing an effective barrier to water and microorganisms. In a typical adult the surface area of skin is between 1.6 and 1.8 sq meters (Bender et al. 1994). In growth and development, skin expands to cover the growing skeleton and soft tissues, being constantly subjected to intrinsic forces of underlying skeletal growth as well as extrinsic mechanical forces from body movements and external supports. Additional intrinsic forces are produced by myofibroblasts after skin injury (Hinz et al. 2001). A better understanding of the mechanobiological environment of skin will be helpful in designing new skin expansion and wound healing therapies. Mechanical forces are also related to cutaneous scarring. Scar formation follows the normal injury repair process in skin. Control and regulation of extrinsic / intrinsic mechanical forces are important to reduce abnormal scarring (Akaishi et al. 2008) (hypertrophic scar or keloid).

2 Mechanobiology of Cutaneous Wound Healing

A. Wound Healing and Mechanical Forces

Cutaneous wound healing normally closes skin gaps and re-establishes an effective epidermal barrier. Complex biochemical events are categorized into four general processes: coagulation, inflammation, proliferation, and remodeling. Coagulation begins immediately after injury, and the inflammatory and proliferative phases start within a few days. The remodeling phase starts within a week of injury and

A. Gefen (Ed.): Bioengineering Research of Chronic Wounds, SMTEB 1, pp. 31–42.
springerlink.com

continues over months. The phases of wound healing, granulation tissue formation, wound contraction and epithelialization are all influenced by mechanical forces (Silver et al. 2003).

In granulation tissue, fibroblasts, myofibroblasts, endothelial and epithelial cells are affected by intrinsic and extrinsic mechanical stimulation. The wound itself contracts by forces produced by myofibroblasts, and is also affected by many extrinsic forces including the natural tension found within skin. Fibroblasts secrete collagen and fibronectin (Couchman et al. 1990) and regulate the volume of extracellular matrix (ECM) by collagenese secretion (Fig.1). Sequential synthesis and enzymatic breakdown of proteins within the extracellular matrix results in remodeling of three-dimensional (3-D) ECM structures (Fig.1). The binding of cells to matrix proteins results in small forces on cells that can cause deformation. If the balance between synthesis and degradation is not carefully maintained, scars can either become hypertrophic or atrophic (Pilcher et al. 1999). In addition to ECM alteration, volume and flow of extracellular fluid (ECF) is increased in the wound due to increased blood vessel permability. The ECM and ECF also produce intrinsic mechanical forces e.g. tension (Gilbert et al. 2007), shear force (Stolberg et al. 2009), osmotic pressure (Denda et al. 2007), and hydrostatic pressure (Lund et al. 1992), and cells convert these mechanical stimuli into electrical signals through mechanoreceptors (mechanosensors) such as mechanosensitive ion channels (Stockbridge et al. 1988), cell adhesion molecules including integrins (Matthews et al. 2006), and actin filaments in the cytoskeleton (Wang et al. 1993, Wu et al. 1999) (Fig.1 and 2). As a result, cell proliferation, angiogenesis and epithelization are accelerated. In addition to these cellular responses, macro-scale tissue response from mechanical stimuli including hypoxia is also closely related to wound healing.

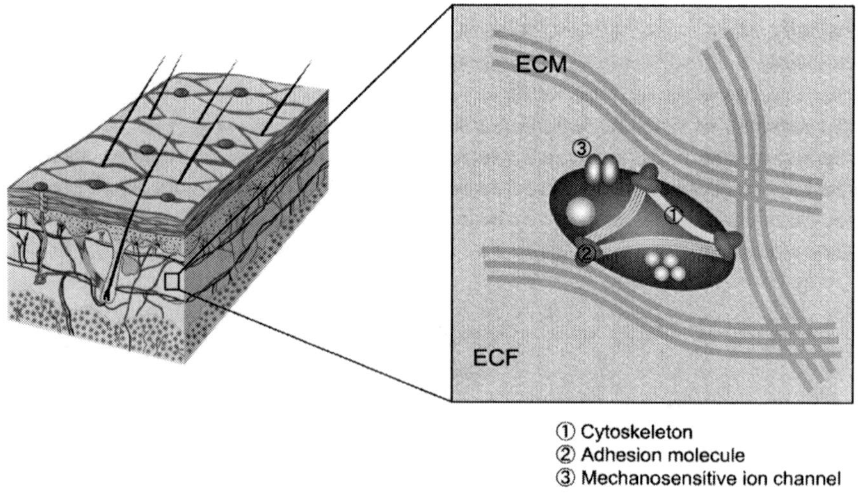

Fig.1. Schema of Cellular Mechanoreceptors (Mechanosensors) in The Skin

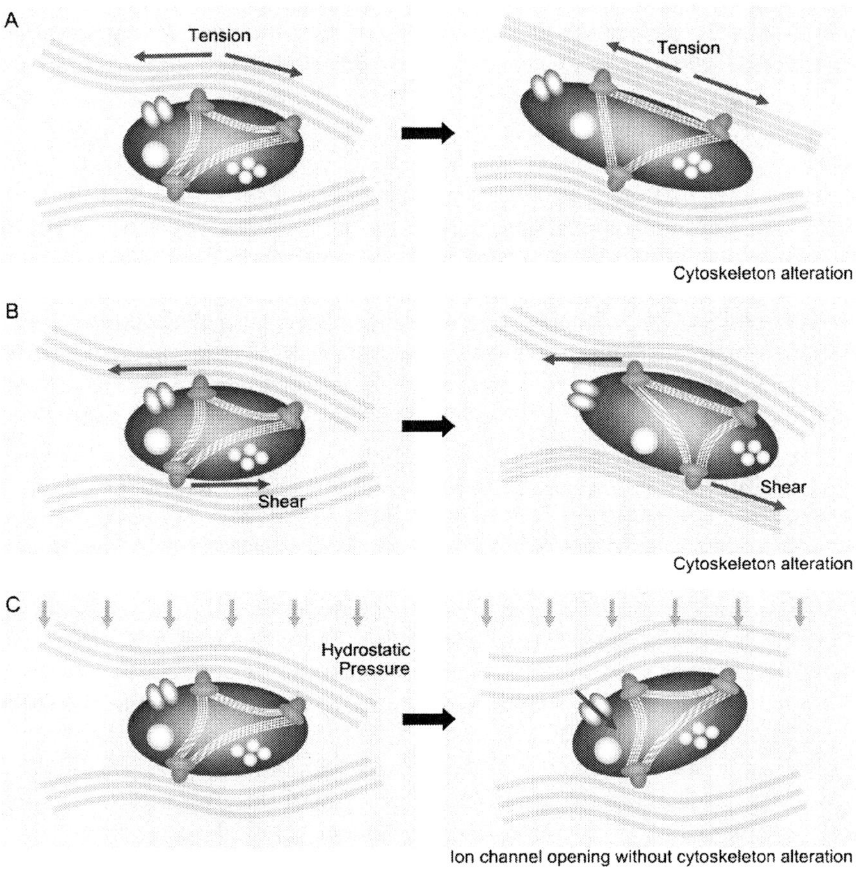

Fig. 2. Schema of Mechanical Forces and Cellular Responses

A. Tensile force
B. Shear force
C. Hydrostatic pressure

The ECM and ECF produce intrinsic mechanical forces e.g. stretching tension, shear force, osmotic pressure, and hydrostatic pressure, and cells convert these mechanical stimuli into electrical signals through mechanoreceptors (mechanosensors) such as mechanosensitive ion channels, cell adhesion molecules including integrins, and actin filaments in the cytoskeleton.

Matrix-related mechanical forces (tension, shear and compression) and fluid-related mechanical forces (hydrostatic and fluid shear forces) contribute to the mechanism of cellular responses.

B. Normal Tissue and Cellular Response of Skin Against Mechanical Forces

With respect to mechanotransduction studies of human dermal fibroblasts *in vitro*, (Parson et al. 1999) reported that procollagen synthesis and procollagen mRNA levels were increased after 48hour of cyclic 20% strain, and the response was increased in presence of transforming growth factor-β1 (TGF-β1). Kessler et al. 2001) also studied human dermal fibroblasts those were cultured within 3-D gels under application of tension. cDNA microarray and Northern blot analysis led to identification of mechano-responsive genes coding for extracellular matrix proteins, fibrogenic growth factors, protease inhibitors, components of focal adhesions, and the cytoskeleton. Mechanical stimulation activate both signaling pathways and mechanosensitive ion channels. The signaling pathways include protein kinases, adaptor proteins, guanidine exchange factors and small GTPases that are recruited to these sites and are directly trigger mitogen-activated protein kinases (MAPK) pathways as well as activation of the NF-κB pathway. G proteins are another family of membrane proteins considered to modulate mechanochemical transduction pathways (Silver et al. 2003). Mechanical stimulation alters the conformation of G protein that leads to growth factor-like changes that initiate secondary messenger cascades leading to cell growth [4)]. With respect to the mechanosensitive ion channels, Ca^{2+} (Giamarchi et al. 2006), K^+ (Hamill 2006), Na^{2+} (Hamill 2006) and Mg^{2+} (Inoue, 2006) have been reported. Ca^{2+} ion involve in phospholipase C (PLC) activation, and PLC leads to protein kinase C (PKC) activation. These PLC and PKC pathways are related to epidermal growth factor (EGF) activation (Silver et al. 2003).

We have focused recent studies on the in *vivo* analyses of tissue and cellular responses against mechanical forces for understanding skin wound healing. Tension is a principle force experienced by skin, the optimal amplitude and waveform may facilitate the growth and expansion of skin. We previously applied static and periodic tensile forces to rat ears and showed vessel dilatation and epidermal proliferation (Pietramaggiori et al 2007). A gene chip analysis performed on this rat model indicated that hypoxia is a possible mechanism in tissue levels for the observed effects Saxena et al 2007). Waveform-specific mechanical loads may accelerate tissue growth by mechanotransduction and as a result of repeated cycles of temporary hypoxia.

C. Clinical Application of Mechanical Forces

In reconstructive surgery, expanded skin due to pregnancy (Riordan et al. 2003, Elwood et al. 2000), implanted tissue expanders (Manders et al. 1984, Gao et al. 2007) or simple tape (Daya and Nair, 2008) has been used. We have demonstrated that cell proliferation and angiogenesis occur in stretched skin and suggest that these biological effects can be modulated by skin stretching devices (Fig.3).

Another important approach is vacuum-assisted closure (VAC) therapy (Fig.3). This is frequently used to treat complex wounds, and we previously demonstrated that wound surface microdeformations promote cell proliferation through mechanotransduction pathways (Saxena et al. 2004, Greene et al, 2006, Scherer et al 2008). Scherer et al. (2008) demonstrated that the foam induces micromechanical

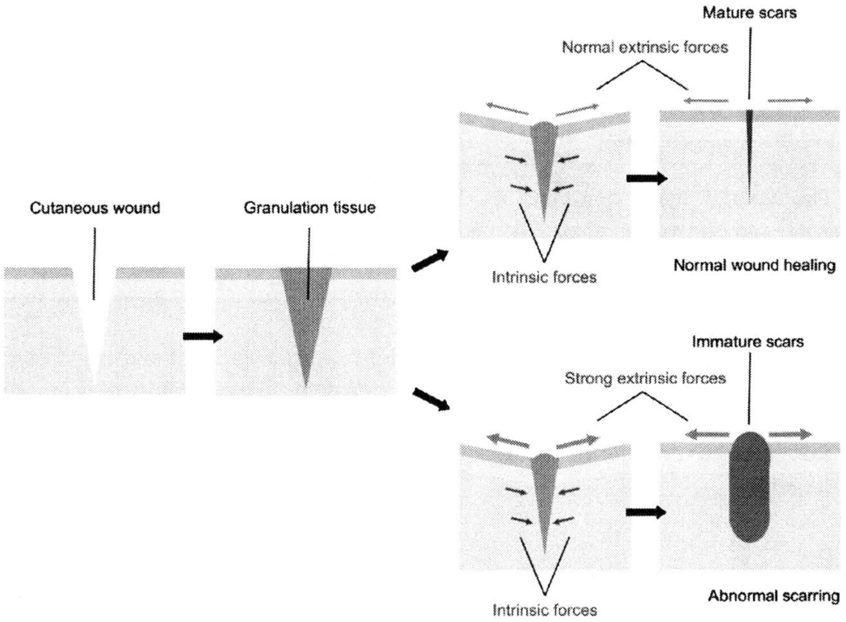

Fig. 3. Strategy of Clinical Application of Mechanical Forces

deformation and hypothesized that the effects of ECF removal and ECM distortion from the VAC may effectively change single-cell morphology and result in various cellular responses such as proliferation and differentiation. Saxena et al. (2004) also demonstrated that the predicted strain within the pores of the foam was visible in the histology cross-sections of treated wounds.

Fluid removal results in hydrostatic pressure gradients and shear forces that may alter genomic expression through mechanosensitive ion channels. Therefore, both fluid-related mechanical forces (hydrostatic and fluid shear forces), and matrix-related mechanical forces (tension and compression) may contribute to the mechanism of action of the VAC device (Fig.2). The magnitude and balance of these force patterns needs to be further studied, but the distribution of strain is compatible with previous results of histological studies. Connective tissues respond to mechanical forces by changing their structure, composition and function. Fibroblasts transduce mechanical forces into biomechanical signals utilizing differentional gene expression and protein secretion.

Mechanical force can be applied for the treatment of intractable wounds such as diabetic wounds and chronic ulcers, effectively. Vacuum-assisted closure (VAC) therapy and skin stretch devices accelerate angiogenesis, nerve growth, epithelization and matrix production.

Mechanobiology of Scarring

A. The Relationship between Scarring and Mechanical Force

It is well known that appropriate intrinsic tension is necessary for incisional wound closure. However, abnormal scarring including hypertrophic scars (HSs) will be generated in the case where there is excess tension in the wound (Fig.4). Generally, HSs and keloids are fibroproliferative disorders (FPDs) of the skin, and mechanical forces promote their growth (Fig.5). We have studied abnormal scarring from the aspect of skin tension as an extreme example of excess wound healing of skin.

The wound itself contracts by intrinsic forces produced by myofibroblast activity, and the wound also affected by many extrinsic forces including the natural tension found within skin. Cell proliferation and angiogenesis are accelerated by strong extrinsic force, this these forces sometime become a cause of abnormal scarring.

Hypertrophic scars (HSs) and keloids are fibroproliferative disorders (FPDs) of the skin, and mechanical forces promote their growth. Theses abnormal scars are an extreme example of excess wound healing of skin. Clinicians have defined that HSs do not grow beyond the boundaries of the original wound, whereas keloids grow horizontally.

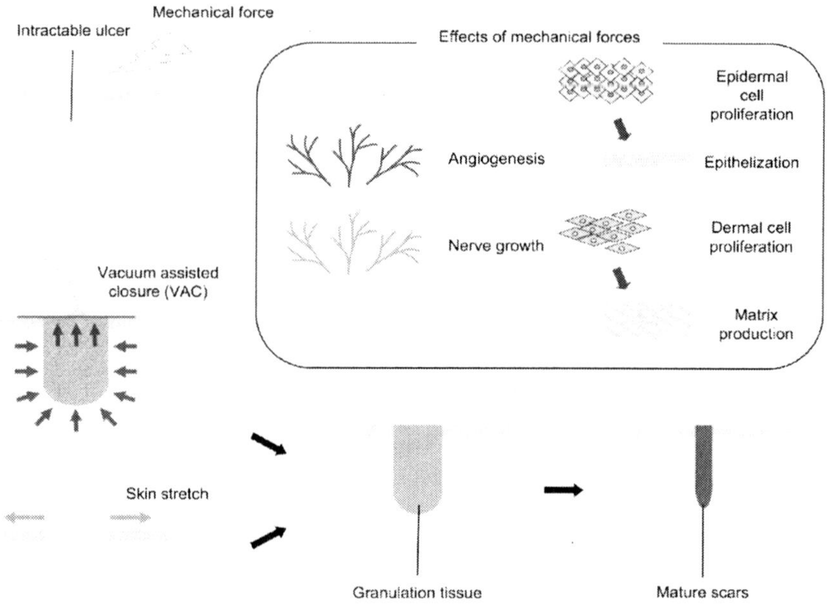

Fig. 4. Schema of Intrinsic and Extrinsic Forces on Cutaneous Wound Healing and Scarring

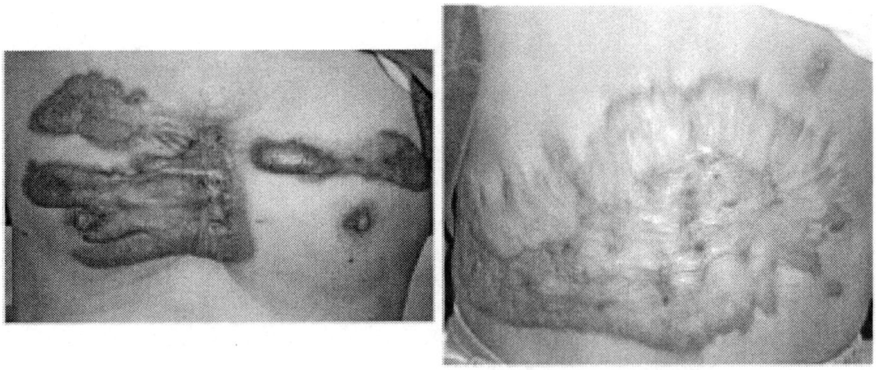

Fig. 5. Keloids and Hypertrophic Scars

1. Site Specificity of HSs/Keloids

Abnormal scars often occur at specific sites such as the anterior chest, shoulder, scapular and the lower abdomen (Ogawa 2003). All of these sites are constantly or frequently subjected to skin stretching due to the movements of the body in daily life. The skin on the anterior chest is constantly being stretched by respiration and upper limb movements, the shoulder and scapular regions are constantly stretched by movements of the upper limbs and the bending motion of the body, and the lower abdomen and suprapubic regions are regularly stretched by seating and standing motions (Akaishi et al. 2008, Ogawa 2008). In contrast, heavy scars rarely occur on the scalp and the anterior lower leg even with abnormal scars on other areas of the body. These sites have in common that bones lie directly under the skin and consequently the skin at these sites is rarely subjected to tension. The site-specificity of HSs/keloids development suggests that mechanical force may not only promote keloid/HS growth, it may also be primarily responsible for their generation (Akaishi et al. 2008, Ogawa 2008).

2. The Relationship between Scar Growing Pattern and the Direction of Stretching Tension

HSs do not grow beyond the boundaries of the original wound and thus only grow vertically. In contrast, keloids grow and spread both vertically and horizontally, similar in many respects to slowly growing malignant tumors. The direction of their horizontal growth results in characteristic shapes that depend on their location. For example, keloids on the anterior chest grow in a pattern that has been likened to a "crab's claw", wheras keloids on the shoulder grow into a "butterfly" shape. These patterns may reflect the predominant directions of skin tension at these sites. Supporting this notion is our computer analysis of the keloids (Fig.6) (Akaishi et al., 2008), which revealed some interesting observations. First, high skin tension was

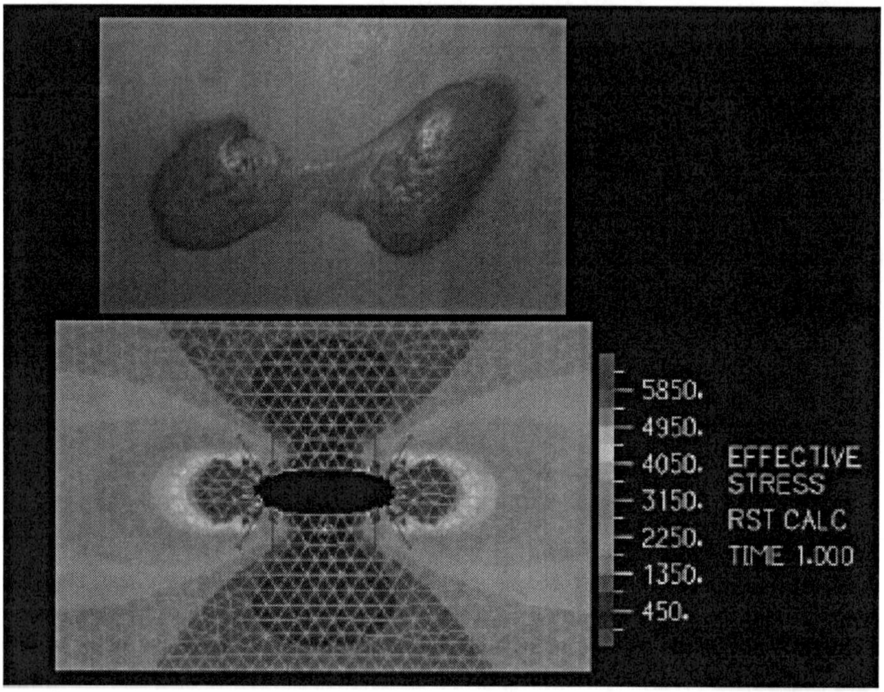

Fig. 6. Computer Simulation of Skin Stretching Tension around Keloids

observed at the edges of the keloid. Second, the keloid centers were regions of lower tension, explaining why keloids generally stop growing in their central regions. Third, the keloid expansion occurred in the direction in which the skin was being pulled, and fourth, the stiffness of the skin at the circumference of a keloid correlates directly with the degree of skin tension at the circumference. These observations strongly support the notion that tension is closely associated with the pattern and degree of keloid growth. Because HSs and normal scars differ in growth characteristics from keloids, they may reflect a difference in responsiveness to skin tension.

Unfortunately, there are no quantitative evaluations of the magnitudes of skin forces at the tissue level that cause abnormal scarring, because the individual difference is large especially in human race.

Finite element model observations suggested the notion that stretching tension is closely associated with the pattern and degree of keloid growth.

3. Animal Models of Hypertrophic Scars

Many authors have attempted to construct suitable animal models of HSs or keloids using mouse, rats and rabbits, however these models seem to be driven more by an acute inflammatory response than the chronic inflammation that actually leads to HSs and keloids (Ogawa and Chin 2008). Aarabi et al. (2007) reported a HS mouse model based on mechanical force loading and showed that scars subjected to tension exhibit less apoptosis. They also showed that inflammatory cells and mechanical forces promote fibrosis. These findings support the well-established notion that mechanical forces strongly modulates cellular behavior.

B. Cellular and Tissue Response against Mechanical Forces on Scarring

Mechanical forces promote growth and/or generate scars. Mechanical forces can be perceived by two different types of receptors in scars. These include cellular mechanoreceptors (also known as mechanosensors) and the receptors on nerve fibers (which include mechanosensitive nociceptors) that produce the somatic sensation of mechanical force. In the cellular level response, cellular mechanoreceptors, including mechanosensitive (MS) ion channels, the cytoskeleton and cell adhesion molecules response against mechanical forces. A hypersensitivity of these structures may be a cause for abnormal scarring.

In tissue level, hyperreactivity or derangement of the mechanosensitive nociceptors of nerve fibers may also cause or contribute to the generation of abnormal scarring. Sensory fibers are considered to be a cause of "neurogenic inflammation", and ineuropeptides released by sensory fibers accelerate the cytokine production by various cells. Neurogenic inflammation is a type of mechanical stress-induced inflammation that is defined as cutaneous antidromic vasodilatation and plasma extravasation. It is mediated by the release of neuropeptides from sensory endings. In our study of neuropeptides after skin stretching of mouse (Chin et al. 2009) gene expression of the neuropeptides, substance P (SP) and calcitonin gene-related peptide (CGRP), their receptors (NK1R and CGRP-RCP), and growth factors (NGF, TGFβ1, VEGF, and EGF) were increased.

Direct evidence for neurogenic inflammation due to the antidromic stimulation of sensory fibers was presented 30 years ago (Jansco et al 1967). The results of neuropeptide activity in skin inflammation include erythema, edema, hyperthermia, and pruritus (Zegarska et al. 2006). We speculate that neurogenic inflammation may also drive or contribute to HS/keloid formation. Mechanosensitive nociceptors on unmyelinated axons (C-fibers and Aδ-fibers) can be stimulated while the skin is stretched (Zheng et al. 2002). In addition, axonal reflexes and the stimulation of antidromic sensory nerves results in the release of vasodilative factors, including neuropeptides such as substance P (SP) and calcitonin gene-related peptide (CGRP), that in turn induce local erythema. These neuropeptides may also up-regulate the gene expression of growth factors such as TGFβ and NGF in various dermal cells, including fibroblasts (Lai et al. 2003). Therefore, neurogenic inflammation is a mediator that can plausibly connect mechanical forces with the development of abnormal scar growth and/or generation.

3 Conclusions

Cutaneous wound healing and scarring are continuous processes of tissue and cellular response. Intrinsic and extrinsic mechanical forces are involved in both wound healing and subsequent scar formation. New knowledge of the mechanisms of mechano-transduction in skin may result in better methods to heal wounds and modulate scarring.

References

1) Aarabi, S., Bhatt, K.A., Shi, Y., Paterno, J., Chang, E.I., Loh, S.A., Holmes, J.W., Longaker, M.T., Yee, H., Gurtner, G.C.: Mechanical load initiates hypertrophic scar formation through decreased cellular apoptosis. FASEB J. 21, 3250–3261 (2007)
2) Akaishi, S., Akimoto, M., Ogawa, R., Hyakusoku, H.: The relationship between keloid growth pattern and stretching tension: visual analysis using the finite element method. Ann. Plast. Surg. 60(4), 445–451 (2008)
3) Akaishi, S., Ogawa, R., Hyakusoku, H.: Keloid and hypertrophic scar: neurogenic inflammation hypotheses. Med. Hypotheses 71(1), 32–38 (2008)
4) Bender, A.E., Bender, D.A.: Body Surface Area. A Dictionary of Food and Nutrition. Oxford University Press, New York (1995)
5) Chin, M.S., Lancerotto, L., Helm, D., Dastouri, P., Prsa, M.J., Ottensmeyer, M., Akaishi, S., Orgill, D.P., Ogawa, R.: Analysis of Neuropeptides in Stretched Skin. Plast. Reconstructive Surg (in Press, 2009)
6) Couchman, J.R., Austria, M.R., Woods, A.: Fibronectin-cell interactions. J. Invest. Dermatol. 94(suppl. 6), S7–S14 (1990)
7) Daya, M., Nair, V.: Traction-assisted dermatogenesis by serial intermittent skin tape application. Plast. Reconstr. Surg. 122(4), 1047–1054 (2008)
8) Denda, M., Nakatani, M., Ikeyama, K., Tsutsumi, M., Denda, S.: Epidermal keratinocytes as the forefront of the sensory system. Exp. Dermatol. 16(3), 157–161 (2007)
9) Elwood, E.T., Ingram, W.L., Carlson, G.W.: Pregnancy as a tissue expander in the repair of a massive ventral hernia. Ann. Plast. Surg. 45(4), 431–433 (2000)
10) Gao, J.H., Ogawa, R., Hyakusoku, H., Lu, F., Hu, Z.Q., Jiang, P., Yang, L., Feng, C.: Reconstruction of the face and neck scar contractures using staged transfer of expanded "Super-thin flaps". Burns 33(6), 760–763 (2007)
11) Giamarchi, A., Padilla, F., Crest, M., Honore, E., Delmas, P.: TRPP2: Ca2+-permeable cation channel and more. Cell. Mol. Biol. (Noisy-le-grand) 52(8), 105–114 (2006)
12) Gilbert, T.W., Stewart-Akers, A.M., Sydeski, J., Nguyen, T.D., Badylak, S.F., Woo, S.L.: Gene expression by fibroblasts seeded on small intestinal submucosa and subjected to cyclic stretching. Tissue Eng. 13(6), 1313–1323 (2007)
13) Greene, A.K., Puder, M., Roy, R., Arsenault, D., Kwei, S., Moses, M.A., Orgill, D.P.: Microdeformational wound therapy: effects on angiogenesis and matrix metalloproteinases in chronic wounds of 3 debilitated patients. Ann. Plast. Surg. 56(4), 418–422 (2006)
14) Hamill, O.P.: Twenty odd years of stretch-sensitive channels. Pflugers Arch. 453(3), 333–351 (2006)
15) Hinz, B., Celetta, G., Tomasek, J.J., Gabbiani, G., Chaponnier, C.: Alpha-smooth muscle actin expression upregulates fibroblast contractile activity. Mol. Biol. Cell 12, 2730 (2001)
16) Inoue, R., Jensen, L.J., Shi, J., Morita, H., Nishida, M., Honda, A., Ito, Y.: Transient receptor potential channels in cardiovascular function and disease. Circ. Res. 99(2), 119–131 (2006)

17) Jancso, N., Janxso-Gabor, A., Szolxsanyi, J.: Direct evidence for neurogenic inflammation and its prevention by denermation and by pre-treatment with capsaicin. Br. J. Pharmacol. 31, 138–151 (1967)
18) Kessler, D., Dethlefsen, S., Haase, I., Plomann, M., Hirche, F., Krieg, T., Eckes, B.: Fibroblasts in mechanically stressed collagen lattices assume a "synthetic" phenotype. J. Biol. Chem. 276(39), 36575–36585 (2001)
19) Lai, X.N., Wang, Z.G., Zhu, J.M., Wang, L.L.: Effect of substance P on gene expression of transforming growth factor beta-1 and its receptors in rat's fibroblasts. Chin. J. Traumatol. 6(6), 350–354 (2003)
20) Lund, T., Onarheim, H., Reed, R.K.: Pathogenesis of edema formation in burn injuries. World J. Surg. 16(1), 2–9 (1992)
21) Manders, E.K., Schenden, M.J., Furrey, J.A., Hetzler, P.T., Davis, T.S., Graham, W.P.: 3rd, Soft-tissue expansion: concepts and complications. Plast. Reconstr. Surg. 74(4), 493–507 (1984)
22) Matthews, B.D., Overby, D.R., Mannix, R., Ingber, D.E.: Cellular adaptation to mechanical stress: role of integrins, Rho, cytoskeletal tension and mechanosensitive ion channels. J. Cell. Sci. 119(Pt 3), 508–518 (2006)
23) Ogawa, R., Chin, M.S.: Animal models of keloids and hypertrophic scars. J. Burn. Care Res. 29(6), 1016–1017 (2008)
24) Ogawa, R., Mitsuhashi, K., Hyakusoku, H., Miyashita, T.: Postoperative electron-beam irradiation therapy for keloids and hypertrophic scars: retrospective study of 147 cases followed for more than 18 months. Plast. Reconstr. Surg. 111(2), 547–553 (2003)
25) Ogawa, R.: Keloid and hypertrophic scarring may result from a mechanoreceptor or mechanosensitive nociceptor disorder. Med. Hypotheses 71(4), 493–500 (2008)
26) Parsons, M., Kessler, E., Laurent, G.J., Brown, R.A., Bishop, J.E.: Mechanical load enhances procollagen processing in dermal fibroblasts by regulating levels of procollagen C-proteinase. Exp. Cell. Res. 252(2), 319–331 (1999)
27) Pietramaggiori, G., Liu, P., Scherer, S.S., Kaipainen, A., Prsa, M.J., Mayer, H., Newalder, J., Alperovich, M., Mentzer, S.J., Konerding, M.A., Huang, S., Ingber, D.E., Orgill, D.P.: Tensile forces stimulate vascular remodeling and epidermal cell proliferation in living skin. Ann. Surg. 246(5), 896–902 (2007)
28) Pilcher, B.K., Wang, M., Qin, X.J., Parks, W.C., Senior, R.M., Welgus, H.G.: Role of matrix metalloproteinases and their inhibition in cutaneous wound healing and allergic contact hypersensitivity. Ann. N. Y. Acad. Sci. 878, 12–24 (1999)
29) Riordan, C., Budny, P., Regan, P.: Pregnancy as an autologous tissue expander for closure of an abdominal-wall defect. Br. J. Plast. Surg. 56(1), 64–66 (2003)
30) Saxena, V., Hwang, C.W., Huang, S., Eichbaum, Q., Ingber, D., Orgill, D.P.: Vacuum-assisted closure: microdeformations of wounds and cell proliferation. Plast. Reconstr. Surg. 114(5), 1086–1096 (2004) discussion 1097-8
31) Saxena, V., Orgill, D., Kohane, I.: A set of genes previously implicated in the hypoxia response might be an important modulator in the rat ear tissue response to mechanical stretch. BMC Genomics 8, 430 (2007)
32) Scherer, S.S., Pietramaggiori, G., Mathews, J.C., Prsa, M.J., Huang, S., Orgill, D.P.: The mechanism of action of the vacuum-assisted closure device. Plast. Reconstr. Surg. 122(3), 786–797 (2008)
33) Silver, F.H., Siperko, L.M., Seehra, G.P.: Mechanobiology of force transduction in dermal tissue. Skin Res. Technol. 9(1), 3–23 (2003) (review)
34) Stockbridge, L.L., French, A.S.: Stretch-activated cation channels ir human fibroblasts. Biophys. J. 54(1), 187–190 (1988)

35) Stolberg, S., McCloskey, K.E.: Can shear stress direct stem cell fate? Biotechnol. Prog. 25(1), 10–19 (2009)
36) Wang, N., Butler, J.P., Ingber, D.E.: Mechanotransduction across the cell surface and through the cytoskeleton. Science 260(5111), 1124–1127 (1993)
37) Wu, Z., Wong, K., Glogauer, M., Ellen, R.P., McCulloch, C.A.: Regulation of stretch-activated intracellular calcium transients by actin filaments. Biochem. Biophys. Res. Commun. 261(2), 419–425 (1999)
38) Zegarska, B., Lelinska, A., Tyrakowski, T.: Clinical and experimental aspects of cutaneous neurogenic inflammation. Pharmacol. Rep. 58(1), 13–21 (2006)
39) Zheng, Z., Lamotte, R.H., Grigg, P.: Comparison of responses to tensile and compressive stimuli in C-mechanosensitive nociceptors in rat hairy skin. Somatosens Mot. Res. 19(2), 109–113 (2002)

Cell Migration along the Basement Membrane during Wound Repair. The Corneal Endothelium as a Model System

Sheldon R. Gordon

Department of Biological Sciences, Oakland University, Rochester, MI 48309-4476 USA
srgordon@oakland.edu

Abstract. The vertebrate corneal endothelium serves as a model system for investigating the role of cell migration during wound repair along the natural basement membrane. As a tissue monolayer, the cells reside on their basement membrane, Descemet's membrane, which serves to modify many cellular behaviors during the injury-induced migratory response. Because the tissue is not directly dependent on either blood or nerve supplies, it is readily amendable to organ culture, where it can be easily maintained and experiments performed on the cells in situ. In addition, results obtained from work using tissue cultured corneal endothelial cells have augmented our understanding of the various pathways that are used for triggering cell movement during in vitro wound repair. The initiation of a migratory response during wound repair involves a multitude of various cellular mechanisms and structural changes that result in the activation of several biochemical pathways and modifications to cell shape as well as the reorganization of the actin cytoskeleton. These changes provide a means for the endothelial cell to translocate along Descemet's membrane into the wound region in order to permit the repopulation of the area with the eventual reestablishment of an intact and functional monolayer. Many of these biochemical and morphological alterations appear to be mediated by various growth factors such as fibroblast growth factor, epidermal growth factor and transforming growth factor-beta. Various biochemical pathways are stimulated after an injury, including phospholipase C, phosphatidylinositol-3 kinase, protein kinase C, Rho, Rac, cdc42 and arachidonic acid metabolism which mediate several events during wound repair such as cell proliferation, epithelial-mesenchymal transition, spreading and the extension of cellular processes. In addition, despite residing on a basement membrane endothelial cells respond to injury by secreting various extracellular matrix components at the cell/basement membrane interface that aids in mediating their movement along Descemet's membrane, which also appears to be regulated in part by protease activity. This review discusses the role of migration in corneal endothelial wound healing. How movement is modified when it occurs during in vitro as opposed to in vivo or organ cultured wound healing will also be examined as will the differences between the cellular responses to large versus small sized wounds.

1 Introduction

Migration is a fundamental cellular process that is inherent in many biological functions. Cell movement occurs during various stages of development (Newgreen

A. Gefen (Ed.): Bioengineering Research of Chronic Wounds, SMTEB 1, pp. 43–84.
springerlink.com

and Erickson 1986; Armstrong 1985) where it plays as important role in organogenesis. Additionally, it is a critical process observed during wound repair in many biological systems and in diseases such as cancer, where cell motility is responsible for the metastasis that characterizes the disease. Furthermore, the initiation of cell movement can be evoked by a variety of mechanisms including the explantation of cells or tissues into culture and responses to chemical gradients that occur during chemotaxis or haptotaxis. Overall, cell movement invokes a multitude of sequential events at the cellular level that together, in a well coordinated fashion, culminates in contributing to the translocation of a cell. In general, cell movement is the result of several processes that include cell signaling, gene expression, cytoskeletal reorganization, the loss of cell-to-cell and/or alterations in cell/substrate contact that result in changes to the morphological appearance of the cell that directly correlates with an active state of motility. In many cultured cells, migration has often been observed to be rather random in its direction (Weiger et al. 2009; Parkinson 1983; Abercrombie and Heaysman 1953) as opposed to migration during wound repair where cell movement is often referred to as "directional" so as to indicate a well focused cell translocation for the purpose of repairing an injury. Evidence in the literature suggests that directed migration involves both changes in organelle distributions and activation of biochemical pathways. Examples of these changes include the reorientation of both the Golgi apparatus (Bershadsky and Futerman 1994) and the microtubule organizing center (Gotlieb et al. 1981; Kupfer, Louvard and Singer 1979) to the front of the cell and the activation of Rho-family GTPases and phosphatidylinositol 3-kinase (PI 3-kinase) (Fukata, Nakagawa and Kaibuchi 2003) (see Table I). In epithelial-type cells, the morphological alterations that correlate with migratory behavior are often referred to as the epithelial-mesenchymal transition (EMT) (Savenger 2001), and reflect those cellular changes that occur from a non-motile to motile state. Some of these cellular alterations involve the loss of cell-cell adhesion, change in the cell polarity, reorganization of the cytoskeleton, migration as isolated cells and the ability of cells to resist anoikis when detached from their substrate (Klymkowsky and Savagner 2009). Both this morphological transformation and the events described above are observed during wound repair in many epithelial cell types, including the corneal endothelium. These observations lend one to conclude that cell migration itself is the result of a well choreographed series of events that sets the stage in order to promote as well as allow cells to traverse from one location to another. The purpose of this chapter is to provide an overview of cell migration during wound repair in the vertebrate corneal endothelium, both in the organ cultured and in vivo tissue. As such, this review will cover several aspects of cellular events that participate in a coordinated fashion for the translocation of endothelial cells in the healing tissue. In addition, where practical, wound healing of cultured corneal endothelial cells is also discussed to further augment an understanding of mechanisms that mediate the repair process, albeit that variations exist between wound repair in cell culture as opposed to the intact tissue layer.

Table 1. List of abbreviations that are used in this article

Abbreviation	Definition
EMT	Endothelial-mesenchymal transition
TGF	Transforming growth factor
EGF	Epidermal growth factor
Csk	C-terminal Src kinase
CMB	Circumferential microfilament bundle
IGF	Insulin-like growth factor
FGF	Fibroblast growth factor
ZO-1	Zonula adherens-1
PDGF	Platelet derived growth factor
PI 3-kinase	Phosphatidylinositol 3-kinase
PKC	Protein kinase C
GTPase	Guanosine triphosphatase
PGE_2	Prostaglandin E_2
SBA	Soybean agglutinin
ECM	Extracellular matrix
GAGs	Glycosaminoglycans
TSP	Thrombospondin

2 The Vertebrate Corneal Endothelium

The posterior aspect of the cornea is covered by a single layer of epithelial cells (Fig. 1), termed the "endothelium", that are derived from neural crest cells (Johnston et al. 1979; Hay and Revel 1969) that migrated onto this surface to form the tissue. During the early embryonic period, migrating cells destined to become the chick endothelium demonstrate active DNA synthesis (Nuttall 1976) and in mammalian systems, these cells continue to exhibit both DNA synthesis and mitosis (Gordon 1990, von Sallman 1961; 1963) as the population grows and cells spread to cover the posterior surface of the cornea. Reneker et al. (2000) have provided evidence that the formation of an intact endothelium is critical for the normal formation of the eye. Using a transgenic mouse model in which the ocular lens expresses either transforming growth factor α (TGFα) or epidermal growth factor (EGF), under the control of the mouse αA-crystallin promoter, results in the failure of neural crest derived mesenchymal cells to migrate along the corneal stroma and differentiate into endothelial cells. This results in multiple defects in

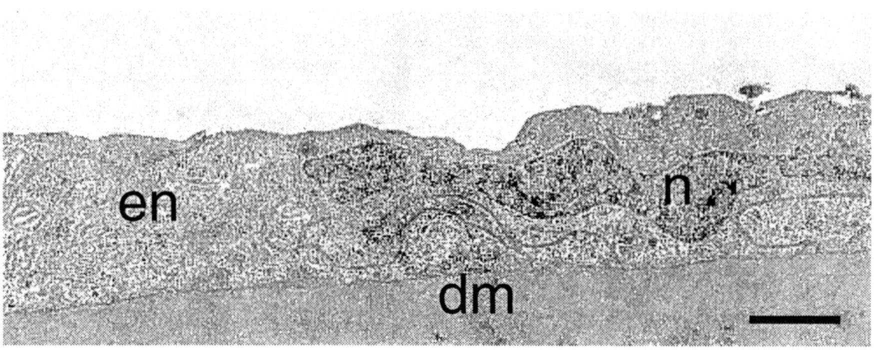

Fig. 1. An electron micrograph of the normal rat corneal endothelium. The endothelial cell (en) is rather flat and resides upon Descemet's membrane (dm). The nucleus (n) observed from this perspective is a somewhat elongated shaped organelle. Bar = 1 μm.

the anterior of the eye, including the attachment of both the lens and iris to the cornea. In a related study, the importance of Src family kinases (SFKs) for proper neural crest mesenchyme migration and subsequent endothelial formation was examined by Takauska and co workers (2008). These investigators knocked out C-terminal Src kinase (Csk), a negative regulator of SFK. using the Protein zero promoter in a Cre-loxP system. Their findings indicate that knocking out Csk prevented neural crest mesenchyme cells from migrating completely across the corneal stroma resulting in the central region of the cornea being devoid of endothelial cells and Descemet's membrane. Those endothelial cells that were present around the tissue periphery had disorganized actin cytoskeletons and exhibited defects in their ability to form stress fibers and lamellipodia when placed in tissue culture. Thus, the establishment of the corneal endothelial monolayer is very dependent on cell migration and this ability of the cells to translocate also plays a major role during wound repair of the tissue.

As gestation proceeds and the morphology of the monolayer begins to form, endothelial cells start to organize much of their actin into circumferential microfilament bundles (CMBs) around their cell peripheries (Gordon 1990). These actin-containing microfilament bundles are associated with the zonula adherens cell junctions (Gordon and Wood 2009; Petroll et al. 1999) and contribute to the maintenance of the monolayer integrity in much the same manner as observed in vascular endothelium (Lee and Gotlieb 2003; Wong and Gotlieb 1986). At about the time that the endothelial monolayer morphology is being established, cells begin to exit the cell cycle and shortly after birth the population becomes a quiescent population of non-cycling cells (Joyce 2003; Joyce, Harris and Zieske 1998; Gao et al. 1993; Gordon 1990). Recent evidence suggests that transforming growth factor β2 (TGF-β2) may function as a mitotic regulator of endothelial cells. Joyce et al. (2002) indicated that the presence of TGF-β2 in the aqueous humor may help to maintain this state of mitotic arrest, but investigations by Nishida et al. (1995)

demonstrate that the endothelium in situ produces mRNA transcripts of all three TGF-β isoforms. Recent studies by Chen et al. (2003) demonstrate that TGF-β2 inhibits endothelial proliferation in cultured cells by enhancing the production of prostaglandin E2, a known mitotic suppressor (Jumblatt 1994). Since TGF-β is known to stimulate the production of extracellular matrix (Schultz and Wysocki 2009; Rahimi and Leof 2007) it is possible that its production by the endothelium or that its presence in the aqueous humor ensures that the cells of the tissue maintain their focus towards the production of extracellular matrix components such as fibronectin and laminin (Usui et al. 1998), since one of the endothelial roles in vivo is the continuous production of Descemet's membrane, which thickens with age (Jun et al. 2006). Although there is some evidence for a stem cell population (Whikehart et al. 2005), under normal conditions, aging in all vertebrates leads to a decreased cell density (Bourne et al. 1997; Møller-Pedersen 1997; Laing et al. 1976) as dying cells appear not to be replaced by the proliferation of the remaining surviving cells of the tissue monolayer. As a means of compensating for their inability to undergo mitosis in the adult endothelium and to ensure a continuous covering exists across Descemet's membrane, cells display a unique propensity to spread themselves out in order to offset cell loss and maintain their cell-to-cell adhesion and integrity of the tissue monolayer (MacCallum et al. 1983; Rahi and Robins 1981).

In addition to the presence of TGF-β, another mechanism that appears to maintain the endothelium in a non-cycling state of quiescence is their adhesion to each other, in which they essentially maintain a form of contact inhibition of growth (Blat, Villaudy and Harel 1994; Martz and Steinberg 1972). When the developing rat endothelium begins to attain an adult-like morphology the formation of the CMB is also observed (Gordon 1990) as is the decline of cycling cells. Gordon (2002) demonstrated that disrupting the CMB of adult endothelial cells in non-injured tissues with cytochalasin B permitted a subsequent mitotic response in organ culture that could be augmented when either insulin or insulin-like growth factor-2 (IGF-2) was presence in the culture medium. Similarly, when pieces of human corneal endothelium were treated with the calcium chelator EDTA and organ cultured in the presence of both epidermal growth factor (EGF) and fibroblast growth factor (FGF), mitotic figures were observed (Senoo, Obara and Joyce 2000). Thus, these results indicate that cell-to-cell and cell/matrix adhesions exhibited in the endothelium, indeed also form a viable basis for the cessation of their cell growth.

Physiologically the corneal endothelium functions as a transport epithelium, concerned with maintaining water balance within the tissue's collagen-rich stroma, thus ensuring corneal deturgesence and transparency. The inward passive diffusion of fluid back into the stroma is, in part, due to the presence of non-continuous tight junctions that allows the tissue to exhibit permeability toward tracer molecules such as horseradish peroxidase (Gordon et al., 1998; Kaye et al., 1973). The maintenance of stromal hydration is accomplished by a mechanism termed the "pump-leak" by Maurice (1972) to characterize the outward active transport of ions that counters the inward passive "leak" of fluid back into the tissue. The non-continuous tight junctions of the tissue constitute the morphological basis for this "leakiness" of the endothelium. The ion pumps that are responsible for maintaining water balance are located in the highly interdigitated lateral cell membranes (Yokota and Waller 1975; Leuenberger and Novikoff 1974; Kaye and Tice 1966) and have been shown to consist of Na^+/K^+ ATPase and bicarbonate pumps as well as several voltage regulated ion channels that move their respective ions into the aqueous humor of the anterior chamber (Mergler and Pleyer 2007; Bonanno 2003).

In vascular endothelial cells, the CMB provides a cellular basis for the tissue's barrier function (Bogatcheva and Verin 2008; Mills and Mandel 1994), restricting the pericellular movement of macromolecules. A similar situation appears to exist in the corneal endothelium. Tight junctions in this tissue do not constitute the "classical" zonula occludens junction around the cell, i.e., they do not form a continuous barrier around endothelial cells, but exist as focal tight junctions. Nevertheless, higher molecular weight cytochemical tracers such as horseradish peroxidase type II and ferritin are excluded from crossing the tissue monolayer either in situ or in vitro (Gordon et al 1998: Raphael and McLaughlin 1990; Kaye and Pappas 1962) and only appear in the endothelial cytoplasm within vesicles or tubules, indicative of absorptive endocytosis. Recent work has implicated the endothelial CMB in the occlusion of such tracers and the maintenance of barrier function. Microtubule depolymerization following nocodazole treatment promotes increases in myosin light chain kinase phosphorylation and RhoA activation producing CMB disruption and alters the staining pattern of ZO-1 (zonula adherens - 1, a cell junction protein) leading to enhanced horseradish peroxidase permeability across the monolayer (Jalimarada et al 2009). Similar mechanisms also appear to govern CMB barrier function in vascular endothelium (Verin et al. 2001; Carbajal et al. 2000). Additional studies by Sakamoto et al. (1995; 1993) demonstrates that thrombin exposure to confluent cultured endothelial cells also produces a breakdown of the CMB pattern with a redistribution of actin filaments into the center of the cells that appears to work through a protein kinase C mechanism. Recent work indicates that both adenosine and extracellular ATP both act to dephosphorylate myosin light chain kinase and restore barrier function (Seinivas et. 2004; Satpathy et al. 2005). Results from these studies support the concept that in the endothelial tissue monolayer the CMB is regulated by contractile mechanisms, which in turn may act as a means of compensating for the lack of a continuous zonula occludens and therefore, under non-injured conditions, it serves as a major contributor toward maintaining the tissue's barrier function.

3 Organ Cultured Corneal Endothelium as a Model System

When studying a complex biological process such as cell migration provoked by an injury, many results can best be achieved by using experimental systems that allow for straightforward observations and analyses of the process. As so aptly stated by Weiss, "The biologist, in dealing with the problems of wound healing, endeavors to single out very simple forms in which the problems can be studied more easily..." (1956). Hence, much of the work carried out in understanding cell movement has come from observations on cell culture systems. Under these conditions investigations have ranged from the early studies of single cell migratory behavior to studies on cell movement during in vitro wound repair processes using model systems such as fibroblasts (Grinnell et al. 2006), vascular endothelial cells (Lee et al. 2001; Gotlieb et al 1991) or the movement of other cell types (Kuriyama and Mayor 2008; Raja et al. 2007). While much solid information has been gleaned from these observations, one must keep in mind that cell culture represents an artificial environment in which to observe cell movement. Often cells translocate across native plastic culture dishes or glass or plastic surfaces that have been coated with either single or multiple combinations of molecules such as polylysine or various extracellular matrix proteins to closer mimic a more natural condition. In other instances, various cell types may be seeded onto extracellular matrices produced by other cells, such as corneal endothelium, to help facilitate a more favorable condition. However, using the organ cultured corneal endothelium allows for investigations of cell movement along a natural basement membrane that is inherent to the tissue since the cells themselves are responsible for its synthesis (Labermeier and Kenney 1983; Murphy et al 1984). Even studies using injury-induced migration in confluent cultures of corneal endothelial cells have certain drawbacks since *in vitro* these cells have been shown to produce an array of extracellular matrix components (MacCallum et al 1982) that are not all present in the native Descemet's membrane.

The corneal endothelium is suited in many ways to be a model system for studying cell movement on a natural basement membrane. The tissue is composed of a homogeneous appearing cellular monolayer of polyhedral shaped squamous epithelial cells (Fig. 2) that closely adhere to Descemet's membrane. It receives neither a direct blood supply nor does it gets innervated from corneal nerve branches. Thus, it is well suited for organ culture based investigations. Cells of the adult tissue are all in a non-proliferating state of the generative cycle, that as previously stated, is thought to be maintained by TGF-β2. The peripheral organization of much of their actin cytoskeletal framework into CMBs is more suited for maintaining endothelial integrity and normal transport functions of the tissue as an intact monolayer than it is for motility and the cells do not display any morphological appearance that would indicate a predisposition toward motile behavior. In short, relative to wound repair and cell movement every cell within the population is essentially at the same position in the starting gate.

A classical approach to endothelial wound repair is to initiate the process by producing an injury such as a circular freeze wound over the central portion of the tissue. This procedure provides an injury to the tissue that results in a region of denuded Descemet's membrane surrounded by the surviving cells (Fig. 3) that will migrate into wound area. Depending on the size of the cornea, which is dependent on the source from which the tissue is secured, the wound can be administered either as a transcorneal freeze injury, such as what my laboratory performs with the rat cornea, or by placing a super-cooled probe directly to the endothelial layer, which is what is done when using corneas from larger sources such as rabbit,

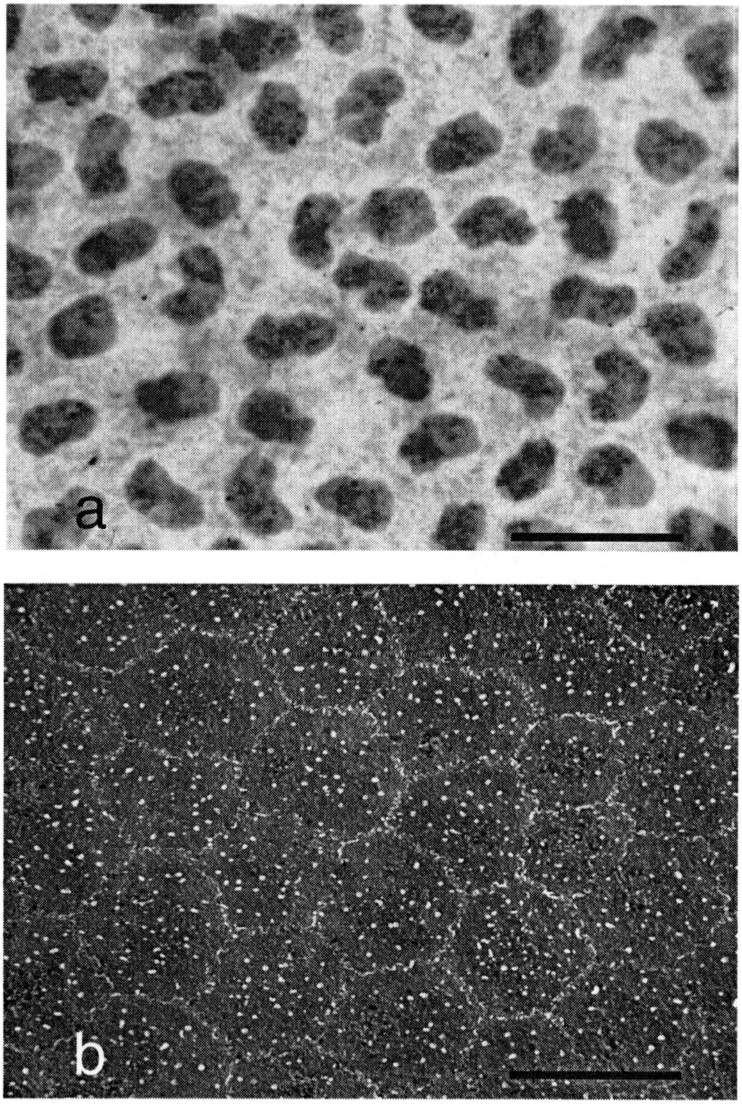

Fig. 2. The general morphology of the endothelium. a. Light microscopy of a flat mount of the rat corneal endothelium. The nuclear morphology ranges from kidney to oval in nature. In the intact tissue, cells form a tight contiguous monolayer. b. Scanning electron micrograph of the rabbit endothelium. Note the hexagonal appearance of the cells and their interleafed lateral membranes. Bars a = 40 μm b = 20 μm.

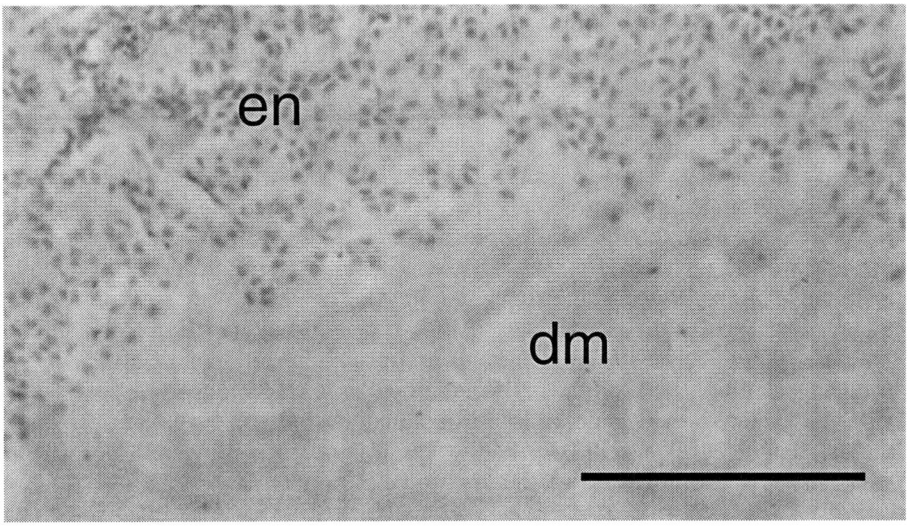

Fig. 3. A light micrograph demonstrating the wound area of the endothelium created as a result of a freeze-injury. Surviving endothelial cells (en) are observed surrounding the wound region, which is now devoid of endothelial cells, exposing bare Descemet's membrane (dm). Bar = 250μm.

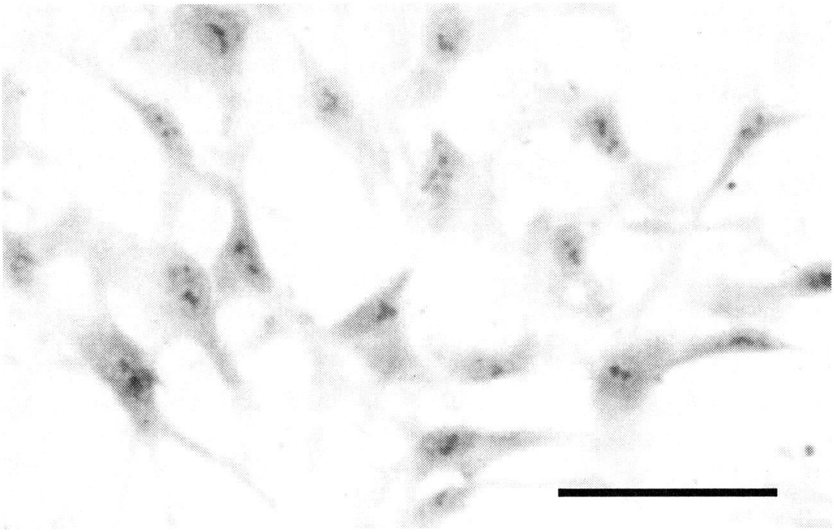

Fig. 4. An example of endothelial cells migrating in response to a freeze injury. The cells have undergone an epithelial mesenchymal transition and now appear fibroblastic and extend long processes to aid in their movement into the wound. Migration is towards the lower left of the micrograph. Bar = 100 μm.

bovines or humans. Using either method produces a well-defined central wound in which the cells surrounding the injury begin to initiate cell cycle progression and undergo directed migration toward the center of the lesion to repopulate the region (Fig. 4).

Following the completion of either in vivo or organ cultured experiments, Descemet's membrane acts as a scaffold to facilitate the stripping of the monolayer and its preparation as a flat mount by simply peeling it and the adhering endothelial cells away from the rest of the corneal tissue. This affords one the opportunity to view wound repair in a manner that allows for the visualization of the entire endothelial tissue. Because the tissue can be easily peeled away from the stroma after fixation with various fixatives such as either Carnoy's or formaldehyde, numerous histological, histochemical, and immunocytochemical procedures can be routinely accomplished and the entire cell population can be observed. In addition, since the isolated tissue is comprised of a homogeneous cell type, the fresh endothelium can be easily separated from the rest of the cornea allowing for biochemical or molecular biological investigations to be performed.

4 Cell Proliferation Is a Component of Endothelial Wound Repair

Injury to the endothelial monolayer produces an injury response mediated mainly by those cells that surround the wounded region of the tissue. Cells immediately adjacent to the circular freeze injury are the first to respond to the insult and are followed by successive tiers of cells until the tissue can organize a suitable wound response to restore the monolayer. These cells undergo a loss in their cell-to-cell contacts, reorganize their actin cytoskeleton, and begin to migrate into the wound area. During this time they also start to undergo macromolecular synthesis (Wang et al. 2002; Ross et al. 2002; Joyce, Mehlir and Zieske 1996; Rothstein and Gordon 1980), DNA synthesis (Fig. 5) and mitosis as part of the repair response (Landshman, Soloman and Belkin, 1989; Tuft, Williams and Coster 1986; Gordon and Rothstein 1978). In contrast to this dynamic response, cells residing in the peripheral areas of the tissue remain static, maintaining their normal morphological appearance and retain their cell-to-cell connections with adjacent neighboring cells. Thus, they maintain their position within the tissue and do not partake in the repair process.

Injury promotes the cell population around the wound region to re-initiate the cell cycle. Although at the moment of wounding all the cells reside in the G_0/G_1 compartment of the cell cycle (Joyce et al. 1996; Gordon and Rothstein 1978) they rapidly lose any cell cycle synchrony relative to macromolecular synthesis and growth initiation. Because of this loss in synchrony in the rat for example, DNA synthesis commences by 24 h post-injury, reaches a peak at 30 h and then exhibits a plateau between 36 to 54 h before declining (Gordon and Rothstein1978). Interestingly, using a bovine corneal endothelial cell culture model system, Miyata et al. (1989) observed a similar timing for DNA synthesis during in vitro wound repair for a circular injury. As one would expect, the mitotic response, although

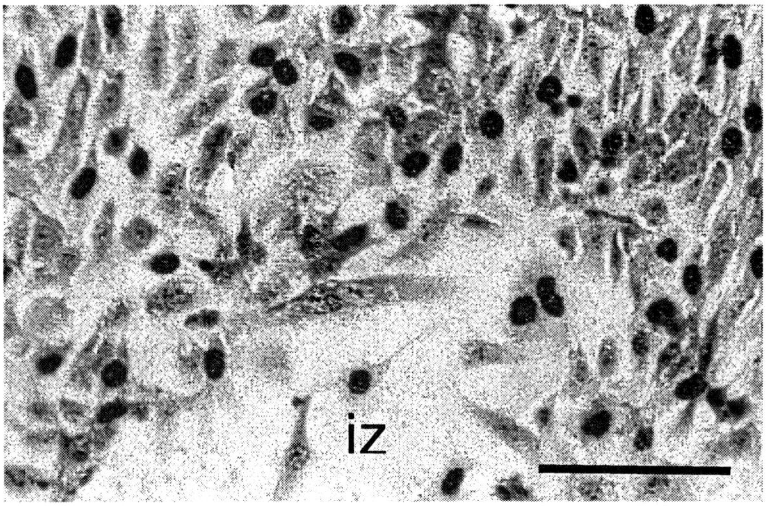

Fig. 5. As a result of injury, rabbit endothelial cells leave their quiescent state and re-initiate the cell cycle as they migrate into the injured area (iz). This autoradiograph demonstrates the incorporation of ^{3}H-thymidine (dark nuclei) characteristic of DNA synthesis. Note that these cells have also undergone an epithelial-mesenchymal transition and appear fibroblast-like in their morphology. Bar = 100 μm.

commencing several hours later, demonstrates a time course that roughly parallels the DNA response. In this case, mitosis is first observed by 36 h, plateaus between 36 and 66 h post-injury and thereafter declines to much lower levels (Gordon and Rothstein 1978). Several investigations have also shown that the mitotic response to injury either in vivo or in organ culture can be augmented by the presence of growth factors such as EGF, FGF or vascular platelet-derived growth factor (Riech et al 1995; Schultz et al 1994; Hoppenreijs et al 1994; 1992; Rich, Hatfield and Louiselle 1992).

Although the significance of the mitotic respond represents a critical need for the reestablishment of cell numbers in order to restore the tissue to full cell density and function, studies indicate that wounds can be repaired in the absence of a proliferative response (Gordon and Rothstein 1982). In such situations, when injured endothelia are exposed to the DNA synthesis inhibitor hydroxyurea, that subsequently prevents cell cycle progression, cell migration and spreading by the cells surrounding the injury area are the main processes used to restore the integrity of the endothelial monolayer. Nevertheless, the importance of restoring endothelial cell numbers remains a tissue priority. This was demonstrated following the cessation of hydroxyurea treatment. In this case, cells that had migrated into the wound and spread out along the once denuded Descemet's membrane, subsequently re-initiated cell cycle progression and completed the mitotic response and restored cell density to values comparable to control (non-hydroxyurea treated) endothelia (Gordon and Rothstein 1982).

5 Migratory Response of Corneal Endothelial Cells during Wound Repair

Epithelial cells tend to migrate into the injury while still maintaining their connections to neighboring cells (Sawada et al. 2008; Tétreault et al. 2008; Dipasquale 1975). This most probably is related to the fact that as an epithelium, they act to maintain their barrier function during wound repair (Zelenka and Arpitha 2008). However, unlike wound repair in classical epithelial healing, corneal endothelial cells responding to a freeze injury tend to break away from their neighboring cells and migrate into the wound region as individual cells, reminiscent of their neural crest origins. The endothelial repair process involves the cells undergoing an epithelial-mesenchymal transition, cell spreading and finally cell migration in order to repair a wound. Changes in the cell morphology and migration during wound repair in vivo or in organ culture along Descemet's membrane mimic those movements of neural crest derived mesenchymal cells during endothelial development. Studies by Nelson and Revel (1975) observe that as these cells migrate across the acellular collagenous corneal stroma during development, the presumptive endothelial cells appear highly flattened, fibroblastic-like in their morphology and display surface ruffles and lamellipodia but neither are well developed and most cellular processes appear to extend outward from the cell periphery. Similar cell features were also noted by Bard, Hay and Meller (1975). In fact ruffling, which is a normal feature of cultured cells but is minimal in endothelial cells migrating along the basement membrane in vivo demonstrates how the nature of the substrate influences the morphological behavior of migrating cells. Though presumptive endothelial cells that are translocating appear elongated, few have their long axis aligned with the radius of the cornea (i.e., in the direction of cell movement) and cells are in contact with adjacent cells mostly due to an overlapping of their cell processes. Nelson and Revel (1975) also point out that the radial inward movement of presumptive endothelial cells during development seems to be guided by "collisions" and contact inhibition with a mechanotaxis input probably supplied by the underlying substratum. This scenario, though it can be modified and somewhat re-interpreted in terms of today's current biological concepts, appears to reflect those same cellular responses observed in cell morphology and migration during the wound repair of the adult tissue.

As is the case for any cell moving along a substrate, endothelial cells must continually make, break and remake connections to the underlying matrix in order to maintain their migration. Because endothelial cells only move along the surface of Descemet's membrane, inhibitors like phenanthroline and phosphoramidon that affect metalloproteases have no effect on their migration. However, leupeptin and antipain, two serine/thiol protease inhibitors, prove effective in retarding migration into the wound area, as does the lysosomotropic amine methylamine, suggesting a possible role for lysosomal proteases in endothelial cell movement (Gordon and DeMoss 1999). Another possible protease that may be involved in endothelial cell movement is urokinase-type plasminogen activator. This enzyme has been associated with cell movement in other systems (Dellas and Loskutoff 2005; Kjøller 2002) and has been shown to be inducible in endothelial cell cultures (Ramsby and Kreutzer 1993a; Fehrenbacher, Gospodarowicz and Shuman 1979). Therefore, it is very likely that endothelial migration in response to injury appears to involve the activation of one or more types of proteases that serve in a permissive role to mediate cell-matrix interactions during movement along Descemet's membrane.

In the corneal endothelium, the size of the injury will dictate both the migratory and mitotic responses observed during wound repair. When small injuries are made that only create small, relatively focal wounds, the endothelial layer behaves in much the same manner as is observed in epithelial wound repair, that is, cell movement occurs without the loss of cell contacts with neighbors. In other words, small endothelial wounds heal by a mechanism similar to the movement of an epithelial cell sheet. Under this condition, cells that border the wound become elongated and begin to move into the denuded region of Descemet's membrane and essentially pull the rest of the cells along with them to cover the defect. If the wound is of a small enough nature only the movement of cells is needed to restore the monolayer and no mitosis will be observed (Honda et al. 1982). If the wound size is increased to a somewhat larger area, mitosis may or may not accompany the migratory response (Matsuda et al 1985), and larger wounds, such as freeze injuries (Landshman et al 1989; Tuft et al. 1986; Gordon and Rothstein 1978) always produce mitotic responses. Recent studies by Grasso et al. (2007) on cultured endothelium suggest, at least for in vitro wound repair, it is not the size or geometry of the wound that dictates the type of healing response, but the presence or absence of the extracellular matrix that defines the cellular mechanism(s) employed for migration.

In the case of small wounds, cells respond by spreading and elongation in the direction of their migration towards the injury (Honda et al. 1982; Matsuda et al. 1985). Using cat endothelium as a model, it was found that in this scenario cells retain contact with neighbors and the observed movements that many of these neighboring cells undergo is described as "cell rearrangements", resulting in variations in the number of lateral surface contacts between cells during the repair process (Honda et al. 1982). Despite the fact that these cells continue to move en mass, individual cells within the sheet exhibit some variability in their direction of migration. Some cells move toward the wound, whereas others show little movement and still others move in a different direction within the cell sheet. Nevertheless, the wound area continues to decrease over time (Matsuda et al. 1985; Honda et al. 1982). Additional observations indicate that during the healing of this type of wound there is a considerable mount of fine motile behavior whereby smaller cell rearrangements and movements continue to take place for extended lengths of time after the initial injury (Honda et al 1982). In confirmation of this work, Matsuda and colleagues (1985) reported that cell migration in their small wound model in rabbit endothelium was more pronounced in the first 12-24 h after injury and slowed so that by 48 h only slight migration was observed. Finally, as the wounds closed, the endothelial cells within the injury zone appeared more spread out and irregular in shape, but began to assume a more hexagonal appearance between 48-72 h after injury.

In response to larger wounds, such as a circular freeze injury, endothelial cells undergo an epithelial-mesenchymal transition in their appearance and move into the wound in a directed radial inward migration toward the wound center until finally repopulating the region. In the case of the rat endothelium, a large wound measures between 2.0-2.5 μm in diameter, resulting in a damaged region of approximately 25-30% of the endothelium. The total time for restoration of the cell population within the damaged area is dependent on the initial size of the wound relative to the entire endothelial area, which usually occurs between two and three days for most injuries, but in some cases, depending on the type of injury, may not occur until 7-10 days or later after the initial wounding (Schubert and Trokel 1984). Cells initially extend out processes as early as two hours after injury in the rat (Gordon, Climie and Hitt 2005) and migration has been reported to commence by 6 – 8 hours after wounding in the rabbit (Schubert and Trokel 1984; Khodadoust and Green 1976). In the rat, by 24 h post-wounding, endothelial cells around the injury border have lost their cell-to-cell contacts with their adjacent neighbors, assumed a fibroblast-like morphology (a result of epithelial-mesenchymal transition) but only demonstrate a small degree of migration into the injury zone. However, soon afterward they commence movement along Descemet's membrane, keeping in close contact with the substrate, and occasionally exhibiting an extension at the leading edge (Fig. 6), eventually repopulating the area by 48 h post-wounding (Gordon and Staley 1990). Initially, after cells have moved into the wound area the region looks rather disorganized and is filled with fibroblast-like cells and mitotic figures (Fig. 7). However, soon these cells cease their migration, begin to spread and reform the monolayer. Interestingly, in these larger wounds, the physiological function of the endothelium, relative to the histological appearance of the healed endothelium, lags behind as measured by corneal thickness (Khodadoust and Green 1976). This suggests that although the tissue layer is reestablished histologically, the barrier function, probably due to the restoration of junctional complexes, occurs at a later time. In a similar study, Gordon and Rothstein (1982) observed wound repair in mitotically inhibited frog endothelia that healed by migration alone. Although the wound area was repopulated solely by migration and a rather normal appearing endothelial layer was obtained by 96 h post-injury, not until six days after the wound did the corneas regain transparency and corneal thickness approached control values. For endothelial cells to uncouple with their neighbors in response to migration there must be a disruption of junctional integrity in order to allow for cell movement. Indeed, in the endothelium, wound response to a freeze injury has been shown to involve the loss of pericellular ZO-1 staining and its re-appearance in a punctate pattern in migrating cells (Gordon and Wood 2009; Petroll et al 1997). Conversely, when wound repair of smaller scrape injuries are examined, the ZO-1 pattern of the responding cell sheet remains intact except at the wound margin where cells extend lamellipodia from their leading edges (Petroll et al 2001; 1995). These studies suggest in a small wound situation, the endothelium by migrating as a cell sheet, may maintain its cell junctions and probably retain some barrier function properties during wound repair as suggested by Zelenka and Arpitha (2008). In contrast, in the case of a larger endothelial injury, the cells have no choice but to dissociate from one another, migrate along Descemet's membrane to reestablish a contiguous monolayer and subsequently restore the barrier function at a later time.

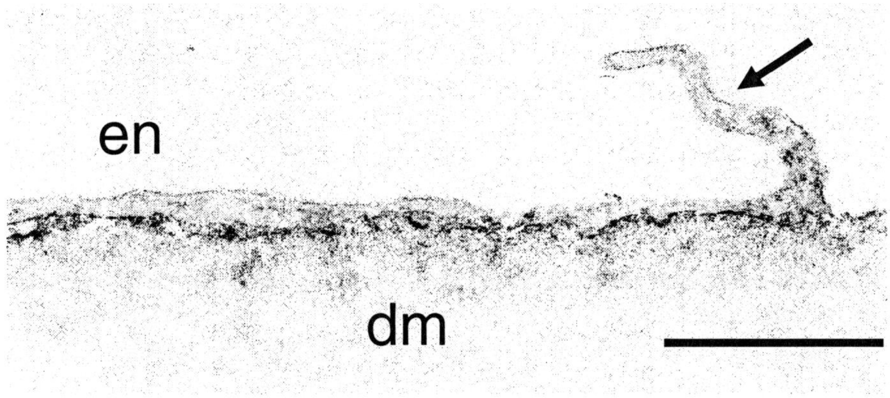

Fig. 6. An electron micrograph depicting a migrating rat endothelial cell (en) moving along Descemet's membrane (dm). The leading edge of the cell is very flat and closely apposed to the basement membrane. This particular cell has a cytoplasmic extension protruding from its leading edge (arrow). The dark material in the micrograph is immunoperoxidase reaction product that detects the presence of fibronectin. Bar = 1 μm.

In many species, for example, the cat, primate and human endothelium (Van Horn et al. 1977; Doughman et al 1976; Van Horn and Hyndiuk 1975) the tissue exhibits a very poor mitotic response and wound healing is accomplished mainly through migratory and not proliferative mechanisms. Thus, there has been a concerted effort involving many investigators to examine a means of increasing endothelial proliferation during wound repair both in vivo and in vitro using various growth factors (see review of Joyce 2003). Interestingly, other studies indicate that endothelial cells can initiate cell cycle traverse when either their cell/matrix associations or cell to-cell contacts are disrupted followed by exposure to growth factors (Gordon 2002; Senoo, Obara and Joyce 2000;). Because of the importance for migration in endothelial repair, investigators have also turned their attention to the effects of growth factors in augmenting cell movement during healing. Using organ cultured endothelia several studies have demonstrated a stimulatory effect of growth factors during wound repair in organ cultured endothelia (Sabatier et al. 1996; Hoppenreijs et al 1994, 1992; Yashida et al. 1989; Weimar, Squires and Knox 1980; Gospodarowicz and Greenburg 1979). In each case, the presence of the growth factor tested (FGF-2, PDGF or EGF), enhanced the migration of the endothelial cells during wound repair. Related in vitro work using endothelial cell cultures have also demonstrated that various growth factors can promote the migratory response during wound repair (Rieck, Cholidis and Hartmann 2001; Schilling-Schön et al. 2000; Joyce, Matkin and Neufeld 1989). In a study by Grant and co-workers (1992) the effect of growth factors on bovine and human endothelial cell migration in vitro was investigated using chemotaxis chambers. It was determined that both cell types positively responded to EGF, FGF and TGF-β. Both

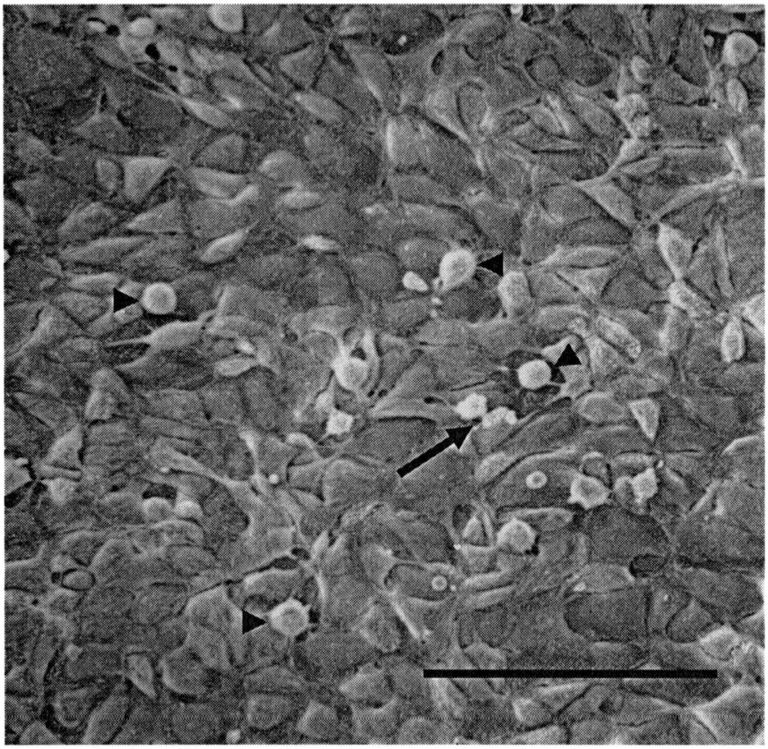

Fig. 7. Scanning electron micrograph of the central region of a freeze injury. Migrating cells have repopulated the wound area by 48 h which, at this time, appears to exhibit some degree of disorganization. Proliferating cells (arrowheads) appear rounded and one cell can be observed in the process of telophase (arrow), the final sub-stage of mitosis. Bar = 100 µm.

human and bovine cells displayed a high sensitivity to TGF-β, responding to concentrations as low as 1pg/ml, though higher concentrations caused a decrease in the migratory response. The importance of understanding the relationship between growth factors and endothelial wound repair is manifested by the fact that basement membranes, including Descemet's membrane act as reservoirs for growth factors (Vlodavsky et al. 1991; Morton et al. 1989; Folkman et al. 1988) that upon tissue injury could be released, bind to their surface receptors on endothelial cells and initiate intracellular signaling pathways that influence the wound repair process by modulating cell shape, growth and migratory activities. Nevertheless, caution must be used when determining how these in vitro studies relate to directed endothelial cell migration during wound repair along Descemet's membrane. In the case of in vivo or organ cultured endothelium, cells migrate toward the center of the wound, yet there is no apparent presence of a growth factor gradient. In vivo, injury promotes the production of secondary aqueous humor that biochemically is indistinguishable from serum (Coulter, Engelke and Eaton 1980;

Weinsieder et al. 1975), and organ cultured endothelium are usually placed in a surrounding environment that contains either serum or growth factors or both of these, such that no distinct chemoattractant gradient is readily apparent. Furthermore, it is rather unlikely that growth factors are deposited in Descemet's membrane in such a manner that they constitute a potential gradient from the periphery to the center of the tissue. In the case of endothelial cells, directed migration may simply be a case of cells finding themselves confronted with a free surface following an injury, and migrating into the wound, being directed (or "squeezed") toward its center due to the physical presence of adjacent neighboring cells that are also moving due to the same stimulus. Recent work has demonstrated that the directed migration of neural crest cells (from which corneal endothelial cells are derived) in vivo is the result of contact inhibition between adjacent cells, whereby cells with free edges become polarized and their interactions with neighboring cells forces them to undergo directional migration (Carmona-Fontaine et al. 2008). Thus, it is not unreasonable to assume that similar interactions occur between adjacent endothelial cells surrounding a wound, resulting in contact inhibition of movement in any direction except for their migration into the wound center. In such a scenario, as the cells converge at the wound center, they now become confronted by adjacent cells on all their surfaces and as a consequence, there is a cessation of migration and the reestablishment of an intact monolayer. From these observations it appears that those mechanisms involved in promoting directed migration during an injury response along the basement membrane may not be necessarily dependent on the presence of a chemoattractant gradient.

Studies on growth factor induced intracellular signaling pathways relative to stimulating cell migration in cultured endothelial cells has shed some light on the mechanisms that regulate motility. Investigations reveal that the stimulation of PI 3-kinase plays an important role for the protrusion of cell processes and shape changes of migrating cells (Lee and Kay 2006a; Lee et al. 2004; Rieck, Cholidis and Hartmann 2001; Gu et al. 1996), and this is a similar situation to that observed in vascular endothelium (Maekawa et al 2003). Recent work has also shown that this stimulation acts in turn to activate p38 (Lee and Kay 2009) a mitogen-activated protein kinase that is known to be involved in motility (Huang, Jacobson and Schaller 2004). In addition, endothelial migration has also been shown to involve protein kinase C (PKC) activation. However, Rieck, Cholidis and Hartmenn (2001) demonstrated that PKC activation was only observed independent of FGF-2 stimulation, whereas, Joyce and Meklir (1992) showed that it was stimulated in the presence of epidermal growth factor (EGF). Continued investigations on these growth factor stimulated pathways will undoubtedly help in elucidating those events that allow for endothelial cell movement during wound repair.

Perhaps the most dramatic event that occurs during endothelial wound repair is the epithelial-mesenchymal transition. This occurrence transforms endothelial cells from their polyhedral shaped morphology into their fibroblast-like morphology, characteristic of migrating cells. Though little is known about the mechanisms regulating this during in vivo repair, Petroll and colleagues (1998) demonstrated that TGF-β isoforms (β_1, β_2, β_3) all promoted epithelial- mesenchymal transition in injured organ cultured cat endothelium, in addition to stimulating actin reorganization and causing the loss of pericellular ZO-1 staining. A related observation was made by Sumioka et al. (2008) who demonstrated that transfecting endothelium with the TGF-β1 receptor antagonist Smad7, with an adenoviral vector, prevented epithelial-mesenchymal transition in wounded rat endothelium in vivo. Under this condition, cells failed to exhibit a fibroblast-like morphology and remained flattened. In addition, recent work indicates that gap junction communication may also be an important mechanism that mediates epithelial-mesenchymal transition during rat endothelial wound repair. In experiments that employed a knockdown of connexin43, a major gap junction protein, endothelial cells fail to undergo change into a fibroblast-like morphology but increased their level of proliferation during wound repair (Nakano et al. 2008). Interestingly, studies have shown that TGF-β signaling may occur by pathways that involve gap junctions (Dai et al 2007; Hirschi et al 2003). Taken together, these studies imply an important role for TGF-β signaling as a mechanism that regulates epithelial-mesenchymal transition in the corneal endothelium on their natural basement membrane.

Additional studies using endothelial cell cultures have shed some partial light on the means by which epithelial-mesenchymal transition made be regulated. However, results of these investigations must be somewhat tempered by the fact that cultured cells are on artificial substrates and not on their natural basement membrane and that cell-matrix interactions are probably somewhat compromised relative to the in vivo situation. This scenario alone, more than likely, influences and possibly modifies the signaling dynamics between these two systems. Nevertheless, investigations on epithelial-mesenchymal transformation in cultured rabbit endothelial cells indicates that FGF-2, working through the phospholipase C - IP3-kinase membrane signaling pathway causes cell protrusions and thus modifies cell shape akin to the morphology of migrating cells (Lee and Kay 2006a; Lee et al. 2004; Gu et al. 1996). As a result of this FGF-2 stimulation, the GTPases Rac and Cdc42 are activated and both function to inhibit Rho activity leading to actin cytoskeletal changes and pseudopodia extension (Lee and Kay 2006b; Lee and Kay 2007). In contrast to what is observed either in organ culture or in vivo, cultured endothelial cells reorganize their actin from stress fibers (characteristic of cells in vitro) into cortical actin bands upon FGF-2 treatment. In the former conditions, when endothelial cells migrate along Descemet's membrane during wound healing, actin is reorganized from cortical bands into distinct stress fibers (Gordon, Climie and Hitt 2005; Ichijima et al. 1993; Gordon, Essner and Rothstein 1982). This difference between actin alterations in these two responses highlights the variances observed between cells in culture and those in vivo. Even so, investigating the epithelial-mesenchymal transition in either system will culminate in our gaining insight into how this dynamic morphological change is brought about.

6 Eicosanoids and Injury-Induced Endothelial Cell Movement

Eicosaniods are twenty-carbon-long fatty acids containing a five member carbon ring that are derived from arachidonic acid released from the cell plasma membrane by the action of phospholipase A_2 that are then processed through either the cyclooxygenase pathway to form prostaglandins or the lipoxygenase pathway into leukotrienes (González-Périz and Clària 2007). Both types of molecules have been implicated in fibroblast wound repair in vitro (Green et al 2004). In Hela cells, the lipoxygenase pathway has been demonstrated to induce cell spreading by reorganizing the actin cytoskeleton (Chun, Auer and Jabcoson 1997) whereas the cyclooxygenase pathway also acts to reorganize microfilaments, but in this case, to mediate cell migration (Glenn and Jacobson 2003).

Little is known about the role of eicosaniods in modulating corneal endothelial migration during wound repair either in vivo or in organ cultured tissue. However, several laboratories have utilized endothelial tissue cultures to determine the function of these molecules during in vitro wound healing. Reich and co-workers (2001) have reported that cessation of leukotriene synthesis by inhibition of the lipoxygenase pathway with NDGA (nordihydroguaiaretic acid) results in a severe restriction of endothelial cell movement following injury. Other studies on cultured endothelial cells have demonstrated that they are capable of synthesizing prostaglandin E_2 (PGE_2), and this activity is abated by indomethacin (Joyce et al 1995; Jumblatt 1994; Neufeld et al. 1986). The initial wounding of corneal endothelial cell cultures promotes an increase in the synthesis of PGE_2, preferentially by the action of cycolooxygenase 2, that after reaching a peak between 6-24 h after injury, reverts to control values as the wound heals by 72 h (Jumblatt and Willer 1996). The action of eicosaniods on endothelial wound repair in vitro has been evaluated either by the direct addition of the appropriate molecule to the culture media or inferred by results obtained with pathway inhibitors. Studies have indicated that in the presence of EGF and indomethacin, an inhibitor of cyclooxygenase, the shape alters from their normal polygonal appearance to one of elongated cells, whereas, when EGF and indomethacin are used together with PGE_2, cells maintain their normal appearance (Neufeld, Matkin and Raymond 1990). Relative to its effects on cell migration during in vitro wound repair, PGE_2 has been reported to induce cell spreading, whereas EGF stimulated individual cell movement (Joyce et al. 1995; Joyce and Meklir 1994). Taken together, these results indicate that during in vitro wound repair, growth factors are responsible for mediating cell migration into the wound, whereas PGE_2, is responsible for cell spreading to re-establish the continuity of the monolayer. Because basement membranes serve as reservoirs for exogenous growth factors, it is not unreasonable to assume a similar type of situation exists during wound repair either in organ culture or in vivo.

7 Actin Cytoskeletal Changes Accompanying Cell Migration

Like all epithelial-type cells, in the normal non-injured corneal endothelium, actin, as noted previously, is mostly organized as apical CMBs that, following actin staining using either fluorescently tagged phallotoxins or immunocytochemistry, appears as a honeycomb pattern (Gordon, Climie and Hitt 2005; Petroll et al. 1999; Barry et al. 1995; Ichijima et al. 1993; Gordon and Staley 1990; Gordon Essner and Rothstein 1982). Studies have shown that the CMB co-localizes with several junctional proteins such as ZO-1, α, β and γ catenins, occludin and cadherins (Petroll et al. 1999; Barry et al. 1995) indicating a close relationship exists between the rabbit endothelial CMB and cell junction. Recently, Gordon and Wood (2009) confirmed that this is also the case in the rat endothelium by demonstrating with electron microscopy the structural association between the CMB and the zonula adherens junction of the endothelium.

Like other transport epithelia residing on natural basement membranes, corneal endothelial cells are highly polarized relative to their apical and basal ends in order to carry out their physiological function. How the CMB structure is maintained as an apical pericellular microfilament band in vivo is not well understood. However, studies employing confluent cultures of endothelium, which form even more elaborate CMBs than those seen in vivo, offer some insights into possible mechanisms. For example, when endothelial cells are allowed to incorporate the monovalent cation channel forming antibiotic gramicidin D or have their extracellular saline composition modified, both of which lead to membrane depolarization, the organization of their CMBs becomes altered. In both cases there is a loss of CMB actin that becomes reorganized as cytoplasmic filaments and results in a disruption of monolayer integrity (Chifflet et al. 2003). In another set of studies, Sakamoto and co-workers (1994) demonstrated that thrombin induced cytoskeletal changes to the endothelial cell CMBs were mediated by the protein kinase C pathway. In this situation, thrombin exposure produced changes in cell morphology and likewise, produced a loss of peripheral actin and an increase in cytoplasmic microfilament bundles orientated with the long axis of the cell, very reminiscent of actin changes described below that are observed in migrating cells during wound repair. Further investigations into these types of mechanisms may provide clues as to how CMB organization is maintained in non-injured tissue and how the cell regulates actin cytoskeletal changes in response to corneal endothelial cell injury.

When corneal endothelial cells translocate during wound repair on the basement membrane they must be able to move into the injury region to repopulate the area and restore an intact monolayer. As pointed out earlier, depending on the type of wound administered (either a scrape or transcorneal freeze injury) will determine the type of cell response. In small scrape wounds, endothelial cells simply spread to restore an intact monolayer, whereas, larger circular freeze wounds initiates cell shape alterations and individual cell translocation into the injury zone (Ichijima et al. 1993). One dramatic change that accompanies this migratory response is a significant change in the appearance of their actin cytoskeleton. In response to an injury, the CMB pattern of rat endothelial cells surrounding the wound rapidly disappears by 2 h after wounding (Gordon, Climie and Hitt 2005). Along with the loss of the CMB, the typical pericellular pattern for ZO-1 also disappears (Fig 8) as cells migrate into the wound (Gordon and Wood 2009; Petroll et al 1995). In contrast, in the case of smaller wounds, as cells spread out to restore monolayer integrity, the pericellular staining for both actin and junctional proteins is maintained (Petroll et al. 2001). Thus, wound size also appears to be a critical determinate in dictating the response of the endothelial cytoskeleton to injury. These results contrast to those obtained by Grasso et al (2007) using in vitro wound repair that indicate the mechanism and organization of the actin

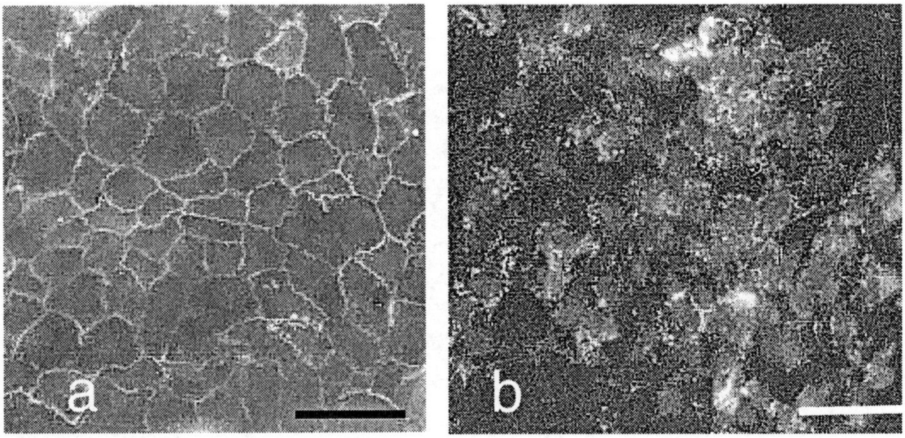

Fig. 8. Fluorescent micrograph demonstrating the distribution of the zonula adherens junctional protein ZO-1 using an anti ZO-1 antibody conjugated to a fluorescent marker. In a, the normal distribution is seen as a fluorescent band around the cell periphery. In b, cells that are migrating into the wound area demonstrate a punctate ZO-1 staining pattern. Bars = 50μm.

cytoskeleton during healing is independent of wound size but is dependent on the presence of absence of an intact extracellular matrix.

As cells begin their early response to the in vivo injury, they extend out processes into the wound, although they remain in contact with their neighbors and their actin cytoskeleton does not demonstrate any other noticeable changes except for the CMB disappearance. By 24 h post-injury, concurrent with their morphological change during epithelial-mesenchymal transformation, endothelial cells surrounding the wound have begun to lose cell-to-cell contact with their neighbors and migrate as individual cells (Ichijima et al. 1995; Gordon 1994) and actin microfilaments now appear distinctly in the cytoplasm in the form of so-called stress fibers (Fig. 9) that are oriented in the direction of movement (Gordon and Buxar, 1997; Ichijima et al. 1993; Gordon, Essner and Rothstein 1982). Similar actin changes have also been noted in other systems, such as organ cultured retinal pigment epithelium (Hergott, Sandig and Kalnins 1989) and aortic endothelium (Rodgers et al. 1989) where cells migrate along a basement membrane in response to wounding. In contrast, during growth factor stimulation of in vitro wound induced cell migration, endothelial cells also assumed shape alterations into fibroblast-like cells, but here their actin is reported to become diffusely distributed (Joyce, Meklir and Nuefeld 1990), similar to what is has been observed in other cell types during in vitro migration (Lewis et al. 1982; Herman, Crisona and Pollard 1981). This stark difference may be accounted for by the fact that in organ culture or in vivo, cell migration is greatly impacted by the chemical composition of Descemet's membrane, which, like all basement membranes, serves to greatly influence all aspects of cell physiology and behavior (See section 8). This fact cannot be overemphasized relative to the actin cytoskeletal organization. When

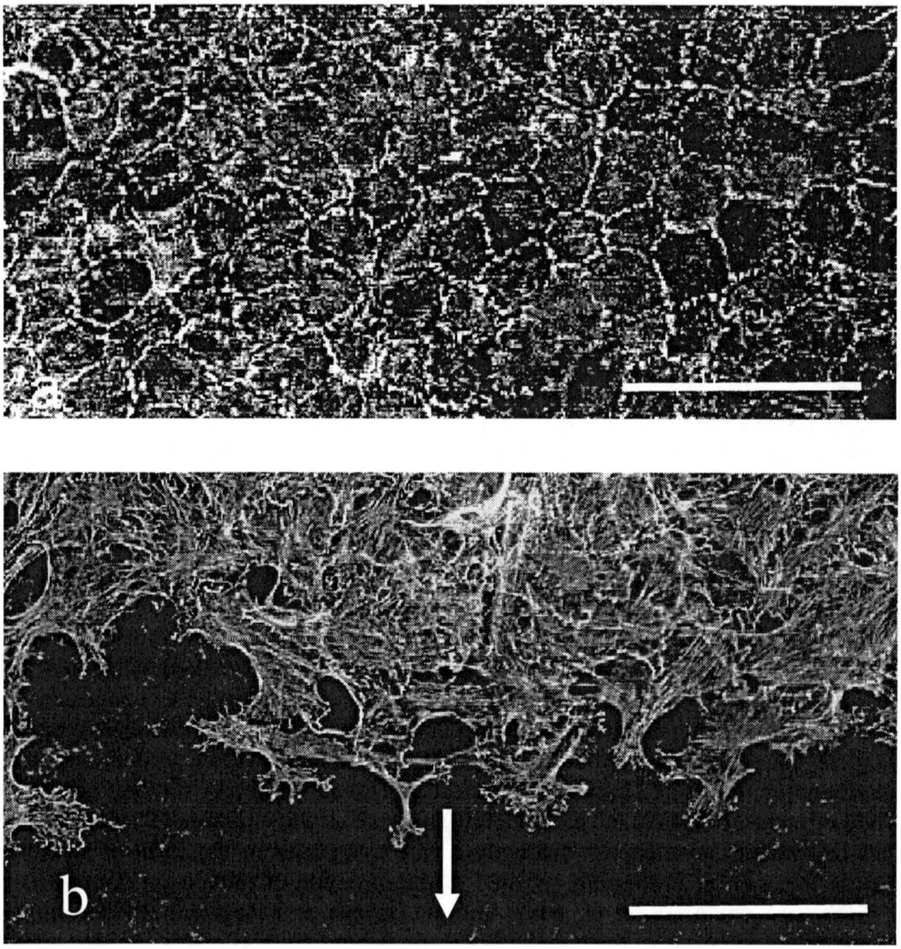

Fig. 9. Fluorescent micrographs that demonstrate changes in the organization of the actin cytoskeleton as a result of a freeze injury. a. In the non-injured tissue, actin is detected as circumferential microfilament bundles (CMBs) that impart the tissue with a honeycomb-like pattern. In b, cells are initiating their migration across denuded Descement's membrane in response to a freeze injury. These cells no longer exhibit a CMB pattern, and their actin has reorganized into stress fibers that now fill the cytoplasm. Arrow indicates the direction of cell movement. Bars = 100 μm.

placed into culture, endothelial cells elaborate a much more definitive actin pattern. As cells move away from their initial explant, they exhibit stress fibers and when they reach confluency, they produce a more elaborate CMB and contain stress fibers within a well spread cytoplasm (Gordon, Essner and Rothstein 1982). The fact that these cells are not on their native substrate, but are on a foreign surface easily accounts for the development of this different pattern in actin organization, thus serving as a dramatic indication of the influence of the basement membrane on cell organization.

Studies by Petroll and colleagues (1996; 1998) on organ cultured cat endothelium demonstrated that during wound healing the addition of TGF-β isoforms (β1, β2, β3) to the culture medium brought about a reorganization of actin into stress fibers with the concomitant loss of pericellular ZO-1 staining to a much greater extent than occurred when tissues were cultivated in either serum or basic fibroblast growth factor (bFGF), suggesting that TGF-β isoforms may act as signals for mediating endothelial actin cytoskeletal changes during wound repair. In addition, in vitro studies with bovine endothelial cell cultures have also examined mechanisms that affect endothelial actin cytoskeletal reorganization. Under culture conditions, it appears that while fibroblast growth factor-2 (FGF2; bFGF) is capable of initiating epithelial-mesenchymal; transition, in doing so, it promotes the loss of stress fibers in cultured rabbit endothelium possibly through the actions of phosphatidylinositol 3-kinases (Lee and Kay 2006c). This is somewhat different to the findings of Petroll et al. (1998) on injured organ cultured corneal buttons cultured in the presence of FGF-2. Under this condition, these cells demonstrated neither epithelial-mesenchymal transition nor stress fiber formation, but did exhibit an increased mitotic response. The difference responses of each study may simply reflect variations between substrates in tissue culture and organ culture. Lee and Kay (2003) also demonstrated that in cultured cells the actin cytoskeleton is influenced through what appears to be a Rho dependent pathway, as determined by use of inhibitors to Rho and ROCK activation. When PI-3 kinase is activated this pathway activates Rac and cdc43 (both GTPases) and leads to the inactivation of Rho with the subsequent reorganization of actin (Lee and Kay 2004). Furthermore, it has also been shown that inhibiting the small GTPase molecule Ras with dominant interfering mutant protein (N17) prevented cell migration in wounded cultured bovine endothelial cells (Sosnowski, Feldman and Feramisco 1993). Thus, the reorganization of the endothelial actin cytoskeleton during wound repair is probably mediated in much the same manner as in many cell types (Pellegrin and Mellor 2007: Schmidt and Hall 2002; Ridley 2001; Hall 1998) and these pathways more than likely are responsible for the actin changes observed during injury-induced cell migration seen in either the organ cultured or in vivo endothelium.

As stress fiber containing endothelial cells move into the wound area, those cells on the peripheral portion of the tissue do not contribute to the healing of the injury and retain their CMBs. The migration of individual cells into the damaged region results in the repopulation of the injury zone with a fair amount of apparent disorganization, a result of all the cells having converged toward the center of the injury. However, between 72-96 h post-injury, this disorganization is beginning to "calm down" and give way to an orderly restoration of a intact monolayer as cells reconnect with their neighbors and reform their CMBs, although the injury region now displays more of pleotypic cell morphology than prior to injury. This type of a wound response is much different than that observed in epithelium whereby cell sheets that undergo wound repair use a "purse-string" closure to repair a defect (Rodriguez-Diaz et al. 2008; Bement, Forscher and Mooseker 1993), which again possibly relates to the neural crest origin of the corneal endothelium.

That stress fibers aid in the facilitation of cell movement was shown by studies that used agents that either directly (cytochalasins), or indirectly (5-fluoruracil), effected microfilament organization and consequently interfered with cell movement and wound repair (Gordon Climie and Hitt 2005; Mohay and McLaughlin 1995; Gordon and Staley 1990; Fujino and Tanishima 1987). Cytochalasins are a group of fungal metabolites that bind to the plus ends of microfilaments and cause their depolymerization. When injured endothelium were organ cultured in the presence of cytochalasin B, wound repair slowed when compared to control tissues. Interestingly, disrupting the actin cytoskeleton in the migrating cells caused alterations in the appearance of their processes, which become rather long and slender. Still, the loss of stress fibers in the cells did not prevent them from moving into the wound but did delay their progress into the region, indicating that these structures are needed in order to allow for efficient migration to occur (Gordon and Staley 1990). In the case of 5-fluorouracil, a pyrimidine analog that acts as an inhibitor of thymidylate synthase and thus prevents DNA synthesis, concentrations ranging between 0.01 and 0.1 mM retard migration and wound repair, whereas, concentrations above 1.0 mM adversely affect microfilament patterns and extensively retard migration (Gordon, Climie and Hitt 2005; Mohay and McLaughlin 1995). The exact mechanism of how fluorouracil promotes its effects on the actin cytoskeleton are at present not understood, but it appears to be of an indirect nature (see below).

Gordon and Buxar (1997) demonstrated that during wound repair the appearance of stress fibers is independent of new actin synthesis, since immunoprecipitated actin failed to incorporate ^{35}S-methionine during wound repair. Actin represents a rather large amount (10%) of the total endothelial protein within the cell and the emergence of stress fibers appears to come about at the expense of dismantling the actin-rich CMB. In other words, actin within the CMB is recruited and reorganized into stress fibers to meet the migratory demands of the cells. When actin reorganization was prevented by stabilizing microfilaments with the mushroom toxin phalloidin (Wieland 1977), a high level of ^{35}S-methionine incorporation now became evident, indicating that only when actin reorganization was prevented, was there any need for the synthesis of new protein (Gordon and Buxar 1997). Similar observations were made by Reuner et al (1996) on rat hepatocytes.

Whether endothelial cells migrating along Descemet's membrane absolutely require actin to translocate is an interesting question based on experimental evidence accumulated with various inhibitors. Indeed, without the presence of organized microfilaments, wound repair is retarded (Gordon, Climie and Hitt 2005; Gordon and Staley 1990), thus indicating that their presence probably allows cells to translocate much more efficiently and effectively when they are intact components of migratory cells. Soybean agglutinin (SBA), a lectin with N-acetyl-D-galactosamine binding specificity, disrupts endothelial microfilament organization in either non-injured or injured tissues (Gordon and Wood 2009; Gordon and Wood 1997). In the presence of this lectin, cells without benefit of an organized actin cytoskeleton (Fig. 10) are still capable of migrating into the wound region, but surprisingly, in about the same time course as control tissues (Gordon and Wood 1997). However, these cells do not extend out long cellular processes but demonstrate much shorter

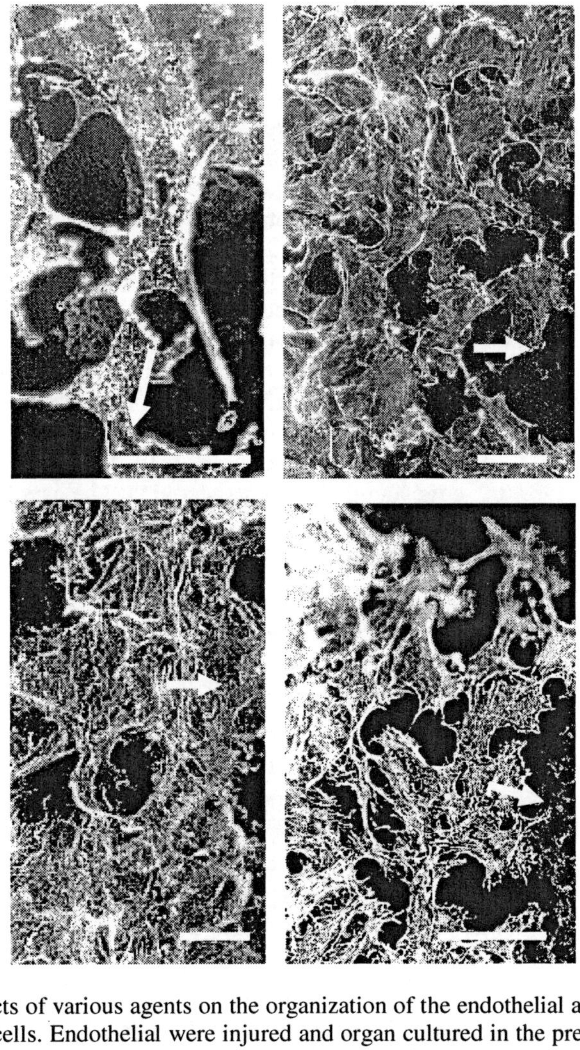

Fig. 10. The effects of various agents on the organization of the endothelial actin cytoskeleton in migrating cells. Endothelial were injured and organ cultured in the presence of $5x10^{-7}$M cytochalasin B (a), 100 µg/ml soybean agglutinin (b) and 0.1mM fluorouracil (c). In all cases the actin cytoskeleton is disrupted and cells fail to exhibit stress fibers, as observed in control tissues (d), but they continue to move into the wound area and subsequently repopulate the injury zone. Arrows indicate the direction of movement. Bars = 50 µm.

and blunter extensions when their actin has been disrupted. Similarly, actin is also disrupted in the presence of 5-fluorouracil and akin to SBA treatment, their cellular extensions are also blunted in migrating cells during wound repair in organ culture (Gordon, Climie and Hitt 2005). The cells nevertheless are still capable of migrating into the wound, albeit, in a more retarded manner than control preparations. In this study, only when the 5-fluorouracil concentration was at 1.0 mM was cell movement essentially inhibited. Using an in vitro culture system, Mohay and McLaughlin (1995) also demonstrated an inhibition of cell migration occurred using a much lower 5-fluorouracil concentration of 0.02 mM. This difference probably represents the difference between cells moving along their basement membrane as opposed to cell migration in an in vitro wound across an artificial substrate. The exact mechanism by which actin becomes disorganized as a result of 5-fluorouracil treatment remains unknown, though it most likely is an indirect affect since actin polymerization in the presence of the drug, as determined by falling ball viscosity and F-actin sedimentation studies, do not differ from control preparations (Gordon, Climie and Hitt 2005). Furthermore, when organ cultured endothelia are returned to normal medium that does not contain 5-fluorouracil, the disrupting effects of the drug are reversed and microfilaments reappear by 24 h afterward and the cells become indistinguishable from their control counterparts by 48 h. Because stress fibers have been shown to exhibit contractility (Pellegrin and Mellor 2007; Katoh et al. 2001), one may think that disrupting them would result in a cessation of cell movement and an inability to heal a wound. However, these results do not exclude the possibility that short actin filaments could still persist adjacent to the leading edge and their actions would still be capable of maintaining cell migration despite the apparent absence of a more extensive and organized actin cytoskeleton.

8 Matrix Proteins and Endothelial Wound Repair

As cells migrate in response to an injury the one physical component that most influences how they migrate is their underlying substrate. The nature of this entity influences many cellular processes including cytoskeletal organization, the expression of cell- matrix adhesion molecules, cell migration, growth and gene expression (Engvall 1995; Lin and Bissel 1993). Basement membranes are complex macromolecular structures (Timpl and Brown 1996; Paulsson 1992) that are synthesized by and resided upon by epithelial cells. Basement membranes are characterized by neither a uniform appearance nor an identical chemical composition, but rather, in a general sense, they are of a similar function, but the specificity of each, relative to their morphology and function is dependent on the location of which they are found (Grant and Leblond 1988). When epithelial cells move in vivo, the substrate which they translocate along is usually an acellular basement membrane or, in the case of fibroblasts, a composite of extracellular matrix (ECM) components. For epithelial cells, residing on a basement membrane means there are also profound effects on their cell organization and polarity. In the case of the corneal endothelium, the apical end of the cell is the site where the CMB is organized and associates with the zonula adherens junction (Gordon and Wood 2009; Petroll et al. 1999) while the lateral cell membranes are the sites of the transport pumps (Leuenberger and Novikoff 1974; Kaye and Tice 1966). This polarity is also demonstrated using biochemical analyses (Soltau, Zhou and McLaughlin 1993) that indicated different protein profiles exist for the apical and basal membranes, as determined by polyacrylamide gel electrophoresis. Furthermore, the morphological appearance of the monolayer differs when observed by scanning electron microscopy between the apical versus the basal surfaces. The apical surface is characterized by the presence of distinct interdigitating lateral membranes that give the tissue monolayer its typical hexagonal-type appearance, whereas, in the basal surface, broad overlapping cell process were seen extending between neighboring cells that did not interdigitate with other processes and the presence of gaps could be detected between them (Sherrard and Ng 1990). In addition, while the apical end of the cell is characterized by the CMB, at deeper regions toward the basal end of the cells, actin is seen in a thread-like pattern where the fibers radiate toward the center (Gordon Essner and Rothstein 1982). Thus, as in other systems, the basement membrane plays an important role in the organization of the cell (Martin-Belmonte and Mostov 2008).

In addition to maintaining corneal hydration and transparency, the corneal endothelium also functions in the formation and continuing synthesis of Descemet's membrane throughout life (Cintron, Covington and Kublin 1988; Johnson, Bourne and Campbell 1982; Wulle 1972). Thus, the cells not only function as transport epithelia, but also act as ECM producers. In this context, when they are placed into tissue culture they demonstrate an exceptional ability to secrete many matrix proteins. These include collagen (Yamaguchi et al. 1991; Sankey et al. 1981; Perlman, Baum and Kaye 1979), glycosaminoglycans (GAGs) (Robinson and Gospodarowicz 1983; Yue and Baum 1976) thrombospondin (Ramsby and Kreutzer 1993), laminin (Gospodarowicz et al. 1981) and fibronectin (Gospodarowicz et al. 1979). Studies by Usui et al. (1998) indicate that at least in the in vitro production of fibronectin and laminin both molecules appear to be regulated by TGF-β and its cell membrane receptor. When maintained for short term cultures endothelial cells secrete many filaments ranging in size from 10-40 nm diameters with some forming hexagonal lattices (Sawasa et al. 1987). After longer periods of culture, cells produce enough material to form basement membrane resembling substrates (MacCallum et al 1982; Sawada, Konomi and Nagai 1984; Perlman and Baum 1974). These observations indicate that in vitro, corneal endothelial cells are capable of producing numerous ECM components and with time, the continued secretion of these products begins to elaborate an organized and complex basement membrane-like substrate.

Endothelial cells not only synthesize ECM components in vivo and in vitro, but like all other cell types, they exhibit various physiological responses to them. For example, when plated onto ECM, cultured endothelial cells are more growth responsive (Blake et al. 1997; Gospodarowicz, Gonzalez and Fujii 1983; Gospodarowicz and Ill; 1980), exhibit phenotypic changes in cell shape (Hsieh and Baum 1985; Beach and Kenney 1982), cell attachment (Scott, Murray and Barnes

1983) and even have altered levels in the synthesis of their ECM components (Underwood and Bennet 1993). Other studies have also show distinct impacts of the ECM on in vitro cell movement during monolayer wounding (Scheef et al. 2007; Grasso, Hernández and Chifflet 2007; Kirkpatrick et al. 1990). These results, if extrapolated to the in vivo situation probably suggests an interesting scenario whereby the synthesis of the tissue's own basement membrane more than likely produces reciprocal effects on the cells, that in turn, assist in determining the physiological status of the tissue monolayer.

In the case of corneal endothelial wounding, either in vivo or in organ culture, a wound produces injury-induced migration along Descemet's membrane. One may conclude that since movement in response to injury entails translocation on a preexisting natural extracellular matrix, cells would simply initiate those processes, other than matrix synthesis and secretion, which are required to allow for movement to proceed and repair to be completed. However, that is not the case. When endothelial cells respond to injury, although they migrate along Descemet's membrane they still demonstrate the need to secrete extracellular matrix materials that facilitate their movement (Sabet and Gordon 1989; Gordon 1988).

In vivo, both laminin and fibronectin (approximately 440 and 800 kD molecular weights, respectively) appear in a pericellular distribution, and similar to actin, produce a honeycomb pattern in the tissue (Gordon 1988). The cytoplasm of endothelial cells shows very little staining for either ECM protein by immunoperoxidase cytochemistry. However, following a freeze wound, cells exhibit very strong cytoplasmic staining by 24 h post-injury (Fig. 11), that is maintained as the wound region becomes repopulated with migrating cells (Gordon 1988). Interestingly, after 14 days, the fibronectin pattern has returned in the repaired tissue but the laminin pattern has not, although cells still show evidence of strong immunoperoxidase cytoplasmic staining. Immunoperoxidase studies at the ultrastructural level reveal that in organ cultured endothelium, fibronectin is deposited along Descemet's membrane and into the interstitial space during wound repair and these localizations are absent in either control samples or in cells around the periphery of wounded tissues (Sabet and Gordon 1989). Furthermore, incubation in colchicines prevents this deposition, indicating a role for microtubules in the cellular transport of matrix material, and puromycin treatment abolishes all immunostaining, indicating that the fibronectin was derived from the endothelial cells themselves and not by a fortuitous deposition from the serum component of the culture medium (Sabet and Gordon 1989). That these changes in immunostaining are the result of an upregulation of gene expression can be inferred from the study of Nickeleit et al. (1996). These investigators demonstrated the presence of low levels of the V and EIIIB fibronectin domains in normal non-injured endothelium using in situ hybridization. As a result of either employing a scrape injury or anterior excimer laser keratectomy to inflict an epithelial injury, the endothelium also responded by upregulating levels of V, EIIIA and EIIIB fibronectin domains. Based on these results, one can infer that the immunoperoxidase staining observed after endothelial wounding, more than likely, is the result of an upregulation of fibronectin mRNA and the subsequent expression of its protein.

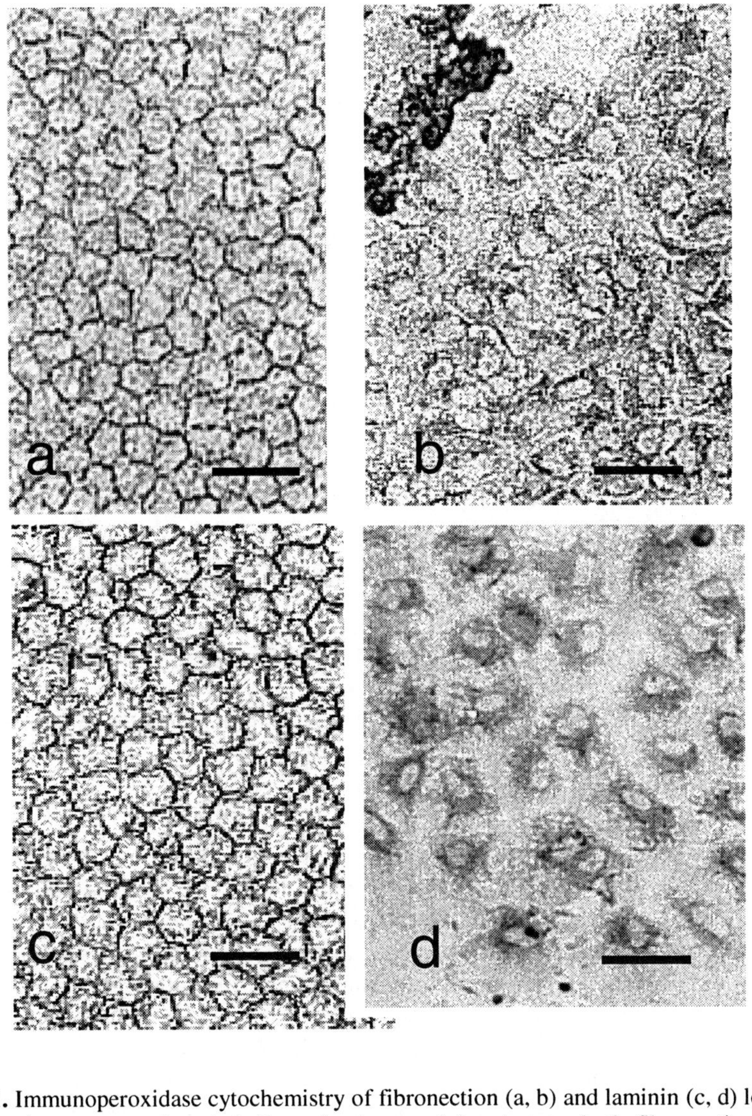

Fig. 11. Immunoperoxidase cytochemistry of fibronection (a, b) and laminin (c, d) localizations in the rat corneal endothelium. In the non-injured state, both fibronectin (a) and laminin (c) show a pericellular distribution of reaction product, producing a honeycomb-like pattern in the tissue, reminiscent of that observed with actin. Following a freeze injury, cells that are migrating into the wound area at 24 h post-injury exhibit increased staining for both fibronectin (b) and laminin (d). The injured area is in the upper left in (b) and lower left in (d). Bars = 50 μm.

In addition to alterations in the appearance of fibronectin and laminin, injury to the endothelium also produces changes in other ECM components. For example, in migrating endothelial cells, from both bovine and human tissues, the levels of cell surface proteoglycans becomes differentially expressed. Both heparan sulfate and chondroitin sulfate levels demonstrated increases, whereas, keratan sulfate levels declined. In non-injured tissues, cell labeling for these components revealed just the opposite situation, i.e., heparan and chondroitin sulfates were low while keratan sulfate was high (Davies et al. 1999; Fullwood et al. 1996). In fact, migrating cells displayed heparan sulfate at the interface between the cell-Descemet's membrane (Davies et al. 1999), strongly suggesting a possible regulatory role in cell movement. This differential appearance of the GAG levels detected between the intact tissue and migrating cells suggests differential roles for these molecules in the non-injured and wounded endothelia.

Another ECM protein that is detected in Descemet's membrane is thrombospondin (TSP) (Hiscott et al 1997; Tucker et al. 1995). TSP is a large (>420 kD molecular weight) protein that is involved in both cell-to-cell and cell-matrix interactions (Hiscott et al. 2006). The TSP family consists of several multifunction proteins (TSP 1-5) that are comprised of two homotrimers (TSP 1 and 2) and three homopentamers (TSP 3, 4 and 5) (Carlson, Lawler and Mosher 2008). In vitro studies have shown that thrombospondin, as is the case with fibonectin and laminin, influence several aspects of cell behavior (Vogel et al. 1993). As would be expected, since TSP is present in Descemet's membrane, it is readily synthesized by cultured endothelial cells in a ratio that indicates it is synthesized by about 8 fold more than laminin, but 6.5 fold less than fibronectin (Munjal et al 1990). In addition, treatment of confluent endothelial cells with in situ polymerized fibrin, a coagulant that can form in the anterior chamber of the eye after injury, also resulted in the induction of thrombospondin synthesis (Ramsby and Kreutzer 1993b).

Recent studies have demonstrated that endothelial cells from throbospondin-1 deficient mice adhere less to fibronectin than vitronectin (another type of ECM protein) and exhibit poor cell migration in response to monolayer wounding (Scheef et al. 2007). In cell monolayers in vitro, TSP localization within the cells appears as a low level of cytoplasmic fluorescence that changes to a punctuate pattern at the level of the extracellular matrix (Munjal et al. 1990) In organ cultured endothelium, the TSP appearance is similar. However, during wound repair, endothelial cells display much more prominent cytoplasmic fluorescence. At the level of Descemet's membrane, TSP localizes as migration tracts, oriented in the direction of cell movement into the wound, suggesting that it is involved in the translocation of these cells along the basement membrane (Munjal et al. 1990). In summary, these results indicate that although the endothelium resides on its own natural basement membrane, the cells are still required to synthesize and deposit ECM molecules to facilitate their migration during wound healing.

9 Summary and Conclusion

The corneal endothelium provides a model system for the study of cell movement during wound repair. The entire cell population resides on Descemet's membrane, their natural basement membrane which influences both cell behavior and response to injury. Because the tissue is neither directly vascularized nor innervated, it is easily amendable for studies that utilize organ culture where the cell/natural matrix relationship remains intact, thus providing a system in which one can observe and investigate injury-induced cell migration, during wound repair, on a natural substrate.

Not surprisingly, endothelial cells respond to an injury defect in much the same manner as other cell types. Wounding the tissue produces both morphological and biochemical changes in the cells that culminate in their migration into the injury area in order to repopulate the region in order to restore an intact and functional tissue monolayer. Migration is mediated by cellular changes in morphology (the epithelial-mesenchymal transition), macromolecular synthesis, reorganization of the actin cytoskeleton into prominent stress fibers, as well as ECM deposition at the cell/matrix interface as cells translocate along the basement membrane into the wound area. Injury also releases the cells surrounding the wound from their state of quiescent and as they initiate movement they also begin to undergo cell cycle progression, entering into DNA synthesis and mitosis. Results from investigations that have studied wound repair in smaller injuries demonstrate that the size of the wound affects the means by which the cells respond. Smaller injuries, depending on their size, may or may not be accompanied by a mitotic response and appear to heal by cell spreading without the loss of cell-to-cell adhesion or the development of actin containing stress fibers. Why cells response to a smaller wound in this manner is still in need of further study.

Many studies on endothelial wound repair have explored several regulatory mechanisms that appear to modulate cell movement. Results obtained from these investigations make it clear that biochemical pathways involving growth factors, PI-3 kinases, Rho-family GTPases and eicosanoid metabolism all function to some extent during wound repair. However, the question that remains is to what degree can the results of the in vitro work be applicable to the in vivo or organ culture system? While most in vitro work will have significant impact on how cells migrate in vivo, a degree of caution must be used when correlating culture data to the in vivo and organ culture systems. Cultured cells reside and migrate on non-native substrates which can affect not only cell/matrix adhesion but also influence different adhesion mediated signaling pathways, thus having the potential to change their behavior. In addition, most experimental studies that use either type of system are carried out in the continual presence of bioactive agents for the duration of the experiment. It is rather unlikely that this is the natural case for the cell. It is more probable that molecules that affect cells are only available for short periods of time and events regulating the cells migratory response are continually being modulated in a "yin-yang" sort of manner. Thus, cell movement should be considered as a multifaceted process that undoubtedly involves the continuous interplay of several morphological and biochemical phenomena that occur repeatedly, again and again in a precise fashion, to mediate the translocation of a cell from one location to another.

References

Abercrombie, M., Heaysman, J.E.M.: Observations on the social behaviour of cells in tissue culture. I. Speed of movement in chick heart fibroblasts in relation to their mutual contacts. Exp. Cell. Res. 5, 111–121 (1953)

Armstrong, P.B.: The control of cell motility during embryogenesis. Cancer Metastasis Rev. 4, 59–79 (1985)

Bard, J.B.L., Hay, E.D., Meller, S.M.: Formation of the endothelium of the avian cornea: A study of cell movement in vivo. Dev. Biol. 42, 334–361 (1975)

Barry, P.A., Petroll, W.M., Andrews, P.M., Cavanagh, H.D., Jester, J.V.: The special organization of corneal endothelial cytoskeletal proteins and their relationship to the apical junctional complex. Invest. Ophthalmol. Vis. Sci. 36, 1115–1124 (1995)

Beach, R.S., Kenney, M.C.: Altered rabbit endothelial phenotypic expression in response to human fibronectin. Biochem. Biophys. Res. Comm. 108, 904–909 (1982)

Bement, W.M., Forscher, P., Mooseker, M.S.: A novel cytoskeletal structure involved in purse string wound closure and cell polarity maintenance. J. Cell. Biol. 121, 565–578 (1993)

Bershadsky, A.D., Futerman, A.H.: Disruption of the Golgi apparatus by brefeldin A blocks cell polarization and inhibits directed cell migration. Proc. Natl. Acad. Sci. USA 91, 5686–5689 (1994)

Blake, D.A., Yu, H., Young, D.L., Caldwell, D.R.: Matrix stimulates the proliferation of human corneal endothelial cells in culture. Invest Ophthalmol. Vis. Sci. 38, 1119–1129 (1997)

Blat, C., Villaudy, J., Harel, L.: Density-dependent inhibition of mouse embryo fibroblast growth: involvement of IGFBP-3. Exp. Cell. Res. 215, 114–118 (1994)

Bogatcheva, N.V., Verin, A.D.: The role of cytoskeleton in the regulation of vascular endothelial barrier function. Microvasc. Res. 76, 202–207 (2009)

Bonanno, J.A.: Identity and regulation of ion transport mechanisms in the corneal endothelium. Prog. Retin. Eye Res. 22, 69–94 (2003)

Bourne, W.M., Nelson, L.R., Hodge, D.O.: Central corneal endothelial cell changes over a ten-year period. Invest. Ophthalmol. Vis. Sci. 38, 779–782 (1997)

Carbajal, J.M., Gratrix, M.L., Yu, C.-H., Schaeffer Jr., R.C.: ROCK mediates thrombin's endothelial barrier dysfunction. Amer. J. Physiol. Cell. Physiol. 279, C195–C204 (2000)

Carlson, C.B., Lawler, J., Mosher, D.F.: Structures of thrombospondins. Cell. Mol. Life Sci. 65, 672–686 (2008)

Chen, K.-H., Hsu, W.-M., Chiang, C.-C., Li, Y.-S.: Transforming growth factor-$\beta 2$ inhibition of corneal endothelial proliferation mediated by prostaglandin. Curr. Eye Res. 26, 363–370 (2003)

Chifflet, S., Hernández, J.A., Grasso, S., Cirillo, A.: Nonspecific depolarization of the plasma membrane potential induces cytoskeletal modifications of bovine corneal endothelial cells in culture. Exp. Cell. Res. 282, 1–13 (2003)

Chun, J., Auer, K.A., Jacobson, B.S.: Arachidonic initiated protein kinase C activation regulates HeLa cell spreading on a gelatin substrate by inducing F-actin formation and exocytotic upregulation of $\beta 1$integrin. J. Cell. Physiol. 173, 361–370 (1997)

Cintron, C., Covington, H.I., Kublin, C.L.: Morphogenesis of rabbit corneal endothelium. Curr. Eye Res. 7, 913–929 (1988)

Coulter III., J.B., Engelke, J.A., Eaton, D.K.: Insulin concentrations in aqueous humor after paracentesis and feeding of rabbits. Invest. Ophthalmol. Vis. Sc. 19, 1524–1526 (1980)

Dai, P., Nakagami, T., Tanaka, H., Hitomi, T., Takamatsu, T.: Cx43 mediates TGF-beta signaling through competitive Smads binding to microtubules. Mol. Biol. Cell. 18, 2264–2273 (2007)

Davies, Y., Lewis, D., Fullwood, N.J., et al.: Proteoglycans on normal and migrating human corneal endothelium. Exp. Eye Res. 68, 303–311 (1999)

Dellas, C., Loskutoff, D.J.: Historical analysis of PAI-1 from its discovery to its potential role in cell motility and disease. Thromb. Haemost 93, 631–640 (2005)

Dipasquale, A.: Locomotive activity of epithelial cells in culture. Exp. Cell. Res. 94, 191–215 (1975)

Doughman, D.J., Van Horn, D.L., Rodman, W.P., et al.: Human cornel endothelial layer repair during organ culture. Arch. Ophthalmol. 94, 1791–1796 (1976)

Engvall, E.: Structure and function of basement membranes. Int. J. Dev. Biol. 39, 781–787 (1995)

Fehrenbacher, L., Gospodarowicz, D., Shuman, M.A.: Synthesis of plasminogen activator by bovine corneal endothelial cells. Exp. Eye Res. 29, 219–228 (1979)

Folkman, J., Klagsbrun, M., Sasse, J., Wadzinski, M., et al.: A heparin-binding angiogenic protein–basic fibroblast growth factor is stored within basement membrane. Am. J. Pathol. 130, 393–400 (1988)

Fujino, Y., Tanishima, T.: Actin in wound-healing of rabbit cor-neal endothelium. II. Study by nitrobenzoxadiazole-phallacidin method. Jpn. J. Ophthamol. 31, 393–404 (1987)

Fukata, M., Nakagawa, M., Kaibuchi, K.: Roles of Rho-family GTPases in cell polarization and directional migration. Curr. Opinion Cell. Biol. 15, 590–597 (2003)

Fullwood, N.J., Davies, Y., Nieduszynski, I.A., et al.: Cell surface-associated keratan sulfate on normal and migrating corneal endothelium. Invest. Ophthalmol. Vis. Sci. 37, 1256–1270 (1996)

Gao, F., Toriyama, K., Ma, H., Nagata, T.: Light microscopic autoradiographic study on DNA synthesis in aging mice corneas. Cell. Mol. Biol. 39, 435–441 (1993)

Glenn, H., Jacobson, B.S.: Cyclooxygenase and cAMP-dependent protein kinase reorganize the actin cytoskeleton for motility in HeLa cells. Cell. Motil. Cytoskeleton. 55, 265–277 (2003)

González-Périz, A., Clària, J.: New approaches to the modulation of the cyclooxygenase-2 and 5-lipoxygenase pathways. Curr. Top Med. Chem. 7, 297–309 (2007)

Gordon, S.R.: Changes in distribution of extracellular matrix proteins during wound repair in corneal endothelium. J. Histochem. Cytochem. 36, 409–416 (1988)

Gordon, S.R.: Changes in extracellular matrix proteins and actin during corneal endothelial growth. Invest. Ophthalmol. Vis. Sci. 31, 94–101 (1990)

Gordon, S.R.: Cytological and immunocytochemical approaches to the study of corneal endothelial wound repair. Prog. Histochem. Cytochem. 28, 1–66 (1994)

Gordon, S.R.: Microfilament disruption in a noncycling organized tissue, the corneal endothelium, initiates mitosis. Exp. Cell. Res. 272, 127–134 (2002)

Gordon, S.R., Climie, M., Hitt, A.L.: 5-Fluorouracil interferes with actin organization, stress fiber formation and cell migration in corneal endothelial cells during wound repair along the natural basement membrane. Cell. Motil. Cytoskeleton. 62, 244–258 (2005)

Gordon, S.R., Czerwinski-Mowers, D., Marchand, J., Shuffett, R.: Endocytosis by the corneal endothelium. I. Regulation of binding and transport of hemeproteins and peroxidase-conjugated lectins across the tissue. Histochem. Cell. Biol. 110, 251–262 (1998)

Gordon, S.R., DeMoss, J.: Exposure to lysosomotropic amines and protease inhibitors retard corneal endothelial cell migration along the natural basement membrane during wound repair. Exp. Cell. Res. 246, 233–242 (1999)

Gordon, S.R., Essner, E., Rothstein, H.: In situ demonstration of actin in normal and injured ocular tissues using 7-nitrobenz-2-oxa-1,3-diazole phallacidin. Cell. Motil. 2, 343–354 (1982)

Gordon, S.R., Rothstein, H.: Studies on corneal endothelial growth and repair. I. Micofluorometric and autoradiographic analyses of DNA synthesis, mitosis and amitosis following freeze injury. Metabol. Ophthalmol. 2, 57–63 (1978)

Gordon, S.R., Rothstein, H.: Studies on corneal endothelial growth and repair. III. Effects of DNA and RNA synthesis inhibitors upon restoration of transparency following wounding. Ophthal. Res. 14, 195–209 (1982)

Gordon, S.R., Wood, M.: Soybean (Glycine max) agglutinin binds to corneal endothelial cells during wound repair and alters their microfilament pattern. Cell. Molec. Biol. 43, 329–336 (1997)

Gordon, S.R., Wood, M.: Soybean agglutinin binding to corneal endothelial cell surfaces disrupts in situ monolayer integrity and actin organization and interferes with wound repair. Cell. Tiss. Res. 335, 551–563 (2009)

Gospodarowicz, D., Gonzalez, R., Fujii, D.K.: Are factors originating from serum, plasma, or cultured cells involved in the growth-promoting effect of the extracellular matrix produced by cultured bovine corneal endothelial cells? J. Cell. Physiol. 114, 191–202 (1983)

Gospodarowicz, D., Greenburg, G.: The effects of epidermal and fibroblast growth factors on the repair of corneal endothelial wounds in bovine corneas maintained in organ culture. Exp. Eye Res. 28, 147–157 (1979)

Gospodarowicz, D., Greenberg, G., Foidart, J.M., Savion, N.: The production and localization of laminin in cultured vscular and corneal endothelial cells. J. Cell. Physiol. 107, 171–183 (1981)

Gospodarowicz, D., Greenburg, G., Vlodavsky, I., et al.: The identification and localization of fibronectin in cultured corneal endothelial cells: cell surface polarity and physiological implications. Exp. Eye Res. 29, 485–509 (1979)

Gospodarowicz, D.: The extracellular matrix and the control of proliferation of corneal endothelial and lens epithelial cells. Exp. Eye Res. 31, 181–199 (1980)

Gotlieb, A.I., Langille, L., Wong, M.K., Kim, D.W.: Structure and function of the endothelial cytoskeleton. Lab Invest. 65, 123–137 (1991)

Gotlieb, A.I., May, L.M., Subrahmanyan, L., Kalnins, V.I.: Distribution of microtubule organizing centers in migrating sheets of endothelial cells. J. Cell. Biol. 91, 589–594 (1981)

Grant, M.B., Khow, P.T., Schultz, G.S., et al.: Effects of epidermal growth factor, fibroblast growth factor and transforming growth factor-β on corneal cell chemotaxis. Invest. Ophthalmol. Vis. Sci. 33, 3292–3301 (1992)

Grant, D.S., Leblond, C.P.: Immunoglod quantitation of laminin, type IV collagen, and heparin sulfate proteoglycan in a variety of basement membranes. J. Histochem. Cytochem. 36, 271–283 (1988)

Grasso, S., Hernández, J.A., Chifflet, S.: Roles of wound geometry, wound size, and extracellular matrix in the healing response of bovine corneal endothelial cells in culture. Am. J. Physiol. 293, C1327–C1337 (2007)

Green, J.A., Stockton, R.A., Johnson, C., Jacobson, B.S.: 5-Lipoxygenase and cyclooxygenase regulate wound closure in NIH?3T3 fibroblast monolayers. Am. J. Cell. Physiol. 287, C373–C383 (2004)

Grinnell, F., Rocha, L.B., Iucu, C., Rhee, S., Jiang, H.: Nested collagen matrices: a new model to study migration of human fibroblast populations in three dimensions. Exp. Cell. Res. 312, 86–94 (2006)

Gu, X., Seong, G.J., Lee, Y.G., Kay, E.P.: Fibroblast growth factor 2 uses distinct signaling pathways for cell proliferation and cell shape changes in corneal endothelial cells. Invest. Ophthalmol. Vis. Sci. 37, 2326–2334 (1996)

Hall, A.: Rho GTPases and the actin cytoskeleton. Science 279, 509–514 (1998)

Hay, E.D., Revel, J.P.: Fine structure of the developing cornea. In: Wolsky, A., Chen, P.S. (eds.) Monographs in developmental biology, Karger, Basal, vol. 1 (1969)

Hergott, G.J., Sandig, M., Kalnins, V.I.: Cytoskeletal organization of migrating retinal pigment epithelial cells during wound healing in organ culture. Cell. Motil. Cytoskel. 13, 83–93 (1989)

Herman, I.M., Crisona, N.J., Pollard, T.D.: Relationship between cell activity and the distribution of cytoplasmic actin and myosin. J. Cell. Biol. 97, 416–424 (1981)

Hirschi, K.K., Burt, J.M., Hirschi, K.D., Dai, C.: Gap junction communication mediates transforming growth factor-beta activation and endothelial-induced mural cell differentiation. Circ. Res. 93, 429–437 (2003)

Hiscott, P., Paraoan, L., Choudhary, A., Ordonez, J.L., Al-Khaier, A., Armstrong, D.J.: Thrombospondin 1, thrombospondin 2 and the eye. Prog. Retin. Eye Res. 25, 1–18 (2006)

Hiscott, P., Seitz, B., Schlötzer-Schrehardt, U., Naumann, G.O.H.: Immunolocalization of thrombospondin 1 in human, bovine and rabbit cornea. Cell. Tissue Res. 289, 307–310 (1997)

Hoppenreijs, V.P., Pels, E., Vrensen, G.F., Oosting, J., Treffers, W.F.: Effects of human epidermal growth factor on endothelial wound healing of human corneas. Invest. Ophthalmol. Vis. Sci. 33, 1946–1957 (1992)

Hoppenreijs, V.P., Pels, E., Vrensen, G.F., Treffers, W.F.: Effects of platelet-derived growth factor on endothelial wound healing of human corneas. Invest. Ophthalmol. Vis. Sci. 35, 150–161 (1994)

Hsieh, P., Baum, J.: Effects of fibroblastic and endothelial extracellular matrices on corneal endothelial cells. Invest. Ophthalmol. Vis. Sci. 26, 457–463 (1985)

Huang, C., Jacobson, K., Schaller, M.D.: MAP kinases and cell migration. J. Cell. Sci. 117, 4619–4628 (2004)

Ichijima, H., Petroll, W.M., Barry, P.A., Andrews, P.M., et al.: Actin filament organization during endothelial wound healing in the rabbit cornea: comparison between transcorneal freeze and mechanical scrape injuries. Invest. Ophthalmol. Vis. Sci. 34, 2803–2812 (1993)

Jalimarada, S.S., Shivanna, M., Kini, V., et al.: Microtubule disassembly breaks down the barrier integrity of corneal endothelium. Exp, Eye Res. 89, 333–343 (2009)

Johnson, D.H., Bourne, W.H., Campbell, R.J.: The ultrastructure of Descemet's membrane. I. Changes with age in normal corneas. Arch. Ophthalmol. 100, 1942–1947 (1982)

Johnston, M.C., Noden, D.M., Hazelton, R.D., Coulombre, J.L., Coulombre, A.J.: Origins of avian ocular and periocular tissues. Exp. Eye Res. 29, 27–43 (1979)

Joyce, N.C.: Proliferative capacity of the corneal endothelium. Prog. Retin. Eye Res. 22, 359–389 (2003)

Joyce, N.C., Harris, D., Mello, D.M.: Mechanisms of mitotic inhibition in corneal endothelium: contact inhibition and TGF-β2. Invest. Ophthalmol. Vis. Sci. 43, 2152–2159 (2002)

Joyce, N.C., Harris, D.L., Zieske, J.D.: Mitotic inhibition of corneal endothelium in neonatal rats. Invest. Ophthalmol. Vis. Sci. 39, 2572–2583 (1998)

Joyce, N.C., Joyce, S.J., Powell, S.M., Meklir, B.: EGF and PGF2: effects on corneal endothelial cell migration and monolayer spreading during wound repair *in vitro*. Curr. Eye Res. 13, 601–609 (1995)

Joyce, N.C., Maklir, B., Joyce, S.J., Zieske, J.P.: Cell cycle protein expression and proliferative status in human corneal cells. Invest. Ophthalmol. Vis. Sci. 37, 645–655 (1996)

Joyce, N.C., Matkin, E.D., Neufeld, A.H.: Corneal endothelial wound closure in vitro. Invest. Ophthalmol. Vis. Sci. 30, 1548–1559 (1989)

Joyce, N.C., Meklir, B., Neufeld, A.H.: In vitro pharmacological separation of corneal endothelial migration and spreading responses. Invest. Ophthalmol. Vis. Sci. 31, 1816–1826 (1990)

Joyce, N.C., Meklir, B.: Protein kinase C activation during corneal endothelial wound repair. Invest. Ophthalmol. Vis. Sci. 33, 1958–1973 (1992)

Joyce, N.C., Meklir, B.: PGE2: a mediator of corneal endothelial wound repair in vitro. Am. J. Physiol. 266, C269–C275 (1994)

Jumblatt, M.M.: Autocrine regulation of corneal endothelium by prostaglandin E2. Invest. Ophthamol. Vis. Sci. 35, 2783–2790 (1994)

Jumblatt, M.M., Willer, S.S.: Corneal endothelial repair. Regulation of prostaglandin E2 synthesis. Invest. Ophthalmol. Vis. Sci. 37, 1294–1301 (1996)

Jun, A.S., Chakravarti, S., Edelhauser, H.F., Kimos, M.: Aging changes of mouse corneal endothelium and Descemet's membrane. Exp. Eye Res. 83, 890–896 (2006)

Katoh, K., Kano, Y., Amano, M., et al.: Rho-kinase–mediated contraction of isolated stress fibers. J. Cell. Biol. 153, 569–584 (2001)

Kaye, G.I., Pappas, G.D.: Studies on the cornea. I. The fine structure of the rabbit cornea and the uptake and transport of colloidal particles by the cornea in vitro. J. Cell. Biol. 12, 457–479 (1962)

Kaye, G.I., Sibley, R.C., Hoefle, F.B.: Recent studies on the nature and function of the corneal endothelial barrier. Exp. Eye Res. 15, 585–613 (1973)

Kaye, G.I., Tice, L.W.: Studies on the cornea. V. Electron microscopic localization of adenosine triphosphatase activity in the rabbit cornea in relation to transport. Invest. Ophthalmol. 5, 22–32 (1966)

Khodadoust, A.A., Green, K.: Physiological function of regenerating endothelium. Invest. Ophthalmol. 15, 96–101 (1976)

Kirkpatrick, C.J., Kampe, M., Rixen, H., et al.: In vitro studies on the expansion of endothelial cell monolayers on components of the basement membrane. Virchows Arch. B. Cell. Pathol. 58, 207–213 (1990)

Kjøller, L.: The urokinase plasminogen activator receptor in the regulation of the actin cytoskeleton and cell motility. Biol. Chem. 383, 5–19 (2002)

Klymkowsky, M.W., Savagner, P.: Epithelial-mesenchymal transition. A cancer researcher's conceptual friend and foe. Am. J. Pathol. 174, 1588–1593 (2009)

Kupfer, A., Louvard, D., Singer, S.J.: Polarization of the Golgi apparatus and microtubule-organizing center in cultured fibroblasts at the edge of an experimental wound. Proc. Natl. Acad. Sci. USA 79, 2603–2607 (1982)

Kuriyama, S., Mayor, R.: Molecular analysis of neural crest migration. Philos. Trans. R. Soc. Lond. B. Biol. Sci. 363, 1349–1362 (2008)

Labermeier, U., Kenney, M.C.: The presence of EC collagen and type IV collagen in bovine Descemet's membranes. Biochem. Biophys. Res. Commun. 116, 619–625 (1983)

Laing, R.A., Sandstrom, M.M., Berrospi, A.R., Liebowitz, H.M.: Changes in the corneal endothelium as a function of age. Exp. Eye Res. 22, 587–594 (1976)

Landshman, N., Solomon, A., Belkin, M.: Cell division in the healing of the corneal endothelium of cats. Arch. Ophthalmol. 107, 1804–1808 (1989)

Lee, T.Y., Gotlieb, A.I.: Microfilaments and microtubules maintain endothelial integrity. Microsc. Res. Tech. 60, 115–127 (2003)

Lee, H.T., Kay, E.P.: FGF-2 induced reorganization and disruption of actin cytoskeleton through PI 3-kinase, Rho, and Cdc42 in corneal endothelial cells. Molec. Vis. 9, 624–634 (2003)

Lee, H.T., Lee, J.G., Na, M., Kay, E.P.: FGF-2 induced interleukin-1β through action of phosphatidylinositol 3-kinase mediates endothelial mesenchymal transformation in corneal endothelial cells. J. Biol. Chem. 279, 32325–32332 (2004)

Lee, J.G., Kay, E.P.: FGF-2-induced wound healing in corneal endothelial cells requires Cdc42 activation and Rho inactivation through the phosphatidlyinositol 3-kinase pathway. Invest. Ophthalmol. Vis. Sci. 47, 1376–1386 (2006a)

Lee, J.G., Kay, E.P.: Cross-talk among Rho GTPases acting downstream of PI 3-kinases induces mesenchymal transformation of corneal endothelial cells mediated by FGF-2. Invest. Ophthalmol. Vis. Sci. 47, 2358–2368 (2006b)

Lee, J.G., Kay, E.P.: FGF-2 mediated signal transduction during endothelial mesenchymal transformation in corneal endothelial cells. Exp. Eye Res. 83, 1309–1316 (2006c)

Lee, J.G., Kay, E.P.: Common and distinct pathways for cellular activities in FGF-2 signaling induced by IL-1β in corneal endothelial cells. Invest. Ophthalmol. Vis. Sci. 50, 2067–2076 (2009)

Lee, T.Y., Noria, S., Lee, J., Gotlieb, A.I.: Endothelial integrity and repair. Adv. Exp. Med. Biol. 498, 65–74 (2001)

Leuenberger, P.M., Novikoff, A.B.: Localization of transport adenosine triphosphate in rat cornea. J. Cell. Biol. 60, 721–731 (1974)

Lewis, L., Verna, J.-M., Levinstone, D., Sher, S., et al.: The relationship of fibroblast translocations to cell morphology and stress fiber density. J. Cell. Sci. 53, 21–36 (1982)

Lin, C.Q., Bissel, M.J.: Multi-faceted regulation of cell differentiation by extracellular matrix. FASEB J. 7, 737–743 (1993)

MacCallum, D.K., Bahn, C.F., Lillie, J.H., Meyer, R.F., Martonyi, C.L.: Evidence for corneal endothelial cell hypertrophy during postnatal growth of the cat cornea. Invest. Ophthalmol. Vis. Sci. 24, 247–250 (1983)

MacCallum, D.K., Lillie, J.H., Scaletta, L.J., et al.: Bovine corneal endothelium in vitro. Elaboration and organization of a basement membrane. Exp. Cell. Res. 139, 1–13 (1982)

Maekawa, H., Oike, Y., Kanda, S., Ito, Y., et al.: Ephrin-B2 induces migration of endothelial cells through the phosphatidylinositol-3 kinase pathway and promotes angiogenesis in adult vasculature. Arterioscler Thromb. Vasc. Biol. 23, 2008–2014 (2003)

Martin-Belmonte, F., Mostov, K.: Regulation of cell polarity during epithelial morphogenesis. Curr. Opinion Cell. Biol. 20, 227–234 (2008)

Martz, E., Steinberg, M.S.: The role of cell-cell contact in "contact" inhibition of cell division: a review and new evidence. J. Cell. Physiol. 79, 189–210 (1972)

Maurice, D.M.: The localization of the fluid pump in the cornea. J. Physiol. (London) 221, 43–54 (1972)

Mergler, S., Pleyer, U.: The human corneal endothelium: new insights into electrophysiology and ion channels. Prog. Retin. Eye Res. 26, 359–378 (2007)
Mills, J.W., Mandel, L.J.: Cytoskeletal regulation of membrane transport events. FASEB J. 8, 1161–1165 (1994)
Miyata, K., Murao, M., Sawa, M., Tanishima, T.: Kinetic study of cell proliferation in a new wound healing model using tissue cultured corneal endothelial cells. Nippon Ganka Gakkai Zasshi 93, 287–293 (1989)
Mohay, J., McLaughlin, B.J.: Corneal endothelial wound repair in normal and mitotically inhibited cultures. Graefes Arch. Clin. Exp. Ophthalmol. 233, 727–736 (1995)
Møller-Pedersen, T.: A comparative study of human corneal keratocyte and endothelial cell density during aging. Cornea 6, 333–338 (1997)
Morton, K., Hutchinson, C., Jeanny, J.C., et al.: Colocalization of fibroblast growth factor binding sites with extracellular matrix components in normal and keratoconus corneas. Curr. Eye Res. 8, 975–987 (1989)
Munjal, I.D., Crawford, D.R., Blake, D.A., Sabet, M.D., Gordon, S.R.: Thrombospondin: biosynthesis, distribution, and changes associated with wound repair in corneal endothelium. Eur. J. Cell. Biol. 52, 252–263 (1990)
Murphy, C., Alvarado, J., Juster, R.: Prenatal and postnatal growth of the human Descemet's membrane. Invest. Ophthalmol. Vis. Sci. 25, 1402–1415 (1984)
Nakano, Y., Oyamada, M., Dai, P., Nakagami, T., et al.: Connexin43 knockdown acclerates wound healing but inhibits mesenchymal transition after corneal endothelial injury in vivo. Invest. Ophthalmol. Vis. Sci. 49, 93–104 (2008)
Nelson, G.A., Revel, J.-P.: Scanning electron microscopic study of cell movements in the corneal endothelium of the avian embryo. Dev. Biol. 42, 315–333 (1975)
Neufeld, A.H., Jumblatt, M.M., Matkin, E.D., Raymond, G.M.: Maintenance of corneal endothelial cell shape by prostaglandin E_2: Effects of EGF and indomethacin. Invest. Ophthalmol. Vis. Sci. 27, L1437–L1442 (1986)
Newgreen, D.F., Erickson, C.A.: The migration of neural crest cells. Int. Rev. Cytol. 103, 89–145 (1986)
Nickeleit, V., Kaufman, A.H., Zagachin, L., Dutt, J.F., et al.: Healing corneas express embryonic fibronectin isoforms in the epithelim, subepithelial stroma, and endothelium. Am. J. Pathol. 149, 549–558 (1996)
Nisha, K., Sotozono, C., Adachi, W., Yamamoto, S., et al.: Transforming growth factor-$\beta 1$, -$\beta 2$ and $\beta 3$ mRNA expression in human cornea. Curr. Eye Res. 14, 235–241 (1994)
Nuttall, R.P.: DNA synthesis during the development of the chick cornea. J. Exp. Zool 198, 193–208 (1976)
Parkinson, W.C.: Motility of mouse fibroblasts in tissue culture. Biophys. J. 42, 17–23 (1983)
Paulsson, M.: Basement membrane proteins: Structure, assembly and cellular interactions. Crit. Rev. Biochem. Mol. Biol. 27, 93–127 (1992)
Pellegrin, S., Mellor, H.: Actin stress fibres. J. Cell. Sci. 120, 3491–3499 (2007)
Perlman, M., Baum, J.: Synthesis of a collagenous basal membrane by rabbit corneal endothelial cells in vitro. Arch. Ophthalmol. 92, 238–239 (1974)
Perlman, M., Baum, J., Kaye, G.I.: Fine structure and collagen synthesis activity of monolayer cultures of rabbit corneal endothelium. J. Cell. Biol. 63, 306–311 (1974)
Petroll, W.M., Hsu, J.K.W., Bean, J., Cavanagh, H.D., Jester, J.V.: The spatial organization of apical junctional complex-associated proteins in feline and human corneal endothelium. Curr. Eye Res. 18, 10–19 (1999)

Petroll, W.M., Jester, J.V., Barry-Lane, P.A., Cavanagh, H.D.: Effects of basic FGF and TGF beta 1 on F-actin and ZO-1 organization during cat endothelial wound healing. Cornea 15, 525–532 (1996)
Petroll, W.M., Jester, J.V., Barry-Lane, P., Cavanagh, H.D.: Assessment of F-actin organization and apical-basal polarity during in vivo cat endothelial wound healing. Invest. Ophthalmol. Vis. Sci. 36, 2492–2502 (1995)
Petroll, W.M., Jester, J.V., Bean, J.J., Cavanagh, H.D.: Myofibroblast transformation of cat corneal endothelium by transforming growth factor-β1,-β2, and –β3. Invest. Ophthalmol. Vis. Sci. 39, 2018–2032 (1998)
Petroll, W.M., Ma, L., Jester, J.V., Cavanagh, H.D., Bean, J.: Organization of junctional proteins in proliferating cat corneal endothelium during wound healing. Cornea 20, 73–80 (2001)
Rahi, A.H., Robins, E.: Human corneal endothelial cell repair in health and disease. Trans. Ophthalmol. Soc. U.K. 101, 30–34 (1981)
Rahimi, R.A., Leof, E.B.: TGF-β signaling: a tale of two responses. J. Cell. Biochem. 102, 593–608 (2007)
Raja, S.K., Garcia, M.S., Isseroff, R.R.: Wound re-epithelialization: modulating keratinocyte migration in wound healing. Front Biosci. 12, 2849–2868 (2007)
Ramsby, M.L., Kreutzer, D.L.: Fibrin induction of tissue plasminogen activator expression in corneal endothelial cells in vitro. Invest. Ophthalmol. Vis. Sci. 34, 3207–3219 (1993a)
Ramsby, M.L., Kreutzer, D.L.: Fibrin induction of thrombospondin in corneal endothelial cells in vitro. Invest. Ophthalmol. Vis. Sci. 34, 165–174 (1993b)
Raphael, B., McLaughlin, B.J.: Adsorptive and fluid phase endocytosis by cultured rabbit corneal endothelium. Curr. Eye Res. 9, 249–258 (1990)
Reneker, L., Silversides, D.W., Xu, L., Overbeek, P.A.: Formation of corneal endothelium is essential for anterior segment development – a transgenic mouse model of anterior segment dysgenseis. Develop. 127, 533–542 (2000)
Reuner, K.H., Dunker, P., van der Does, A., Wiederbold, M., et al.: Regulation of actin synthesis in rat hepatocytes by cytoskeletal rear-rangements. Eur. J. Cell. Biol. 69, 189–196 (1996)
Rich, L.F., Hatfield, J.M., Louiselle, I.: Influences of basic fibro-blast growth factor on cat corneal endothelial wound repair in vivo. Curr. Eye Res. 11, 719–725 (1992)
Ridley, A.J.: Rho GTPases and cell migration. J. Cell. Sci. 114, 2713–2722 (2001)
Rieck, P.W., Cholidis, S., Hartmann, C.: Intracellular signaling pathway of FGF-2-modulated corneal endothelial cell migration during wound healing in vitro. Exp. Eye Res. 73, 639–650 (2001)
Rieck, P., Oliver, L., Engelmann, K., Fuhrmann, G., Hartmann, C., Courtois, Y.: The role of exogenous/endogenous basic fibroblast growth factor (FGF2) and transforming growth factor beta (TGF beta-1) on human corneal endothelial cells proliferation in vitro. Exp. Cell. Res. 220, 36–46 (1995)
Robinson, J., Gospodarowicz, D.: Glycosaminoglycans synthesized by cultured bovine corneal endothelial cells. J. cell. Physiol. 117, 368–376 (1983)
Rodgers, K.A., Dandig, M., McKee, N.H., Kalnins, V.I.: The distribution of microfilament bundles in rabbit endothelial cells in the intact aorta during wound healing in situ. Biochem. Cell. Biol. 67, 553–562 (1989)
Rodriguez-Diaz, A., Toyama, Y., Abravanel, D.L., Wiemann, J.M., et al.: Actomyosin purse strings: renewable resources that make morphogenesis robust and resilient. HFSP J. 2, 220–237 (2008)

Ross, L.L., Danehower, S.C., Prioa, A.D., Sontag, M., et al.: Coordinated activation of corneal wound response genes in vivo as observed by in situ hybridization. Exp. Eye Res. 61, 435–450 (1995)
Rothstein, H., Gordon, S.R.: Studies on corneal endothelial growth and repair. II. Increased transcription as detected by incorporation of 3H-uridine and 3H-actinomycin D. Tiss. Cell 12, 547–659 (1980)
Sabatier, P., Rieck, P., Daumer, M.L., et al.: Effet du basic Fibroblast Growth Factor (bFGF) recombinant humain sur la cicatrisation endothéliale en organoculture de la cornée humaine. J. Fr. Ophthalmol. 19, 200–207 (1996)
Sabet, M.D., Gordon, S.R.: Ultrastructural immunocytochemical localization of fibronectin deposition during corneal endothelial wound repair. Evidence for cytoskeleton involvement. Bio. Cell 65, 171–179 (1989)
Sakamoto, T., Hinton, D.R., Sakamoto, H., Gopalakrishna, R., et al.: Thrombin induced cytoskeletal change in cultured bovine corneal endothelial cells mediated via protein kinase C pathway. Curr. Eye Res. 14, 35–45 (1994)
Sankey, E.A., Bown, F.E., Morton, L.F., et al.: Analysis of the collagen types synthesized by bovine corneal endothelial cells in culture. Biochem. J. 198, 707–710 (1981)
Satpathy, M., Gallagher, P., Jin, Y., Srinivas, S.P.: Extracellular ATP opposes thrombin-induced myosin light chain phosphorylation and loss of barrier function. Exp. Eye Res. 81, 183–192 (2005)
Savenger, P.: Leaving the neighborhood: molecular mechanisms involved during epithelial-mesenchyme transition. BioEssays 23, 912–923 (2001)
Sawada, H., Furthmayr, H., Konomi, H., Nagai, Y.: Immunoelectronmicroscopic localization of extracellular matrix components produced by bovine corneal ensothelial cells in vitro. Exp. Cell. Res. 171, 94–109 (1987)
Sawada, H., Konomi, H., Nagal, Y.: The basement membrane of bovine corneal endothelial cells in culture with β-aminopropionitrile: biosynthesis of hexagonal lattices composed of a 160 nm dumbbell-shaped structure. Eur. J. Cell. Biol. 35, 226–234 (1984)
Sawada, H., Tanaka, H., Ono, M.: Ultrastructure of tracheal epithelial cells migrating in an in vivo environment. Arch. Histol. Cytol. 71, 223–234 (2008)
Scheef, E.A., Huang, Q., Wang, S., Sorenson, C.M., Sheibani, N.: Isolation and characterization of corneal endothelial cells from wild type and thrombospondin-1 deficient mice. Mol. Vis. 13, 1483–1495 (2007)
Schilling-Schön, A., Pleyer, U., Hartmann, C., Rieck, P.W.: The role of endogenous growth factors to support corneal endothelial migration after wounding in vitro. Exp. Eye Res. 71, 583–589 (2000)
Schmidt, A., Hall, A.: Guanine nucleotide exchange factors for Rho GTPases: turning on the switch. Genes Dev. 16, 1587–1609 (2002)
Schubert, H.D., Trokel, S.: Endothelial repair following Nd:YAG laser injury. Invest. Ophthalmol. Vis. Sci. 25, 971–976 (1984)
Schultz, G., Khaw, P.T., Oxford, K., MaCauley, S., et al.: Growth factors and ocular wound healing. Eye 8, 184–187 (1994)
Schultz, G.S., Wysocki, A.: Interactions between extracellular matrix and growth factors in wound healing. Wound Rep. Reg. 17, 153–162 (2009)
Scoot, D.M., Murray, J.C., Barnes, M.J.: Investigations on the attachment of bovine corneal endothelial cells to collagens and other components of the subendothelium. Role of fibronectin. Exp. Cell. Res. 144, 472–478 (1983)
Senoo, T., Obara, Y., Joyce, N.C.: EDTA: a promoter of proliferation in human corneal endothelium. Invest. Ophthalmol. Vis. Sci. 41, 2930–2935 (2000)

Sherrerd, E.S., Ng, Y.L.: The other side of the corneal endothelium. Cornea 9, 48–54 (1990)

Soltau, J.B., Zhou, L.X., McLaughlin, B.J.: Isolation of plasma membrane domains from bovine corneal endothelial cells. Exp. Eye Res. 56, 115–120 (1993)

Sosnowski, R.G., Feldman, S., Feramisco, J.R.: Interference with endogenous Ras function inhibits cellular responses to wounding. J. Cell. Biol. 121, 113–119 (1993)

Srinivas, A.P., Satpathy, M., Gallagher, P., Larivière, E., Van Driessche, W.: Adenosine induces dephosphorylation of myosin II regulatory light chain in cultured bovine corneal endothelial cells. Exp. Eye Res. 79, 543–551 (2004)

Sumioka, T., Ikeda, K., Okada, Y., et al.: Inhibitory effect of blocking TGF-β/Smad signal on injury-induced fibrosis of corneal endothelium. Mol. Vis. 14, 2272–2281 (2008)

Takatsuka, A., Yagi, R., Koile, M., Oneyama, C., et al.: Ablation of Csk in neural crest lineages causes corneal anomaly by deregulating collagen fibril organization and cell motility. Dev. Biol. 315, 474–488 (2008)

Tétreault, M.P., Chailler, P., Beaulieu, J.F., Rivard, N., Ménard, D.: Specific signaling cascades involved in cell spreading during healing of micro-wounded gastric epithelial monolayers. J. Cell. Biochem. 105, 1240–1249 (2008)

Timpl, R., Brown, J.C.: Supramolecular assembly of basement membranes. BioEssays 18, 123–132 (1996)

Tucker, R.P., Hagios, C., Chiquet-Ehrismann, R., Lawler, J.: In situ localization of thrombospondin-1 and thrombospondin-3 transcripts in the avian embryo. Dev. Dyn. 208, 326–337 (1997)

Tuft, S.J., Williams, K.A., Coster, D.J.: Endothelial repair in the rat cornea. Invest. Ophthalmol. Vis. Sci. 27, 1199–1204 (1986)

Underwood, P.A., Bennett, F.A.: The effect of extracellular matrix molecules on the in vitro behavior of bovine endothelial cells. Exp. Cell. Res. 205, 311–319 (1993)

Usui, T., Takase, M., Kaji, Y., Suzuki, K., et al.: Extracellular matrix production regulation by TGF-β in corneal endothelial cells. Invest. Ophthalmol. Vis. Sci. 39, 1981–1989 (1998)

Van Horn, D.L., Hyndiuk, R.A.: Endothelial repair in primate corneas. Exp. Eye Res. 21, 113–124 (1975)

Van Horn, D.L., Sendele, D.D., Seidman, S., Buco, P.J.: Regenerative capacity of the corneal endothelium in rabbit and cat. Invest. Ophthalmol. Vis. Sci. 16, 597–613 (1977)

Verin, A.D., Birukova, A., Wang, P., Liu, F., et al.: Microtubule disassembly increases endothelial cell barrier dysfunction: role of MLC phosphorylation. Amer. J. Physiol. Lung. Cell. Mol. Physiol. 281, L565–L574 (2001)

Vlodavsky, I., Fuks, Z., Ishai-Michaeli, R., et al.: Extracellular matrix-resident basic fibroblast growth factor: implication for the control of angiogenesis. J. Cell. Biochem. 45, 167–176 (1991)

Vogel, T., Guo, N.H., Krutzsch, H.C., Blake, D.A., et al.: Modulation of endothelial cell proliferation, adhesion, and motility by recombinant heparin-binding domain and synthetic peptides from the type 1 repeats of thrombospondin. J. Cell. Biochem. 53, 74–84 (1993)

von Sallman, L., Caravaggio, L.L., Grimes, P.: Studies on the corneal endothelium of the rabbit. I. Cell division and growth. Aner. J. Ophthalmol. 51, 955–966 (1961)

von Sallman, L., Grimes, P., McElvain, M.: Studies on the corneal endothelium of the rabbit. II. The generative cycle of the cell. Arch. Ophthalmol. 69, 815–823 (1963)

Wang, D.-A., Haiming, D., Jaggar, J.H., Brindley, D.B., et al.: Injury-elicited differential transcriptional regulation of phospholipids growth factor receptors in the cornea. Am. J. Physiol. Cell. Physiol. 283, C1646–C1654 (2002)
Weiger, M.C., Wang, C.C., Krajcovic, M., et al.: Spontaneous phosphoinositide 3-kinase signaling dynamics drive spreading and random migration of fibroblasts. J. Cell. Sci. 122(Pt 3), 313–323 (2009)
Weimar, V.L., Squires, E.L., Knox, R.J.: Acceleration of healing of rabbit corneal endothelium by mesodermal growth factor. Invest. Ophthalmol. Vis. Sci. 19, 350–361 (1980)
Weinsieder, A., Briggs, R., Reddan, J., et al.: Induction of mitosis in ocular tissue by chemotoxic agents. Exp. Eye Res. 20, 33–44 (1975)
Weiss, P.: Symposium on wound healing and tissue repair. In: Patterson, W.B. (ed.) Wound healing and tissue repair. Univ. Chicago Press, Chicago (1956)
Whikehart, D.R., Parikh, C.H., Vaughn, A.V., Mishler, K., Edelhauser, H.F.: Evidence suggesting the existence of stem cells for the human corneal endothelium. Mol. Vis. 11, 816–824 (2005)
Wieland, T.: Modification of actins by phallotoxins. Naturwissenschaften 64, 303–309 (1977)
Wong, M.K., Gotlieb, A.I.: Endothelial cell monolayer integrity. I. Characterization of dense peripheral band of microfilaments. Arteriosclerosis 6, 212–219 (1986)
Wulle, K.G.: Electron microscopy of the fetal development of the corneal endothelium and Descemet's membrane of the human eye. Invest. Ophthalmol. 11, 817–904 (1972)
Yamaguchi, N., Sato, N., Ko, J.S., Ninomiya, Y.: Cloning of $\alpha 1$(IV) and $\alpha 2$(IV) collagen cDNAs from rabbit corneal endothelial cell RNA. Invest. Ophthalmol. Vis. Sci. 32, 2924–2930 (1991)
Yokota, S., Waller, W.K.: Electron microscopic localization of carbonic anhydrase (CA) activity in rabbit cornea. Albrecht. Von. Graefes Arch. Klin. Exp. Ophthalmol. 197, 145–152 (1975)
Yoshida, A., Laing, R.A., Joyce, N.C., Neufeld, A.H.: Effects of EGF and indomethacin on rabbit corneal endothelial wound closure in excised corneas. Invest. Ophthalmol. Vis. Sci. 30, 1991–1996 (1989)
Yue, B.Y.J.T., Baum, J.: The synthesis of glycosaminoglycans by cultures of rabbit and corneal endothelial and stromal cells. Biochem. J. 158, 567–573 (1976)
Zelenka, P.S., Arpitha, P.: Coordinating cell proliferation and migration in the lens and cornea. Semin. Cell. Dev. Biol. 19, 113–124 (2008)

The Importance of the Microenvironment of Support Surfaces in the Prevalence of Pressure Ulcers

Steven I. Reger and Vinoth K. Ranganathan

Department of Physical Medicine and Rehabilitation, Cleveland Clinic, Cleveland, Ohio, USA

Abstract. Soft tissue breakdown is a major cause of disablement in the United States. External pressure has been the most frequently discussed stress factor in the formation of ulcers. Analysis of published data on the prevalence of pressure ulcers and interface pressures at various anatomic sites indicate a nearly non-existent or slightly negative correlation between prevalence and interface pressure for the general and the spinal cord injured populations respectively. This lack of direct relation suggests the major influence of environmental factors in addition to mechanical factors (pressure, shear strain and friction, etc.,) on ulcer formation and indicates the need for control of the support surface microenvironment. However, most of the reported results from studies evaluating support surfaces focus mainly on pressure relief and neglect to address adequately the environmental contributing factors of ulcer formation. Studies directly relating primary stress factors and tissue viability with prevalence and incidence of pressure ulcers are needed to better understand the benefits of pressure relieving support surfaces and to improve the effectiveness of prevention and treatment interventions. The effects of microenvironment on support surface function are reviewed here to aid the healthcare providers in the choosing of the most appropriate support surface to meet the patient's needs.

1 Introduction

1.1 Pressure Ulcer Prevalence and Incidence, Economic Cost

Soft tissue breakdown is a major cause of disablement in the United States affecting an estimated 1.3 to 3 million patients [1,2] and the total cost of associated health care is $8 billion/yr [3]. Miller and Delozier [4] estimated that treating patients with pressure ulcer cost $1.335 billion (average charge was $21,675) in 1992. The average charge for treating pressure ulcers was nearly $37,800 or approximately $17.2 billion in total cost for the year 2003. The prevalence (the number of patients with ulcers divided by the number of patients at risk for ulcers) and incidence (the number of patients who develop a pressure ulcer after admission to a hospital) of pressure ulcers in the acute care settings has remained steady around 15.3% and 7.6% respectively for the period 1999-2004 [5] It is universally accepted that a reduction of pressure between the body and the support surface

A. Gefen (Ed.): Bioengineering Research of Chronic Wounds, SMTEB 1, pp. 85–100.
springerlink.com

interface will reduce or prevent the occurrence of pressure ulcer. Landis' [6] observation (using a microinjection method) of 32mmHg capillary pressure has been interpreted as a threshold above which pressure ulcer occurs and is often used by the healthcare industry as a guideline for testing the effectiveness of a support surface. The healthcare industry and healthcare providers have been very active to develop new products and treatment and prevention guidelines to reduce the complexity and occurrence of pressure ulcers. The objective of this chapter is to present the relationship between interface pressure and the occurrence of pressure ulcers at various anatomic sites from published literature and briefly review the role played by factors other than pressure in the development of pressure ulcers.

1.2 Current Definition of PU

A pressure ulcer is localized injury to the skin and/or underlying tissue usually over a bony prominence, as a result of pressure, or pressure in combination with shear and/or friction [7].

1.3 Major Factors

External pressure has been the most frequently discussed stress factor in the formation of ulcers. Other primary stress factors associated with ulcer formation are shear, friction and the resulting deformation of the soft tissues.

1.4 Current Methods/Procedures to Prevent PUs

Currently, there are more than 200 support surfaces [8] available that aim to either redistribute or reduce interface pressures below the 32mmHg threshold. These pressure relieving and pressure reducing support surfaces have been widely applied with the hope of reducing the frequency of ulcers. The effectiveness of these myriad support surfaces is usually tested by their ability to reduce interface pressures, usually in a healthy population. Many studies [9-19] have been published by researchers, clinicians and the support surface industry to aid the healthcare staff in choosing the right support surface for their patients. These studies mostly evaluate or compare the different support products using interface pressure measurements alone, mainly because they are readily available, easy to measure, non-invasive and reasonably repeatable. Randomized control trials are costly, limited in number of similar subjects participating and require long duration to complete.

1.4.1 Status of These Methods

Despite all these efforts by the healthcare providers and the healthcare industry, the rate of occurrence of hospital stays with pressure sores has increased by 63% from 280,000 cases in 1993 to 455,000 cases in 2003 [20]. 72% of patients with pressure ulcer were 65 years of age and older, and 19% of patients were between 45 and 64 years of age [20]. The five most common conditions for pressure sore related hospital stays were septicemia, pneumonia, urinary tract infection, aspiration pneumonitis and congestive heart failure. The common concomitant conditions for those admitted primarily for pressure sores were paralysis, spinal cord injury, substance abuse, malnutrition, multiple sclerosis, stroke and senility. This has led many researchers to revisit the issue of the role played by pressure in the development of pressure ulcers and the effectiveness of the pressure relieving support surfaces.

1.4.2 Reasons for Limited Success

However, due to paucity of other relevant data, healthcare providers continue to rely on research studies reporting mainly interface pressure to select pressure relief support devices and fail to adequately address all the other factors contributing to the formation of ischemic necrosis [21-23]. Berlowitz and Brienza reviewed the literature relevant to the pathophysiology and pathogenesis of pressure ulcer and noted that pressure ulcers are the result of deep tissue damage and that eliminating of pressure is not necessarily the highest priority for prevention and treatment of deep tissue damage [24]. As a result, these expensive support surfaces provide mixed, or in some cases, no benefit at all to the patients. Correlating the interface pressure measured for the various pressure relieving support surfaces with the relevant prevalence or incidence information will be very useful in understanding the role of pressure in reducing pressure ulcer and the effectiveness of the pressure relieving systems. Bader and White also pointed out that tissue viability could be compromised in elderly patients making them more susceptible to pressure sores during surgical procedures [25].

2 Relationship between the Prevalence of Pressure Ulcers and Interface Pressures

At different anatomic regions the occurrence of the ulcers and the corresponding body interface pressures are highly variable depending on the tissue health, thickness, support characteristics, and the method of measurement. Figure 1 shows the distribution of pressure ulcers at various anatomic locations for the general population and the SCI patients. The overall prevalence of pressure ulcer in the acute care setting was reported to be stable around 15% for the years 1999-2004 [5]. Except for occiput, elbow and heel locations, the distribution of pressure ulcers was similar between the two populations.

The weighted average interface pressure by the number of subjects at various anatomic locations as reported by the surveyed publications is shown in Figure 2. The interface pressure for the healthy volunteers and the general population is lower compared to the SCI population except for the occiput [27]. The higher interface pressure in SCI patients is due to a lack of innervation, resulting in disuse atrophy, rapid bottoming out, more tissue deformation, total compression of the paraplegic tissues and obstruction of blood flow. During weight bearing, the rapid reduction of paraplegic soft tissue thickness (bottoming out) creates higher pressure gradients and larger shear effect in the thinner soft tissues between the support surface and the bony prominence. [26]

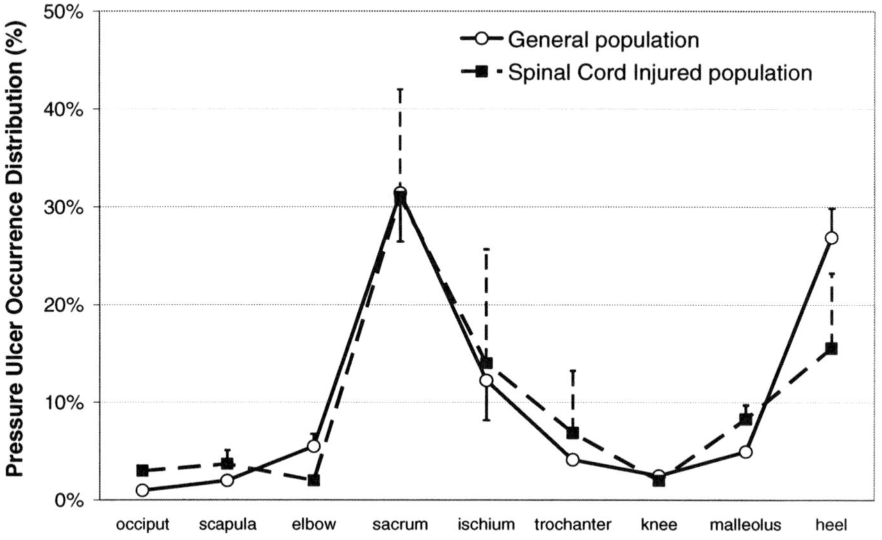

Fig. 1. The graph shows pressure sore occurrence distribution (weighted average) at various anatomic locations for the general and SCI population. Standard deviations are not shown if the data was from a single study [27]. (Used with permission).

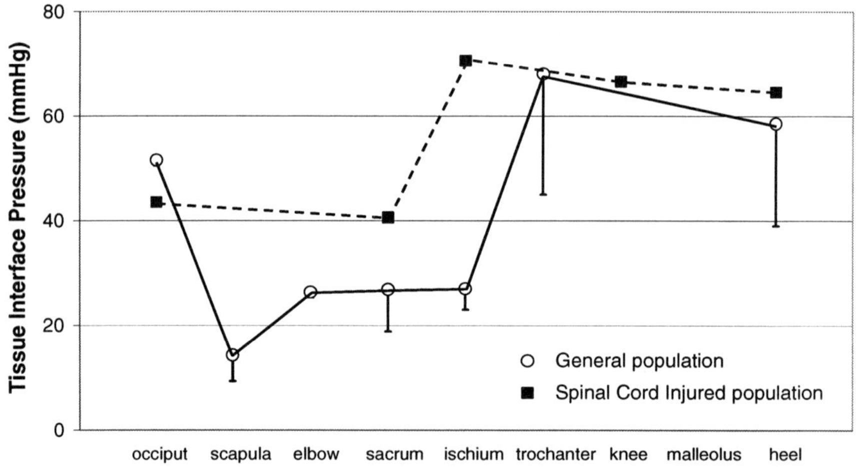

Fig. 2. The graph shows tissue interface pressure (weighted average) at various anatomic locations for the general and SCI population. Standard deviations are not shown if the data was from a single study [27]. (Used with permission).

Table 1. Correlation analysis between reported interface pressure and pressure ulcer occurrence distribution (weighted averages) for the general population and SCI patients at selected anatomic locations [27]. (Used with permission).

	General Population		Spinal Cord Injury Patients	
Location	Occurence in % (range)	Interface Pressure in mmHg (range)	Occurrence in % (range)	Interface Pressure in mm Hg
Occiput	1	49 (23-51)	3	44
Sacrum	31 (26-37)	24 (12-43)	31 (14-37)	41
Ischium	12 (8-17)	26 (22-29)	14 (9-35)	71
Heel	27 (23-30)	61 (14-102)	16 (11-26)	65
Correlation Coeff	**-0.118**		**-0.191**	

A correlation analysis was done between the interface pressures and the prevalence for 4 locations where the data were available for both the general population and the SCI patients (Table 1) [27]. The authors found a slightly negative correlation for both the general population of ulcer patients and the SCI patients. Both groups had the same distribution for pressure ulcer at sacrum (31%) but the SCI patients had a 70% higher interface pressure compared to general population of ulcer patients. In the case of heel ulcers, the general population had a much higher occurrence (27% vs. 16%) compared to SCI patients while the interface pressure was actually smaller (61mmHg vs. 65mmHg) [27].The differences between the two populations indicate the change in weight bearing ability of soft tissues in the absence of innervations.

The review article acknowledged the various limitations and shortcomings of using data from multiple sources, years, surfaces, patient groups, investigators, measurement technologies, missing data, small sample size and lack of standardization regarding support surfaces, methodology and reported data [27]. A good analysis would require a long duration randomized controlled study with continuous or at least frequent measurement of interface pressure at various anatomic sites along with other confounding factors (temperature, moisture, friction, immersion etc.) for each of the support device to be tested and relate it to the rate of ulcer formation.

3 Microenvironment

3.1 What Constitutes Microenvironment?

There are more than one hundred biomechanical and pathophysiological factors identified to contribute to the formation of ischemic necrosis of the skin and soft tissues [1]. External pressure has been the most frequently discussed but other primary stress factors associated with ulcer formation are shear [28], friction and the resulting deformation of the soft tissues. The secondary or environmental factors important in bed immobility are temperature, moisture, duration of the applied load, tissue atrophy, and posture. Microclimate is the effect of temperature, humidity and airflow at the patient support contact. These factors influence tissue quality by reducing the strength and the stiffness of soft tissues and increase the coefficient of the friction of the skin.

The data used for analysis (shown in Table 1) has many significant limitations and indicates the need for better controlled studies investigating the relationship between stress, environmental factors and pressure ulcer prevalence or incidence for the support surfaces currently used. While the published data has many limitations, the analysis presented here indicates no direct or positive relationship between interface pressure and the distribution of pressure ulcers at the anatomic locations. This lack of direct relation between interface pressure and the frequency of ulcers at anatomic sites suggests the major influence of secondary factors at these anatomic sites on ulcer formation and mandates the control and reporting of these factors at the microenvironment of the support surface.

3.2 Shear Stress

3.2.1 Effect of Shear

Shear stress is generated in the soft tissues by the tangential force component of the body weight gravity force on the contact area externally and by the parallel and opposite tangential force on the bony prominence internally (Figure 3). Tangential forces acting on the skin develop shear stress in the tissues through friction and cause the tissue layers to slide with respect to each other. The amount of sliding depends on the looseness of the connective fibers between the tissue layers and create tissue deformation and stretching. If the fibers are tight, the skin and

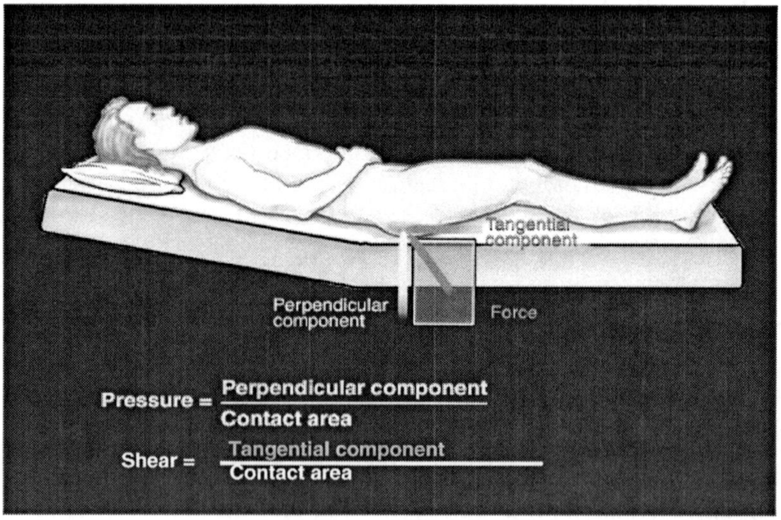

Fig. 3. Illustration of force components of pressure and shear stress on the body. (Courtesy of Cleveland Clinic).

subcutaneous tissue will be subjected to higher shear stress and more stretching; if the fibers are loose, there will be more sliding and lower shear stress and less stretching.

3.2.2 How Is It Minimized?

Shear stress is reduced by a decrease in the tangential force and an increase in the contact area. Loose covers and increased immersion in the support medium also increase contact area and further reduce shear stresses. When shear induced tissue sliding is present, vessels approaching the skin surface perpendicularly will bend and occlude at the connective layers between the tissue planes. Thus, shear will increase the effect of pressure in reducing flow through the blood vessels [29]. Conversely, if shear stress is reduced, tissues can tolerate higher pressures without blood flow occlusion.

3.2.3 Limitations / Difficulties

The shear occlusion of perforating deeper perpendicular blood vessels will create larger ischemic volume of soft tissues than pressure occlusion of deep surface vessels. There are no good proven and validated methods for measuring shear stress in a patient care environment.

3.3 Pressure

3.3.1 Effect of Pressure

The pressure stress in the soft tissues arise from the gravity force component perpendicular to the external contact area and from the body weight acting through the nearest bony prominence (Figure 3). Sustained pressure can stretch soft tissues and blood vessels, causing multiple microthrombi around the point of maximum compression [30]. In the design of support surfaces the objective is to increase contact area by greater "immersion" allowing the body to sink deeper into the support, distributing the force and reducing the pressure. The cyclic transfer of weight from high-pressure areas is the guiding principle underlying alternating pressure support systems. The alternating pressure creates pressure gradients, which are related to shear stress stretching and may damage adipose tissues and capillaries which are lacking tensile strength [28,31]. Interestingly, Knox reported in his article that a only a weak negative relationship existed between skin surface temperature and interface pressure [32].

3.3.2 How Is It Minimized?

The effect of pressure and shear stress are similar on the blood flow to the tissues. Both stresses reduce blood flow and occlude tissue perfusion [33]. In the absence of shear, tissues can tolerate nearly twice as much surface pressure before the loss of tissue perfusion [31]. Thus, a simple reduction of tissue stretching can nearly double the tissues' ability to withstand pressure without the development of ischemia. Many support surfaces take advantage of this fact and reduce shear, thus improving the weight-bearing tolerances of the soft tissues [34]. Lim et al [35] have introduced a promising concept of "Off-Loading" sitting by partially removing the ischial support to reduce excessive pressure under buttocks.

3.3.3 Limitations / Difficulties

Two groups of patients, those with neurological disorders (spinal cord injury, stroke, head trauma) and the elderly, account for most of the reported prevalence of pressure ulcers. The similarity is not surprising since the loss of muscle strength, skin and muscle proteins, and muscle mass with age are similar to the losses observed with neurological disorders. The changes in the mechanical strength and susceptibility to external loads of the skin and subcutaneous tissues in neurological disorders indicate a significant reduction of tissue viability from normally innervated tissues. With the ageing of the population, these soft tissue changes are increasing in importance and may override the benefits of pressure reduction alone.

3.3.4 Tissue Deformation Due to Shear Stress and Pressure

Compression of soft tissues and their deformation under externally applied forces are the limiting factors in the fitting of body support systems in bed supports and wheelchair seating. Tissue pressure and strain are also important determinants of comfort on beds and mattresses. Studies have reported information about the significant effect of externally applied pressure on the dimension changes of muscle, fat and skin and the compatibility of tissues and cushion for load transfer [26]. Figure 4A shows the unloaded ("free hanging") muscular and subcutaneous tissues in a normal subject and Figure 5A the unloaded atrophied tissues of a paraplegic subject due to disuse. The extent of tissue atrophy is more clearly indicated by the increased reduction of tissue thickness upon loading of the paraplegic shown in figure (5B) from that of the normal loaded tissue, illustrated in figure (4B). Reduction of tissue tone in paraplegia is also indicated by the extensive lateral tissue bulge near the trochanter in Figure (5B), suggest increased coupling of vertical load to transverse elongation in the loaded tissues indicating tension and shear loading in the lateral direction. This lateral "bulging" and transverse distortion of the paraplegic tissues under vertical compressive loading indicates the need for precise matching of shape and material of the support surface to the unloaded tissue contours to prevent the lateral "bulge" [26].

Rapid reduction in tissue thickness was evident with increasing surface pressures for both normal and the denervated tissues. The paraplegic tissues were maximally compressed or indented and bottomed out under their own body weight at the bony prominences without additional weight. The additional external weight caused no further reduction of tissue thickness; it increased surface and deep pressures only. On the normal tissues, however, the additional weight could cause further reduction of thickness at all locations except at the male trochanter. The transverse images of the soft tissues and bony prominences showed thicker tissues in normal subjects at every pressure, leading to the realization that higher pressure gradients from skin to bone must exist in the paraplegic than in the normal tissues [26]. Recent studies by Linder-Ganz et al [36]. using open-MRI scans also reported that internal tissue loads are significantly higher in paraplegics. Bottoming out of the paraplegic tissues showed the increased risk of tissue trauma caused by total compression and the resulting obstruction of blood flow.

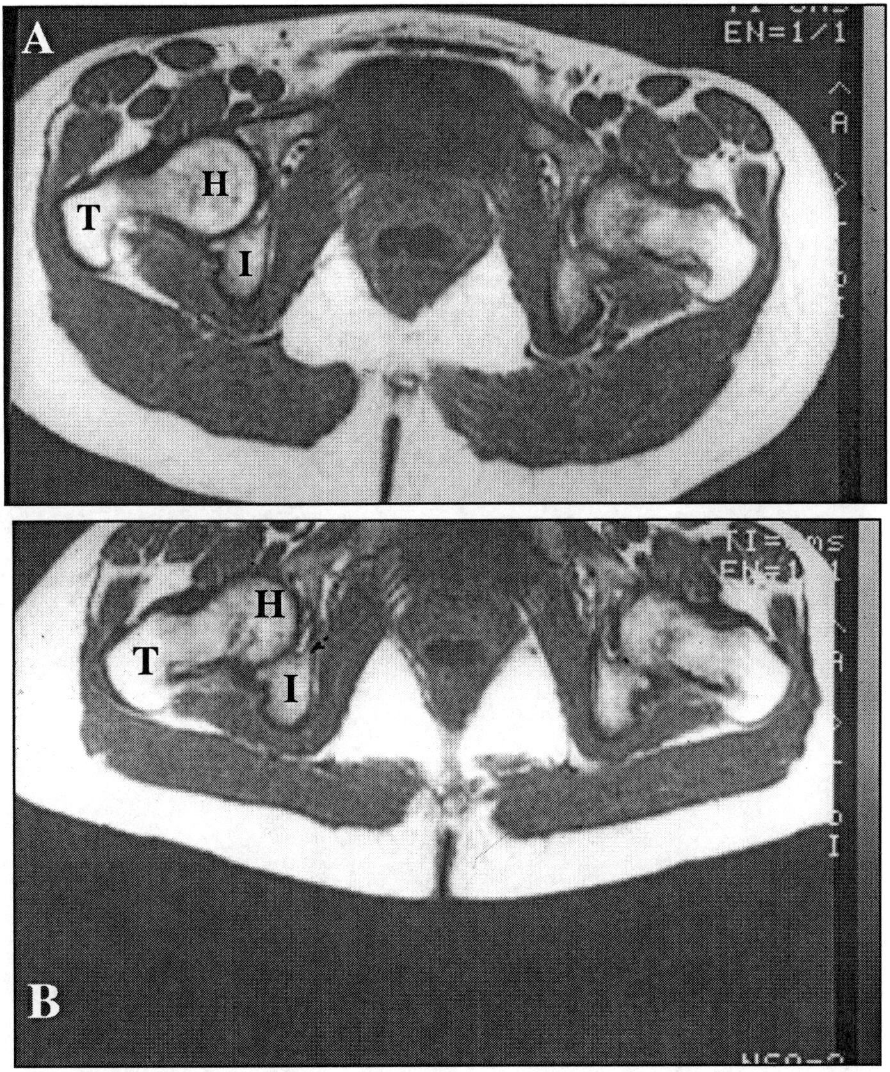

Fig. 4. Transverse magnetic resonance image of normal female near ischial tuberosities [26]. (A) Tissues without external support, "free hanging". (B) Loaded tissues on the right side with flat cushion support below. T: Trochanter of femur; H: head of femur; I: Ischium.

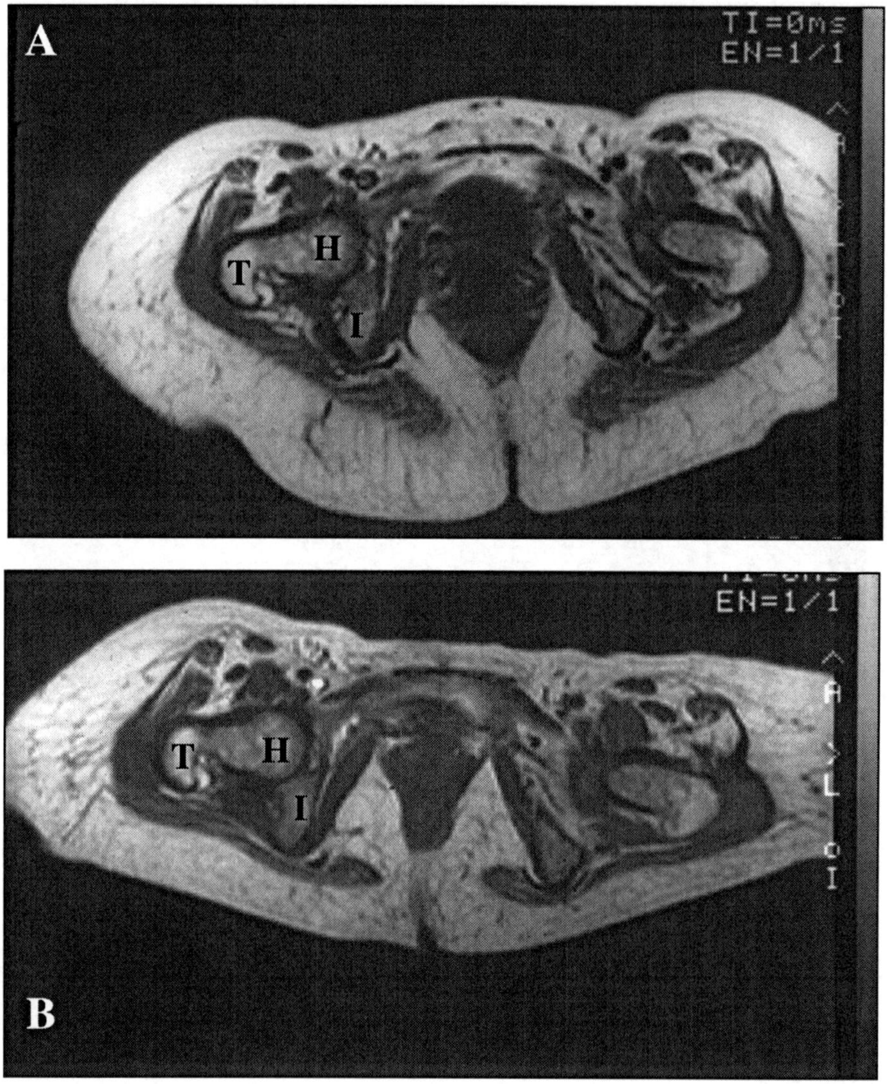

Fig. 5. Transverse magnetic resonance image of paraplegic female near ischial tuberosities [26]. (A) Tissues without external support, "free hanging". (B) Tissues on the right side with flat cushion support below. Note blood vessel distortion by the mechanical environment. T: Trochanter of femur; H: head of femur; I: Ischium.

Sae-Sia et al studied the effect of pressure loading over a 2hr period on the sacral skin blood flow (SBF) and skin temperature in acute spinal cord injury patients [37]. They observed a negative change in SBF during pressure loading along with a higher initial positive slope of the SBF reactive hyperemia response compared to patients with orthopedic trauma and healthy subjects. This suggests microvascular dysfunction in acute SCI subjects and indicates that turning interval guidelines need to be revisited for acute SCI patients.

3.4 Friction

3.4.1 Effect of Friction

Friction is a phenomenon, which describes the surface's ability to prevent motion due to forces tangential to the contact area (Figure 6). The tangential or friction force depends on the perpendicular force and the coefficient of friction at the contact and it is independent of the contact area. When the cover of the support surface is designed to allow movement over its foundation, due to lower coefficient of friction, the slippage occurs between the cover and the bed and not within the tissue layers and the tension in the skin is decreased without stretch and occlusion of blood vessels (Figure 6) [27].

3.4.2 How Is It Minimized?

The tangential or shear force is reduced most effectively by decreasing the coefficient of friction on the support surface (Figure 6A). The effect of high coefficient of friction is shown in Figure 6B [27].The effectiveness of properly inflated air, water, and viscous fluid or gel supports rests on these principles. Combinations of these biomechanical principles are commonly used in modern support surfaces to create a better physical environment for tissue survival.

3.4.3 Limitations / Difficulties

Measuring friction in a patient care environment is not easy. If the interface surface is frictionless, patient stability will be affected as patients may slide or be unable to maintain posture.

3.5 Temperature

3.5.1 Effect of Temperature

Elevating body temperature increases the metabolic activity of the tissues by 10% per degree Celsius of temperature rise, thus increasing the need for oxygen and energy source at the cellular level. If the patient has impaired circulation from local pressure and shear then the tissues will starve and release contents of lysozymes inducing autodigestion of cytoplasma. The metabolic activity may also cease from lack of energy and the accumulation of waste products. It has been shown [29] that pressure induced tissue injury accelerates with increasing body temperature (see Table 2). An equally significant effect of increases in skin temperature is the induction of the sweat response and the potential accumulation of moisture in the skin at the skin-support interface. This local perspiration increases friction, shear, maceration and may weaken skin in the area [38]. Also, the adverse effects of fever should not be overlooked [30]. Interestingly, Knox observed that a negative correlation existed between core body temperature and skin surface temperature [32].

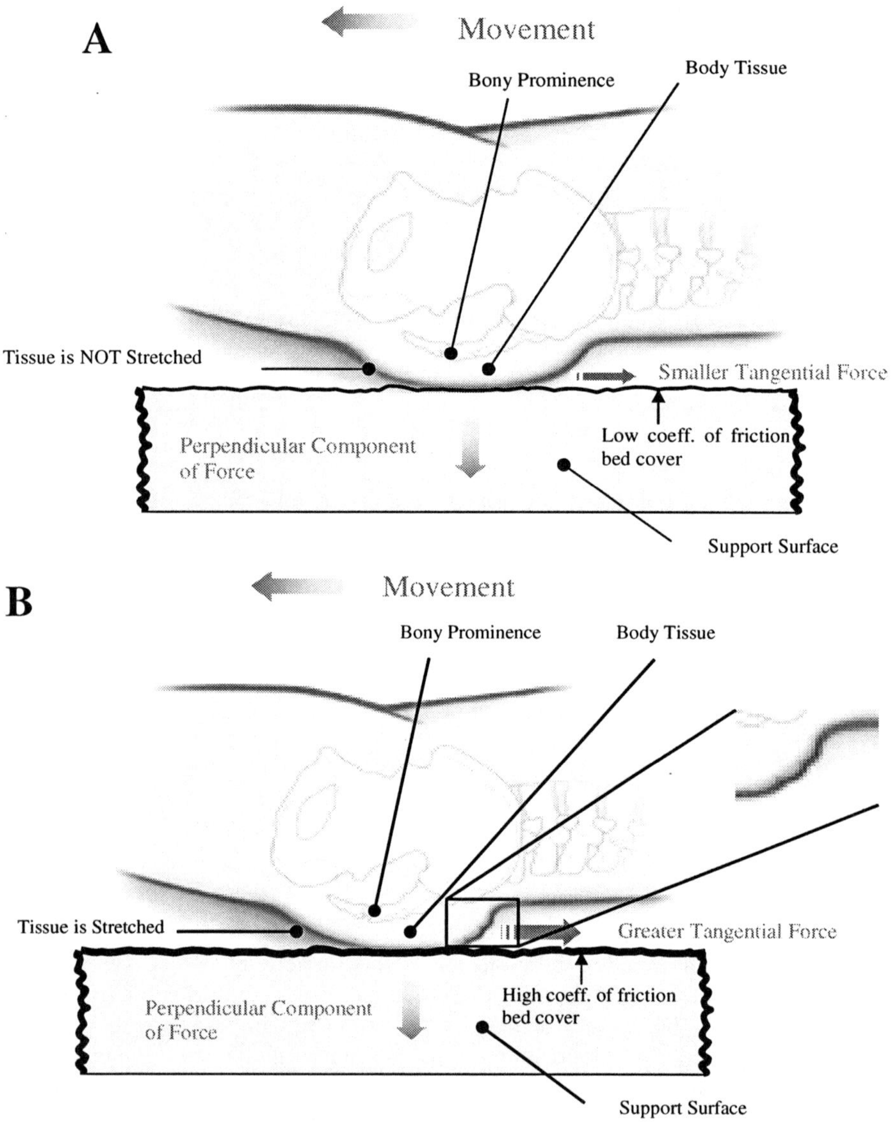

Fig. 6. (A) The illustration shows the lack of skin stretch in the case of low coefficient of friction between the cover and the bed. (B) The illustration shows the skin stretched on one side and folding of tissues on the side opposite to the direction of movement (see inset) in the case of high coefficient of friction between the cover and the bed. Force components on the support [27]. (Used with permission).

Table 2. Pressure Induced Tissue Injury Accelerates With Increasing Body Temperature [29] (Used with permission)

Experimental pig model	@ 100 mm Hg 5 hrs
25°C	no break down
35°C	partial thickness injury
45°C	full thickness breakdown

3.5.2 How Is It Minimized?

Temperature is usually maintained by varying the airflow at the patient/support surface interface area. Low and High-air loss mattresses are able to maintain optimal body temperature by cooling the skin. Reducing body temperature aids in preventing or minimizing infection and fever. Studies have suggested that a 5 degrees C reduction in skin temperature would have an effect similar to that of the most expensive support surface available [38].

3.5.3 Limitations / Difficulties

Excess cooling at the interface may cause some discomfort to the patient and may also enhance muscle tension.

3.6 Moisture

3.6.1 Effect of Moisture

Hydration of the weight-bearing skin opens a new set of destructive influences on skin integrity. Moisture from sweating or from urinary or fecal incontinence will hydrate the skin, dissolve the molecular collagen crosslinks of the dermis and soften the stratum corneum (maceration). Skin maceration results in the reduction of the stiffness, the near complete loss of the connective tissue strength and in the erosion of the dermis under the action of shear forces. Another result of skin hydration is the rapid increase of the coefficient of friction of the epidermis, which promotes adhesion of the skin to the support surface and produce elevated shear, easy sloughing and ulceration. Compounding the destructive effect of stress is hydration diluting the natural skin acidity, reducing antibacterial properties of the epidermis leading to easier sepsis. A recent study [39] attempted to delineate the differences in pathophysiology and histopathology in patients with both incontinence and pressure ulcer lesions. They observed that there is no justification for singling out moisture lesions from pressure ulcer lesions.

3.6.2 How Is It Minimized?

The clinician has two excellent technologies for controlling the microclimate at the skin-support surface interface. The dynamic low- and high-air-loss [40] and the air-fluidized support systems are designed to reduce stress and temperature, evaporate moisture and prevent heat accumulation and cool the interface with the support surface. The evaporation of one kilogram of water from the skin at the support surface will remove 580 Kcal of heat from the body through the "latent heat of vaporization" [41]. Thus the cooling power at the rate of the total water loss through the skin for an average person with 1.8m^2 skin surface and at the normal sweat rate of 26.7g/m^2/hr, the heat loss is 27.9Kcal/hr. With proper design and nursing care aimed to maintain physiologic water balance, dynamic air loss supports are able to control interface pressure, shear, friction, and the temperature and moisture of the support environment.

3.6.3 Limitations / Difficulties

The operating cost of these air-loss systems can be significant and are not usually reimbursable in the home-care environment.

4 Summary and Conclusion

It is not the intention of the authors to discount the usefulness of clinical support surfaces. On the contrary, the authors wish to remind the healthcare providers that the usefulness of any pressure relieving support surface depends on the control of the microenvironment and not just pressure alone. It is therefore important for healthcare providers to understand the influence of the microenvironment in the development of pressure ulcers when selecting and applying a support surface for their patients. More research is still needed to understand the interactions between the microenvironment and pressure sore occurrence.

4.1 Areas for Future Research

A recent study has proposed endothelial dysfunction as an additional risk factor for pressure ulcer [42]. Endothelial dysfunction is defined as impaired vasodilation to specific stimuli which results in reduced vasodilation, proinflamatory state and prothrombotic state. Statins and angiotensin receptor blockers improve endothelial function indicating good potential for future research.

References

1. Lyder, C.H.: Pressure ulcer prevention and management. JAMA 289(2), 223–226 (2003)
2. Allman, R.M.: Pressure sores among the elderly. N. Engl. J. Med. 320(13), 850–853 (1989)
3. Kuhn, B.A., Coulter, S.J.: Balancing the pressure ulcer cost and quality equation. Nurs. Econ. 10(5), 353–359 (1992)
4. Miller, H., Delozier, J.: Cost implications of the pressure ulcer treatment guideline. Columbia, MD: Center for Health Policy Studies; Contract No. 282-91-0070. Sponsored by the Agency for Health Care Policy and Re-search (1994)
5. Whittington, K.T., Briones, R.: National Prevalence and Incidence Study: 6-year sequential acute care data. Adv. Skin Wound Care 17(9), 490–494 (2004)
6. Landis, E.M.: The microinjection studies of capillary blood pressure in human skin. Heart 15, 209–228 (1930)

7. National Pressure Ulcer Advisory Panel, http://www.npuap.org/pr2.htm (accessed May 12, 2009)
8. Russell, L.: Overview of research to investigate pressure-relieving surfaces. Br. J. Nurs. 10(21), 1421–1426 (2001)
9. Lilla, J.A., Friedrichs, R.R., Vistnes, L.M.: Floatation mattresses for preventing tissue breakdown. Geriatrics 30(9), 71–75 (1975)
10. Boorman, J.G., Carr, S., Kemble, J.V.: A clinical evaluation of the air-fluidised bed in a general plastic surgery unit. Br. J. Plast. Surg. 34(2), 165–168 (1981)
11. Berijan, R.A., Douglas, H.O., Holyoke, E.D., et al.: Skin pressure measurements on various mattress surfaces in cancer patients. Am. J. Phys. Med. 62(5), 217–226 (1983)
12. Bar, C.A.: Evaluation of cushions using dynamic press measurement. Prosthet. Orthot. Int. 15(3), 232–240 (1991)
13. Allen, V., Ryan, D.W., Murray, A.: Repeatability of subject/bed interface pressure measurements. J. Biomed. Eng. 15(4), 329–332 (1993)
14. Hedrick-Thompson, J., Halloran, T., Strader, M.K., McSweeney, M.: Pressure-reduction products: making appropriate choices. J. ET. Nurs. 20(6), 239–244 (1993)
15. Allen, V., Ryan, D.W., Murray, A.: Measurements of interface pressure between body sites and the surfaces of four specialized air mattresses. British Journal of Clinical Practice 48(3), 125–129 (1994)
16. Collier, M.E.: Pressure-reducing mattresses. Journal of Wound Care 5(5), 207–211 (1996)
17. Sachse, R.E., Fink, S.A., Klitzman, B.: Multimodality evaluation of pressure relief surfaces. Plastic and Reconstructive Surgery 102(7), 2381–2387 (1998)
18. Fontaine, R., Risley, S., Castellino, R.: A quantitative analysis of pressure and shear in the effectiveness of support surfaces. J. Wound Ostomy Continence Nurs. 25(5), 233–239 (1998)
19. Hardin, J.B., Cronin, S.N., Cahill, K.: Comparison of the effectiveness of two pressure-relieving surfaces: low-air-loss versus static fluid. Ostomy Wound Manage 46(9), 50–56 (2000)
20. Russo, C.A., Elixhauser, A. (2006) Hospitalizations Related to Pressure Sores. HCUP Statistical Brief #3. Agency for Healthcare Research and Quality, Rockville, MD (2003), http://www.hcup-us.ahrq.gov/reports/statbriefs/sb3.pdf (accessed May 2, 2006)
21. Gefen, A., Levine, J.: The false premise in measuring body-support interface pressures for preventing serious pressure ulcers. J. Med. Eng. Technol. 31(5), 375–380 (2007)
22. Makhsous, M., Rowles, D.M., Rymer, W.Z., et al.: Periodically relieving ischial sitting load to decrease the risk of pressure ulcers. Arch. Phys. Med. Rehabil. 88(7), 862–870 (2007)
23. Nicosia, G., Gliatta, A.E., Woodbury, M.G., Houghton, P.E.: The effect of pressure-relieving surfaces on the prevention of heel ulcers in a variety of settings: a meta-analysis. Int. Wound J. 4(3), 197–207 (2007)
24. Berlowitz, D.R., Brienza, D.M.: Are all pressure ulcers the result of deep tissue injury? A review of the literature. Ostomy Wound Manage 53(10), 34–38 (2007)
25. Bader, D.L., White, S.H.: The viability of soft tissues in elderly subjects undergoing hip surgery. Age Ageing 27(2), 217–221 (1998)

26. Reger, S.I., McGovern, T.R.F., Chung, K.C.: Biomechanics of tissue distortion and stiffness by magnetic resonance imaging. In: Bader, D. (ed.) Pressure Sores - Clinical Practice and Scientific Approach, ch. 14, pp. 177–190. MacMillan, London (1990)
27. Reger, S.I., Ranganathan, V.K., Sahgal, V.: Support surface interface pressure, microenvironment, and the prevalence of pressure ulcers: an analysis of the literature. Ostomy Wound Manage 53(10), 50–58 (2007)
28. Reichel, S.M.: Shear force as a factor in decubitus ulcers in paraplegics. J. Am. Med. Assoc. 166(7), 762–763 (1958)
29. Kokate, J.Y., Leland, K.J., Held, A.M., et al.: Temperature-modulated pressure ulcers: a porcine model. Arch. Phys. Med. Rehabil. 76(7), 666–673 (1995)
30. Parish, L.C., Lowthian, P., Witkowski, J.A.: The decubitus ulcer: many questions but few definitive answers. Clin. Dermatol. 25(1), 101–108 (2007)
31. Bennett, L., Kavner, D., Lee, B.Y., Trainor, F.S.: Shear vs. pressure as causative factors in skin blood flow occlusion. Arch. Phys. Med. Rehabil. 60, 309–314 (1979)
32. Knox, D.M.: Core body temperature, skin temperature, and interface pressure. Relationship to skin integrity in nursing home residents. Adv. Wound Care 12(5), 246–252 (1999)
33. Cherry, G.W., Ryan, T.J.: Pathophysiolgy. In: Parish, L.C., Witkowski, J.A., Crissey, J.T. (eds.) The decubitus ulcer in clinical practice, pp. 33–43. Springer, Berlin (1997)
34. Jay, R.: Pressure and Shear: their effects on support surface choice. Ostomy Wound Manage 4(8), 36–45 (1995)
35. Lim, D., Lin, F., Hendrix, R.W., Moran, B., Fasanati, C., Makhsous, M.: Evaluation of a new sitting concept designed for prevention of pressure ulcer on the buttock using finite element analysis. Med. Biol. Eng. Comput. 45(11), 1079–1084 (2007)
36. Linder-Ganz, E., Shabshin, N., Itzchak, Y., Yizhar, Z., Siev-Ner, I., Gefen, A.: Strains and stresses in sub-dermal tissues of the buttocks are greater in paraplegics than in healthy during sitting. J. Biomech. 41(3), 567–580 (2008)
37. Sae-Sia, W., Wipke-Tevis, D.D., Williams, D.A.: The effect of clinically relevant pressure duration on sacral skin blood flow and temperature in patients after acute spinal cord injury. Arch. Phys. Med. Rehabil. 88(12), 1673–1680 (2007)
38. Lachenbruch, C.: Skin cooling surfaces: estimating the importance of limiting skin temperature. Ostomy Wound Manage 51(2), 70–79 (2005)
39. Houwing, R.H., Arends, J.W., Canninga-van Dijk, M.R., Koopman, E., Haalboom, J.R.: Is the distinction between superficial pressure ulcers and moisture lesions justifiable? A clinical-pathologic study. Skin med. 6(3), 113–117 (2007)
40. Reger, S.I., Adams, T.C., Maklebust, J.A., Sahgal, V.: Validation test for climate control on air-loss supports. Arch. Phys. Med. Rehabil. 82(5), 597–603 (2001)
41. Scott, J.W.: The body temperature. In: Best, C.H., Taylor, B. (eds.) The Physiological Basis of Medical Practice, p. 895. The Williams and Wilkins Co., Baltimore (1961)
42. Struck, B.D., Wright, J.E.: Pressure ulcers and endothelial dysfunction: is there a link? J. Nutr. Elder. 26(3-4), 105–117 (2007)

Partial Differential Equations for Modelling Wound Geometry

Hassan Ugail

School of Computing, Informatics and Media
University of Bradford, Bradford BD7 1DP
United Kingdom
h.ugail@bradford.ac.uk

Abstract. Wounds arising from various conditions are painful, embarrassing and often requires treatment plans which are costly. A crucial task, during the treatment of wounds is the measurement of the size, area and volume of the wounds. This enables to provide appropriate objective means of measuring changes in the size or shape of wounds, in order to evaluate the efficiency of the available therapies in an appropriate fashion. Conventional techniques for measuring physical properties of a wound require making some form of physical contact with it. We present a method to model a wide variety of geometries of wound shapes. The shape modelling is based on formulating mathematical boundary-value problems relating to solutions of Partial Differential Equations (PDEs). In order to model a given geometric shape of the wound a series of boundary functions which correspond to the main features of the wound are selected. These boundary functions are then utilised to solve an elliptic PDE whose solution results in the geometry of the wound shape. Thus, here we show how low order elliptic PDEs, such as the Biharmonic equation subject to suitable boundary conditions can be used to model complex wound geometry. We also utilise the solution of the chosen PDE to automatically compute various physical properties of the wound such as the surface area, volume and mass. To demonstrate the methodology a series of examples are discussed demonstrating the capability of the method to produce good representative shapes of wounds.

1 Introduction

Wounds, especially those related to chronic conditions such as pressure ulcers, venous ulcers, and diabetic ulcers are painful, embarrassing and costly. A given wound may typically range in size from around 0.5 cm to 10 cm across and is of variable depth. The wound can be deeply undermined with the cavities ranging in size up to several cubic centimetres. Figure 1 shows two examples of wounds, one showing a deep foot ulcer relating to a chronic condition and the other showing several wounds on a human arm shape.

In the USA the estimated annual cost of pressure ulcer treatment alone range between \$3.5 and \$7 billion [1,2,3]. In the UK it is conservatively estimated that wounds relating to venous ulceration alone affects around 580,000 individuals at any one time of which a significantly high proportion are the elderly [4,5]. This

A. Gefen (Ed.): Bioengineering Research of Chronic Wounds, SMTEB 1, pp. 101–125.
springerlink.com

prevalence imposes a significant burden on healthcare systems, with an approximate cost of between £300 and £600 million per year. A single Vascular Unit in England sees some 300-400 patients weekly, at an annual cost of £0.25million. The cost is not just financial: wounds take a toll in missed workdays and chronic unemployment because painful, open wounds disqualify people from many lines of work and many patients suffer damaging psychological side-effects. Many patients are treated at home by community nurses. For example, over 60% of treatment costs are for community nursing services [6]. Community nurses may spend 44% of their time dressing and providing care [7], yet treatment often remains ineffective [8,9] and considerable uncertainty remains regarding the most effective treatments [10].

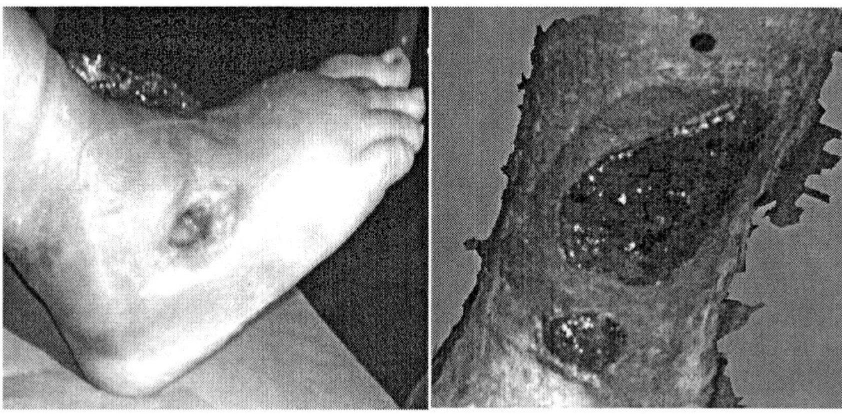

Fig. 1. Example wounds, a deep foot ulcer related to a chronic condition and deep wounds on an arm

There exist a number of modalities to treat wounds. At any given setting these days, for wound care it requires rapid, highly reliable, low cost as well as accurate methods of assessing wound dimensions. From a diagnostic point view in order to properly assess a given wound it requires an understanding of the 3-dimentional geometric shape of the wound. Without any accurate and objective means of measuring changes in size or shape of wound, it is difficult or impossible to properly evaluate the efficiency of the available therapies [11,12,13].

A crucial task, during the treatment of wounds, especially those that are considered chronic, is the measurement of their area and volume [11,13]. This enables to provide appropriate objective means of measuring changes in the size or shape of wounds, in order to evaluate the efficiency of the available therapies in an appropriate fashion. At the moment there appears to exist a number of measurement methods which are used clinically. Of which the most widely used methodology is the direct contacting measurements which are not accurate, carry a risk of infection and uncomfortable to the patient.

Conventional techniques for measuring the area and volume of wounds depend on making physical contact with the wound, for example by drawing around the periphery on an acetate sheet or by making an alginate cast of the wound [6]. For commonly used objective measures of wound healing the use of planimetric and two dimensional methods are very common. Methods of measuring wound dimensions often involve the direct measurements with the use of a ruler [14]. The ruler enables a quick and easy way to measure the wound whereby quantities such as the length, width and the depth of the wound is estimated [15]. Other common methods include the hand tracing of the perimeter of the wound with use of a grid to calculate wound area. Computer based wound documentation systems such as MAVIS (Measurement of Area and Volume Instrument System, Photometrix Imaging Ltd, Mid Glamorgan, UK), allow digital wound tracing [16]. Here the tracing of the wound boundary is undertaken by following the wound edge with a mouse cursor on a digital photograph of the wound.

There is currently significant interest in developing non-invasive measurement systems using optical methods such as 'structured light' (a technique that projects stripes on to a surface and infers the shape from changes in the linearity of the reflected stripe) or stereophotogrammetry [9,16,17]. The availability of high-resolution 3D digital cameras, increasing computing power and the development of software techniques for manipulating three-dimensional information has benefited this area. However, equipment associated with these sorts of measurement methods is not often portable and is often costly, thus making the prohibitive for routine medical use. Nevertheless this is probably the most exhaustive mechanism for representation of the geometry of wound surface which is based on the reconstruction of the surface geometry in 3-dimensional space. This enables accurate calculation of the surface area and volume of the wound.

Here we address a possible approach that would enable to model and measure the geometry of complex wound geometry. The idea behind is based on shape parameterisation. In parametric design the basic approach is to develop a generic description of an object or a class of objects in which the shape is controlled by the values of a set of design variables or parameters. A new design, created for a particular application, is obtained from this generic template by selecting particular values for the design parameters so that the item has particular properties suited to that application. Thus, one could envisage a scenario where a generic parameterised 3-dimensional shape of a wound is available for a user where the user would be then able to create a fine tuned model based on a handful of direct measurements taken from a patient.

There exists a wide variety of methods that can be utilised to generate the geometry of wounds. Examples include boundary based methods such as polygon based design [18], extrusions and surface of revolution [19] and polynomial patches [20], procedural modelling such as implicit surfaces [21] and fractals [22] and volumetric models such as constructive solid geometry [23] and subdivision [24]. Many of these techniques, especially polygon based design and polynomial patches, would be appropriate for modelling the geometry of wounds.

Of the above mentioned techniques many commercial Computer Aided Design (CAD) systems today employ the conventional polynomial geometry modelling schemes. One of them is spline based schemes now dominated by Non Uniform Rational B-Splines (NURBS) [25,26]. There has been a considerable amount work undertaken in the area of geometry representation using NURBS. An example of geometry representation using spline based methods has been developed by Piegl [27, 28]. The example describes how the geometry of a brush handle can be designed by using a series of cross sectional profile curves. For this problem, the points per cross section were taken between 10 to 48 and the number of control points after merging all the surface patches were nearly 4000. In addition to the 4000 control points, one also has to take into account the associated weights of the spline functions that are used. In a similar fashion Pottmann [29] describes the approximation of a ruled cylindrical surface using NURBS. Here a surface of bi-degree with 7×25 control points were utilsed.

Thus, conventional geometry modelling techniques based on splines are not suitable to overcome geometry parameterisation problems since such methods typically use polynomial patches (e.g. Non Uniform Rational B-Splines or NURBS), which often require hundreds of control points in order to represent a realistic object. Furthermore, such "spline" based patches often do not exactly meet at the boundaries and consequently need to be 'trimmed' or stitched in order to close the geometry of the object in question.

Triangular meshes or subdivision schemes [30,31,32] for geometry representation has recently been popular as an alternative to spline based techniques. For example, Hubeli and Gross [33] describe a geometric surface in which Doo-Sabin subdivision scheme is applied on a two manifold surface to represent its geometry. In order to reconstruct the geometry of the shape they have utilized around 50 smoothing steps using the subdivision method. Similarly, Catmull [34] reconstruct a teapot using his subdivision algorithm. Although being much more flexible than spline based techniques, triangular meshes also have restrictions and disadvantages. For example, when applying an extreme deformation to a triangle mesh, certain triangles exhibit strong stretching which leads numerically and visually undesirable triangles that have to be overcome.

Geometry based on Partial Differential Equations (PDEs) have recently emerged as a powerful tool for geometric shape modelling [35,36,37,38,39]. Using this methodology, a surface is generated as the solution to an elliptic Partial Differential Equation (PDE) using a set of boundary conditions. The PDE method is efficient in the sense that it can represent complex three-dimensional geometries in terms of a relatively small set of design variables.

Mathematical boundary-value problems such as the use of elliptic Partial Differential Equations (PDEs) are well known for their application in many engineering problems including stress/stain analysis, fluid flows and electromagnetism. From a geometric design point of view such boundary-value problems have found their way into various application areas such as surface design, geometric mesh smoothing and fairing. To this end it is noteworthy to refer to Bloor and Wilson's PDE method [40, 37, 41], for intuitive shape generation. This method is based on the solution of the Biharmonic PDE with appropriately chosen boundary conditions. We also take note of the work by Schneider and Kobbelt and others on geometric mesh fairing [42, 43, 44], where the properties of the Biharmonic operator is used to fair triangular meshes. Other PDE based techniques in geometric design include level set formulations for image processing, computer graphics and computer vision [45,46,47] geometric mesh processing [48, 49] and Radial Basis Functions [50]. Such PDE based techniques form well posed problems as well as possess appealing functionality in terms of their solution being smooth functions whereby such solutions can be controlled by the specified boundary conditions.

The advantage of the geometry based on PDEs is that it can represent complex geometry in terms of a small set of design variables [51], instead of many hundreds of variable using traditional modelling techniques. In broad terms this is because its boundary-value approach means that PDE geometry are defined by data distributed around just their boundaries, instead of data distributed over their surface area, e.g. control points. The method defines a shape in terms of a number of surface patches that collectively describe the object's surface. However, unlike spline-based approaches, the method produces surface patches that exactly meet at their mutual boundaries, without the need for complicated trimming. Thus, a PDE model, when changed by altering the values of its design parameters, remains continuous; there is no need for a designer to intervene in order to close up any holes that might appear at patch boundaries. In the present context, this means that the PDE geometry can be made to adapt automatically to changes in the shape of the wound in question.

2 Elliptic Partial Differential Equations for Shape Modelling

In very simple terms, one can describe Partial Deferential Equations (PDEs) as a mathematical tool that can be used to describe a given physical phenomena. This description is given in the form of a mathematical relation between different rates of change of the phenomena in study with respect to different variables such as the 3-dimentional physical coordinates or time. The range of application areas that form the use of PDEs vary from physical phenomena such as fluid flow applications such as weather forecasting to economics and finance.

The subject of PDEs holds an exciting and special position in applications relating to modelling physical phenomena. PDEs emerged as a subject in the 18th century due to the failure of ordinary differential equations to describe physical phenomena. The subject of PDEs has been developed by major names in applied mathematics such as Euler, Legendre, Laplace and Fourier. For example, Euler and Laplace developed the idea of potential theory [52] while Fourier developed series expansions for heat equation. Many advances in modern science have been based on the discovery of the underlying PDE for the process in question. Examples include Maxwell's equations [53] which describe a unified theory between electricity and magnetism, the Schrödinger's equation for quantum mechanics and Navier-Stokes' equation [54] for modelling fluid flows with applications to weather forecasting, motion of stars inside galaxies and flow around aerofoil wings.

Mathematically speaking, these rates of change are known as derivatives and in particular, these rates are known as partial derivatives when the function that is being differentiated depends on two or more variables. For instance, assume that a function F depends on x, y and t; that is, $F(x, y, t)$ where $0 \le x \le 1$, $0 \le y \le 1$ and $t \ge 0$. The rate of change of F with respect to x denoted as $\frac{\partial F}{\partial x}$ which represents the partial derivative of F with respect to x. Now if we are to represent the rate of change of $\frac{\partial F}{\partial x}$ with respect to y we can denote this as $\frac{\partial^2 F}{\partial x \partial y}$.

Assuming that a given physical phenomenon can be mathematically modelled using F where the phenomenon is governed by the relation between its rate of change with respect to the variables x and y and a function $G(x, y, t)$, we can write such a PDE as,

$$\frac{\partial F}{\partial x} + \frac{\partial^2 F}{\partial x \partial y} = G(x, y, t). \tag{1}$$

2.1 Classification of Partial Differential Equations

It is common practice in mathematics to use a quantity called the discriminant for determining the nature of the roots associated with a given second order algebraic equation. The discriminant could tell if both the roots are real and different, if there is only one root or if they are complex. Similarly, PDEs can be classified into different types of equations depending on the value of the discriminant. If one assumes that the general second order partial deferential equation in two variables is given by,

$$A\frac{\partial^2 F}{\partial^2 x} + B\frac{\partial^2 F}{\partial x \partial y} + C\frac{\partial^2 F}{\partial^2 y} + \ldots, \tag{2}$$

then the discriminant of the Equation (2) can be written as,

$$B^2 - 4AC. \tag{3}$$

The classification is divided into three major groups. i.e.,

1. $B^2 - 4AC < 0$

PDEs fulfilling this condition are regarded as elliptic partial differential equations. An example of elliptic partial differential equations is the Laplace equation. The solutions of these types of PDEs are generally given in terms of harmonic functions and are smooth within the domain in which they are solved. Moreover, if the coefficients multiplying the terms involving the unknown function and its derivatives are separable, the solution can be found using Fourier transforms. The aim of this work is to enhance the use of elliptic PDEs as a surface and solid generation technique relating to wound geometry. Thus, further details of specifically selected low order elliptic PDEs would be the main theme of discussion of this paper.

2. $B^2 - 4AC > 0$

Any second order PDE satisfying this condition is classified as a parabolic PDE. The heat equation is an example of a parabolic equation. Parabolic PDEs are typically related to evolution problems such as heat diffusion. For that reason they are also known as evolution equations since they describe how a physical property changes through time across a given domain. Generally, the solution to this type of an equation is less stable than those to elliptic PDE i.e. they reach a singularity as they evolve with time.

3. $B^2 - 4AC = 0$

Any PDE satisfying the above condition is called a hyperbolic PDE. An example of such a type PDE is the wave equation.

2.2 The Biharmonic Equation

In this work we are placing special emphasis on the Biharmonic PDE which is utilised to be the basic building block for modelling the wound geometry. The Biharmonic PDE falls into the category of elliptic PDEs of the type commonly denoted by,

$$(\nabla^2)^k F = 0, \tag{4}$$

where ∇^2 represents the Laplace operator and $k \geq 1$. The form of Laplace operator in a Cartesian 3-dimentional coordinate system is denoted as,

$$\nabla^2 = \frac{\partial^2 F}{\partial^2 x} + \frac{\partial^2 F}{\partial^2 y} + + \frac{\partial^2 F}{\partial^2 z}. \tag{5}$$

The case when $k = 2$ Equation (4) is known as the Biharmonic equation. Its solution is given in terms of functions whose fourth partial derivatives are continuous and satisfy the Biharmonic condition. Examples in which this equation has played an important role describing physical phenomena are the case of stokes flows in fluid dynamics, where the Biharmonic equation is used and the stream function describing the flow or in continuum mechanics where it is used to find the Airy or Love stress functions to describe plane stress or pane strain problems in order to find the displacement function.

In order to solve the Biharmonic PDE, or for that matter any PDE, a number of boundary conditions are usually required. The nature of the boundary conditions required to solve the equation depends on the type of problem and in general there are two types of boundary conditions. They are Dirichlet boundary conditions and he Neumann boundary conditions. The Dirichlet boundary conditions specify the value of F on the boundary of the region in which the solution wants to be found. The Neumann boundary conditions specify the value of the normal derivative of the function F at the boundary of the domain.

2.3 The Solution of the Biharmonic Equation

The solution to the Biharmonic equation can be found through a number of techniques. Full analytic solution can be found in some cases but it also has been solved using numerical techniques, which in general tend to provide very stable results. This is attributed to the fact that the discrete formulation of the Laplace equation is regarded as an averaging process.

The task of finding a solution to the Biharmonic equation is by no means trivial. Sometimes with a suitable choice of boundary conditions and restriction placed on the domain of the solution one can obtain a simple analytic solution to the equation. However, in the vast majority of cases this is not the case and often approximate solutions based on numerical analysis are sought. Here we outline the common methods that can be employed to find solution of the Biharmonic equation subject to a given set of boundary conditions.

2.3.1 Analytic Methods

Among the popular analytic methods available for finding the solution of the Biharmonic equation involves the use of separation of variables. This method is generally used to solve linear PDEs and consists of expressing the unknown function in terms of a product of a series of functions, each of which depend only and only on one of the independent variables.

2.3.2 Spectral Methods

These methods usually express the solution of the PDE in terms of its Fourier series, which is then substituted in the PDE itself in order to obtain a system of ordinary differential equations. This simplifies the problem. However it is often necessary to employ numerical techniques to find the solution to each ordinary differential equation.

2.3.3 Numerical Methods

The most common of numerical methods employed in solving Biharmonic PDE is the finite differences, finite element method and element of boundary. Below, a brief description of each of these techniques is given.

- Finite difference method

Finite difference methods are based on grid-type discretisation of the unknown function in the domain in which the solution to the PDE is to be found. The derivatives involved in the Biharmonic equation are then expressed in terms of these discrete points according to well established rules at every point in the grid and as many neighbouring points as required by the order of the derivative in turn. It is worth mentioning that the value of the function and its derivatives are also expressed in the same manner but are somehow compensated with the boundary conditions. Then, all the corresponding expressions are substituted in the original PDE, leading to a system of algebraic equations that can be easily solved. Finite difference methods can be further categorised into explicit, implicit and semi-implicit methods. The type of method selected to solve a particular PDE depends on criteria often related to the inner stability of the method.

- Finite element method

The working principle of this technique consists of approximating the original Biharmonic equation into a system of ordinary differential equations that can be integrated numerically using well known methods. The main challenge when applying this technique to consist of approximating the original equation in order to ensure the solution is stable.

- Boundary element method

This method consists of finding a suitable set of boundary values from the integral equation resulting from formulating the original Biharmonic equation in integral form. The boundary values found can then be used to calculate the numerical solution of the original PDE. Finite element methods are regarded as accurate; however, they usually lead to very large matrix systems and therefore their computational cost is quite high.

3 Modelling Wound Geometry

In this paper we utilise the elliptic PDE known as the Biharmonic equation in order to model the geometry of wound shapes. The aim here is to show how a boundary-value approach can be adopted to model complex geometry corresponding to wound shapes intuitively. The Biharmonic equation when solved subject to suitable boundary conditions enables to define complex geometry in terms of a small set of design variables [51]. This is because the boundary-value adopted here surfaces are defined by data distributed around just their boundaries, instead of data distributed over their geometry.

In the current context the PDE geometry we want to model can be regarded as a parametric surface patch $\underline{X}(u,v)$, defined as a function of two parameters u and v on a finite domain $\Omega \subset R^2$, by specifying boundary data around the edge region of $\partial\Omega$. Typically the boundary data are specified in the form of $\underline{X}(u,v)$ and a number of its derivatives on $\partial\Omega$. Here one should note that the coordinate of a point u and v is mapped from that point in Ω to a point in the physical space. To satisfy these requirements the surface $\underline{X}(u,v)$ is regarded as a solution of a PDE of the form,

$$\left(\frac{\partial^2}{\partial u^2}+\frac{\partial^2}{\partial v^2}\right)^2 \underline{X}(u,v)=0, \tag{6}$$

The PDE given in Equation (6) is of fourth order. Therefore, in order to solve the Equation, four boundary conditions are required. Here four positional curves are taken as the four boundary conditions.

Let Ω be a finite domain defined as $\{ \Omega : 0 \le u \le 1, 0 \le v \le 2\pi \}$ such that,

$$\begin{aligned} \underline{X}(0,v) &= \underline{P}_0(v), \\ \underline{X}(s,v) &= \underline{P}_s(v), \\ \underline{X}(t,v) &= \underline{P}_t(v), \\ \underline{X}(1,v) &= \underline{P}_1(v), \end{aligned} \tag{7}$$

where $\underline{P}_0(v)$ and $\underline{P}_1(v)$ define the edges of the surface at $u = 0$ and $u = 1$ respectively. $\underline{P}_s(v)$ and $\underline{P}_t(v)$ are the positions of the second and third functions as described in Equation (7). Here s and t are the positions of the interior curves such that, $0 \le s < t$, and $s < t \le 1$.

Using the method of separation of the variables, the explicit solution of Equation (6) can be written as,

$$\underline{X}(u,v) = \underline{A}_0(u) + \sum_{n=1}^{\infty} \{\underline{A}_n(u)\cos(nv) + \underline{B}_n(u)\sin(nv)\}, \tag{8}$$

where,

$$\underline{A}_0 = \underline{a}_{00} + \underline{a}_{01}u + \underline{a}_{02}u^2 + \underline{a}_{03}u^3, \tag{9}$$

$$\underline{A}_n(u) = \underline{a}_{n1}e^{nu} + \underline{a}_{n2}ue^{nu} + \underline{a}_{n3}e^{-nu} + \underline{a}_{n4}ue^{-nu}, \tag{10}$$

$$\underline{B}_n(u) = \underline{b}_{n1}e^{nu} + \underline{b}_{n2}ue^{nu} + \underline{b}_{n3}e^{-nu} + \underline{b}_{n4}ue^{-nu}, \tag{11}$$

where $\underline{a}_{00}, \underline{a}_{01}, \underline{a}_{02}, \underline{a}_{03}, \underline{a}_{n1}, \underline{a}_{n2}, \underline{a}_{n3}, \underline{a}_{n4}, \underline{b}_{n1}, \underline{b}_{n2}, \underline{b}_{n3}$ and $\underline{b}_{n4}$ are vector valued constants, whose values are determined by the imposed boundary conditions at $u = 0, u = s, u = t$ and $u = 1$.

For a given set of boundary conditions, in order to define the various constants in the solution, it is necessary to perform Fourier analysis of the boundary conditions and identify the various Fourier coefficients. For a finite number of Fourier modes N (e.g. $4 \le N \le 10$) the approximate surface solution can be defined as,

$$\underline{X}(u,v) = \underline{A}_0(u) + \sum_{n=1}^{N} \{\underline{A}_n(u)\cos(nv) + \underline{B}_n(u)\sin(nv)\} + \underline{R}(u,v), \tag{12}$$

where $\underline{R}(u,v)$ is called a remainder function defined as,

$$\underline{R}(u,v) = \underline{r}_1(v)e^{wu} + \underline{r}_2(v)ue^{wu} + \underline{r}_3(v)e^{-wu} + \underline{r}_4(v)ue^{-wu}, \tag{13}$$

where $\underline{r}_1, \underline{r}_2, \underline{r}_3, \underline{r}_4$ are obtained by considering the difference between the original boundary conditions and the boundary conditions satisfied by the function,

$$\underline{F}(u,v) = \underline{A}_0(u) + \sum_{n=1}^{N} \{\underline{A}_n(u)\cos(nv) + \underline{B}_n(u)\sin(nv)\}, \tag{14}$$

and $w = \dfrac{1}{N+1}$.

The remainder function $\underline{R}(u,v)$ is calculated by means of the difference between the original boundary conditions and the boundary conditions satisfied by the function $\underline{F}(u,v)$ therefore it guarantees that the chosen boundary conditions are exactly satisfied [55]. Note that the terms $\underline{r}_j$ are vector valued functions depending on the parametric coordinate v. Another thing to point out is that it is necessary to express the boundary conditions defining the particular solution to Equation (12) in terms of the corresponding Fourier series.

Once the boundary conditions are expressed in terms of a Fourier series, the values of a_{ij} and b_{ij} can be found. For this purpose, a series of systems of linear equations need to be solved. These systems are of the form,

$$Mx = c, \tag{15}$$

where M represents the matrix associated with the system, x denotes the unknown of such a system and c is the independent value of each algebraic equation in the system. In particular, the system to be solved to find the solution to Equation (12) is given by,

$$\begin{aligned} M_0 x_{A0} &= c_{A0}, \\ M_{0n} x_{An} &= c_{An}, \\ N_n x_{Bn} &= c_{Bn}, \end{aligned} \tag{16}$$

where M_0, M_n and N_n are matrices. The subscripts in these matrices denote the system with which they are associated with. For instance, M_0 is the matrix associated with the term A_0 in Equation (9), whilst M_n and N_n are linked to the terms A_n and B_n respectively. The vectors x_{A0}, x_{An} and x_{Bn} represent the unknowns in the system associated with each respective term in Equation (10) and (11). Finally, the independent vectors c_{A0}, c_{An} and c_{Bn} determined by the boundary conditions.

The matrices are thus given by,

$$M_0 = \begin{pmatrix} 1 & 0 & 0 & 0 \\ 0 & 1 & 0 & 0 \\ 1 & 1 & 1 & 1 \\ 0 & 1 & 2 & 3 \end{pmatrix}, \tag{17}$$

$$M_n = N_n \begin{pmatrix} 1 & 0 & 1 & 0 \\ n & (1+n) & -n & (1-n) \\ e^n & e^n & e^{-n} & e^{-n} \\ ne^n & (1+n)e^n & -ne^{-n} & (1-n)e^{-n} \end{pmatrix}. \tag{18}$$

The vectors x_{A0}, x_{An} and x_{Bn} and are defined by,

$$\begin{aligned} x_{A0} &= (a_{00}, a_{01}, a_{02}, a_{03}), \\ x_{An} &= (a_{n1}, a_{n2}, a_{n3}, a_{n4}), \\ x_{Bn} &= (b_{n1}, b_{n2}, b_{n3}, b_{n4}). \end{aligned} \tag{19}$$

As mentioned before, the independent vectors c_{A0}, c_{An} and c_{Bn} are determined by the boundary conditions; that is, they are associated with the Fourier series of all the boundary conditions. Let ca_{i1}, $ca_{i2}, ca_{i3}, ca_{i4}$ cb_{i1}, cb_{i2}, cb_{i3} and cb_{i4} represent the coefficients associated with the Fourier series of each of the boundary conditions respectively. Thus, the independent vectors are defined by,

$$\begin{aligned} c_{A0} &= (ca_{00}, ca_{01}, ca_{02}, ca_{03}), \\ c_{An} &= (ca_{n1}, ca_{n2}, ca_{n3}, ca_{n4}), \\ c_{Bn} &= (cb_{n1}, cb_{n2}, cb_{n3}, cb_{n4}). \end{aligned} \tag{20}$$

The next and final step to find the solution to Equation (12) is therefore to solve the systems of equations previously defined. This can be done using readily available algorithms [56]. Note that the number of systems to be solved in this fashion is equal to $2N+1$. Although the constants required to determine $R(u,v)$ are calculated in a similar manner, the number of systems to be solved depend on the resolution assigned to the v direction. for instance, if the solution of Equation (12) is to be calculated over a uv mesh of 20 by 20 points, then 20 systems would be solved to obtain the full solution. This is due to the dependency of the remainder term on v.

3.1 Examples of PDE Geometry

In this section we discuss some of examples which show how the geometry based on solutions of PDEs can be generated. We choose the Biharmonic equation given in (6) and the boundary conditions are taken in the format described in Equation (7).

As a first example we show how a fourth order PDE surface is generated where all the conditions are taken to be function conditions. Fig. 2(a) shows the boundary conditions to solve the Biharmonic PDE. Fig. 2(b) shows the shape of a surface generated by the fourth order PDE where the conditions are specified in terms of the curves shown in Fig. 2(a). In particular, the conditions are such that, $X(0,v)=c_1(v)$, $X(1/3,v)=c_2(v)$, $X(2/3,v)=c_3(v)$ and $X(1,v)=c_4(v)$. Since we are taking four function conditions to solve the fourth order PDE, all the curves in this case lie on the resulting surface. Thus, in this particular case the resulting PDE surface is a smooth interpolation between the given set of functional conditions.

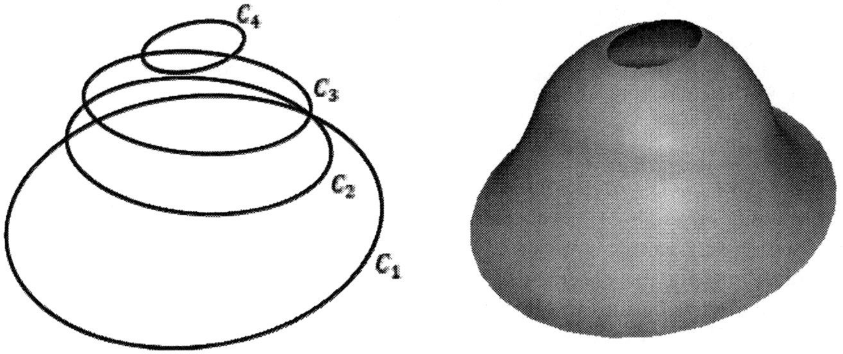

(a) (b)

Fig. 2. The shape of a surface generated by the fourth order PDE where the conditions are all taken to be function conditions. (a) The conditions defined in the form of curves in 3-space. (b) The resulting surface shape.

As a second example we show how a four sided patch can be generation as the solution to the Biharmonic equation. Again here all the conditions are taken to be function conditions. Here Fig. 3(a) shows the necessary boundary conditions to solve PDE. Fig. 3(b) shows the shape of a surface generated by the fourth order PDE where the conditions are specified in terms of the curves shown in Fig. 3(a). In particular, the conditions are such that, $X(0,v)=c_1(v)$,

$X(1/3, v) = c_2(v)$, $X(2/3, v) = c_3(v)$, $X(1, v) = c_4(v)$. Note here the parameter range for u and v such that, $0 \le u \le 1$ and $0 \le v \le \pi$. Since we are taking four function conditions to solve the fourth order PDE, all the curves in this case lie on the resulting surface. Thus, in this particular case the resulting PDE surface is a smooth interpolation between the given set of functional conditions.

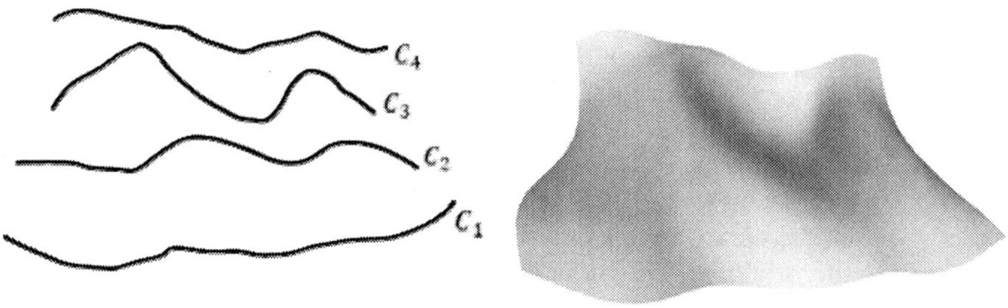

(a) (b)

Fig. 3. A four-sided surface patch generated by the fourth order PDE where the conditions are all taken to be function conditions. (a) The conditions defined in the form of curves in 3-space. (b) The resulting surface shape.

The above two examples how solutions of PDEs based on the 4th order Biharmonic equation enables to generate surface shapes. As discussed above, the basic idea here is to generate a series of curves that can be utilised to define the boundary conditions for the chosen PDE. As seen in the examples, the resulting surface shape can always be intuitively predicted from the shapes of the chosen curves.

3.2 Modelling Geometry of Wound Shapes

Here we show how the above describe technique can be utilised to generate a wide range of shapes which correspond to wound geometries. The idea here is to show how generic shapes of wounds (both the surface of the wound and the interior part of the wound) can be generated based on the above techniques. In what follows we first discuss how the surface of a wound can be modelled using the Biharmonic equation. We then adopt the technique for generating enclosed surface which correspond to the interior shape of the wound.

3.2.1 Modelling Surface Geometry of Wounds

In this subsection we describe how the surface of a would can be created using the solution to the Biharmonic equation as described earlier. As usual, we generate four boundary curves through which we seek an interpolation (corresponding to the surface of the wound) based on the Biharmonic equation. The curves are defined as splines of the form $P_i = \sum_i C_i B_i$ where B_i is a cubic polynomial and C_i are the corresponding control points. Thus, the curves once defined through spline form are discretely sampled in order to obtain their Fourier representations so that the analytic solution outlined previously can be utilised.

Fig. 4(a) shows typical boundary conditions for the chosen PDE. Fig. 4(b) shows an example wound surface where Fig. 4(a) shows the corresponding curves utilised to generate the surface shapes. Here the boundary conditions are defined as periodic functions such that $X(0,v) = c_1(v)$, $X(0.3,v) = c_2(v)$, $X(0.7,v) = c_3(v)$ and $X(1,v) = c_4(v)$. Note that the curve $c_4(v)$ in this case is a point in 3-space whose Fourier representation is utilised as one of the boundary conditions for the PDE.

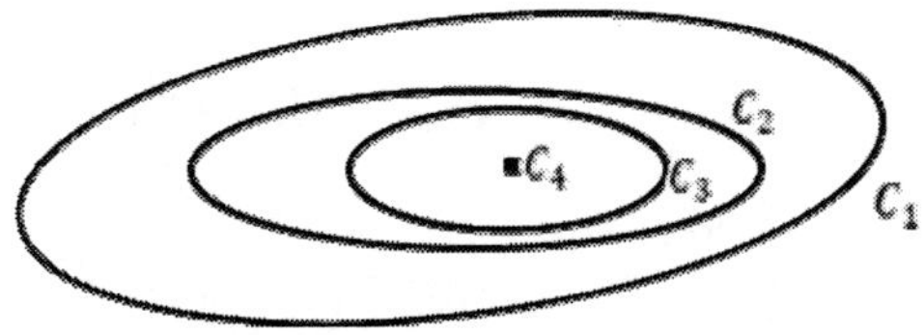

(a)

(b)

Fig. 4. An example wound surface generated by the fourth order PDE where the conditions are all taken to be function conditions. (a) The conditions defined in the form of curves in 3-space. (b) The resulting wound surface shape.

Fig. 5 shows further examples of wound surface geometry generated using the above described methodology. In each case the boundary conditions are similar to that shown in Fig. 4(a) where each of the boundary conditions is defined as a spline function. The corresponding control points of the spline are manipulated in real-time in order to generate alternative shapes of boundary curves resulting alternative shapes of wound surface geometry.

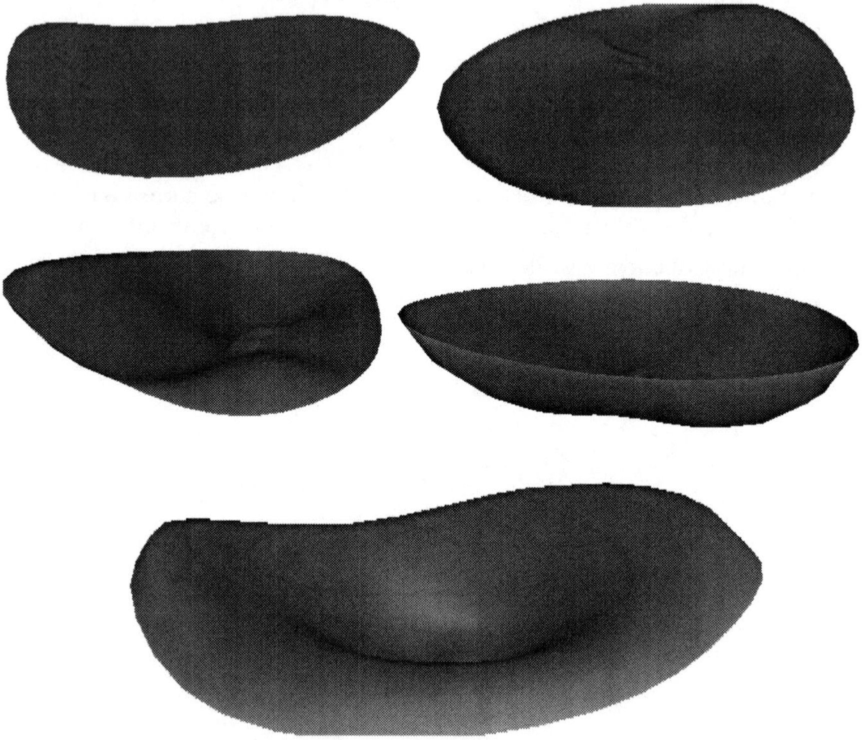

Fig. 5. Example wound surface geometry generated by using the Biharmonic equation. In each case the surface geometry is generated by manipulating the boundary conditions through the control points of the spline functions.

3.2.2 Modelling the Interior Geometry of Wounds

In this section we show how the PDE based methodology can be utilised to generate the interior geometry of the wound shapes. Like the surface of the wounds, the interior shapes are equally important and therefore being able to model the interior part of a wound is paramount. Fig. 6(a) shows the boundary conditions required to solve the PDE. Fig. 6(b) shows a generic shape of wound which has been created using two fourth order Biharmonic surface patches with a common boundary. Fig. 6(a) shows the boundary curves corresponding to this shape. Here we generate two surface patches, one for the curves $c_1(v)$, $c_2(v)$, $c_3(v)$ and $c_4(v)$ and the other for the curves $c_4(v)$, $c_5(v)$, $c_6(v)$ and $c_7(v)$. Thus, $c_4(v)$ is taken to be a common boundary where both the surface patches meet. Again the curves $c_1(v)$

and $c_7(v)$ are taken to be point in 3-space. Note also that all the curves are defined as spline functions whose control points can be utilised to manipulate the curves in 3-space.

With the above formulation, the analytic solution is utilised to generate the geometry shown in Fig. 6(b).

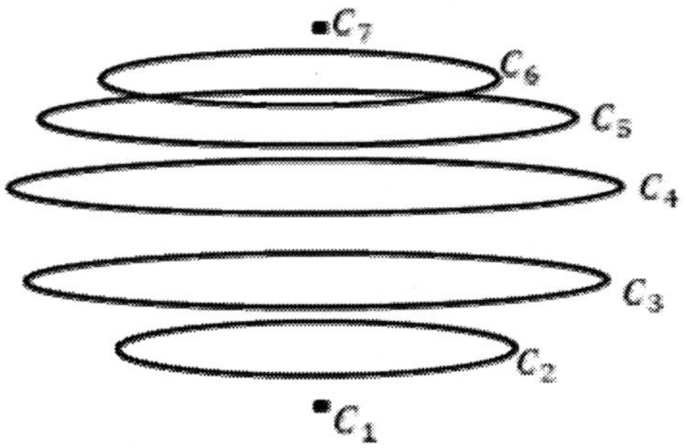

(a)

(b)

Fig. 6. An example geometry corresponding to the interior of a wound generated as a two blended fourth order PDE surface patches. (a) The boundary conditions defined in the form of curves in 3-space. (b) The resulting wound geometry shape.

Figure 7 shows further examples of wound geometry generated using the above described methodology. In each case the boundary conditions are similar to that shown in Fig. 6(a) where each of the boundary conditions are defined as spline functions. The corresponding control points of the spline are manipulated in real-time in order to generate alternative shapes of boundary curves, resulting in the shapes of wound geometry as shown.

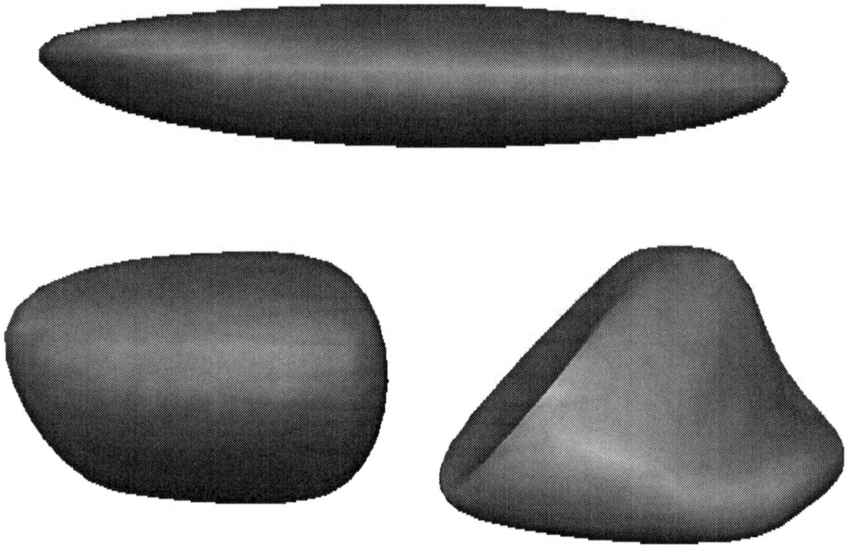

Fig. 7. Example geometries corresponding to the interior of typical wounds. In all cases the geometry is generated by means of manipulating the boundary conditions defined Fig (a), in real-time.

4 Measuring Properties of Wounds

In this section we discuss how PDE based geometry modelling techniques can be coupled with computational techniques in order to measure various properties of wounds. Properties such as the surface area, enclosed volume and mass of the wound may be of use by clinicians in order to characterise the wound and estimate the healing properties of various therapies. We show how the geometry generated by PDE surfaces can be directly utilised to compute physical properties of the wounds.

From the previous sections it is clear that we can generate a complex geometric shape of a wound as an analytic solution of a fourth order Biharmonic PDE. This means that we can define the geometry of a complex wound shape as a mathematical function. This function can thus be utilised to compute the physical properties of the wound.

4.1 Surface Area and Volume of the Wound Shape

As described in detail above, a surface $\underline{X}(u,v)$ defined over a finite domain Ω is utilised to describe the shape of the wound where $\underline{X}(u,v)$ is an analytic function defined as a solution of the Biharmonic equation subject to a suitable set of boundary conditions.

Then the surface area of the wound shape is given by,

$$A_w = \iint dudv, \tag{21}$$

and the volume enclosed by the wound is given by,

$$V_w = \iint dV, \tag{22}$$

where

$$dV = \underline{X}_u \wedge \underline{X}_v dudv. \tag{23}$$

Note here the subscript at $\underline{X}_u$ and $\underline{X}_v$ denote the partial differentiations.

The product in Equation (23) can be easily computed by means of the inverse of the Jacobian of the transformation J_v defined by,

$$J_v^{-1} = \begin{vmatrix} x_u & x_v \\ y_u & y_v \\ z_u & z_v \end{vmatrix}. \tag{24}$$

Hence, provided we are able to represent the wound shape closely by means of the choosing appropriate boundary data the surface area and volume of the wound can be automatically computed using the above technique. This eliminates invasive mechanisms for measuring properties of the wound such as the use of square based grids for estimating the surface area of the wound.

4.2 Mass Properties of the Wound

In some cases it may also be necessary to compute the mass properties of the actual wound. Here we propose a mechanism for computing the mass of the wound.

Assuming the wound under consideration is solid of uniform density we can model the solid geometry of the wound shape by means of the PDE,

$$\left(\frac{\partial^2}{\partial u^2} + \frac{\partial^2}{\partial v^2} + \frac{\partial^2}{\partial w^2}\right)^2 \underline{X}(u, v, w) = 0. \tag{25}$$

Equation (25) can be seen as a mapping from a solid cube parameterised by u, v and w to a solid. Subject to suitable boundary conditions, which can be determined from the surface of the wound geometry, Equation (25) can be solved to create the solid geometry of the wound. However, unlike the previous case of the Biharmonic equation defined over the parameter space u and v Equation (25) does not possess an analytic solution. Hence the equation now has to be solved numerically.

Given the surface geometry of a wound shape, suitable boundary conditions can be chosen for Equation (25) in order to generate the solid geometry corresponding to the wound. Given the function $\underline{X}$ defining the solid geometry of the wound one can compute the mass of the wound by the volume integral,

$$M_w = \iiint dxdydz. \tag{26}$$

As before, rather than calculating this quantity over the physical space we can transform it to the parametric domain where an element of volume is given by,

$$dV = (\underline{X}_u \wedge \underline{X}_v).\underline{X}_w dudvdw, \tag{27}$$

where, as before, the subscripts denote the partial differentiations. The triple product in Equation (27) is the inverse of the Jacobian of the transformation J_v defined by,

$$J_v^{-1} = \begin{vmatrix} x_u & x_v & x_w \\ y_u & y_v & y_w \\ z_u & z_v & z_w \end{vmatrix}. \tag{28}$$

Thus, the mass of the wound can be measured by means of evaluation the integral,

$$M_w = \iiint J_v^{-1} dudvdw. \tag{29}$$

It is noteworthy that in most cases the integral in Equation (29) has to be compute numerically with J_v being evaluated at suitable points over a unit cube.

5 An Example of Wound Modelling

In this section we show how a given wound geometry can be modelled using the methodology discussed above. Here we use 4th order PDE surface patches in order to model a wound shape by means of utilising real data.

Fig 8(a) shows the 3-dimentional geometry corresponding to the deep foot chronic ulcer shown in Figure 1. The surface data of the geometry was acquired using multiple-camera photogrammetry by means of the DSP400 system from 3dMD Ltd. This commercial technology has been widely used for acquiring medical images, especially in the USA, and captures data in a few milliseconds. The surface resolution (i.e. the separation of data points) is approximately 2mm with a positional accuracy of approximately 0.2mm.

In this example we show how we can generate the geometry corresponding to the foot and the corresponding wound shown in Figure 1. In order to generate a representative PDE surface shape, we extract a series of curves along the profile of the geometry model. For this purpose we import the scanned geometry mesh model into an interactive graphical environment through which we can examine and interact with the model. We then manually identify a number of regions on the scanned geometry model. These regions are then utilised to determine the number of PDE surface patches required to produce a good representative model.

The criteria in determining the number of PDE surface patches required is purely based on the required degree of accuracy by which one wants to approximate the scanned geometry using PDE surface patches. i.e. the more the number of surface patches the higher the degree of accuracy. Once the number of surface

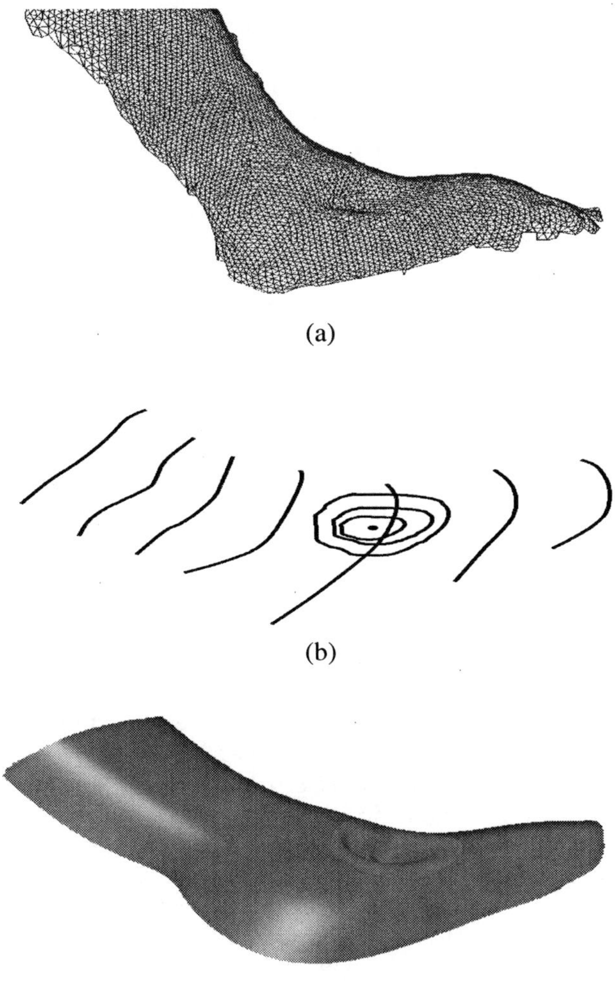

(a)

(b)

(c)

Fig. 8. Example wound geometry modelling using PDEs

patches required is decided then appropriate number of curves for each surface patch is extracted from the scanned geometry data. To do this we create a series of free-form cubic spline curves within the interactive environment. The spline curves are then projected onto the scanned geometry at the positions where the PDE boundary curves are to be extracted.

Fig. 8(b) shows the curves that have been extracted. First the curves corresponding to main foot is utilised to generate a smooth surface for the geometry of the foot (without the wound shape). In this case the foot surface is generated using two 4th order PDEs with a common boundary. A portion of the geometry corresponding to the wound shape is then trimmed out from the main foot surface in order to accommodate the wound shape. The trimming process which we utilised for this purpose is outlined in [37]. Finally an additional 4th order patch is generated for the wound shape as shown in Fig. 8(c).

The above example demonstrates how one can develop a practical system for measuring both the 3-dimentional shape and its associated properties which can be useful by clinicians and other medical staff working in the related areas. Such a system should have templates of generic PDE based wound shapes whereby the user can input a handful of key measurements taken from the wound. These key measurements should enable the system to generate appropriate boundary conditions to generate a close approximation of the wound shape and provide the relevant physical properties to the user.

6 Conclusions and Future Work

Modelling wound geometry is a crucial task in order to evaluate the efficiency of the available therapies in an appropriate fashion. A crucial task, during the treatment of wounds, is the measurement of the size area and volume of the wounds.

Here we have presented a method to model a wide variety of geometries of wound shapes. The shape modelling is based on formulating mathematical boundary-value problems relating to solutions of Partial Differential Equations (PDEs). In order to model a given geometric shape of the wound a series of boundary functions which correspond to the main features of the wound are selected. These boundary functions are then utilised to solve an elliptic PDE whose solution results in the geometry of the wound shape. Thus, here we show how low order elliptic PDEs, such as the Biharmonic equation subject to suitable boundary conditions can be used to model complex wound geometry. We also utilise the solution to automatically compute various physical properties of the wound such as the surface area, volume and mass. To show the practical ability of the methodology a series of examples are discussed demonstrating the capability of the method to produce good representative shapes of wounds.

In this work we have mainly utilised low order elliptic PDEs where the chosen PDE is solved subject to suitable boundary conditions. The work has potential for developing software tools whereby efficient geometry parameterisation of wounds can be undertaken by means of taking a small set of key measurements from given wounds. A description of a particular wound can be obtained from this generic template by assigning values for the parameters so that the model is matched to the given wound geometry. The assigned values – and hence the 'tailor-made model' – would be calculated from a small number of simple measurements taken from the patient, somewhat similar to the way in which a tailor or dressmaker obtains key measurements that determine the fit of bespoke clothing.

Another potential area to investigate would be the study of time-dependent PDEs where the geometric shape of the wound can be associated with its healing kinetics in order to model the healing patterns of wounds subject to given set of therapies.

References

Bansal, C., Scott, R., Stewart, D., Cockerell, C.J.: Decubitus Ulcers: A Review of the Literature. International Journal of Dermatology 44, 805–810 (2005)

Pressure Ulcers in America: Prevalence, Incidence and Implications for the Future. An Executive Summary of the National Pressure Ulcer Advisory Panel Monograph. Advanced Skin Wound Care 14(4), 208–215 (2001)

Stausberg, J., Kroger, K., Maier, I., Schneider, H., Niebel, W.: Interdisciplinary Decubitus Project. Pressure Ulcers in Secondary Care: Incidence, Prevalence and Relevance. Advanced Skin Wound Care 18, 140–145 (2005)

Doherty, D., Ross, F., Yeo, L., Uttley, J.: Leg Ulcer Management in an Integrated Service, The South Thames Evidence Based Practice (STEP) Project, Report (6), Kingston University (2000)

Fletcher, A.: Common Problems of Wound Management in the Elderly. In: Harding, K.G., Leaper, D.L., Turner, T.D. (eds.) Proceedings of the First European Conference on the Advances in wound management, pp. 25–29. Macmillan, London (1992)

Kings Fund Grant to Help Treat Leg Ulcers. Br. Med. J. 297, 1412 (1988)

Carroll, S.R.: Hydrocolloid Wound Dressings in the Community. Dermatol. Pract. 8, 24–26 (1990)

Bosanquet, N.: Costs of Venous Ulcers: From Maintenance Therapy to Investment Programmes. Phlebology 7(suppl. 1), 44–46 (1992)

Krouskop, T.A., Baker, R., Wilson, M.S.: A Noncontact Wound Measurement System. Journal of Rehabilitation Research and Development 39(3), 337–346 (2002)

Plassmann, P., Melhuish, J.M., Harding, K.G.: Methods of Measuring Wound Size: A Comparative Study. WOUNDS: A compendium of clinical research and practice 6(2), 54–61 (1994)

Goldman, R.J., Salcido, R.: More than one way to Measure a Wound: An Overview of Tools and Techniques. Advanced Skin Wound Care 15, 236–243 (2002)

Houghton, P.E., Kincaid, C.B., Campbell, K.E., Woodbury, M.G., Keast, D.H.: Photographic Assessment of the Appearance of Chronic Pressure and Leg Ulcers. Ostomy Wound Manage 46(4), 28–30 (2000)

Lucas, C., Classen, J., Harrison, D., De, H.: Pressure Ulcer Surface Area Measurement using Instant Full-Scale Photography and Transparency Tracings. Advanced Skin Wound Care 15, 17–23 (2002)

Dyson, M., Suckling, J.: Stimulation of Tissue Repair by Ultrasound: A survey of the mechanisms involved. Physiotherapy 64, 105–108 (1978)

Maklebust, J.: Pressure Ulcer Assessment. Clinics in Geriatric Medicine 13(3), 455–481 (1997)

Plassmann, P., Jones, T.D.: MAVIS: A Non-Invasive Instrument to Measure Area and Volume of Wounds, Measurement of Area and Volume Instrument System. Medical Engineering Physics 20(5), 332–338 (1998)

Thali, M.J., Braun, M., Dirnhofer, R.: Optical 3D Surface Digitizing in Forensic Medicine: 3D Documentation of Skin and Bone Injuries. Forensic Science International 137(2-3), 203–208 (2003)

Hoschek, J., Lasser, D.: Computer Aided Geometric Design. A K Peters, Wellesley (1993)

Mortenson, M.E.: Geometric Modelling. Wiley-Interscience, New York (1985)

Farin, G.: Curves and Surfaces for Computer Aided Geometric Design, A Practical Guide. Morgan-Kaufmann, San Francisco (2001)

Bajaj, C., Blinn, J., Bloomenthal, J., Cani-Gascuel, M., Rock-wood, A., Wyvill, B., Wyvill, G.: Introduction to Implicit Surfaces. Morgan-Kaufmann, San Francisco (1997)

Szeliski, R., Terzopoulos, D.: From Splines to Fractals. In: Proceedings of the 16th annual conference on Computer Graphics and Interactive Techniques, pp. 51–60. ACM, New York (1989)

Rappoport, A., Spitz, S.: Interactive Boolean Operations for Conceptual Design of 3-d Solids. In: Proceedings of the 24th Annual Conference on Computer Graphics and Interactive Techniques, pp. 269–278. ACM, New York (1997)

DeRose, T., Kass, M., Truong, T.: Subdivision Surfaces in Character Animation. In: Proceedings of SIGGRAPH 1998, pp. 85–94. Addison-Wesley, Reading (1998)

Faux, I.D., Pratt, M.J.: Computational Geometry For Design And Manufacture. Halsted Press, Wiley (1979)

Farin, G.: From conics to NURBS, A Tutorial and Survey. IEEE Computer Graphics and Applications 12(5), 78–86 (1992)

Piegl, L.: Recursive Algorithms for the Representation of Parametric Curves and Surfaces. Computer Aided Design 17(5), 225–229 (1985)

Piegl, L.: The NURBS Book. Springer, New York (1997)

Pottmann, H., Leopoldseder, S., Hofer, M., Steiner, T., Wang, W.: Industrial Geometry: Recent Advances and Applications in CAD. Computer Aided Design 37(7), 751–766 (2005)

Catmull, E., Clark, J.: Recursively Generated B-Spline Surfaces on Arbitrary Topological Meshes. Computer Aided Design 10(6), 350–355 (1978)

Loop, C.T.: Smooth Subdivision Surfaces based on Triangles, Master's Thesis, Department of Mathematics, University of Utah (1987)

Dyn, N., Levine, D., Gregory, J.A.: A Butterfly Subdivision Scheme for Surface Interpolation with Tension Control. ACM Transactions on Graphics 9(2), 160–169 (1990)

Hubeli, A., Gross, M.: A survey of Surface Representations for Geometric Modeling, Computer science department, ETH Zurich, Swizerland, CS Technical report No. 335 (2000)

Catmull, E.: Subdivision Algorithm for the Display of Curved Surfaces, PhD thesis, University of Utah (1974)

Bloor, M.I.G., Wilson, M.J.: Generating Blend Surfaces using Partial Differential Equations. Computer Aided Design 21(3), 165–171 (1989)

Du, H., Qin, H.: A Shape Design System using Volumetric Implicit PDEs. Computer Aided Design 36(11), 1101–1116 (2004)

Ugail, H., Bloor, M.I.G., Wilson, M.J.: Techniques for Interactive Design Using the PDE Method. ACM Transactions on Graphics 18(2), 195–212 (1999)

You, L., Comninos, P., Zhang, J.J.: PDE Blending Surfaces with C2 Continuity. Computers and Graphics 28(6), 895–906 (2004)

Monterde, J., Ugail, H.: A General 4th-Order PDE Method to Generate Bézier Surfaces from the Boundary. Computer Aided Geometric Design 23(2), 208–225 (2006)

Bloor, M.I.G., Wilson, M.J.: Using Partial Differential Equations to Generate Freeform Surfaces. Computer-Aided Design 22, 202–212 (1990)

Ugail, H.: On the Spine of a PDE Surface. In: Wilson, M.J., Martin, R.R. (eds.) Mathematics of Surfaces. LNCS, vol. 2768, pp. 366–376. Springer, Heidelberg (2003)

Schneider, R., Kobbelt, L.: Geometric Fairing of Irregular Meshes for Free-Form Surface Design. Computer Aided Geometric Design 18(4), 359–379 (2001)

Kim, B., Rossignac, J.: Localized Bi-Laplacian Solver on a Triangle Mesh and its Applications, Georgia Institute of Technology, GVU Technical Report Number: GIT-GVU-04-12 (2004)

Schneider, R., Kobbelt, L., Seidel, H.-P.: Improved Bi-Laplacian Mesh Fairing. In: Lyche, T., Schumaker, L.L. (eds.) Mathematical Methods for Curves and Surfaces, Oslo, 2000. Innovations in Applied Mathematics Series, pp. 445–454. Vanderbilt Univ. Press, Nashville (2001)

Rudin, L., Osher, S., Fatemi, E.: Nonlinear Total Variation Based Noise Removal Algorithms. Physica D 60(1-4), 259–268 (1992)

Paragios, N., Deriche, R.: A PDE-Based Level Set Approach for Detection and Tracking of Moving Objects, INRIA Technical Report, INRIA, France (1997)

Vese, L.A., Chan, T.F.: A Multiphase Level Set Framework for Image Segmentation using The Mumford and Shah Model. International Journal of Computer Vision 50(3), 271–293 (2002)

Schneider, R., Kobbelt, L.: Generating fair meshes with g1 boundary conditions. In: Proceedings of the Geometric Modelling and Processing 2000, pp. 251–261. IEEE, Washington (2000)

Desbrun, M., Meyer, M., Schroder, P., Barr, A.H.: Implicit Fairing of Irregular Meshes using Diffusion and Curvature Flow. In: Proceedings of SIGGRAPH 1999, pp. 317–324. ACM, New York (1999)

Powell, M.J.D.: The theory of radial basis function approximation in 1990. In: Proceedings of 4th Summer School, Advances in numerical analysis. Wavelets, subdivision algorithms, and radial basis functions, vol. 2, pp. 105–210. Oxford University Press, Oxford (1992)

Ugail, H., Wilson, M.J.: Efficient Shape Parameterisation for Automatic Design Optimisation using a Partial Differential Equation Formulation. Computers and Structures 81(29), 2601–2609 (2003)

Kellogg, O.D.: Foundations of Potential Theory. Dover Publications, New York (1969)

Jackson, J.D.: Classical Electrodynamics. Wiley, Chichester (1999)

Acheson, D.J.: Elementary Fluid Dynamics. Oxford University Press, Oxford (1990)

Bloor, M.I.G., Wilson, M.J.: Spectral Approximations to PDE Surfaces. Computer-Aided Design 28, 145–152 (1996)

Press, W.H., Teukolsky, S.A., Vetterling, W.T., Flannery, B.P.: Numerical Recipes in C. Cambridge University Press, Cambridge (1992)

A Suite of Continuum Models for Different Aspects in Wound Healing

F.J. Vermolen[a] and E. Javierre[b,c,d]

[a] Delft Institute of Applied Mathematics, Delft University of Technology, Mekelweg 4, Delft, 2628 CD, The Netherlands
F.J.Vermolen@tudellft.nl
[b] CIBER-BBN Centro de Investigación Biomédica en Red en Bioingeniería, Biomateriales y Nanomedicina, Zaragoza, Spain
[c] Group of Structural Mechanics and Materials Modelling, Aragón Institute of Engineering Research (I3A), University of Zaragoza, 50018 Zaragoza, Spain
[d] Aragón Health Sciences Institute, Zaragoza, Spain
Etelvina.Javierre@unizar.es

Abstract. Wound healing proceeds through a sequence of partly overlapping processes that can be classified into three phases: inflammation, proliferation and remodeling. Among them, wound closure, angiogenesis and wound contraction play a significant role. First, the connective tissue that is present initially is replaced with fibrous tissue. The production of fibrous tissue by fibroblasts is initiated by fibroblast migration into the wound region and enhanced by the proliferation of these cells. Since the fibroblasts start pulling on the surrounding tissue, the wound starts contracting. Mathematically, the visco-elastic equations are solved in combination with a set of diffusion-convection-reaction equations. Once connective tissue has been replaced with fibrous tissue, a capillary network is established. The capillary network formation is enhanced by a macrophage derived growth factor, of which its production is initiated by a lack of oxygen. Several models exist and are discussed here for this partial process. For the wound re-epithelialization, in which the wound is actually closed by a layer of epidermal cells, several models are used. In this work, we will consider a simplified mathematical model that tracks the epidermal cell density and a generic growth factor that either inhibits or activates this process. Further, an alternative model is discussed, which contains a discontinuous switch mechanism, based on the assumption that the wound edge moves as a result of its curvature and of the concentration of a generic growth factor. In this work, finite element results will be presented, as well as parts of the performed mathematical analysis.

1 Introduction

Bone regeneration, bone remodeling, bone ingrowth and wound healing are very complicated processes from a biological point of view. The first processes involves cell migration and a chain of differentiations of several cell-types triggered by bone fracture. Under certain conditions, cartilage formation, mineralization and formation of fibrous tissue take part in the callus in which the bone heals. Some complicated mathematical models for bone regeneration have been reported in the literature, [Huiskes *et al.* 1997, Ament and Hofer 2000, Andreykiv 2006, Bailon-Plaza and van der Meulen 2001, Adam 1999, Calvo 2008, Martínez *et al.* 2006,

A. Gefen (Ed.): Bioengineering Research of Chronic Wounds, SMTEB 1, pp. 127–168.
springerlink.com

Davies 2003, Chen *et al.* 2007, Carpenter and Carter 2008, García *et al.* 2005, Doblaré *et al.*, Doblaré and Garciá 2001, García *et al.* 2002, Fornells *et al.* 2007, Doblaré and García, García *et al.* 2007, Sanz-Herrera *et al.* 2008], to mention just a few. The later process, wound healing or soft tissue regeneration, involves cell migration, the production and decay of growth factors and a (re-)establishment of the vascular network surrounding the area with an increased mitotic activity. Experimental validation of the models of both complicated biological processes is indispensable. The present chapter focuses on the regeneration of tissue and healing of (deep) wounds.

When a wound occurs, blood vessels are cut and blood enters the wound. Due to blood coagulation, the wound is temporarily closed and as a result the blood vessels adjacent to the wound are also closed. In due course contaminants will be removed from the wounded area and the blood vessel network will be restored, but initially due to insufficient blood supply, there will be a low concentration of nutrients which are necessary for cell division and wound healing. Wound healing, if it occurs, proceeds by a combination of several processes: wound contraction (due to pulling forces caused by fibroblasts entering the wound area underneath the epidermal cells), chemotaxis (movement of cells induced by a concentration gradient), neo-vascularization, synthesis of extracellular matrix proteins, and scar remodeling. Previous models incorporate cell mitosis, cell proliferation, cell death, capillary formation, oxygen supply and growth factor generation, including studies by [Sherratt and Murray 1991, Filion and Popel 2004, Maggelakis 2003, Gaffney *et al.* 2002], to mention just a few. A recent work devoted to mathematical biology has been written by [Murray 2004], in which the issue of wound healing is also treated. The wound healing process can roughly be divided into the following partially overlapping consecutive stages:

1. Formation of a blood clot on the wound to prevent undesired chemicals from entering the tissue of the organism (blood clotting phase);
2. Wound contraction caused by fibroblasts entering the wounded area under the epidermal cells. The fibroblasts pull the tissue underneath if their density falls between certain limits (see page 506, [Murray 2004]), which causes wound contraction (proliferative phase);
3. Formation of a network of tiny arteries (capillaries) for blood flow to supply the necessary nutrients for wound healing (angiogenesis);
4. Division and growth of epidermal cells (mitosis), taking place during the actual healing of the wound (proliferative phase).

Once the gap of a deep wound has been filled with connective tissue, fibroblasts move into the wound gap underneath the epidermis. Once the fibroblasts enter the wound gap, they produce extracellular matrix (ECM) and restore the dermis. Furthermore, they also start pulling on the ECM and this traction force increases up to a certain maximum due to saturation of the stress as the fibroblast density increases. More on this topic, can be found in [Murray 2004] among many others. At the same time angiogenesis takes place in order to supply the active cells with oxygen, nutrients and constituents necessary for their function (mitosis, growth and cytokynes release). For this purpose tiny capillaries are formed during the process of

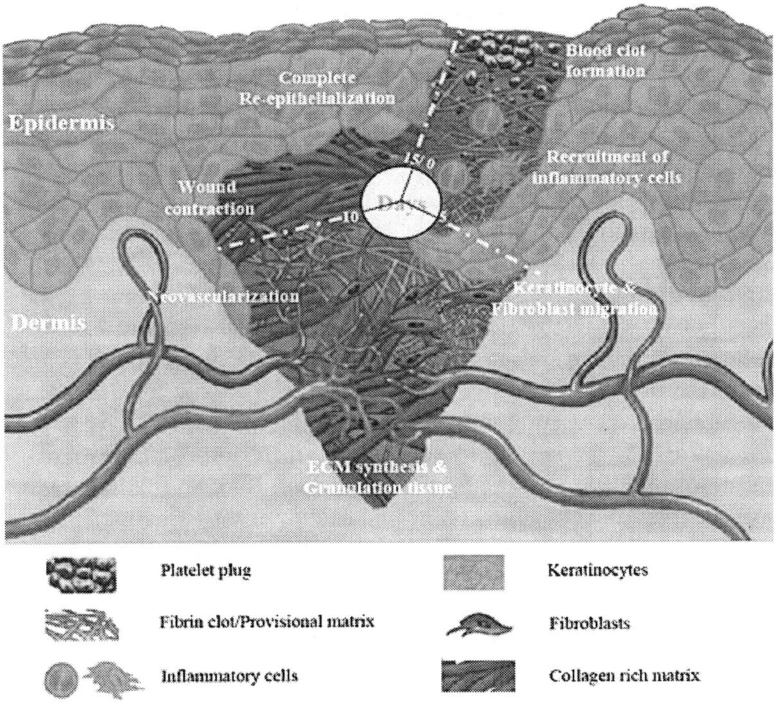

Fig. 1. A schematic of the events during wound healing. The dermis, epidermis and blood clot are illustrated. Fibroblasts move into the blood clot occupied area. The picture was taken from http://www.bioscience.org/2006/v11/af/1843/figures.htm

angiogenesis. Some models for capillary network formation have been proposed by [Gaffney *et al.* 2002] and [Maggelakis 2004]. A schematic of wound healing, incorporating the aforementioned processes, is shown in Figure 1.

Epidermal wound closure is modeled by [Sherratt and Murray 1991] who consider cell division and growth factor generation simultaneously for healing of epidermal wounds. Their model consists of a system of reaction-diffusion equations. Among many others, an alternative model, based on an active layer at the wound edge was proposed by [Adam 1999]. We also refer to [Vermolen *et al.* 2007] for a description of several models for wound healing and several aspects of wound healing, such as wound closure and angiogenesis. Until now, the conditions for wound healing were only analyzed for geometries where only one spatial co-ordinate could be used. As far as we know, in all the mathematical studies mentioned before, either neo-vascularization or wound closure is modeled. Hence, these processes are considered to be sequential. However, according to the medical literature [Stadelman *et al.* 1997], these sequential processes partially overlap. A first attempt to combine the effects of angiogenesis and wound closure is made by [Vermolen and Adam 2007] and [Vermolen 2008]. In the first aforementioned paper, the influence of angiogenesis on wound closure is dealt with and the influence of wound closure on angiogenesis was not taken into account. This last issue is covered in the second paper: the increased density of the vascular network travels with the moving wound edge. Hence, the key innovations in the second study were the following: Finite Element solutions that can be applied to arbitrary wound geometries; and a combination of models for angiogenesis and wound closure as partially overlapping consecutive processes. In the present chapter, the model is extended to deal with wound contraction, which follows angiogenesis and occurs at the same stage as wound closure.

We are aware of the literature existing of studies that couple wound closure or neo-vascularization with mechanical strains and stresses, for instance [Murray 2003, Murray 2004] and [Olsen *et al.* 1995]. These mechanical influences are important for the modeling of deeper wounds and it is a topic for the present study. In these models a convection term for the cell density appears due to the rate of displacement (passive convection). Further, the diffusion coefficients will depend on local strain and become directionally dependent. Another example of the significance of the mechanical loading on wound healing is the experimentally sustained observation that wounds aligned with the lines of skin tension tend to heal with better results. In the present study, only a few growth factors are taken into account. The present model is a simplification of reality. Since, in reality a large number of growth factors have to be taken into account and the picture is even more complicated than that. In [Wearing and Sherratt 2000], the signaling of keratinocyte growth factors, originating from mesenchymal cells, acting on epithelial cells is considered. This is crucially important for the building blocks of the micro-vascular network and skin consisting of layers of cells (epithelium). They develop a mathematical model for which traveling wave solutions are constructed as well as one-dimensional finite difference solutions. An interesting review on the modeling of angiogenesis in the context of tumor growth was written by [Alarcon *et al.* 2006]. In their paper, it is claimed that the modeling of the coupling between angiogenesis and tumor growth is crucially important. Furthermore, the assumption that the local concentration of oxygen is proportional to the density of the endothelial cells is debated by them. Another issue concerns the heterogeneities of the vascular system, resulting into a heterogeneous blood flow. In the present chapter an attempt is made to couple angiogenesis with tissue regeneration. The issues covered in the present chapter, involve chemistry and mechanical issues. Some papers that treat angiogenesis in relation with tumor growth are due to [Balding and McElwain 1985, Mantzaris *et al.* 2004, Alarcon *et al.* 2006].

Since wounds possibly occur after a traumatic event or surgery, a good and efficient healing is crucial. Several treatments are known to enhance wound healing. In order to design alternative and hopefully more efficient treatments, a thorough knowledge of the process is indispensable. Furthermore, a good mathematical model for wound healing could be useful for surgeons to determine how a post-operative wound heals. From a mathematical point of view, the setting up of such a model in terms of a nonlinear set of partial differential equations and boundary conditions is a challenge. Another challenge is a parameter sensitivity analysis, which reveals the most significant parameters of the model with respect to the model results. The mathematical models can be used to examine the effects of wound geometry on the healing time, or on the local injection of certain hormones, *i.e.* certain growth factors, to enhance wound healing. For these purposes, a calibrated mathematical model can provide quicker insights than animal experiments. Another issue concerns wound therapies related to strain lines or addition of drugs. The models can be helpful to investigate the impact of certain treatments on wound healing.

The present chapter is organized as follows. First, two existing models for wound contraction are presented, with some computations in two-dimensional wound domains. Second, two models for angiogenesis are described. This is followed by the treatment of two models for wound closure. Then, a recently accepted model for wound closure in which the angiogenesis process triggered by a shortage on oxygen is presented. We continue with the description of the numerical method for the nonlinearly coupled set of partial differential equations. We end up with some concluding remarks.

We would like to emphasize that the present study is an attempt to describe a qualitative mathematical models for tissue regeneration, and an overview of some important mathematical models for several partial processes taking place during healing of a dermal wound. A quantitative model taking into account the numerous growth factors is beyond the scope of this chapter. In a future study, we will deal with the combination of wound contraction, angiogenesis and wound closure. The phenomenon of tissue healing itself is still being investigated. A rather complete state-of-the-art picture from the medical literature can be found in the thesis due to [Lamme 1999].

2 Methodology

In this manuscript, several problems will be solved: the Fisher-Kolmogorov equation, nonlinear reaction-diffusion problems and the visco-elastic equations. In this section, we will briefly discuss the solution procedures that are employed.

2.1 The Fisher-Kolmogorov Equation

The Fisher-Kolmogorov (FK) equation models spatial spread of a certain biological species in an infinite habitat. The proliferation is modeled by a logistic term: the species increases in number up to a certain equilibrium value. We will use traveling wave analysis for solutions of the FK-equation. The treatment will be intuitive, rather than in mathematical rigor. The FK-equation is given by

$$\frac{\partial u}{\partial t} = \frac{\partial^2 u}{\partial x^2} + u(1-u), \tag{1}$$

where the first term and second term in the right hand side of the above equation respectively account for diffusive transport and proliferation (mitosis). The left hand side describes the overall cell accumulation. This equation admits traveling wave solutions of the kind $u(x,t) = v(s) = v(x-ct)$ under certain conditions. We assume that $u(x,0) \in [0,1]$, and deal with the cases

$$\lim_{x \to -\infty} u(x,t) = u_L, \qquad \lim_{x \to \infty} u(x,t) = u_R, \tag{2}$$

where $u_L \neq u_R$. Considering traveling wave solutions, we obtain

$$\frac{d^2v}{ds^2} + c\frac{dv}{ds} + v - v^2 = 0. \tag{3}$$

Using reduction of order with $w := v'$, gives

$$\begin{cases} v' = w, \\ w' = v(v-1) - cw. \end{cases} \tag{4}$$

It is clear that $(v,w) = (0,0)$ and $(v,w) = (1,0)$ are equilibria. The associated Jacobian is given by

$$J(v,w) = \begin{pmatrix} 0 & 1 \\ 2v-1 & -c \end{pmatrix}, \tag{5}$$

and hence for the equilibria, we have the following eigenvalues

- $(v,w) = (0,0) \, : \, \lambda_{1,2} = \dfrac{-c \pm \sqrt{c^2-4}}{2}$ (wounded state),
- $(v,w) = (1,0) \, : \, \lambda_{1,2} = \dfrac{-c \pm \sqrt{c^2+4}}{2}$ (unwounded state).

In order to have biologically relevant solutions, we must have non-oscillatory solutions, which exclude negative values for v. This implies that the eigenvalues for the origin must be real, and hence we require $|c| \geq 2$. The origin becomes either an attractor if $c > 0$ with eigenvector $[1 \;\; -1]^T$ or a repellor if $c < 0$ with eigenvector $[1 \;\, 1]^T$. The point $(v,w) = (1,0)$ is always a saddle point. For the cases $c \leq -2$ and $c \geq 2$, we sketch the trajectory connecting $(0,0)$ with $(1,0)$, in Figures 2 and 3. From these arguments, we can draw the following conclusions:

- $c \geq 2$: There is a trajectory going from $(1,0)$ to $(0,0)$ with increasing s by going through the second quadrant ($v > 0$ and $w < 0$), where $w = v' < 0$. Hence this trajectory corresponds to *decreasing* values of v for increasing s;
- $c \leq -2$: There is a trajectory going from $(0,0)$ to $(1,0)$ with increasing s by going through the first quadrant ($v > 0$ and $w > 0$), where $w = v' > 0$. Hence this trajectory corresponds to *increasing* values of v for increasing s;
- There is a relation between the wave speed and the maximum derivative (that is the point of inflection of v, $v''(s) = 0$). [Sachdev 1987] motivates that $c \approx \frac{L}{4}$, where L is the thickness of the profile (where $0 < u < 1$), to an error of $O(\frac{1}{c^4})$. Suppose that the initial profile has a thickness smaller than $L = 8$, then the solution cannot be a traveling wave immediately.

One can conclude that the motion of the front will always be directed towards decreasing values of $u(x,t)$.

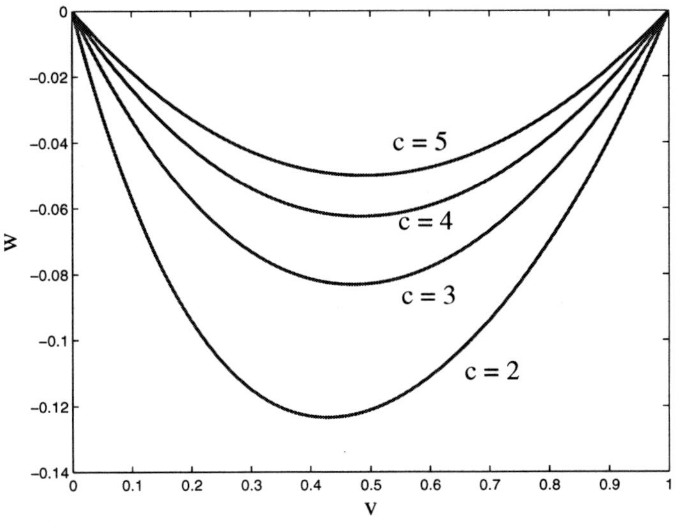

Fig. 2. The trajectories for $c \geq 2$ for several values of wave speed c

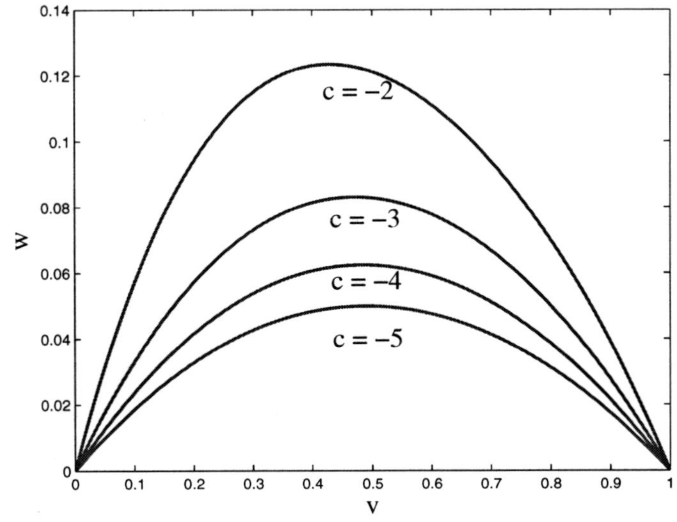

Fig. 3. The trajectories for $c \leq -2$ for several values of wave speed c

The trajectory from (1,0) to (0,0) in the v,w-space defines the solution. The equation of the trajectory in the v,w-space is defined by

$$\frac{dw}{dv} = \frac{v^2 - v - cw}{w}. \tag{6}$$

After some scaling, by setting $\overline{w} = cw$ and defining $\varepsilon = \frac{1}{c^2}$, a perturbation method can be used to obtain the solution $w(v)$, by setting

$$\overline{w}(v,\varepsilon) = g_0(v) + \varepsilon g_1(v) + \varepsilon^2 g_2(v) + \dots,$$

where the functions g_0, g_1, g_2, etc. can be solved by considering equal powers of ε. More on this issue can be found in [Sachdev 1987] or [de Vries *et al.* 2006] for instance. Most important is that the minimal wave speed is given by $c = 2$. In [Sachdev 1987], a discussion about the actual wave speed and steepness of the profile can be found. We note that the traveling wave solutions are valid in the (intermediate) long term.

Equation (3) can be solved directly using a perturbation series

$$u(s,\varepsilon) = u_0(s) + \varepsilon u_1(s) + \varepsilon^2 u_2(s) + \dots,$$

where $\varepsilon := \frac{1}{c^2}$. The above expression is substituted into equation (3) and equal powers of ε are considered.

2.2 A Diffusion-Reaction Equation for Chemotaxis

We consider a very simple model for chemotaxis in order to illustrate the numerical method that we use to approximate the solutions of the equations. Here, we only look at a scaled model for chemotaxis, which reads as:

$$\begin{aligned} \frac{\partial n}{\partial t} &= D_n \Delta n - \alpha \nabla \cdot (n \nabla c) + rn(1-n), \\ \frac{\partial c}{\partial t} &= D_c \Delta c + s\left(\frac{n}{1+n} - c\right). \end{aligned} \tag{7}$$

Here n and c denote the cell density and a chemical that influences cell transport. The left hand sides of the above equations represent the total accumulation of the cells and of the chemical. With respect to the first equation, the first, second and third terms, respectively, account for diffusive transport, chemotaxis (transport induced by the gradient of the chemical) and proliferation. For the second equation, the first term accounts for diffusion. The second term contains production of this chemical by the cells and a decay term. Further, α, D_n, r and s are treated as constants. The initial condition is given and as boundary conditions, we use natural boundary conditions, saying that the flux at the boundary is prescribed, which read as:

$$
\begin{aligned}
&\mathrm{d}_n\frac{\partial n}{\partial \nu}-\alpha n\frac{\partial c}{\partial \nu}=0, \text{ on } \partial\Omega\\
&D_c\frac{\partial c}{\partial \nu}=0, \qquad\qquad \text{on } \partial\Omega.
\end{aligned}
\tag{8}
$$

Here ν denotes the unit outward normal vector on the boundary $\partial\Omega$.

The system of partial differential equations admit a traveling wave structure. An example can be found in the book of [Murray 2004]. In the current approach, we use the finite element method. To this extent, the partial differential equations are multiplied by a test-function and integrated over the domain of computation, where integration by parts is used to minimize the order of the spatial derivatives. By doing so, we arrive at:

$$
\begin{aligned}
&\int_\Omega \frac{\partial n}{\partial t}\phi d\Omega+\int_\Omega D_n\nabla n\cdot\nabla\phi d\Omega-\alpha\int_\Omega n\nabla c\cdot\nabla\phi d\Omega=r\int_\Omega n(1-n)\phi d\Omega,\\
&\int_\Omega \frac{\partial c}{\partial t}\psi d\Omega+\int_\Omega D_c\nabla c\cdot\nabla\psi d\Omega=\int_\Omega s\left(\frac{n}{1+n}-c\right)\psi \mathrm{d}\Omega.
\end{aligned}
\tag{9}
$$

At each time t, we have to find $n,c\in H^1(\Omega)$ such that the above integral equations hold for any choice $\phi,\ \psi\in H^1(\Omega)$. To this extent, the solutions are written as a linear combination of basis functions, that is

$$
n(\mathbf{x},t)=\sum_{j=1}^{k} n_j(t)\phi_j(\mathbf{x}), \qquad c(\mathbf{x},t)=\sum_{j=1}^{k} c_j(t)\phi_j(\mathbf{x}),
$$

which, when substituted into equation (9), yield

$$
\begin{aligned}
&\sum_{j=1}^{k} n'_j\int_\Omega \phi_j\phi_i d\Omega+\sum_{j=1}^{k} n_j\int_\Omega D_n\nabla\phi_i\cdot\nabla\phi_j d\Omega-\alpha\sum_{j=1}^{k} c_j\int_\Omega n(\mathbf{x},t)\nabla\phi_i\cdot\nabla\phi_j d\Omega=\\
&r\sum_{j=1}^{k} n_j\int_\Omega (1-n(\mathbf{x},t))\phi_i\phi_j d\Omega,\\
&\sum_{j=1}^{k} c'_j\int_\Omega \phi_i\phi_j d\Omega+\sum_{j=1}^{k} c_j\int_\Omega D_c\nabla\phi_i\cdot\nabla\phi_j d\Omega=\\
&s\sum_{j=1}^{k}\left[n_j\int_\Omega\left(\frac{1}{1+n(\mathbf{x},t)}\right)\phi_i\phi_j d\Omega-c_j\int_\Omega \phi_i\phi_j d\Omega\right].
\end{aligned}
\tag{10}
$$

All the integral expressions in the above equation denote matrix entries. We evaluate the expressions for linear triangular elements and approximate the integrals using Newton-Cotes quadrature, then we arrive at:

$$
\begin{aligned}
M^{ij} &:= \int_{\Omega} \phi_i \phi_j d\Omega = \frac{|\Delta|}{6}\delta_{ij},\\
S^{ij}_{nn} &= \int_{\Omega} D_n \nabla\phi_i \cdot \nabla\phi_j d\Omega = \frac{|\Delta|}{6}(\beta_i\beta_j + \gamma_i\gamma_j)\sum_{p=1}^{3} D_n(\mathbf{x}_p),\\
S^{ij}_{nc} &= -\alpha\int_{\Omega} n(\mathbf{x},t)\nabla\phi_i \cdot \nabla\phi_j d\Omega = -\alpha\frac{|\Delta|}{6}(\beta_i\beta_j + \gamma_i\gamma_j)\sum_{p=1}^{3} n(\mathbf{x}_p,t),\\
S^{ij}_{cn} &= \int_{\Omega} D_c \nabla\phi_i \cdot \nabla\phi_j d\Omega = \frac{|\Delta|}{6}(\beta_i\beta_j + \gamma_i\gamma_j)\sum_{p=1}^{3} D_c(\mathbf{x}_p),\\
M^{ij}_{nn} &:= \int_{\Omega} (1-n(\mathbf{x},t))\phi_i\phi_j d\Omega = \frac{|\Delta|}{6}(1-n(\mathbf{x}_i,t))\delta_{ij},\\
M^{ij}_{cn} &:= \int_{\Omega} \frac{1}{1+n(\mathbf{x},t)}\phi_i\phi_j d\Omega = \frac{|\Delta|}{6}\frac{1}{1+n(\mathbf{x}_i,t)}\delta_{ij}.
\end{aligned}
\tag{11}
$$

In the above expressions, we use the Kronecker Delta

$$
\delta_{ij} = \begin{cases} 1, & i = j,\\ 0, & i \neq j, \end{cases}
$$

and $\phi_i(\mathbf{x}) = \alpha_i + \beta_i x + \gamma_i y$. After the semi-discretization, the time integration is done using a backward Euler integration method, in which the integral (that is the matrices) are evaluated at the previous time step. Hence, the time integration method is of IMEX (IMplicit EXplicit) type. We also note that the point $(n,c) = (0,0)$ and $(n,c) = (1,1/2)$ are equilibria being an attractor and a saddle point respectively. For our applications, the IMEX method has sufficient stability properties. For a more concise treatment of the numerical time integration of diffusion-reaction problems, we refer to [Hundsdorfer 2003].

2.3 The Visco-elastic Equation

Next, we consider the visco-elastic equations. Here, we deal with a weak solution of

$$
-\nabla\cdot\sigma = -\mathbf{f},
$$

where

$$
\sigma(\mathbf{u}) = \mu_1\frac{\partial\varepsilon(\mathbf{u})}{\partial t} + \mu_2(\nabla\cdot\frac{\partial\mathbf{u}}{\partial t})\mathbf{I} + \frac{E}{1+\nu}\left(\varepsilon(\mathbf{u}) + \frac{\nu}{1-2\nu}(\nabla\cdot\mathbf{u})\mathbf{I}\right).
$$

Here, we defined $\mathbf{u} = [u,v]^T$. The above PDE represents a force balance. The first two terms in the above equation for $\sigma(\mathbf{u})$ account for viscous effects of the soft tissue. The second term set of two terms deal with elastic effects of the tissue. Further,

f is an internal body force. We use homogeneous Dirichlet conditions for the displacements and homogeneous natural boundary conditions for the force. Also, an appropriate initial condition has to be specified for $\mu_1\varepsilon(\mathbf{u}) + \mu_2(\nabla\cdot\mathbf{u})\mathbf{I}$. We denote the domain of computation by Ω and its boundary by Γ. The boundary is decomposed $\Gamma = \Gamma_1 \cup \Gamma_2$. To this extent, we introduce the following sets of functions:

$$U_0 := \{u \in H^1(\Omega):\ u = 0 \text{ on } \Gamma_1\},$$

$$V_0 := \{v \in H^1(\Omega):\ v = 0 \text{ on } \Gamma_2\},$$

$$U := C^1((0,T],U_0) \cup C^0([0,T],U_0),$$

$$V := C^1((0,T],V_0) \cup C^0([0,T],V_0).$$

This gives the following weak formulation:

$$\mathbf{u} \in U \times V: \int_\Omega \sigma(\mathbf{u}) : \varepsilon(\phi) d\Omega = -\int_\Omega \mathbf{f}\cdot\phi d\Omega, \qquad \forall\phi \in U_0 \times V_0. \tag{12}$$

In the above expression, we used the matrix inner product defined by

$$A : B := \sum_{i=1}^{m}\sum_{j=1}^{m} A_{ij}B_{ij},$$

where A, B are $m\times m$-matrices. Further, we used the scalar inner product

$$\mathbf{u}\cdot\phi := u\phi_1 + v\phi_2, \text{ where } \phi = [\phi_1,\phi_2]^T.$$

The relation between the strain tensor ε and the displacements, u and v, is given by

$$\varepsilon(\mathbf{u}) := \begin{pmatrix} \frac{\partial u}{\partial x} & \frac{1}{2}\left(\frac{\partial u}{\partial y} + \frac{\partial v}{\partial x}\right) \\ \frac{1}{2}\left(\frac{\partial u}{\partial y} + \frac{\partial v}{\partial x}\right) & \frac{\partial v}{\partial y} \end{pmatrix}.$$

To determine the entries of the strain tensor for each element we use an averaging procedure over the adjacent elements inspired from Galerkin's method. To illustrate this procedure, we consider

$$\varepsilon_{xx} = \frac{\partial u}{\partial x} \quad \text{, in } \Omega.$$

We write a weak form of the above equation to obtain

$$\int_\Omega \varepsilon_{xx}\phi d\Omega = \int_\Omega \frac{\partial u}{\partial x}\phi d\Omega.$$

The partial differential equations are solved using the Finite Element method with triangular elements and piecewise linear basis functions. For the time integration of the nonlinear partial differential equations, we use an backward (implicit) Euler method such that a toilsome stability criterion is circumvented. We write the displacement and strain as a linear combination of the basis functions

$$\sum_j \varepsilon_{xx,j}^k \int_\Omega \phi_i \phi_j d\Omega = \sum_j u_j^k \int_\Omega \phi_i \frac{\partial \phi_j}{\partial x} d\Omega, \quad \forall i \in \{1,\ldots,N\},$$

where k denotes the time index. At each time-step this system of equations is solved. Here, Newton-Cotes integration lumps the matrix to a diagonal matrix. The other terms for the strain tensor are treated similarly. One can demonstrate, using a combination of Grönwall's Lemma and Korn's inequality, that a given initial displacement, will vanish as t tends to infinity, if $\mathbf{f} = \mathbf{0}$. If $\mathbf{f} \neq \mathbf{0}$, then the solution converges to a steady-state. These assertions can be observed from experiments. The rigorous proof will be presented in a more mathematical study.

3 Wound Contraction

3.1 The Model Due to Tranquillo

For the modeling of wound contraction, we use the model due to [Tranquillo and Murray 1992], also described in [Murray 2004]. After coagulation of blood, the wound is closed and connective tissue fills the wound gap. At the consecutive stage fibroblasts enter the wound gap and start proliferating up to a equilibrium density in a logistic manner. The transport is modeled by a diffusive flux. The incoming fibroblasts start producing a tissue matrix on which they exert a contractile force. The fibroblast balance becomes

$$\frac{\partial n}{\partial t} + \text{div}\,(\mathbf{u}_t n) = \nabla \cdot (D \nabla n) + rn(n_0 - n). \tag{13}$$

Here n, D, $\mathbf{u}$, r and n_0 respectively denote the fibroblast density, motility coefficient, displacement, proliferation rate and fibroblast equilibrium density as in the unwounded state. The second term on the left-hand side follows from a passive convection of the cells due to the deformation of the structure. The first and second terms of the right hand side, respectively, account for diffusion and cell proliferation. Note that the above equation is of Fisher type, which in the absence of passive convection admits solutions with a traveling wave structure. The production of extra cellular matrix (ECM) by the fibroblasts is modeled by

$$\frac{\partial \rho}{\partial t} + \text{div}\,(\mathbf{u}_t \rho) = bn(\rho_0 - \rho). \tag{14}$$

Here b, ρ and ρ_0 respectively represent the ECM production rate, ECM density and equilibrium ECM density. The last term in the right-hand side stands for production of the ECM, and is based on the assumption that the production rate is proportional to the number of fibroblasts and the difference between the equilibrium ECM density and the ECM density itself. The second term on the left-hand side deals with passive convection due to deformation of the tissue. It can be shown that $\rho = \rho_0$ and $n = n_0$ are stable steady-state solutions under div $\mathbf{u}_t = 0$. Following [Murray 2004], the motility coefficient depends on the local strain in the following way

$$\mathbf{D} = \frac{D}{2} \cdot \begin{pmatrix} 2+\varepsilon_{xx}-\varepsilon_{yy} & 2\varepsilon_{xy} \\ 2\varepsilon_{xy} & 2+\varepsilon_{yy}-\varepsilon_{xx} \end{pmatrix}. \tag{15}$$

For the force equilibrium, we have the following equation

$$-\text{div}\,\sigma = \rho \mathbf{F}, \tag{16}$$

where σ denotes the stress and $\mathbf{f}$ represents a body force resulting from a spring type reaction to the pulling forces. The stress contains the following components: visco-elasticity (the first three terms, the first two representing viscous forces and the third term resulting from linear elasticity (Hooke's Law)) and cell traction, which is proportional to the ECM content and the fibroblast density.

$$\sigma = \mu_1 \varepsilon_t + \mu_2 (\nabla \cdot \mathbf{u}_t)\mathbf{I} + \frac{E}{1+\nu}\left[\varepsilon + \frac{\nu}{1-2\nu}(\nabla \cdot \mathbf{u})\mathbf{I}\right] + \frac{\tau n \rho}{1+\lambda n^2}\mathbf{I}, \tag{17}$$

Here μ_1, μ_1, E and ν respectively denote viscosity (the dynamic and kinematic viscosity), Young's modulus and Poisson's ratio. Except for the last term in the right hand side of the above equation, all terms are standard and were defined earlier in this chapter. The last term takes into account the contractile forces exerted by the fibroblasts on the ECM. The traction saturation constant is denoted by λ. This parameter guarantees the existence of a fibroblast density for which the traction is maximized. [Olsen *et al.* 1995] multiplies the λ parameter by ρ instead. This will be treated in the next subsection. The spring force, $\mathbf{F}$, acting as a body force is given by

$$\mathbf{F} = -s\mathbf{u}, \tag{18}$$

where s denotes the tethering elasticity coefficient.

The domain of computation is assumed to be divided into the wound region Ω_w and the undamaged region Ω_u, that is $\Omega = \Omega_w \cup \Omega_u \cup (\overline{\Omega}_w \cap \overline{\Omega}_u)$ and Ω_w is embedded within Ω_u. The closures of Ω_w and Ω_u respectively are denoted by $\overline{\Omega}_w$ and $\overline{\Omega}_u$. Note that these areas change as the wound heals. Initially, we assume the wound region to be occupied by connective tissue and hence the fibroblast and ECM densities are set equal to zero. Therefore, we use

$$n(\mathbf{x},0) = \begin{cases} 0, & \text{for } \mathbf{x} \in \Omega_w, \\ n_0, & \text{for } \mathbf{x} \in \Omega_u, \end{cases} \tag{19}$$

for the fibroblasts, and

$$\rho(\mathbf{x},0) = \begin{cases} 0, & \text{for } \mathbf{x} \in \Omega_w, \\ \rho_0, & \text{for } \mathbf{x} \in \Omega_u, \end{cases} \tag{20}$$

for the ECM density. Furthermore, we assume that the previous wound healing processes have not deformed the ECM, and hence we impose zero displacement as initial solution for $\mathbf{u}$. Finally, the domain of computation is taken sufficiently large so that no-flux for fibroblasts and zero displacements are biologically justifiable

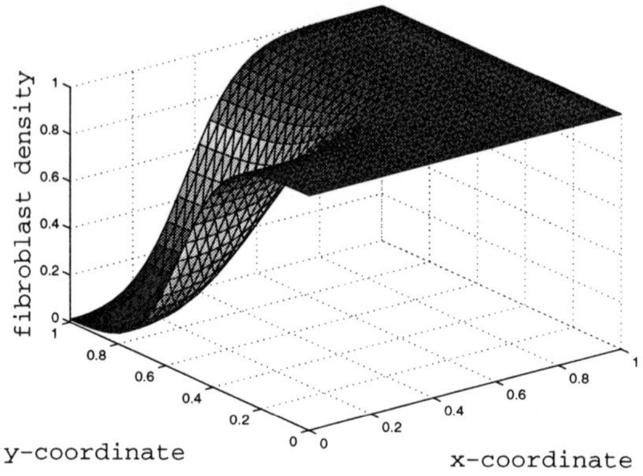

Fig. 4. The fibroblast profile in a dermal wound gap during the inflammatory phase

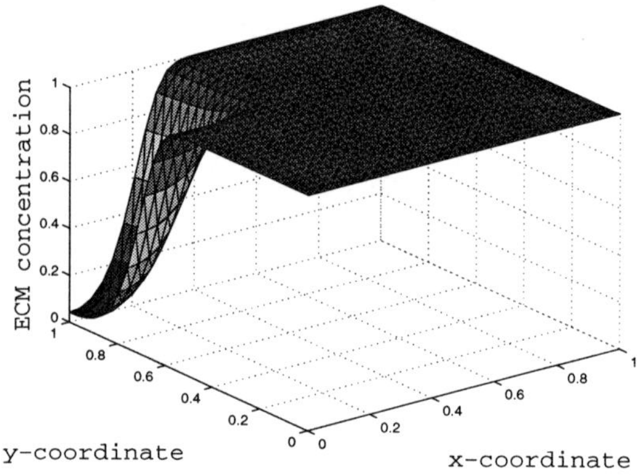

Fig. 5. The ECM profile in a dermal wound gap during the inflammatory phase

boundary conditions. The ECM density satisfies a first order hyperbolic equation (due to passive convection) of which the solution is determined by characteristics evolving from $t = 0$ in the $(t, \mathbf{x})$-space. Hence, for the ECM density, no boundary condition can be prescribed. To illustrate the contraction phase, that is the reparation of the dermis, we consider a wound gap on the top left of the dermis. Fibroblasts enter the gap region and start proliferating. Further, they start producing, and pulling on ECM, which gives a contractile behavior. In Figures 4 and 5 the fibroblast density and ECM concentration are shown at a time in the proliferative phase, in which the dermis underneath the epidermis is repaired. In Figure 6, we show the displacement of mesh nodes near the wound. It can be seen that the upper part of the dermis is bulged as it gets swollen. At lower parts in the dermis, the nodes are contracted, which is not well visible. A magnification of the swollen upper part of the dermis

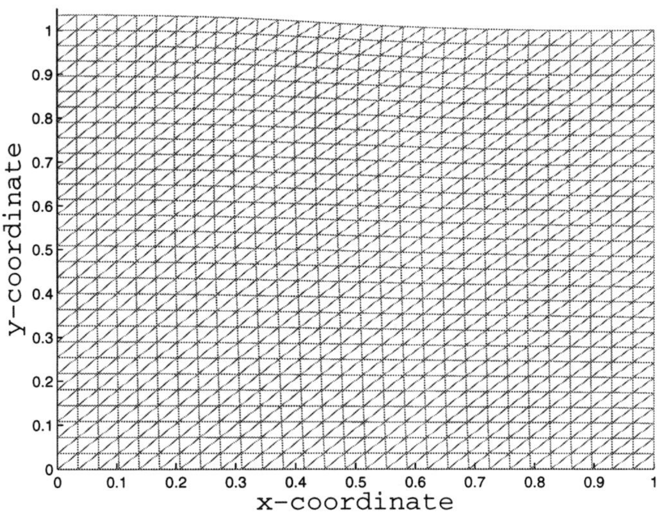

Fig. 6. The contraction of the mesh near the dermal gap. In the current simulations, the mesh is uniform in the entire domain of computation.

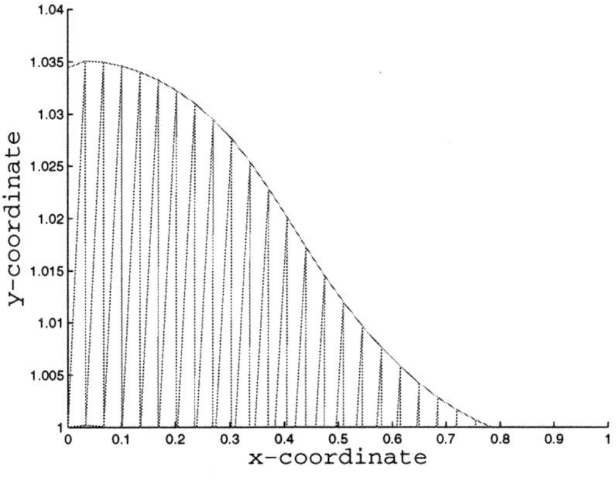

Fig. 7. The contraction of the mesh near the dermal gap, magnification in the top region of the dermis (which borders the epidermis)

is shown in Figure 7. The results that we have shown so far have been obtained using Finite Element simulations. In the current results, the simplification has been made that the upper part of the dermis, which borders the epidermis, is a boundary at which there are no external forces. Since the wound occurs at the top left position, the fibroblast exert a pulling force which has a component directed to top-left. This is why, the upper part of the dermis gets slightly bulged.

We are aware of the fact that we neglect that Young's modulus and Poisson ratio may depend on the composition of the tissue. We assume these dependences to have a secondary effect on the results.

3.2 The Model Due to Olsen *et al*

The model proposed by [Olsen *et al.* 1995] differs from the one due to [Tranquillo and Murray 1992] in two fundamental hypotheses: (1) there is a chemical or growth factor present at the wound site which triggers the contraction process, and (2) myofibroblasts differentiate from fibroblasts to amplify and transmit the traction forces generated by the migration of fibroblasts to the surrounding ECM. Some small differences are likewise found in the governing equations of the common variables. Thus, fibroblast balance according to [Olsen *et al.* 1995] becomes

$$\frac{\partial n}{\partial t} + \nabla \cdot \Big(-D_n \nabla n + \frac{a_n}{(b_n + c)^2} n \nabla c + n \frac{\partial \mathbf{u}}{\partial t} \Big) =$$
$$\Big(r_n + \frac{r_{n,max} c}{C_{1/2} + c} \Big) n \Big(1 - \frac{n}{K} \Big) - \frac{k_{1,max} c}{C_k + c} n + k_2 m - d_n n, \tag{21}$$

where a chemotaxis term has been added in the third term of the left hand side, with respect to [Tranquillo and Murray 1992]. Further, with respect to [Tranquillo and Murray 1992], the myofibroblasts have been taken into account as well. Here n, m and $\mathbf{u}$, respectively, denote the fibroblast density, myofibroblast density and displacement vector. Furthermore, fibroblasts differentiation into myofibroblasts is chemically modulated through $\frac{k_{1,max} c}{(C_{1/2}+c)}$, in which c denotes the concentration of a growth factor accumulated at the wound site during the inflammatory phase and produced by the (myo)fibroblasts. The terms of the right hand side account for (growth factor stimulated) proliferation, differentiation to and from myofibroblasts and cell death. Myofibroblasts are non motile cells, which yields the following balance

$$\frac{\partial m}{\partial t} + \nabla \cdot \Big(m \frac{\partial \mathbf{u}}{\partial t} \Big) = \varepsilon_r \Big(r_n + \frac{r_{n,max} c}{C_{1/2} + c} \Big) m \Big(1 - \frac{m}{K} \Big)$$
$$+ \frac{k_{1,max} c}{C_k + c} n - k_2 m - d_m m. \tag{22}$$

Here the right hand side contains proliferation, differentiation from fibroblasts, differentiation to fibroblasts and cell death. Collagen deposition is due to both fibroblasts and myofibroblasts and it is chemically enhanced, modeled by

$$\frac{\partial \rho}{\partial t} + \nabla \cdot \Big(\rho \frac{\partial \mathbf{u}}{\partial t} \Big) = \Big(r_\rho + \frac{r_{\rho,max} c}{C_\rho + c} \Big) \frac{n + \eta_b m}{R_\rho^2 + \rho^2} - d_\rho (n + \eta_d m) \rho. \tag{23}$$

Here ρ denotes the collagen density. The first and second terms in the right hand side denote growth factor amplified collagen production and decay. Hence, the kinetics of the chemical c clearly determine of the contraction progress. In order to sustain the contraction further in time, the diffusible chemical is produced by fibroblasts and myofibroblasts, which yields the following balance

$$\frac{\partial c}{\partial t} + \nabla \cdot \left(-D_c \nabla c + c \frac{\partial \mathbf{u}}{\partial t} \right) = \frac{k_c (n + \zeta m) c}{\Gamma + c} - d_c c. \tag{24}$$

Here, the second term in the left hand side represents diffusion. The right hand side contains growth factor production by the fibroblasts and myofibroblasts, and a decay term of the growth factor. The composite material formed of ECM and cells is modelled as a linear, isotropic viscoelastic continuum for which balance of linear momentum gives

$$\nabla \cdot \sigma + \mathbf{f}_{ext} = \mathbf{0}, \tag{25}$$

where σ denotes the stress tensor accounting for the contributions from the ECM and the cells, that is

$$\sigma = \sigma_{ecm} + \sigma_{cell}, \tag{26}$$

and $\mathbf{f}_{ext}$ represents the external forces acting on the tissue.

Cell traction stresses are modelled as isotropic, generated by fibroblasts, transmitted (and amplified) within the ECM by myofibroblast and inhibited at high collagen concentrations (*i.e.* contact inhibition). Hence

$$\sigma_{cell} = \frac{\tau_0 (1 + \xi m) n \rho}{R_\tau^2 + \rho^2} \mathbf{I}. \tag{27}$$

For the rest, the mechanical balance is similar to the one as proposed by [Tranquillo and Murray 1992]. The interested reader is referred to [Javierre *et al.* 2009].

In Figures 8 and 9, we show profiles of the myofibroblast and fibroblast density, respectively. The myofibroblasts are cells between the state of muscle cells

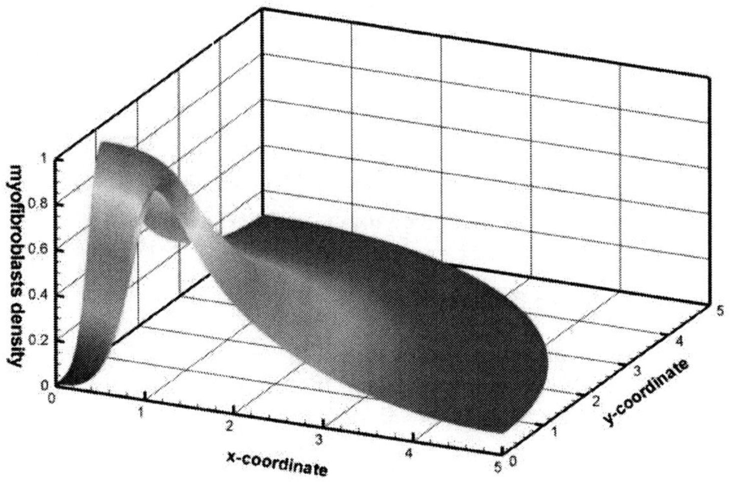

Fig. 8. Finite element solutions for the myofibroblasts using the Olsen model for wound contraction

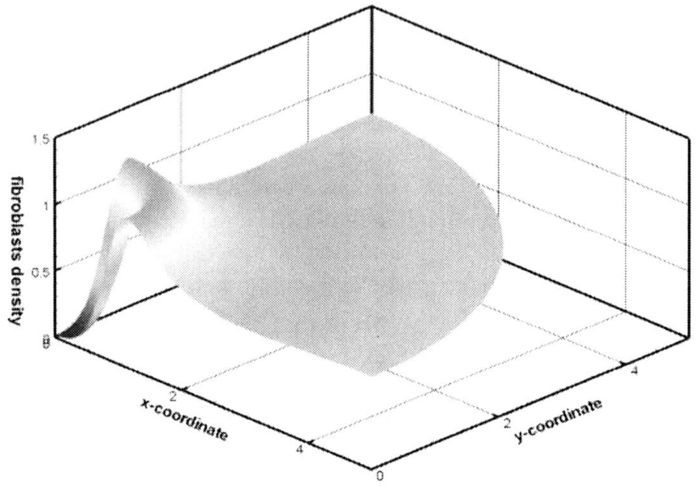

Fig. 9. Finite element solutions for the myofibroblasts using the Olsen model for wound contraction

and fibroblasts, and hence they provide a significant additional pulling force. One of their known functions with respect to wound repair is that they help contracting the wound edge to speed up healing. Initially, myofibroblasts are absent and appear at the wound site due to differentiation from fibroblasts. Since they are nonmotile cells, the profile of the myofibroblasts is elevated near the wound edge only. The fibroblasts are initially present in the undamaged region, but not in the wound area. However, as healing proceeds the fibroblast density increases in the wound region due to proliferation, transport and transformation from myofibroblasts. Since the latter cells are only present in significant numbers near the wound edge, the fibroblast density is elevated there, compared to the model of Tranquillo.

4 Angiogenesis

We refer the interested reader to [Javierre *et al.* 2009] for more details on the mechano-chemical model of wound contraction described here.

4.1 The Model Due to Gaffney *et al*

[Gaffney *et al.* 2002] present a simple model for angiogenesis and investigate one-dimensional finite difference solutions and asymptotic techniques. The capillaries are cylinders of endothelial cells, enveloped by a basement membrane. Angiogenesis is initiated by the degradation of the membrane at localized sites facing the wound. Then, endothelial cells migrate out of the vessel, such that a tube, or *sprout*, extends from the parent vessel. Cells behind the newly formed capillary tips proliferate such that the sprout extends further towards the wound. The tips can branch and join to form a circuit through which blood can flow. The sprouting process starts again from these new capillaries. In the model by [Gaffney *et al.* 2002], a two variable formalism is proposed and analyzed. The variables are the tip concentration, $n(\mathbf{x},t)$, and endothelial cell density (which is a building block for the blood capillaries), $b(\mathbf{x},t)$. For these two quantities, the following set of partial differential equations is considered:

$$\begin{aligned}\frac{\partial n}{\partial t} &= \nabla \cdot \{D_1 \nabla n + D_2 n \nabla b\} + f(n,b), \qquad \text{in } \Omega,\ t > 0,\\ \frac{\partial b}{\partial t} &= \lambda_1 \nabla \cdot \{D_1 \nabla n + D_2 n \nabla b\} + g(n,b), \ \text{in } \Omega,\ t > 0,\end{aligned} \tag{28}$$

where λ_1 represents the average number of endothelial cells in a capillary tip. Further, cell and tip migration takes place as a result of a biased random walk. The terms in the right hand side in the above equation account for transport and production or decay, which will be explained in some more detail later in this section. The second term in the migration part models an additional migration mechanism towards a decreasing blood vessel density (that is, towards a decreasing density of endothelial cells). It replaces effects of chemotaxis, mechanotaxis or haptotaxis appearing in alternative models. In the above equation, the functions $f(n,b)$ and $g(n,b)$ are defined by

$$\begin{aligned} f(n,b) &= \lambda_2 n - \lambda_3 n^2 - \lambda_4 nb,\\ g(n,b) &= \lambda_6 ab(b_0 - b) + \lambda_6 \chi nb(b_1 - b) + \lambda_5(\lambda_3 n^2 + \lambda_4 nb).\end{aligned} \tag{29}$$

First, we consider the function $f(n,b)$. The first term denotes tip branching: the splitting of tips into new tips. The number of new tips generated increases with the number of existing tips. This dependence is assumed to be linear. The second term is based on the fact that the probability of a tip to 'meet' another tip at a certain position is proportional to n^2. In probability terms: the probability that at a certain location there is a tip is proportional to n, which is the tip density. Then the probability that another tip is present at this location, is proportional to n^2, in which it is assumed that the events of the tips to be at this location are independent. When two tips join, they form a closed loop effectively reducing the tip density and increasing the density of capillaries (and hence of endothelial cells). Another reason for tips to 'vanish' is the joining with a capillary. Using a similar probability argument, one arrives at a loss term that is proportional to $n\ b$. For the function $g(n,b)$, we note the following. The first term contains proliferation due to ordinary logistic growth of the endothelial cell concentration. The second term contains an increased logistic growth due to the presence of tips that are built of endothelial cells. Besides, tip carry endothelial cells, two tips that join form a bridge of endothelial cells. This contribution is taken into account by the $\lambda_5 \lambda_3 n^2$-term, and the joining of a tip with a capillary gives an endothelial cell built construction as well. This lastmentioned process is modeled by the $\lambda_5 \lambda_4 nb$-term.

Far away from the wound, we will assume no transport of the tips and the endothelial cells. Hence,

$$D_1 \frac{\partial n}{\partial \nu} + D_2 n \frac{\partial b}{\partial \nu} = 0, \text{ on } \partial\Omega, \tag{30}$$

where ν denotes the outward pointing unit normal vector on the boundary $\partial\Omega$. As initial conditions, it is assumed that there are neither capillaries, nor tips in the wound space. Outside the wound, the endothelial cell density, due to the undamaged network of capillaries, is given by b_0, that is

$$b(\mathbf{x},0) = \begin{cases} 0, & \text{in } \Omega_w \subset \Omega, \\ b_0, & \text{in } \Omega \setminus \Omega_w. \end{cases} \tag{31}$$

Here Ω_w is referred to as the initial wound space. Further, the initial condition for the capillary tip concentration is given by

$$n(\mathbf{x},0) = \begin{cases} 0, & \text{in } \Omega_w, \\ n_0, & \text{in } \Omega_\delta := \{\mathbf{x} \in \Omega \setminus \Omega_w : \text{dist}(\mathbf{x}, \Omega_w) < \delta\}, \\ 0, & \text{in } \Omega \setminus (\Omega_w \cup \Omega_\delta). \end{cases} \tag{32}$$

For the numerical solutions, it is necessary to smooth the lastmentioned step-function. [Gaffney *et al.* 2002] determine traveling wave solutions in the form of $n,\ b(x,t) = \overline{n},\ \overline{b}(x+ct)$. They come up with a minimal wave speed. They consider three asymptotic regions which are matched. Finite element results to the aforementioned equations are given in Figure 10 and 11. The first figure displays the

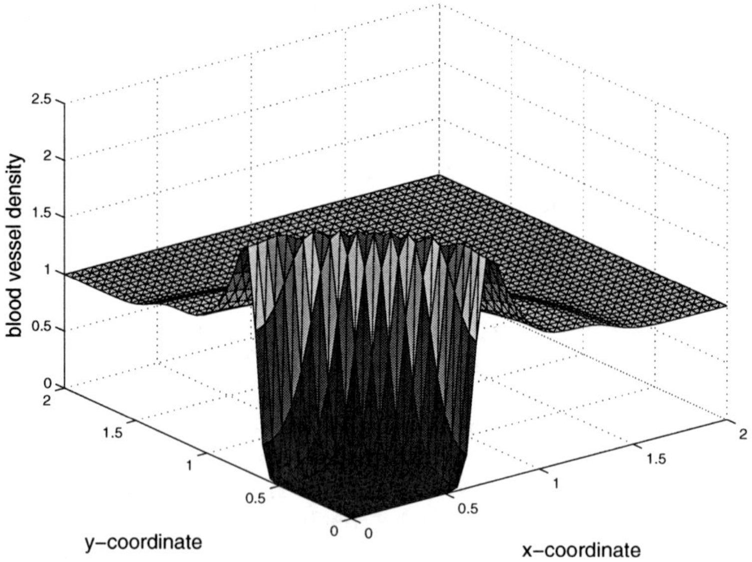

Fig. 10. Finite element solutions for a spherical initial wound. The endothelial cell density has been displayed.

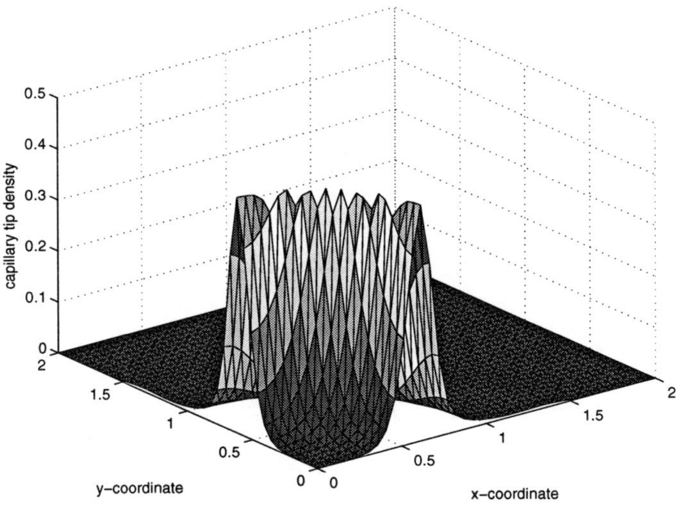

Fig. 11. Finite element solutions for a spherical initial wound. The capillary tip density has been displayed.

endothelial cell density and the second figure displays the capillary tip density. The finite element method was used to solve the set of partial differential equations. The time integration was done using an IMEX method. Linear elements were used with Newton-Cotes quadrature.

Figure 11 reveals that at the wound boundary, there is an increased capillary tip density. This increased tip density is necessary, since at the wound boundary epidermal cells divide and are mobile, moving into the wound region. This activity is supported by the high amount of oxygen and high number of nutrients due to the abundant number of capillary tips. We note that the model due to [Gaffney *et al.* 2002] was inspired from the studies due to [Pettet *et al.* 1996]. The lastmentioned authors include, next to vessel- and tip density, a chemo-attractant in their model. Further, low oxygen content areas are identified with areas of a low blood vessel density. The model due to [Maggelakis 2003] actually uses the lack of oxygen as a trigger of blood vessel formation.

4.2 The Model Due to Maggelakis

Maggelakis constructed this model of angiogenesis in 2003 [Maggelakis 2003]. Angiogenesis is a crucial process for tissue regeneration and for tumor growth [Rossiter *et al.* 2004]. It is assumed that the tips of the capillaries act as the only sources for oxygen supply. Due to the injury, the microvascular network is damaged in the wound area and as a result the oxygen concentration decreases there. This lack of oxygen triggers the activation of macrophages, which among other tasks like being scavengers to remove harmful bacteria, start producing the macrophage

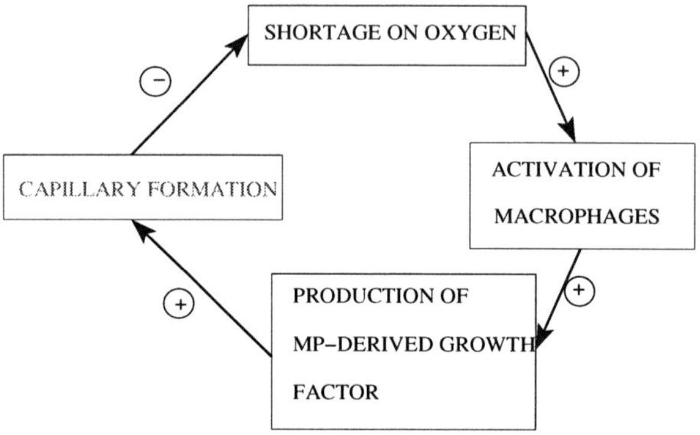

Fig. 12. A schematic of the negative feedback mechanism for the model for angiogenesis due to Maggelakis. The +/- signs denote that the increase has a positive/negative effect on the quantity. For instance in the flow chart, a large number of capillaries has a negative influence on the shortage of oxygen, since the oxygen level will increase due to transport from the capillaries.

derived growth factors (MDGF), such as Vascular Endothelial Growth Factors (VEGF). These growth factors trigger the regeneration of capillaries and hence the network is restored. In the work of [Rossiter *et al.* 2004], it is shown that loss of VEGF's causes a tremendous delay in healing time of deep wounds due to the appearance of blood vessel-free zones. Their work is experimental, in which mice have been used, but it illustrates the importance of the presence of VEGF's and capillaries.

With the production of capillaries, the oxygen concentration increases, by which the production of new capillaries is stopped. The flow chart of this system with a negative feedback mechanism is sketched in Figure 12. Let u_1 and u_3 respectively denote the oxygen concentration and the capillary density and let them be functions of time t and space within the domain of computation Ω; then a mass balance for the oxygen concentration results into the following partial differential equation (PDE):

$$\begin{aligned} &\frac{\partial u_1}{\partial t} = D_1 \Delta u_1 - \lambda_{11} u_1 + \lambda_{13} u_3, \text{ for } \mathbf{x} \in \Omega, \\ &\text{subject to } u_1(\mathbf{x},0) = \begin{cases} 0, & \text{for } \mathbf{x} \in \Omega_w, \\ u_1^{eq}, & \text{for } \mathbf{x} \in \Omega_u. \end{cases} \end{aligned} \tag{33}$$

The first, second and third terms in the right hand side of this equation, account for diffusive transport, oxygen consumption and oxygen increase due to transport by and through the capillaries. It is assumed that that there is no transport of oxygen over the boundary of the domain of computation (the outer boundary of Ω) due to symmetry or due to the far distance from the wound. This results into a homogeneous Neumann boundary condition. The above equation is based on the assumption that the oxygen supply and oxygen consumption depend linearly (affinely) on the capillary density and oxygen concentration respectively. When the oxygen level is low, macrophages start releasing MDGF's (macrophage derived growth factors) which enhance the growth of blood vessels and hence help restore the capillaries that provide the skin with the necessary nutrients and oxygen for cell division needed for wound closure. An assumption in the model is that MDGF is produced if the oxygen level is below a threshold value, say u_θ. The production rate, Q, is assumed to depend linearly on the lack of oxygen, that is

$$Q = Q(u_1) = \begin{cases} 1 - \dfrac{u_1}{u_\theta}, & \text{if } u_1 < u_\theta, \\ 0, & \text{if } u_1 \geq u_\theta. \end{cases} \tag{34}$$

The mass balance of MDGF's, its concentration being denoted by u_2, results into the following PDE's in the wound region Ω_w and out of the wound region Ω_u:

$$\begin{aligned} \frac{\partial u_2}{\partial t} &= D_2 \Delta u_2 + \lambda_{21} Q(u_1) - \lambda_{22} u_2, \quad \text{for } \mathbf{x} \in \Omega_w, \\ \frac{\partial u_2}{\partial t} &= D_2 \Delta u_2 - \lambda_{22} u_2, \quad \text{for } \mathbf{x} \in \Omega_u. \end{aligned} \tag{35}$$

The first terms of the right hand side account for diffusive transport of the growth factor. The second term in the top equation accounts for regeneration of the macrophage derived growth factor due to a lack of oxygen, and the $\lambda_{22}u_2$-term accounts for decay of the MDGF concentration. The initial MDGF concentration is assumed to be zero in the entire domain of computation Ω and a homogeneous Neumann boundary condition is used. The capillary density is assumed to grow as a result of the MDGF's in a logistic manner, that is

$$\frac{\partial u_3}{\partial t} = D_3 \Delta u_3 + \lambda_{33} u_2 u_3 (1 - \frac{u_3}{u_3^{eq}}), \quad \text{for } \mathbf{x} \in \Omega, \tag{36}$$

where u_3^{eq} denotes the equilibrium capillary density of the undamaged tissue. In the above equation, the first and second terms, respectively, stand for diffusive transport of capillaries and proliferation of capillaries induced by the MDGF's. The capillary density is assumed to satisfy the following initial condition

$$u_3(\mathbf{x}, 0) = \begin{cases} 0, & \text{for } \mathbf{x} \in \Omega_w, \\ u_3^{eq}, & \text{for } \mathbf{x} \in \Omega_u, \end{cases} \tag{37}$$

which states that capillaries are removed from the wound area at the moment of injury. A homogeneous Neumann boundary condition is used for u_3. We assume the capillary tips to migrate via a random walk process, opposed to the bias of migration as in Gaffney's model. Maggelakis sets in a nonzero artificial starting value for the capillary density to have the capillary density to increase up to the equilibrium value. The assumption that capillary tips migrate by a random walk is also a key-assumption in the work due to [Plank and Sleeman 2003] and [Plank and Sleeman 2004]. Analytic solutions are considered in [Maggelakis 2003] for the above equations under the assumptions that the diffusion of oxygen and growth factors is instantaneous and that $D_3 = 0$. The induced growth of MDGF's due to a lack of oxygen can be classified as a negative feedback mechanism. We finally note that the model due to [Gaffney *et al.* 2002] captures the increase of the capillary tip and capillary density at the wound edge in a natural way. However, the well-known issue concerning the oxygen level is ignored in their model, but it is incorporated in Maggelakis' model. A combination of these two formalisms could be very illustrative.

5 Wound Closure

5.1 The Model Due to Sherratt and Murray

The mechanism for wound closure is cell division and growth (mitosis). This mechanism is triggered by a complicated system of growth factors. In the present model we use the simplification that wound closure is influenced by one generic growth factor only, which is a severe simplification of reality. The growth factor concentration influences the production of epidermal cells and the growth factors are produced by the epidermal cells. If the number of epidermal cells is low, then, the production of growth factors is excessively high. Whereas, as the healed state is reached, then, the production decreases such that the healed cell concentration is stable: the mitotic rate is balanced by the decay rate of the cells. This results into a nonlinear feedback system. Following [Sherratt and Murray 1991], we assume that the main source of growth factors is the epidermal cells. Growth factors diffuse through the tissue region and the concentration decays due to reactions with other chemicals present in the tissue. Let u_4 and u_5 respectively denote the epidermal cell density and growth factor concentration, then the expression of Sherratt and Murray where the accumulation of the epidermal cells is determined by diffusive transport, proliferation, and cell death(see the first, second and third terms of the right hand side, respectively) is given by

$$\begin{aligned} &\frac{\partial u_4}{\partial t} = D_4 \Delta u_4 + s(u_5) u_4 (2 - \frac{u_4}{u_4^{eq}}) - \lambda_{44} u_4, \qquad \mathbf{x} \in \Omega, \\ &\text{subject to } u_4(\mathbf{x},0) = \begin{cases} 0, & \text{for } \mathbf{x} \in \Omega_w, \\ u_4^{eq}, & \text{for } \mathbf{x} \in \Omega_u. \end{cases} \end{aligned} \tag{38}$$

The function $s = s(u_5)$ is a nonlinear function of the growth concentration describing the mitotic rate, see [Murray 2004, Sherratt and Murray 1991]. For the growth factor accumulation a similar relationship due to diffusive transport, production and decay (see the first, second and third terms in the right hand side) is obtained:

$$\begin{aligned} &\frac{\partial u_5}{\partial t} = D_5 \Delta u_5 + f(u_4) - \lambda_{55} u_5, \qquad \mathbf{x} \in \Omega, \\ &\text{subject to } u_5(\mathbf{x},0) = \begin{cases} 0, & \text{for } \mathbf{x} \in \Omega_w, \\ u_5^{eq}, & \text{for } \mathbf{x} \in \Omega_u. \end{cases} \end{aligned} \tag{39}$$

In the above equation $f(u_4)$ denotes a nonlinear relation for the growth factor regeneration. In the model due to Sherratt and Murray, two different types of growth factors are considered: 1. activators; and 2. inhibitors, both with their characteristic functions for s and f, given by

$$s(u_5) = \frac{2c_m(h-\beta)u_4}{c_m^2+u_4^2} + \beta, \quad \beta = \frac{1+c_m^2-2hc_m}{(1-c_m)^2}, \quad f(u_4) = \frac{u_4(1+\alpha^2)}{u_4^2+\alpha^2}, \tag{40}$$

for the activator case and by

$$s(u_5) = \frac{(h-1)u_5+h}{2(h-1)u_5+1}, \quad f(u_4) = u_4, \tag{41}$$

for the inhibitor. Here h, β and c_m are considered as known constants, and we refer to [Murray 2004] for more details. We want the initial state to be unstable and the boundary conditions to be stable, so that the functions u_4 and u_5 converge to the values of the boundary conditions u_4^{eq} and u_5^{eq} as $t \to \infty$. In other words, the unwounded state is stable with respect to small perturbations, whereas the wounded state is unstable.

[Sherratt and Murray 1991] conducted a detailed traveling wave analyses to obtain insight into the characteristics of the solution. By this analysis they found a lower bound for the traveling wave speed, which is reached in the intermediate stages of the simulation. In Section 5.3 of this chapter, we will propose a combination of the models due to [Maggelakis 2004], [Sherratt and Murray 1991].

We note that Sherratt & Murray report a qualitative agreement with experiments conducted on rabbit's ears in [Murray 2004, Sherratt and Murray 1991]. Their comparison is qualitative and their model predicts the right behavior. To get a more quantitative agreement between the model and experiments, probably a regression procedure is required to get the appropriate values for the parameters involved. In Figures 13 and 14, examples of profiles for the epidermal cell density and epidermal growth factor concentration are shown respectively. It can be seen that both profiles exhibit an elevation near the wound edge. This is consistent with the traveling wave analysis carried out by Sherratt and Murray. This behavior also confirms the need for an increased capillary tip density there. It also supports assumptions concerning an increased activity at the wound edge.

5.2 The Model Due to Adam

In this section the model based on the ideas of [Adam 1999] is presented. Firstly, the model for the regeneration, decay and transport of the growth factor is given, and subsequently the healing process as a result of the presence of the growth factor is described (see [Vermolen *et al.* 2006]). Finally, a description of the coupling of the two processes is described. Further, we assume that healing takes place if and only if the concentration of the growth factor at the wound edge, exceeds a threshold value $\hat{c}$.

We use Ω_1, Ω_2 and Ω_3 to denote the wound itself, the active layer and the outer tissue respectively. Since the wound is healing, the areas $\Omega_1(t)$, $\Omega_2(t)$ and $\Omega_3(t)$ are

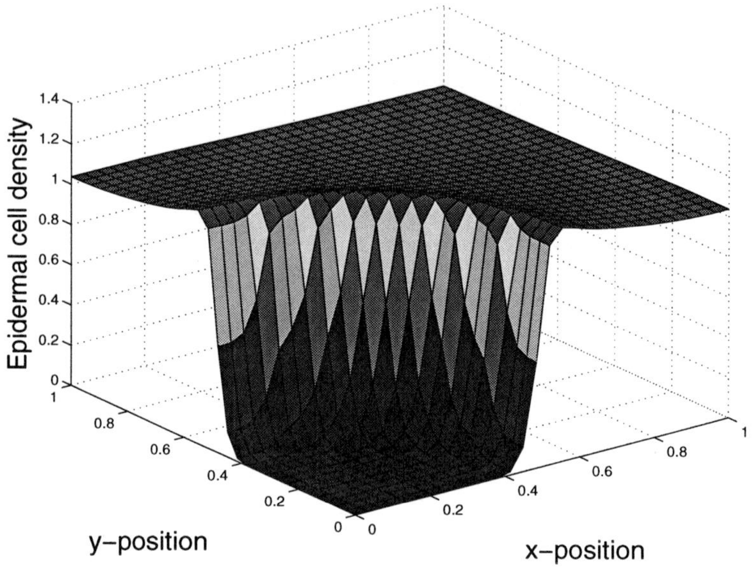

Fig. 13. The epidermal cell density around a circular wound

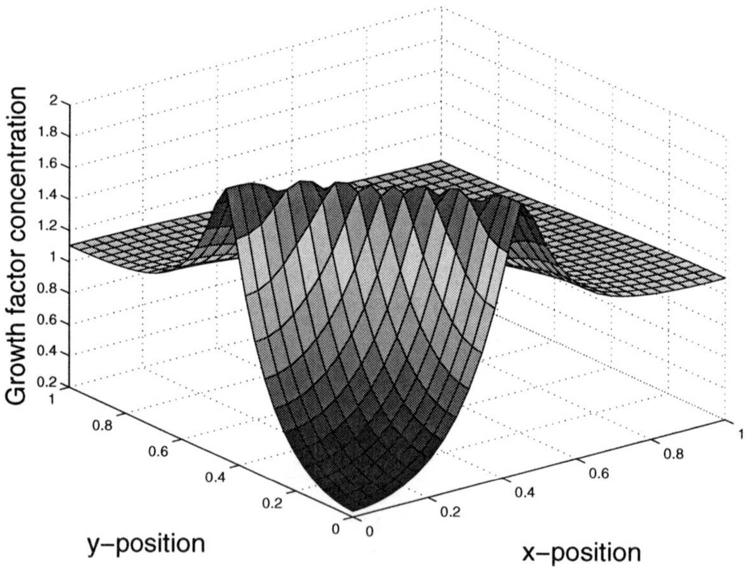

Fig. 14. The epidermal growth factor concentration around a circular wound in case of an activator

functions of time. These areas are part of the solution. Far away from the wound, that is at the boundary of the domain of computation, $\partial\Omega$, we assume that there is no transport of growth factor. The wound edge, the interface between the wound (Ω_1) and the active layer (Ω_2), is indicated by $W(t)$ (*i.e.* $W = \overline{\Omega}_1 \cap \overline{\Omega}_2$).

Let the overall domain of computation be given by Ω, then following [Adam 1999], the partial differential equation for the growth factor concentration, c, is given by

$$\frac{\partial c}{\partial t} - \operatorname{div} D \operatorname{grad} c + \lambda c = P\mathbf{1}_{\Omega_2}(\mathbf{x}), \text{ for } (t,\mathbf{x}) \in (0,T] \times \Omega, \tag{42}$$

$$\frac{\partial c}{\partial \nu} = 0, \text{ for } (t,\mathbf{x}) \in (0,T] \times \partial\Omega, \tag{43}$$

$$\text{where } \mathbf{1}_{\Omega_2(t)}(\mathbf{x}) = \begin{cases} 1, & \text{for } \mathbf{x} \in \Omega_2(t) \\ 0, & \text{for } \mathbf{x} \in \Omega_1(t) \cup \Omega_3(t) \end{cases}, \tag{44}$$

As the initial condition, we have

$$c(0,\mathbf{x}) = 0, \text{ for } \mathbf{x} \in \Omega. \tag{45}$$

In the equations D, P and λ denote the constant diffusion coefficient, production rate constant and the decay coefficient of the growth factor. In equation (42), the second and third terms in the left hand side account for diffusive transport and decay of the growth factor, respectively. The right hand side models production of the growth factor. The nature of this production is motivated by the significant increase of mitotic and motile activity around the edge of the wound with respect to other positions away from the wound edge. These constants are non-negative in our parabolic PDE. The growth factor concentration, c, is to be determined. Equation (43) represents the boundary condition and the indicator function $\mathbf{1}_{\Omega_2(t)}(\mathbf{x})$ accounts for the growth factor production taking place in the active layer only. In [Vermolen and Javierre 2009], it is demonstrated that if we use $\mathbf{1}_{\overline{\Omega}_2}$ instead of $\mathbf{1}_{\Omega_2}$ there is an inconsistency as $D \to 0$. [Adam 1999] considers the derivation of a *critical size defect*, which is the smallest wound that does not heal, where the time derivative in the diffusion reaction equation was not taken into account.

Healing at a certain location of the interface implies that the inward normal component of the velocity, v_ν, of the interface W is positive. In the present chapter, we use the assumption from [Adam 1999] that the interface moves if and only if the growth factor concentration exceeds a threshold concentration $\hat{c}$, hence

$$\begin{aligned} &v_\nu \neq 0 \text{ if and only if } c(t,\mathbf{x}) \geq \hat{c} \text{ for } (t,\mathbf{x}) \in (0,T] \times W(t), \\ &\text{else } v_\nu = 0. \end{aligned} \tag{46}$$

This implies that in order to determine whether the wound heals at a certain location on W at a certain time t, one needs to know the growth factor concentration there.

Adam considers analytic expressions for the time independent case for several geometries: planar (linear) geometry [Adam 1999], a circular wound on a spherical surface [Adam 2004(1)], a circular wound on a planar surface [Adam 2004(2)].

A wound in spherical symmetry is considered in terms of analytic expressions by [Arnold 2001].

As it has been motivated in [Vermolen *et al.* 2006], we assume that the healing rate is proportional to the local curvature of the wound. Hence, in agreement with equation (46), the velocity component in the outward (from Ω_1, that is the wound) normal direction is given by

$$v_v = -(\alpha + \beta\kappa)w(c(t,\mathbf{x}) - \hat{c}), \text{ for } (t,\mathbf{x}) \in (0,T] \times W(t), \tag{47}$$

where κ is the local curvature and $\alpha, \beta > 0$ are considered as non-negative constants, prohibiting growth of the wound if $\kappa \geq 0$. Further, the function $w(s)$ falls within the class of Heaviside functions, that is $w(s) \in H(s)$, where $H(.)$ represents the family of Heaviside functions, for which we have

$$H : s \to \begin{cases} 0, & \text{if } s < 0, \\ \in [0,1], & \text{if } s = 0, \\ 1, & \text{if } s > 0. \end{cases} \tag{48}$$

Some models with the same principles as the active layer and / or the discontinuous switch condition can be found in [Hogea *et al.* 2006, Adam 1987, Adam 1999, Shymko and Glass 1976, Britton and Chaplain 1993].

Fig. 15. Healing of a star-fish shaped wound simulated with Adam's model, using the level set method. The parameters are $\alpha = 1$, $\beta = 0.05$ and $\hat{c} = 0$.

In [Vermolen *et al.* 2006], some preliminary computations are done by the use of Finite elements and with the moving interface in one dimension (for a planar and circular wound). In [Javierre *et al.* 2008], a Finite Element in more dimensions is presented in combination with the moving wound edge. The moving boundary problem is solved using a level set method to track the wound edge. Further, some clinical implications are given concerning the wound geometry, active layer thickness *etc.*. In [Vermolen and Javierre 2009], it is demonstrated that solutions exist in $H^1(\Omega)$ at each time t. In the lastmentioned work, it is also demonstrated that the discontinuous switch condition can give rise to delayed healing under conditions in which built-up of the growth factor level is slow with respect to the actual healing speed that would follow from the interface velocity parameters α and β. Also analytic solutions are constructed in the right function set, and bounds for the waiting time before healing sets in are determined in the lastmentioned work. Currently, a cut-cell method to improve tracking of the moving boundary has been implemented by [Zemskov *et al.* 2009]. In Figure 15, we show the healing behavior of a starfish wound. The wound edge immediately starts moving since the threshold concentration has been chosen equal to zero. The geometry gradually changes from starfish-like to a more rounded configuration, prior to the complete healed state. In Figure 16, we show the same simulation with different conditions, here the influence of curvature is more pronounced. It can be seen that in regions of negative curvature

Fig. 16. Healing of a star-fish shaped wound simulated with Adam's model, using the level set method. The parameters are $\alpha = 1$, $\beta = 0.1$ and $\hat{c} = 0$.

Fig. 17. Healing of a star-fish shaped wound simulated with Adam's model, using the level set method. The parameters are $\alpha = 1$, $\beta = 0.1$ and $\hat{c} = 10^{-3}$.

the wound edge retreats during the initial stages, which makes the wound even more rounded than in the previous run. The wound finally heals in a circular shape. We show the same computation, under different conditions, in which the wound breaks up on its way to complete healing in Figure 17. Here, the GF concentration has to accumulate before the wound edge starts moving. Since the time for the growthfactor concentration to reach the threshold concentration is smallest at the regions of negative curvature, the wound edge starts retreating first at these positions. Subsequently, as the wound edge moves at all positions, the wound breaks up during the later stages of the healing process. Hence, the qualitative behavior as predicted by the model, depends on the values of the parameters.

5.3 A Combination of Angiogenesis and Wound Closure

In this subsection, we use the same notation as in subsections 3.2. and 4.1. for the oxygen, MDGF, capillary tip, epidermal cell and epidermal growth factor concentration. The idea is that a sufficient amount of nutrients and oxygen is needed for the mitotic activity of the epidermal cells. In the conventional wound closure models, it is assumed that the required amount of nutrients is always there. So the levels of nutrients and oxygen are assumed to be instantaneous and angiogenesis is commonly assumed to have completed. On the other hand, the models for angiogenesis like the formalism due to Gaffney, assume angiogenesis to be the rate-determining step. In the lastmentioned formalism, the mechanism of wound closure by migration and proliferation of epidermal cells is assumed to follow the vascularization pattern instantaneously. In the present model, we try to link the gradual build-up of oxygen and nutrients with the actual wound closure process. Furthermore, the capillary density is adjusted according to the wound closure. A first attempt of coupling neo-vascularization and wound closure was made by [Vermolen and Adam 2007]. In the lastmentioned work, the influence of the capillary density on wound closure was taken into account. However, it was assumed that there was no influence of wound closure on the formation of capillaries. This debatable assumption is relaxed here. This implies that the angiogenesis model needs its input from the wound closure model. Hence, in the wound region, an excessive capillary density is needed to supply all necessary nutrients. This advocates for an increased equilibrium capillary density in the wound site, where the epidermal cell concentration is low. Therewith, it follows that the equilibrium capillary density, u_3, is a decreasing function of the epidermal cell density, u_4. We assume that if $u_4 > u_4^{eq}$, that is, if the epidermal cell density is larger than the undamaged cell concentration, then, the equilibrium capillary density is no longer dependent on epidermal cell density, hence it is constant. The function $u_3^{eq} = u_3^{eq}(u_4)$ should be chosen such that the healed state is a stable equilibrium. Let $q > 1$ be a scalar parameter determining the enhancement factor ψ_q of the equilibrium capillary density inside the wound during the wound healing. Then in order to have the healed state as a stable equilibrium, we introduce a function $\psi_q(m)$

$$\psi_q = \psi_q(u_4) = \begin{cases} q(1 - \frac{u_4}{u_4^{eq}}) + \frac{u_4}{u_4^{eq}}, & \text{if } u_4 \leq u_4^{eq}, \\ 1, & \text{if } u_4 > u_4^{eq}. \end{cases} \tag{49}$$

This function is used to adjust the equilibrium capillary density u_3^{eq}.

Furthermore, the formation of epidermal cells and their ability to produce the epidermal cell mitosis regulating growth factor are determined by the amount of oxygen and nutrients supplied. The mitotic rate is assumed to increase as the capillary density increases. Since, the cell division and growth rate are finite, there is a maximum division rate at which the division rate is no longer sensitive with respect to an increase of nutrients and oxygen supplied. Hence, this advocates for the existence of a maximum mitotic rate. To model this, we introduce a function ϕ_p representing the rate-function in the mitosis rate with the capillary density, which is given by

$$\phi_p(u_3) = \begin{cases} \frac{u_3}{u_3^{eq}}, & \text{if } u_3 \leq p u_3^{eq}, \\ p, & \text{if } u_3 > p u_3^{eq}. \end{cases} \tag{50}$$

The quantity p denotes the increase of the mitotic activity of the tissue cells as a result of the number of capillaries that deliver the needed nutrients and oxygen. The functions ψ_q and ϕ_p are substituted into the PDE's (36), (38) and (39) of the models due to [Sherratt and Murray 1991] and [Maggelakis 2004]. The scheme

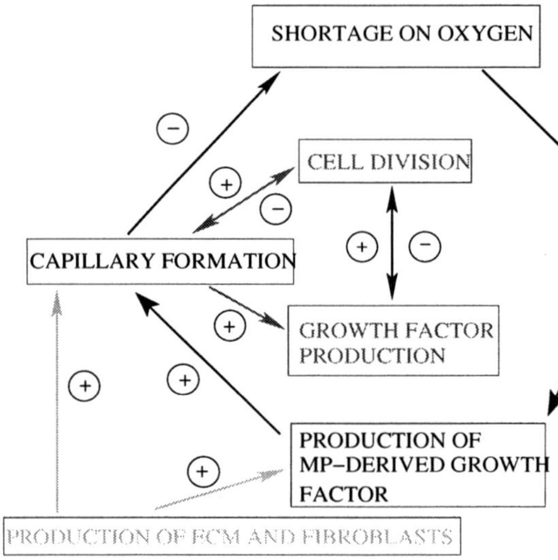

Fig. 18. Schematic of the feedback mechanism of the production of epidermal cells, growth factors and neo-vascularization. The +/- signs denote that the increase has a positive/negative effect on the quantity. For instance in the flow chart, a large number of capillaries has a negative influence on the shortage of oxygen, since the oxygen level will increase due to transport from the capillaries.

for the influences between the various densities and concentrations is depicted in Figure 18. After the adaptations, one obtains:

$$
\begin{aligned}
\frac{\partial u_1}{\partial t} &= D_1 \Delta u_1 - \lambda_{11} u_1 + \lambda_{13} u_3, \\
\frac{\partial u_2}{\partial t} &= D_2 \Delta u_2 + \lambda_{21} Q(u_1) - \lambda_{22} u_2, \\
\frac{\partial u_3}{\partial t} &= D_3 \Delta u_3 + \lambda_{33} u_2 \left(1 - \frac{u_3}{u_3^{eq} \psi_q(u_4)}\right) u_3, \\
\frac{\partial u_4}{\partial t} &= D_4 \Delta u_4 + \phi_p(u_3) s(u_5) u_4 \left(2 - \frac{u_4}{u_4^{eq}}\right) - \lambda_{44} u_4, \\
\frac{\partial u_5}{\partial t} &= D_5 \Delta u_5 + \phi_p(u_3) \lambda_{55} f(u_4) - \lambda_{55} u_5.
\end{aligned}
\tag{51}
$$

The above PDE's hold on Ω, except for the second PDE, which holds in Ω_w only. Its counterpart in Ω_u is the same as PDE (35). The functions ψ_q and ϕ_p, defined in a simple manner, are chosen such that the wound state is unstable while the undamaged state is stable. Further, the meaning of all the terms of the above partial

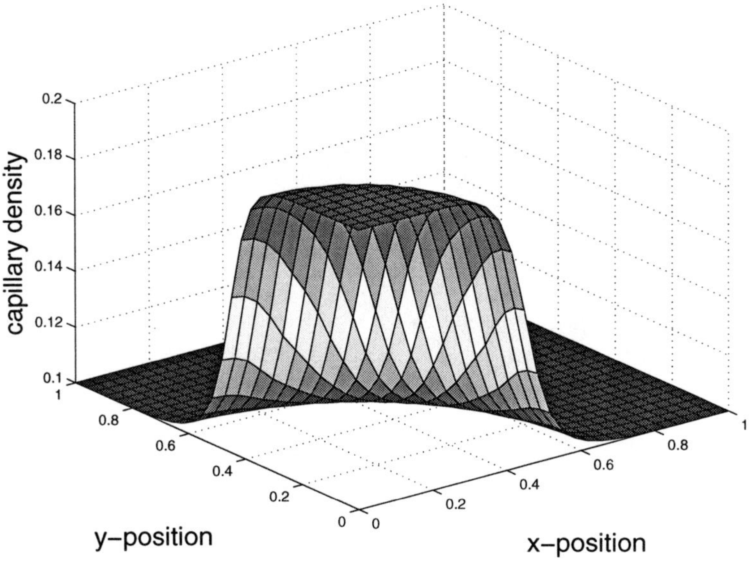

Fig. 19. The capillary density profiles for the model which combines angiogenesis and wound closure

differential equation, can be looked for in the earlier sections of this chapter. In the basis set of the parameter choice, p and q have been assigned equal values.

In Figure 19, an example of a plot of the capillary density profile is shown with an increase in the wound region. This increase facilitates the necessary increased delivery of oxygen and nutrients in order to enhance the wound closure process. Figure 20 shows the case of wound closure in which angiogenesis is assumed to be instantaneous (for the faster case), and in which angiogenesis is a rate-ccontrolling process. To illustrate the influence of the angiogenesis rate on wound closure in terms of several aangiogenesisparameters, we give some results in which the capillary movement rate, increase of capillary equilibrium density, oxygen regeneration coefficient and capillary regeneration rate coefficient are varied.

An increase of the capillary diffusivity increases the capillary transport and hence vessels that are generated within the region underneath the wound 'move' all over the domain. Whereas, if the diffusivity is low, then the newly generated vessels only extend and 'move' locally, that is in the wound region. This is why an increase of the vascular mobility decreases the overall rate of wound closure. This behavior is shown in Figure 21.

The next parameter concerns the increase of the normalized equilibrium capillary density where the number of epidermal cells is low, that is underneath the wound region. For an increased number of capillaries, more oxygen and nutrients reach the epidermal layer such that the healing process is accelerated. This is shown in Figure 22.

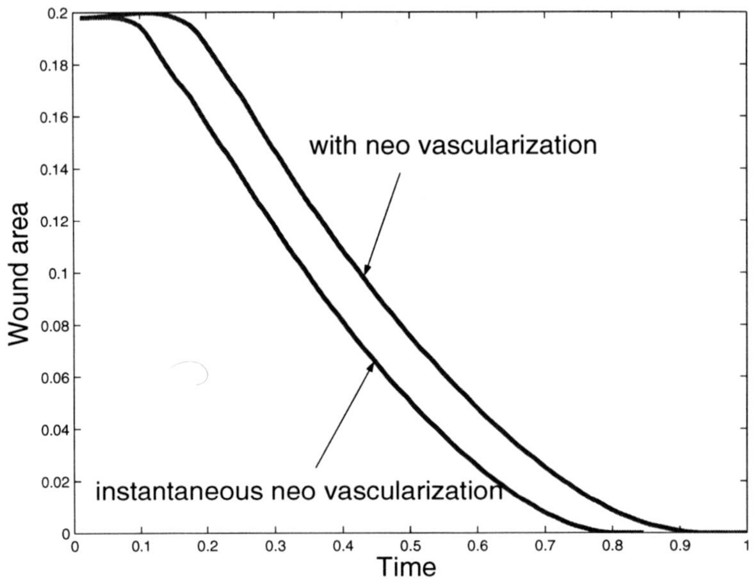

Fig. 20. The wound area as a function of time when the kinetics of angiogenesis are taken into account and when angiogenesis is assumed to proceed instantaneously

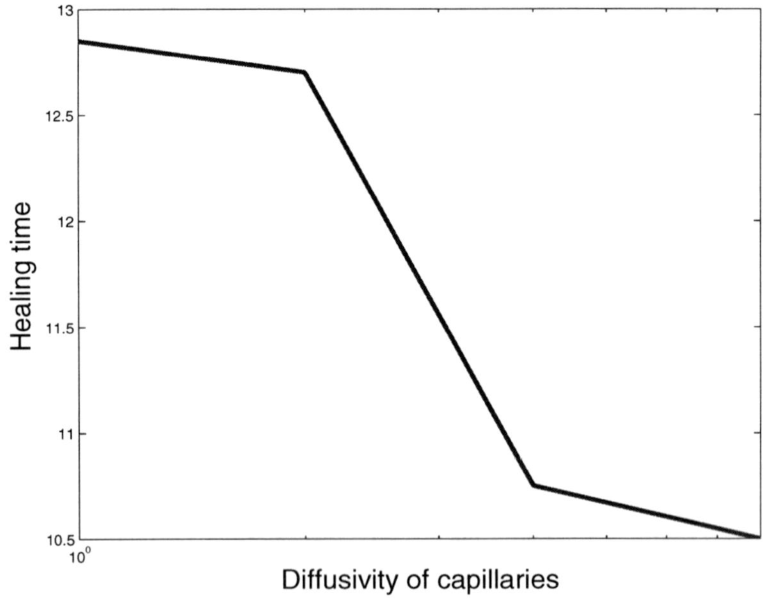

Fig. 21. The nondimensional healing time as a function of the diffusion coefficient of the capillaries

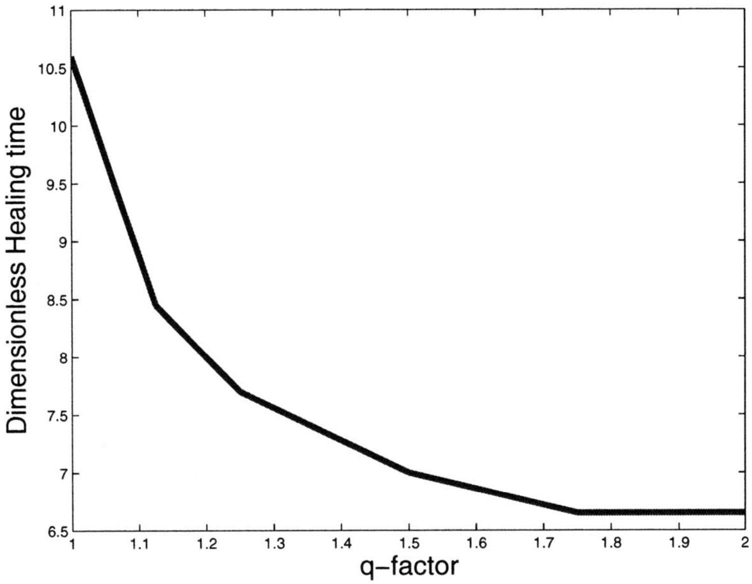

Fig. 22. The nondimensional healing time as a function of the interaction parameter q

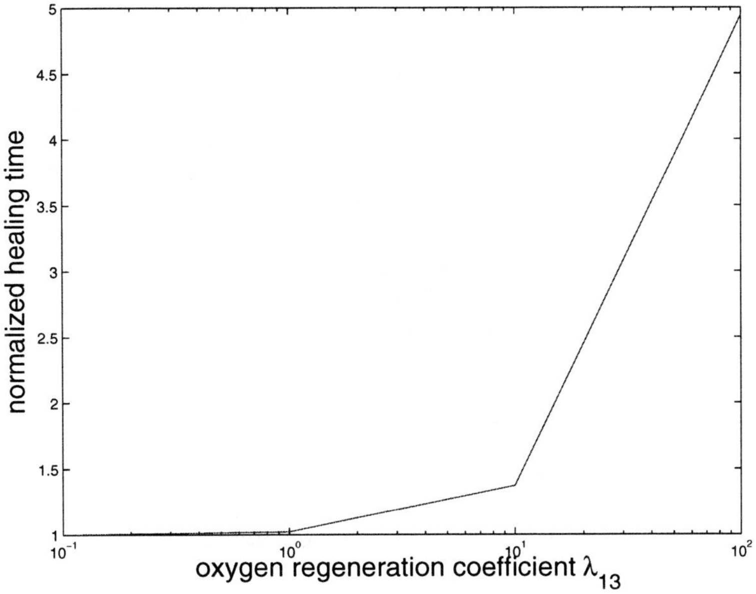

Fig. 23. The nondimensional healing time as a function of the oxygen regeneration coefficient with respect to the capillaries

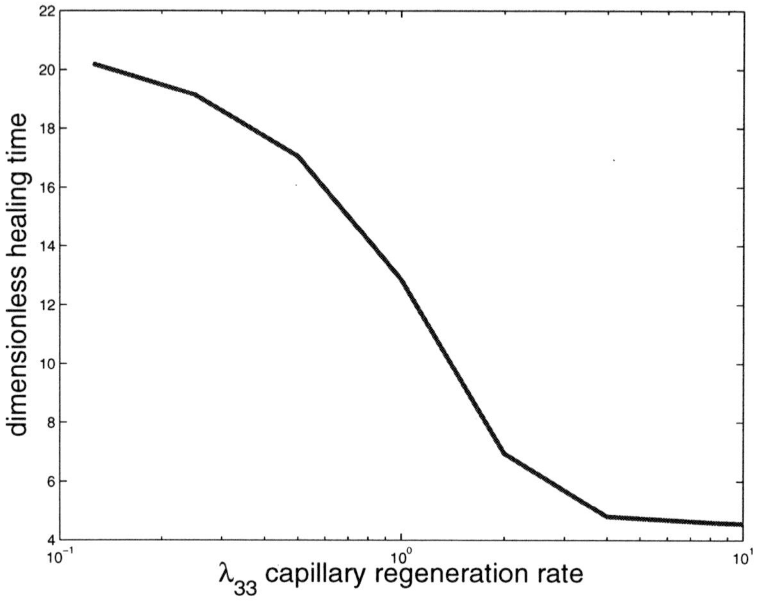

Fig. 24. The nondimensional healing time as a function of the capillary regeneration coefficient

Subsequently, we deal with the oxygen regeneration rate coefficient due to the oxygen flow through the capillaries. The more oxygen is transported and reaches the epidermal layer, the fewer MDGF's are generated and herewith the production rate of capillaries is reduced. Because he throughflow of nutrients will be impaired, wound closure is delayed in the end. This behavior is shown in Figure 23.

Finally, we show some results in which the capillary regeneration rate coefficient is varied. As the capillary regeneration rate increases, the necessary chemicals will enter the wound region easily and quickly, and hence wound closure is accelerated, see Figure 24 for Finite Element results.

6 Discussion and Conclusions

A suite of mathematical models has been described taking into account several partial processes in wound healing. We eemphasizethat the description is far from complete. For instance, Keratinocyte signaling is missing, while this is an important process as well [Wearing and Sherratt 2000]. This process incorporates chemotaxis, that is cell movement as a result of the gradient of a chemical species, which is missing in the current models. The only model that incorporates a chemotaxis-like structure with respect to endothelial cells, is the angiogenesis model due to [Gaffney *et al.* 2002]. The insights from the two-dimensional representation of the wound contraction model due to [Tranquillo and Murray 1992], where only one spatial dimension is taken into account, with respect to bulging of the top part of the skin are new. Further, the coupling between angiogenesis and wound closure is described in [Vermolen 2008] for the first time, as far as we know. In the present formalism, the capillary density increases in the wound region. One could also advocate for an increase around the wound edge only. This is actually modeled in [Gaffney *et al.* 2002]. The combination of Gaffney's model, which naturally gives an increase of the capillary tip density at the wound edge, and Maggelakis' angiogenesis model, which incorporates the lack of oxygen that triggers the vascularization mechanism, probably is an interesting way to go, since both formalisms incorporate crucial partial mechanisms. This will be an issue in future studies on mathematical models for wound healing.

To summarize, we draw the following conclusions:

- The very simplified wound contraction model due to [Tranquillo and Murray 1992] predicts bulging of the upper part of the skin tissue.
- The active layer in the discontinuous switch model due to Adam [Adam 1999] is sustained, or at least not refuted, by the increased capillary tip density resulting, as predicted by [Gaffney *et al.* 2002]. The increased tip density provides an abundant transport of oxygen and of other chemicals for the cells in the vicinity of the wound edge to become active in proliferation and growth factor production.
- The simple model due to Adam possesses the most relevant features of wound healing in terms of wound edge movement.

In a current study, not mentioned in this manuscript, we analyze the relations between wound contraction, angiogenesis and wound closure. Further, we want to improve the formalism of angiogenesis. This process plays a crucial role in wound healing, bone fracture healing and in the growth of tumors.

Another important comment is the lack of experimental validation of the models. In [Murray 2004], a qualitative comparison of simulations of wound contraction due to [Tranquillo and Murray 1992] with experiments is reported on page 510 in [Murray 2004]. The agreement is good from a qualitative point of view. According to [Murray 2004], the agreement could be improved if plasticity effects would be introduced. The model for wound closure due to [Sherratt and Murray 1991] was validated qualitatively with experiments on ears of rabbits in several studies. This comparison was reported in [Sherratt and Murray 1991]. The agreement is reasonable from a qualitative point of view.

For all these models, at least a qualitative validation is necessary. Since the number of parameters involved is very large, and since the values of certain parameters are hard to get, it is desirable to use inverse modeling or optimization in a least squares sense to get a quantitative agreement between experiments and simulations. The optimization strategy needs to be efficient, which also requires an efficient solution of the partial differential equations in the model. For an optimization strategy, model reduction (combining a detailed model with a very simplified model to get biological parameters) might be useful. This will require the use of reliable simplified models for the description of wound healing.

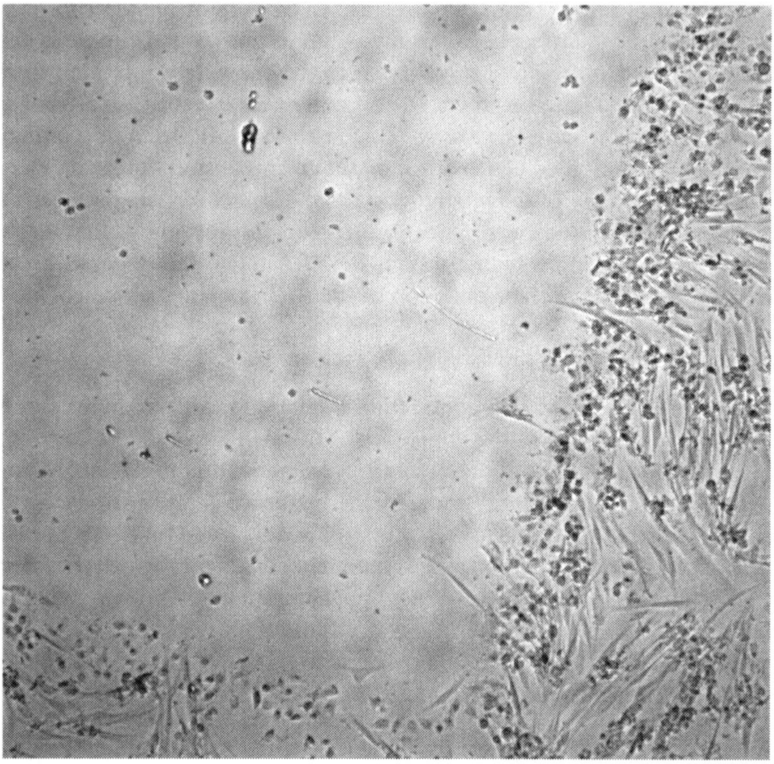

Fig. 25. Photo of a fibroblast cell culture from which cells in a square center were removed. The picture of the in vitro model was taken 16h 40min post-wounding. The fibroblasts near the wound edge get motile and move towards the center of the wound.

One of the reasons of a lack of a validation of the models with experiments, is that wound healing is an extremely complicated biological process. The current models only describe partial processes that occur within wound healing, such as wound closure, wound contraction, and angiogenesis. And even for these partial processes, the models represent a simplified representation. That is why it is hard to validate each model as such. Therefore, we are aiming at a formalism incorporating several partial processes and a coupling between them. To this extent, it is crucially important to understand the current models. In Figure 25, a culture of migrating fibroblasts near a squared wound (24 h after wounding, obtained by impeding the placing of fibroblasts in the wound area during the culturing phase), from a preliminary study, is shown. This work is ongoing research in the experimental laboratory of the Group of Structural Mechanics and Materials Modelling of the Aragón Institute of Engineering Research (I3A) and the University of Zaragoza.

Acknowledgement. The financial support by Senternovem, an agency within the Dutch Ministry of Economic Affairs, in the framework of the IOP self-healing materials is gratefully acknowledged. This support enables the fruitful collaboration between the researchers from the Netherlands and Spain. Further, the authors want to acknowledge the experimental laboratory of the Group of Structural Mechanics and Materials Modelling of the Aragón Institute of Engineering Research (I3A) and the University of Zaragoza for supplying the picture of the in vitro model used to illustrate the healing process (Figure 25).

References

[Huiskes *et al.* 1997] Huiskes, R., van Driel, W.D., Prendergast, P.J., Søballe, K.: A biomechanical regulatory model for periprosthetic fibrous-tissue differentiation. J. Mater. Sc.: Materials in medicine 8, 785–788 (1997)

[Ament and Hofer 2000] Ament, C., Hofer, E.P.: A fuzzy logic model of fracture healing. J. Biomech. 33, 961–968 (2000)

[Andreykiv 2006] Andreykiv, A.: Simulation of bone ingrowth, Thesis at the Delft University, Faculty of Mechanical Engineering (2006)

[Bailon-Plaza and van der Meulen 2001] Bailon-Plaza, A., van der Meulen, M.C.H.: A mathematical framework to study the effect of growth factors that influence fracture healing. J. Theor. Biol. 212, 191–209 (2001)

[LaCroix and Prendergast 2002] LaCroix, D., Prendergast, P.J.: A mechano-regulation model for tissue differentiation during fracture healing: analysis of gap size and loading. J. BioMech. 35(9), 1163–1171 (2002)

[Calvo 2008] Calvo, P.M.: Mathematical modeling and computational simulation of the mechanobiological behavior of bone implants interfaces, thesis at the University of Zaragoza, Spain (2008)

[Martínez *et al.* 2006] Martínez, G., García-Aznar, J.M., Doblaré, M., Cerrolaza, M.: External bone remodeling through boundary elements and damage mechanics. Mathematics and Computers in Simulation 73, 183–199 (2006)

[García *et al.* 2005] García-Aznar, J.M., Rueberg, T., Doblaré, M.: A bone remodeling model coupling microdamage growth and repair by 3D BMU-activity. Biomechanical Models in Mechanobiology 4, 147–167 (2005)

[Doblaré *et al.*] Doblaré, M., García-Aznar, J.M.: On the numerical modeling of growth, differentiation and damage in structural living tissues. Archives of Computatational Methods in Engineering 11(4), 1–44 (2005)

[Doblaré and Garciá 2001] Doblaré, M., Garciá-Aznar, J.M.: Application of an anisotropic bone-remodeling model based on a damage-repair theory to the analysis of the proximal femur before and after total hip replacement. Journal of Biomechanics 34, 1157–1170 (2001)

[García *et al.* 2002] García-Aznar, J.M., Doblaré, M., Cegoñino, J.: Bone remodeling simulation: a tool for implant design. Computational Materials Science 25, 100–114 (2002)

[Fornells *et al.* 2007] Fornells, P., Garciá-Aznar, J.M., Doblaré, M.: A finite element dual porosity approach to model deformation-induced fluid flow in cortical bone. Annals of Biomedical Engineering 35(10), 1687–1698 (2007)

[Doblaré and García] Doblaré, M., García, J.M.: Aisotropic bone remodeling model based on a continuum damage-repair theory. Journal of Biomechanics 35, 1–17 (2002)

[García *et al.* 2007] García, J.M., Martínez, M.A., Doblaré, M.: An anisotropic internal-external bone adaptation model based on a combination of CAO and continuum damage mechanics technologies. Computer Methods in Biomechanics and Biomedical Engineering 4(4), 355–377 (2007)

[Sanz-Herrera *et al.* 2008] Sanz-Herrera, J.A., García-Aznar, J.M., Doblaré, M.: Micro-macro numerical modeling of bone regeneration in tissue engineering. Computational Methods in Applied Mechanical Engineering 197, 3092–3107 (2008)

[Davies 2003] Davies, J.E.: Understanding peri-implant endosseous healing. Journal of Dental Education 67(8), 932–949 (2003)

[Chen *et al.* 2007] Chen, G., Pettet, G.J., Pearcy, M., McElwain, D.L.S.: Modeling external bone adaptation using evolutionary structural optimization. Biomechanical Models in Mechanobiology 6, 275–285 (2007)

[Carpenter and Carter 2008] Carpenter, R.D., Carter, D.R.: The mechanobiological effects of periosteal surface loads. Biomechanical Models in Mechanobiology 7, 227–242 (2008)

[Sherratt and Murray 1991] Sherratt, J.A., Murray, J.D.: Mathematical analysis of a basic model for epidermal wound healing. J. Math. Biol. 29, 389–404 (1991)

[Filion and Popel 2004] Filion, J., Popel, A.P.: A reaction diffusion model of basic fibroblast growth factor interactions with cell surface receptors. Annals of Biomed. Eng. 32(5), 645–663 (2004)

[Maggelakis 2003] Maggelakis, S.A.: A mathematical model for tissue replacement during epidermal wound healing. Appl. Math. Modell. 27(3), 189–196 (2003)

[Gaffney *et al.* 2002] Gaffney, E.A., Pugh, K., Maini, P.K.: Investigating a simple model for cutaneous wound healing angiogenesis. J. Math. Biol. 45(4), 337–374 (2002)

[Plank and Sleeman 2003] Plank, M.J., Sleeman, B.D.: A reinforced random walk model of tumour angiogenesis and anti-angiogenic strategies. Mathem. Medic. and Biol. 20, 135–181 (2003)

[Plank and Sleeman 2004] Plank, M.J., Sleeman, B.D.: Lattice and non-lattice models of tumour angiogenesis. Bull. Mathem. Biol. 66, 1785–1819 (2004)

[Murray 2004] Murray, J.D.: Mathematical biology II: spatial models and biomedical applications. Springer, New York (2004)

[Maggelakis 2004] Maggelakis, S.A.: Modeling the role of angiogenesis in epidermal wound healing. Discr. and Cont. Sys. 4, 267–273 (2004)

[Adam 1999] Adam, J.A.: A simplified model of wound healing (with particular reference to the critical size defect). Math. and Comput. Modell. 30, 23–32 (1999)

[Vermolen and Adam 2007] Vermolen, F.J., Adam, J.A.: A Finite Element Model for Epidermal Wound Healing. In: Shi, Y., van Albada, G.D., Dongarra, J., Sloot, P.M.A. (eds.) ICCS 2007. LNCS, vol. 4487, pp. 70–77. Springer, Heidelberg (2007)

[Vermolen *et al.* 2007] Vermolen, F.J., van Rossum, W.G., Javierre, E., Adam, J.A.: Modeling of self-healing of skin tissue. In: Self-healing materials an alternative approach to 20 centuries of materials science, pp. 337–364. Springer, Dordrecht (2007)

[Stadelman *et al.* 1997] Stadelman, W.K., Digenis, A.G., Tobin, G.R.: Physiology and healing dynamics of chronic cutaneous wounds. The American Journal of Surgery 176(2), 265–385 (1997)

[Olsen *et al.* 1995] Olsen, L., Sherratt, J.A., Maini, P.K.: A mechanochemical model for adult dermal wound closure and the permanence of the contracted tissue displacement role. J. Theor. Biol. 177, 113–128 (1995)

[Murray 2003] Murray, J.D.: On the mechanochemical theory of biological pattern formation with application to vasculogenesis. Biol. Model. 326, 239–252 (2003)

[Wearing and Sherratt 2000] Wearing, H.J., Sherratt, J.D.: Keratinocyte growth factor signalling: a mathematical model of dermal-epidermal interaction in epidermal wound healing. Math. Biosc. 165, 41–62 (2000)

[Rossiter *et al.* 2004] Rossiter, H., Barresi, C., Pammer, J., Rendl, M., Haigh, J., Wagner, E.F., Tschachler, E.: Loss of vascular endothelial growth factor A activity in murine epidermal keratinocytes delays wound healing and inhibits tumor formation. Cancer Research 64, 3508–3516 (2004)

[Alarcon *et al.* 2006] Alarcon, T., Byrne, H., Maini, P., Panovska, J.: Mathematical modeling of angiogenesis and vascular adaptation. In: Paton, R., McNamara, L. (eds.) Studies in multidisciplinary, vol. 3, pp. 369–387 (2006)

[Balding and McElwain 1985] Balding, D., McElwain, D.L.S.: A mathematical model of tumour-induced capillary growth. Journal of Theoretical Biology 114, 53–73 (1985)

[Mantzaris *et al.* 2004] Mantzaris, N.V., Webb, S., Othmer, H.G.: Mathematical modeling of tumor-induced angiogenesis. Journal of Mathematical Biology 49, 111–187 (2004)

[Vermolen 2008] Vermolen, F.J.: A simplified finite element model for tissue regeneration with angiogenesis. ASCE Journal of Engineering Mechanics 135(5) (2009)

[Lamme 1999] Lamme, E.N.: Artificial skin and tissue regeneration, Thesis, The University of Amsterdam, the Netherlands (1999)

[Friesel & Maciang 1995] Friesel, R.E., Maciang, T.: Molecular mechanisms of angiogenesis: fibroblast growth factor signal transduction. FASEB J. 9, 919–925 (1995)

[Stoletov *et al.* 2002] Stoletov, K.V., Ratcliffe, K.E., Terman, B.I.: Fibroblast growth factor receptor substrate 2 participates in vascular endothelial growth factor-induced signaling. FASEB J. 16, 1283–1285 (2002)

[Adam 1999] Adam, J.A.: A simplified model of wound healing (with particular reference to the critical size defect). Math. and Comput. Modell. 30, 23–32 (1999)

[Adam 2004(1)] Adam, J.A.: The effect of surface curvature on wound healing in bone. Applied Mathematics Letters 15, 59–62 (2002)

[Adam 2004(2)] Adam, J.A.: Inside mathematical modelling: building models in the context of wound healing in bone. Discrete and continuous dynamical systems-series B 4(1), 1–24 (2004)

[Arnold 2001] Arnold, J.S.: A simplified model of wound healing III: The critical size defect in three dimensions. Mathematical and Computer Modelling 34, 385–392 (2001)

[Javierre *et al.* 2008] Javierre, E., Vermolen, F.J., Vuik, C., van der Zwaag, S.: A mathematical approach to epidermal wound closure: Model Analysis and Computer Simulations. J. Mathematical Biology (2008) (to appear), `http://www.springerlink.com/content/w4j6633345j7228k/fulltext.pdf`, doi:10.1007/s00285-008-0242-7

[Vermolen *et al.* 2006] Vermolen, F.J., van Baaren, E., Adam, J.A.: A simplified model for growth factor induced healing of circular wounds. Mathematical and Computer Modeling 44, 887–898 (2006)

[Britton and Chaplain 1993] Britton, N.F., Chaplain, M.A.J.: A qualitative analysis of some models of tissue growth. Mathematical Biosciences 113, 77–89 (1993)

[Adam 1987] Adam, J.A.: A simplified model for tumor growth. Mathematical Biosciences 87, 229–244 (1987)

[Shymko and Glass 1976] Shymko, R.M., Glass, L.: Cellular and geometric control of tissue growth and mitotic instability. Journal of Theoretical Biology 63, 355–374 (1976)

[Hogea *et al.* 2006] Hogea, C.S., Murray, B.T., Sethian, J.A.: Simulating complex tumor dynamics from avascular to vascular growth using a general level-set method. Journal of Mathematical Biology 53, 86–134 (2006)

[Vermolen and Javierre 2009] Vermolen, F.J., Javierre, E.: On the construction of analytic solutions for a diffusion-reaction equation with a discontinuous switch mechanism. Journal of Computational and Applied Mathematics (to appear, 2009), doi:10.1016/j.cam.2009.05.022

[Zemskov *et al.* 2009] Zemskov, S., Vermolen, F.J., Javierre, E.: An advanced finite element method for wound healing and crack closure (in progress, 2009)

[Hundsdorfer 2003] Hundsdorfer, W., Verwer, J.G.: Numerical solution of time-dependent advection-diffusion-reaction equations. Series in Computational Mathematics. Springer, Heidelberg (2003)

[Javierre *et al.* 2009] Javierre, E., Moreo, P., Doblaré, M., García-Aznar, M.J.: Computational Modelling of wound contraction. In: Proceedings of the Congreso de Métodos Numéricos en Ingeniería 2009, Barcelona. SEMNI (2009)

[de Vries *et al.* 2006] de Vries, G., Hillen, T., Lewis, M., Müller, J., Schönfisch, B.: A course in mathematical biology: Quantitative modeling with mathematical and computational methods. SIAM, Philadelphia (2006)

[Sachdev 1987] Sachdev, P.L.: Nonlinear diffusive waves. Cambridge University Press, Cambridge (1987)

[Tranquillo and Murray 1992] Tranquillo, R.T., Murray, J.D.: Continuum model of fibroblast-driven wound contraction inflammation-mediation. Journal of Theoretical Biology 158(2), 135–172 (1992)

[Pettet *et al.* 1996] Pettet, G.J., Byrne, H.M., McElwain, D.L.S., Norbury, J.: A model of wound healing angiogenesis in soft tissue. Mathematical Biosciences 136, 35–63 (1996)

MRI Integrated with Computational Methods for Determining Internal Soft Tissue Loads as Related to Chronic Wounds

Sigal Portnoy[1], Nogah Shabshin[2], Itzhak Siev-Ner[3], Anat Kristal[3], and Amit Gefen[1]

[1] Department of Biomedical Engineering, Tel Aviv University
[2] Department of Diagnostic Imaging
[3] Department of Orthopaedic Rehabilitation, Chaim Sheba Medical Center
gefen@eng.tau.ac.il

Abstract. Excessive and prolonged internal soft tissue loads are one of the main factors responsible for the initiation of internal injuries that may, if ignored or untreated, escalate into chronic wounds. Since internal tissue loads cannot be measured *in vivo*, computational methods that incorporate the actual anatomy of the living body, are currently the best available resource for acquiring internal mechanical knowledge. In this chapter we discuss various methods that use computational modeling integrated with anatomical data, scanned by magnetic resonance imaging (MRI) in order to determine internal soft tissue loads. Specifically we will elaborate on linear and non-linear finite element (FE) methods and hyperelastic warping.

1 Introduction

Since first discovering X-rays in 1895, imaging techniques have greatly evolved, expanding our insight by answering common daily questions ranging from what is in our airport luggage to the existence and whereabouts of a brain tumor. With the growing accessibility to advanced imaging technologies, accurate non-invasive visualization of internal structures of the living body became a widely-used stepping stone for clinical diagnostics, semi-invasive surgeries and subject-specific modeling in research facilities. In the research community, imaging techniques such as computerized tomography (CT), ultrasound (US) and magnetic resonance imaging (MRI) are routinely incorporated into studies of the of interaction between hard tissue, i.e. bones and soft tissues, e.g. muscle, fat and connective tissues. These imaging techniques enable three-dimensional (3D) modeling of the anatomy of the scanned body, therefore allowing the creation of an exact subject-specific computational model of an organ or a body structure. Mimicking real-life anatomy and physiology in the virtual environment enables a unique and powerful insight into mechanical processes *in vivo* and prediction of future outcomes of possible surgical procedures.

Magnetic resonance imaging is less commonly used in research laboratories, mainly because of cost issues. However the advantages of MRI are numerous.

A. Gefen (Ed.): Bioengineering Research of Chronic Wounds, SMTEB 1, pp. 169–180.
springerlink.com

Unlike most imaging techniques, e.g. X-ray and CT, MRI does not involve ionizing radiation. Furthermore, when compared to other unharmful imaging methods, e.g. US, the accuracy and resolution of MRI surpasses them all.

Early studies that focused on the aetiology of chronic wounds, e.g. pressure ulcers (PU) and deep tissue injury (DTI), aspired to investigate contact surface pressures as well as internal soft tissue loads. Pressure related wounds are caused mainly by excessive and prolonged tissue deformation and can initiate at the skin surface, i.e. PU, or in soft tissues in the vicinity of bony prominences, i.e. DTI. When investigating internal pressure-related complications such as DTI or disk wear in a joint, measurements of contact pressures on the skin interface are futile. Accordingly, early investigations of internal injuries were mostly limited to physical internal pressure measurements, acquired by means of pressure sensors embedded in anesthetized animals (Le et al, 1984), synthetic phantoms or cadavers. Later, computational capabilities enabled limited simplistic computational models with poor geometric fidelity. Fortunately, current studies have greatly evolved in the last few years, producing publications of highly validated methods for calculating internal soft tissue loads.

In this chapter we will present several prevalent MRI-based methods for evaluating internal loads in soft tissues as related to chronic wounds. A schematic description of the experimental-computational approach to assess the internal mechanical soft tissue loads is outlined in Fig. 1: the MRI scans may be acquired in the common tube-like clinical MRI construction, usually with the field strength of up to 3T. Another option is the open-MRI setting that incorporates an opening at both sides of the MRI table, where the scanning takes place, and is usually dedicated to brain surgery and claustrophobic patients. The open-MRI is best suitable for trials that require natural sitting or standing during the scan. The field strength of the open-MRI is usually 1.5T. There are MRI research systems that have field strengths reaching 9.4T for human studies and 21T for animal studies. For research purposes, scans are acquired with and without applied body weight (Linder-Ganz et al, 2007) or in various positions, e.g. different knee flexion angles (Besier et al, 2008), or creating a load by an external indenter (Stekelenburg et al, 2006) ultimately resulting in "loaded" and "unloaded" images in the studied region of interest (ROI). These images are used as a post-processing mean to create a 3D geometrically-accurate model of the ROI. Each tissue in the model is then attributed data regarding its mechanical behavior, usually literature-based data relying on empirical observations. The applied load / deformation are used as stimulation for the model and the distribution of internal loads is obtained. Validation of the results is performed by comparing deformations produced in the model by a given load to the measured deformations seen in the MRI scans or vice versa, by comparing surface loads produced in the model by a given displacement of the bones or an external indenter, to measured interface loads.

The finite element (FE) method is currently the most predominant tool for approximation of the mechanical conditions in 3D anatomically-accurate models. The FE method is basically a numerical computation tool of the mechanical state of a complex object, which is broken apart into a grid of geometrically simple

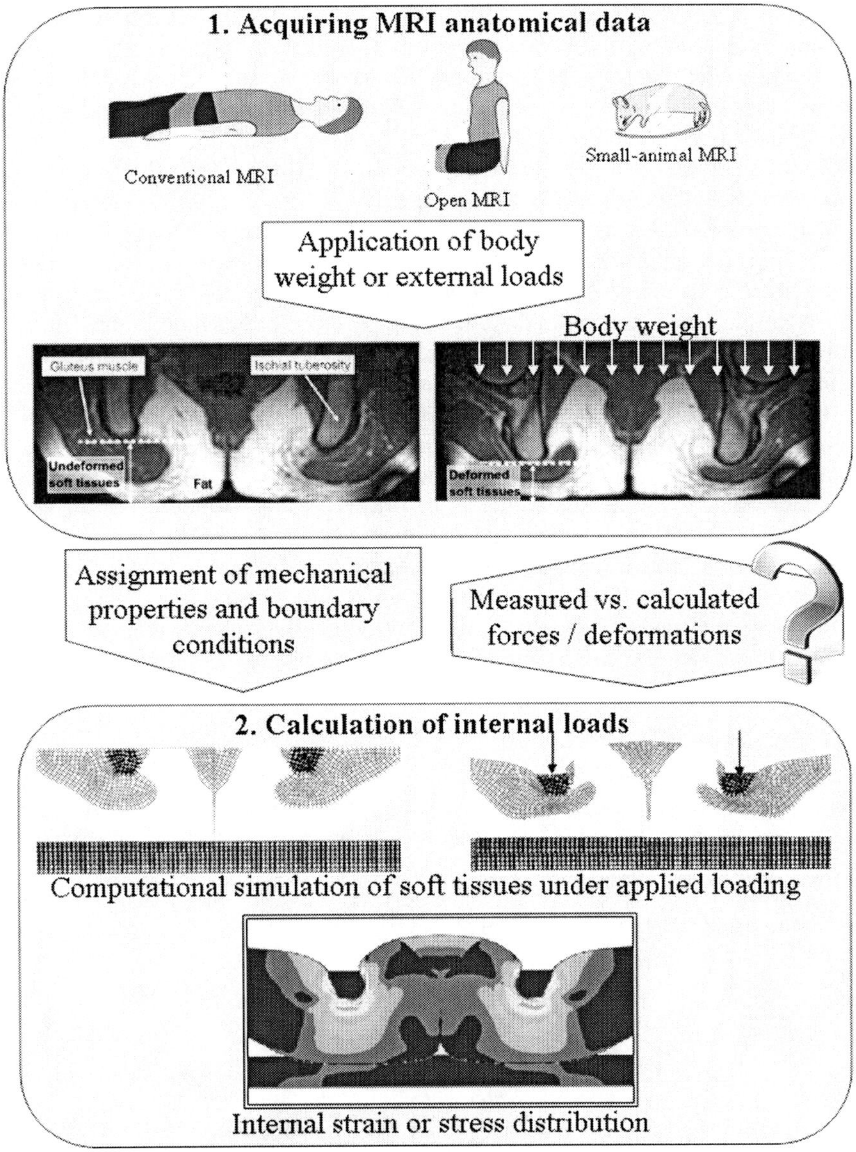

Fig. 1. Schematic description of the MRI-based experimental-computational methods for evaluating internal mechanical loads in soft tissues

elements, called a mesh. The geometry of the model is usually predesigned in mechanical computer-aided-design (CAD) programs, e.g. SolidWorks and CATIA. Solid models can then be exported to a commercial FE software, e.g. ABAQUS, NASTRAN and ANSYS, and assigned be material properties and boundary conditions. The FE method, previously restricted to eronautics, automotive and defense industries due its requirement of high computational power, has in the last decade become attainable for biomedical engineering research. Accordingly, studies incorporating MRI-based geometry with FE analyses (termed coupled MRI-FE; Linder Ganz et al, 2008) have set out to investigate internal loads in cartilages and disks (Knecht et al, 2008; Besier et al, 2008; Tanaka et al, 2004; Kumaresan et al, 2009), internal shear stresses in blood vessels (Funamoto et al, 2009), cardiac deformation (Papademetris et al, 2002) and more (Noakes et al, 2008). Specifically several animal studies (Stekelenburg et al, 2006) as well as human studies (Gefen 2002; Spears et al, 2007; Linder-Ganz et al 2007; Makhsous et al 2007) demonstrated the evaluation of *in vivo* internal mechanical states that inflict internal injuries in soft tissues, i.e. muscle, fat and fascia. Theses studies hold in their grasp the future of subject-specific clinical evaluation and surgical planning, expanding our understanding of numerous pathologies and physiological states.

Although greatly evolved, the methods used today for analyses of internal soft tissue strains and stresses are not yet fully perfected. When concerning the soft tissue mechanics, an optimal computational model should take into account the following considerations (Gallagher et al, 1982): large deformations of the soft tissues, their viscoelastic behavior and the interactive nature of the macroscopic behavior with the constituents at the microscopic level (Breuls et al, 2002). Since generally, these conditions cannot be fully met with current computational tools, assumptions are made. The next sections introduce the fundamentals of the linear and non-linear coupled MRI-FE methods and the concept of hyperelastic warping.

2 Coupled MRI-FE Linear Models

The field equations for solid mechanics are given by equilibrium behavior, constitutive equations and boundary conditions. For small deformations, where the strains are $|\varepsilon_{ij}| << 1$ and the rotation $|\omega_{ij}^2| << \|\varepsilon_{ij}\|$ the displacement gradient is expressed by (Zienkiewicz and Taylor, 2005):

$$\frac{\partial u_i}{\partial x_j} = \varepsilon_{ij} + \omega_{ij} \tag{2.1}$$

where λ and μ are Lamé elastic paramenters and δ_{ij} is the Kronecker delta function. The strains arising from sources other then displacement are denoted by $\varepsilon_{ij}^{(0)}$.

Using a strain energy density given by:

$$W = \frac{1}{2}\varepsilon_{ij} C_{ijkl} \varepsilon_{kl} - \varepsilon_{ij} C_{ijkl} \varepsilon_{kl}^{(0)} \tag{2.2}$$

where C_{ijkl} are elastic moduli, we can obtain the Cauchy stresses:

$$\sigma_{ij} = C_{ijkl}\left(\varepsilon_{kl} - \varepsilon_{kl}^{(0)}\right) \tag{2.3}$$

Assuming isotropic linear elastic material, the Cauchy stresses are:

$$\sigma_{ij} = \lambda\delta_{ij}\left(\varepsilon_{kk} - \varepsilon_{kk}^{(0)}\right) + 2\mu\left(\varepsilon_{ij} - \varepsilon_{ij}^{(0)}\right) \tag{2.4}$$

Studies incorporating actual anatomical MRI-based data with small deformation, linear calculations of internal tissue loads, may be applicable for comparing between postures or studying the effect of an isolated parameter in the model. Unfortunately, the absolute values of the loads obtained by small deformation analyses in biological tissues are unreliable due to the high strains that reside in soft tissue structure, and the viscoelasticity that governs their behavior. In order to analyze deep tissue loads more accurately, large deformation non-linear analyses must be employed.

3 Coupled MRI-FE Non-linear Models

Recently, the coupled MRI-FE linear models have been upgraded towards consideration of the non-linear behavior of biological tissues and the large deformations occurring in soft tissues. The most general type of nonlinear elastic behavior is the hyperelastic model, in which it is assumed that there is a strain energy density, W, from which stresses S_{ij} are defined as (Zienkiewicz and Taylor, 2005):

$$S_{ij} = \frac{\partial W}{\partial E_{ij}} \tag{3.1}$$

where E_{ij} is the tensor of large strains.

Assuming an isotropic material, the strain energy density of the hyperelastic material can generally be calculated as:

$$W = \sum_{i+j=1}^{N} C_{ij}(I_1 - 3)^i (I_2 - 3)^j + \sum_{i=1}^{N} \frac{1}{D_i}(J-1)^{2i} \tag{3.2}$$

with $N \le 6$, where D_i are the temperature-dependant material parameters, and J is the relative volume change so that $J = \prod_{i=1}^{3} \lambda_i$. Also, I_1 and I_2 are the first and second deviatoric strain invariants defined using the principal stretched λ_i so that $I_1 = \sum_{i=1}^{3} \lambda_i^2$ and $I_2 = \sum_{i=1}^{3} \lambda_i^{-2}$.

Several common forms of the hyperelastic strain energy densities are deduced from equation 3.2. For example, when substituting N=1 in equation 3.2, the Mooney-Rivlin form is obtained (Mooney, 1940).

Owing to the increasing computer power, anatomically-accurate 3D FE modeling became a powerful tool, which is available in research facilities worldwide.

Our research group has been applying the MRI-FE non-linear method for the purpose of studying internal soft tissue loads that arise in wheelchair-users and lower-limb prosthetic-users. When investigating the internal load distribution under the ischial tuberosities of wheelchair-users, early linear coupled MRI-FE models, e.g. Todd and Tacker (1994), paved the way for MRI-based non-linear FE models (Linder-Ganz et al, 2007; Linder-Ganz et al, 2008; Elsner and Gefen, 2008) that evaluated internal strains and stresses in skeletal muscle tissue at the proximity of the ischial tuberosities, where the highest tissue loads accumulate. The successful characterization of internal strain and stress distributions in the soft tissues of the buttock motivated application of the MRI-based non-linear 3D FE modeling towards understanding of the load patterns in the residuum of trans-tibial amputation (TTA) prosthetic-users (Portnoy et al, 2008). The implementation of the MRI-FE methods to evaluate deep tissue loads in the residual limbs of prosthetic-users will be detailed herein to illustrate its use.

The muscle flap in the TT residuum is compressed between the truncated bones, i.e. the tibia and fibula, and the hard socket of the prosthetic leg. The residuum is therefore prone to injuries resulting from the prosthetic socket/residuum interaction, e.g. blisters and PU, and also DTI at the distal end of the truncated bones. The injuries may progress undetected by the prosthetic-user in cases where neuropathic comorbidity is present. While commercial prosthetic accessories and aids are plentiful, tools that analyze the internal mechanical conditions in the TTA residuum during load-bearing are scarce. The residuum of each TTA patient differs anatomically and mechanically due to different surgical techniques, level of amputation, and muscle atrophy. The inter-patient variability of the TTA residual limbs affects DTI onset and progression rate. Accordingly, the reality of the different TTA surgical outcomes necessitates advanced computational tools that evaluate patient-specific internal mechanical states in soft tissues of the residuum.

For this purpose, we employed a patient-specific MRI-based non-linear 3D FE modeling of the residual limbs of two male traumatic longtime prosthetic-users, with no known co-morbidities or limb pain (see Table 1 for patient information). Helsinki approval (#4302/06 from Sheba Medical Center, Ramat-Gan, Israel) and informed consents were obtained before each trial. The experimental-computational method is thoroughly detailed in Portnoy et al (2008) and will be briefly discussed herein to illustrate the computational method described in this section.

In order to acquire 3D anatomical data of the residuum during load-bearing, we created a plaster cast replica of the residual limb of each TTA patient (Fig. 2a). The cast was used as a metallic-free container for the residuum, mimicking the prosthetic socket, during the MRI scanning procedure. Trials took place in an open-MRI ("Signa SP'' model, General Electric Co., CT, USA) where two large openings at the sides of the scanning table enable erect patient positions during sitting and standing. In our trials, the patients stood on the MRI scanning table (Fig. 2a) without applying load to their residuum. Then the patients were instructed to apply comfortable unpainful bodyweight loading to their residual limb. The resulting interface pressures (see Table 1) were measured using thin and flexible MRI-compatible force sensors (FlexiForce, Tekscan Co. MA, USA).

Two 3D solid models were constructed (SolidWorks 2009, SolidWorks, MA, USA) using the 4mm-spaced axial MRI scans (Fig. 2a,b). The 3D solid models consisted of the tibia, fibula, muscle flap, fat and cast container. The tibial length, residuum length and tibial bevelment measurements were acquired from these models (Table 1).

Table 1. Patient information for both participants, geometrical characteristics of the residuum, the measured downward displacements of the bones relative to the most distal point of the limb at the scans with and without load-bearing, and the measured interface pressures applied by each patient.

	Patient	
	#1	#2
Age [yrs]	49	63
Weight [kg]	70	64
Time since transtibial amputation [yrs]	29	41
Downward displacement of tibia [mm]	1.8	1.2
Measured interface pressures [kPa]	92±9	330±74
Tibial length [cm]	5.8	6.8
Residuum length [cm]	8.1	7.6
Tibial bevelment [°]	0	18

The 3D models were then exported to a non-linear large deformation FE solver (ABAQUS v. 6.8, SIMULIA, RI, USA). A 2mm thick skin membrane was added to the outer soft tissue layer. All the 3D bodies of the FE model were meshed (Fig. 2b) using second-order 10-node modified quadratic tetrahedron elements ("C3D10M" in ABAQUS). The skin was meshed with 6-node quadratic triangular membrane elements (''M3D6"). The coronal MRI scans of the loaded and unloaded residual limb were used to measure the downward displacement of the bones, relative to the most distal point of the limb (Table 1). This displacement was used as a boundary condition applied to the FE bones. The exterior surface of the cast was constrained for all translations and a friction coefficient of 0.7 was set between the skin and the cast. All of the internal surfaces, i.e. fat to muscle and muscle to bone interfaces, were tied together.

The internal compression strains [%] in an oblique view, capturing both the tibia and the fibula, are depicted in Fig. 2c. For both TTA patients, the compression strains under the tibia were higher than compression strains under the fibula. Patient #1 had a flat tibial end (0° bevelment) and a bulkier muscle flap. The compression strains in the residuum of patient #1 dispersed to the distal outlying area of the residuum. The compression strains under the beveled tibia of patient #2 with the thinner soft tissue cushioning under his truncated bones showed a narrow pattern of high strains in a smaller region directly under the tibia (Fig. 2c).

The limitations of this study are mostly related to the neglect of scar and fibrotic tissues that affect the internal strain distribution. Also, the truncated bones in the TTA residuum may slide on top of the muscle flap. This sliding was not

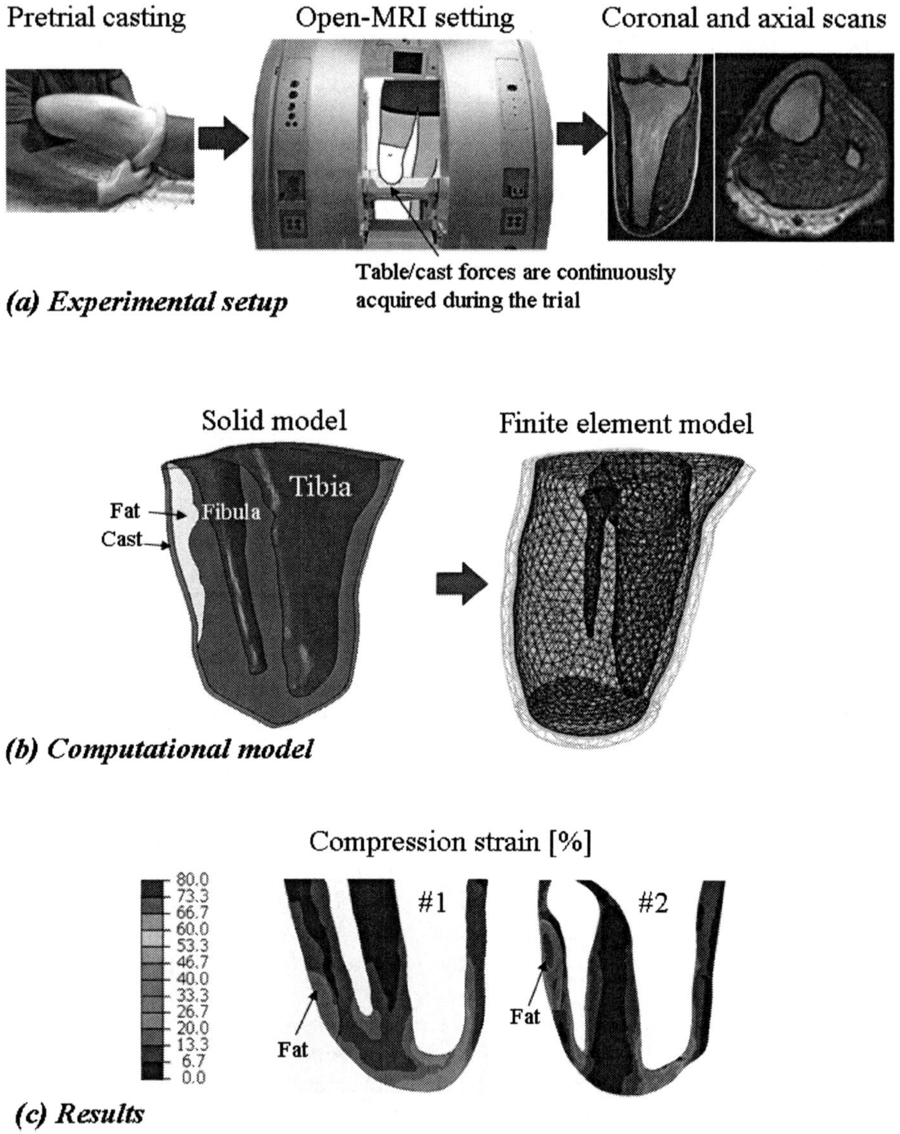

Fig. 2. (a) Axial and coronal MRI scans of the confined transtibial amputation residuum with and without load bearing were obtained in an open-MRI setting. The scans were used to create the (b) geometry of the residuum, and define boundary conditions for the finite element models for (c) calculation of internal soft tissue strains.

included in our analyses as the constraints of this surface were set with a "tie" connection. Finally, the plaster cast replica of the residuum, used in each of our trials, does not fully represent the TTA prosthetic socket. Most TTA patients use a patella-tendon-bearing (PTB) prosthesis, which is indented in the patellar tendon and behind the knee in order to allow shifting of loads from the distal end of the residual limb towards the more pressure-tolerant areas around the knee. The use of the uncertified plaster cast may have resulted in overestimation of the strains.

In summary, it appears that the non-beveled tibia of patient #1 may have created a more uniform distribution of internal strains, thereby theoretically reducing the risk for DTI. The more inclined tibia of patient #2 caused a dangerously high strain concentration in its proximity. Nevertheless, this finding may be considered relevant only to vertical loading of the residual limb, in the standing position studied herein.

4 Hyperelastic Warping

Hyperelastic warping was first introduced by Rabbitt et al (1995) as a FE tool that combines image-based data with the mechanics of a deforming hyperelastic material under unknown forces. This research group implemented MRI-based warping technique to predict left ventricular strain distributions during diastole (Veress et al, 2005) and during systole (Phatak et al, 2009) as well as for determining medial collateral knee ligament strains during knee flexion (Phatak et al, 2007). The full algorithm of hyperelastic warping is detailed in Veress et al (2005) and its outline will be briefly described in the following example. In one of their recent publication, Phatak et al (2007) performed MRI scans to ten cadaveric knees, during various loading scenarios. The geometry of the medial collateral ligament was manually segmented from the MRI scans and meshed in an FE software (TrueGrid, XYZ Scientific, Livermore, CA). A transversely isotropic hyperelastic constitutive model was used to describe the ligament, containing collagen fibers embedded in an isotropic matrix.

The warping algorithm was integrated into the FE code. The concept of the algorithm is to optimize the alignment of a FE deformed state of the initial "template" model with its resulting fixed "target" image. The deformed coordinates are described by $\varphi(X) = X + u(X)$ where $u(X)$ is the displacement field. The deformation gradient $F(X)$ is:

$$F(X) = \frac{\partial \varphi(X)}{\varphi X} \tag{4.1}$$

Accordingly, the Cauchy-Green deformation tensor is $C = F^T F$. The optimization is achieved by minimizing an energy functional E, that consist of an image-based potential energy density U, and the hyperelastic strain energy energy W:

$$E(\varphi) = \int_\beta W(X, C)dv + \int_\beta U(T(X), S(\varphi))dv \tag{4.2}$$

The image intensities of the template and target images are $T(X)$ and $S(\varphi)$, respectively. The template image deforms with respect to the target. The image energy is represented as:

$$U(T(X), S(\varphi)) = \frac{\lambda}{2}(T(X) - S(\varphi))^2 \tag{4.3}$$

where λ is a penalty parameter that enforces reference to target alignment. The image-based force depends on the template and target intensities and their gradients. Finally, the Cauchy stress tensor is obtained:

$$\sigma = \frac{1}{J} F \frac{\partial W}{\partial C} F^T \tag{4.4}$$

where $J = \det(F) = \rho_o / \rho$, and ρ_0, ρ are densities in the reference and deformed configurations, respectively. Model validation of the aforementioned example was accomplished by comparing the warping algorithm strain predictions with measurements obtained from experimental measurements. For this purpose, contrast markers were attached along the visible fiber direction and the strains were measured by a 3D motion analysis system, resulting in overall correlation of R^2=0.81. Sensitivity analyses yielded no change in the code prediction as a result of different fiber stiffness values. The authors reported stretch from 0.97 to 1.05 during passive knee flexion from 0º to 90º and concluded that hyperelastic warping is applicable for noninvasive evaluation of *in vivo* soft tissue loads.

5 Discussion

Experimental-computational methods that incorporate MRI into precise subject-specific models are increasingly improved for the evaluation of internal soft tissue loads. Despite the great potential and proven validity of this powerful tool, MRI-based models are not used in current clinical care facilities for evaluation of the condition of patients and for discerning between different treatment options. This is mainly due to the time-consuming non-automatic current methods for constructing these models. The necessity of an automotive pre-processor that can rapidly manufacture an input file of a FE mesh using MRI data was previously addressed by Todd and Wang (1996) who wrote a Visual Basic code that automatically generated a 3D FE mesh from 2D MRI scans. Current commercial softwares exist (e.g. Mimics or Simpleware) that process MRI images into solid 3D models and export the models to an FE software.

Currently, the creation of an MRI-based subject-specific FE model requires mechanics-trained personnel to build the model and perform the analysis, which can take several hours to days. Additionally, clinical decision-makers should be guided regarding the visualization and comprehension of the analyses results. The implications of clinically-known pathologies on internal soft tissue loads must be thoroughly studied in order to fully characterize guidelines for patient diagnostics that are based on internal strain quantification.

Although greatly evolved, there still exists the challenge of perfecting the MRI-based computational models to represent the soft tissues in the ROI, for all its diversity, to included muscle fibers representation and orientation, subject-specific assignment of mechanical properties of soft tissues, inhomogeneity and so on.

A deeper understanding of the formation of internal loads in these soft tissues is essential for both clinical evaluation of the patient and devising the treatment outline to sustain the health and viability of internal soft tissues. The loads calculated using these models can be further utilized for extrapolation of the biologic damage they inflict. Future work will undoubtedly use the methods described in this chapter to eventually determine a soft tissue injury threshold that will improve clinical decision-making and enable better prevention guidelines in the field of chronic wounds.

References

Besier, T.F., Gold, G.E., Delp, S.L., Fredericson, M., Beaupre, G.S.: The influence of femoral internal and external rotation on cartilage stresses within the patellofemoral joint. J. Orthop. Res. 26, 1627–1635 (2008)

Breuls, R.G., Sengers, B.G., Oomens, C.W., Bouten, C.V., Baaijens, F.P.: Predicting local cell deformations in engineered tissue constructs: a multilevel finite element approach. J. Biomech. Eng. 124, 198–207 (2002)

Elsner, J.J., Gefen, A.: Is obesity a risk factor for deep tissue injury in patients with spinal cord injury? J. Biomech. 41, 3322–3331 (2008)

Funamoto, K., Suzuki, Y., Hayase, T., Kosugi, T., Isoda, H.: Numerical validation of MR-measurement-integrated simulation of blood flow in a cerebral aneurysm. Ann. Biomed. Eng. 37, 1105–1116 (2009)

Gallagher, R.H., Simon, B.R., Johnson, P.C., Gross, J.F.: Finite elements in biomechanics. Wiley, New York (1982)

Gefen, A.: Stress analysis of the standing foot following surgical plantar fascia release. J. Biomech. 35, 629–637 (2002)

Knecht, S., Luechinger, R., Boesiger, P., Stussi, E.: MRI-Based inverse finite element approach for the mechanical assessment of patellar articular cartilage from static compression test. Biomed. Tech. 53, 285–291 (2008)

Kumaresan, S., Yoganandan, N., Pintar, F.A., Maiman, D.J.: Finite element modeling of the cervical spine: role of intervertebral disc under axial and eccentric loads. Med. Eng. Phys. 21, 689–700 (1999)

Le, K.M., Madsen, B.L., Barth, P.W., Ksander, G.A., Angell, J.B., Vistnes, L.M.: An in-depth look at pressure sores using monolithic silicon pressure sensors. Plast. Reconstr. Surg. 74, 745–756 (1984)

Linder-Ganz, E., Shabshin, N., Itzchak, Y., Gefen, A.: Assessment of mechanical conditions in sub-dermal tissues during sitting: a combined experimental-MRI and finite element approach. J. Biomech. 40, 1443–1454 (2007)

Linder-Ganz, E., Shabshin, N., Itzchak, Y., Yizhar, Z., Siev-Ner, I., Gefen, A.: Strains and stresses in sub-dermal tissues of the buttocks are greater in paraplegics than in healthy during sitting. J. Biomech. 41, 567–580 (2008)

Makhsous, M., Lim, D., Hendrix, R., Bankard, J., Rymer, W.Z., Lin, F.: Finite element analysis for evaluation of pressure ulcer on the buttock: development and validation. IEEE Trans. Neural Syst. Rehabil. Eng. 15, 517–525 (2007)

Mooney, M.: A theory of large elastic deformation. J. Appl. Phys. 11, 582–592 (1940)

Noakes, K., Pullan, A., Bissett, I., Cheng, L.: Subject specific finite elasticity simulations of the pelvic floor. J. Biomech. 41, 3060–3065 (2008)

Papademetris, X., Sinusas, A.J., Dione, D.P., Constable, R.T., Duncan, J.S.: Estimation of 3-D left ventricular deformation from medical images using biomechanical models. IEEE Trans. Med. Imaging 21, 786–800 (2002)

Phatak, N.S., Maas, S.A., Veress, A.I., Pack, N.A., Di Bella, E.V., Weiss, J.A.: Strain measurement in the left ventricle during systole with deformable image registration. Med. Image Anal. 13, 354–361 (2009)

Phatak, N.S., Sun, Q., Kim, S.E., Parker, D.L., Sanders, R.K., Veress, A.I., Ellis, B.J., Weiss, J.A.: Noninvasive determination of ligament strain with deformable image registration. Ann. Biomed. Eng. 35, 1175–1187 (2007)

Portnoy, S., Yizhar, Z., Shabshin, N., Itzchak, Y., Kristal, A., Dotan-Marom, Y., Siev-Ner, I., Gefen, A.: Internal mechanical conditions in the soft tissues of a residual limb of a transtibial amputee. J. Biomech. 41, 1897–1909 (2008)

Rabbitt, R.D., Weiss, J.A., Christensen, G.E., Miller, M.I.: Mapping of hyperelastic deformable templates using the finite element method. Proc. SPIE (Vision Geometry IV) 2573, 252–265 (1995)

Spears, I., Miller-Young, J., Sharma, J., Ker, R., Smith, F.: The potential influence of the heel counter on internal stress during static standing: A combined finite element and positional MRI investigation. J. Biomech. 40, 2774–2780 (2007)

Stekelenburg, A., Oomens, C.W., Strijkers, G.J., Nicolay, K., Bader, D.L.: compression-induced deep tissue injury examined with magnetic resonance imaging and histology. J. App. Physiol. 100, 1946–1954 (2006)

Tanaka, E., del Pozo, R., Tanaka, M., Asai, D., Hirose, M., Iwabe, T., Tanne, K.: Three-dimensional finite element analysis of human temporomandibular joint with and without disc displacement during jaw opening. Med. Eng. Phys. 26, 503–511 (2004)

Todd, B.A., Thacker, J.G.: Three-dimensional computer model of the human buttocks, in vivo. J. Rehabil. Res. Dev. 31, 111–119 (1994)

Todd, B.A., Wang, H.: A visual basic program to pre-process mri data for finite element modeling. Comput. Biol. Med. 26, 489–495 (1996)

Veress, A.I., Gullberg, G.T., Weiss, J.A.: Measurement of strain in the left ventricle during diastole with cine-mri and deformable image registration. J. Biomech. Eng. 127, 1195–1207 (2005)

Zienkiewicz, O.C., Taylor, R.L.: The finite element method for solid and structural mechanics. Elsevier, Oxford (2005)

A Finite-Element Biomechanical Model for Evaluating Buttock Tissue Loads in Seated Individuals with Spinal Cord Injury

Mohsen Makhsous[1,2,3,4] and Fang Lin[1,2,4]

[1] Department of Physical Therapy and Human Movement Sciences
[2] Department of Physical Medicine & Rehabilitation
[3] Department of Orthopaedic Surgery, Northwestern University, Chicago, Illinois, USA
[4] Department of Sensory Motor Performance Program, Rehabilitation Institute of Chicago, Chicago, Illinois, USA
m-makhous2@northwestern.edu

Abstract. Pressure ulcer remains to be a serious problem in today's healthcare, which has a detrimental impact on the quality of life of the wheelchair users, especially for those with spinal cord injury, and imposes a tremendous economic burden on health care. Development of effective prevention strategies and treatment modalities of pressure ulcer calls for improved understanding of soft tissue response to external loading. Initial evidences suggest that the stress in deep tissue surrounding a bony prominence may reach multiple times higher than that at the superficial skin, however, interface pressure still is the only available clinical tool to assess sitting load, which provides little to no insight to the situation in deep tissues. Moreover, experimental evaluation of the load transfer, from superficial through multiple layers of soft tissues to the load bearing bones, remains nearly impossible. Therefore, finite element models are a powerful tool to help examine biological structures and predict outcomes from potential interventions. However, to date, successful finite element models for simulating buttock tissue response in sitting are rare. In this chapter, the authors present a comprehensive description of the methodology building and validating a 3-dimensional finite element model for human buttocks with the consideration of the joint configuration and realistic boundary conditions in a sitting posture. In addition, with the validated model, sitting induced soft tissue deformation, internal pressure, and von-Mises stress were computed in a practical application evaluating a novel seating system designed to prevent pressure ulcer formation and promote healing of existing pressure ulcer. Along with the methodology of establishing and validating the buttock finite element model for sitting research, and the application example, the authors provide explicit opinions for various weaknesses usually seen in this field and suggest possible future directions for deepening our knowledge and advancing our technologies for performing successful and practical finite element simulation for buttock tissue.

1 Introduction

1.1 Unsolved Questions and Gaps in Pressure Ulcer Research Call for Mathematical Simulation

An improved understanding of soft tissue response to external loading such as in sitting may afford improved pressure ulcer (PU) prevention and treatment

A. Gefen (Ed.): Bioengineering Research of Chronic Wounds, SMTEB 1, pp. 181–205.
springerlink.com

strategies. Although there are concerns and initial evidences that the local tissue stress in deep tissue layers may have been 3~9 times higher than that at the superficial skin (Bouten et al. 2003; Linder-Ganz et al. 2004; Sun et al. 2005), currently, interface pressure still remains the only available clinical tool to assess sitting load, which provides little to no insight to the situation in deep tissues. Moreover, a full-field experimental evaluation of the load transfer, from the superficial layer to the load bearing bones through multiple layers of soft tissues, remains nearly impossible. It is, therefore, anticipated that those difficulties associated with experimental measurements can be overcome by computational modeling.

1.2 Why Finite Element Modeling in PU Research

Mathematical finite element (FE) models, which allow integration of anatomy, physiology, and clinical observations, can be a powerful tool to help examine biological structures (Mow et al. 1984; Bradley et al. 1997; Keyak et al. 1998; May-Newman et al. 1998; Ferrant et al. 1999; Hagemann et al. 1999; Miga et al. 1999; Sonka et al. 2000) and predict outcomes from potential interventions (Bro-Nielsen et al. 1996; Bielser et al. 1999; Mazella et al. 1999). To analyze soft tissue response for a complex anatomical structure and address non-linear material behavior, it is indispensable to use computer FE modeling.

An improved understanding of the response of deep muscle tissue to external compressive load may afford improved prevention and treatment strategies. However, an experimental evaluation of such response on human remains impossible. Therefore, related studies rely on computational modeling (Todd et al. 1994; Ragan et al. 2002; Linder-Ganz et al. 2004; Sun et al. 2005; Lin et al. 2006) and animal models (Salcido et al. 1994; Dupont-Versteegden et al. 1998; Bosboom et al. 2001; Bosboom et al. 2003; Gefen et al. 2005).

1.3 History of FE Modeling Related to PU Research

There have been several attempts in using FE analysis to evaluate sitting induced mechanical response in deep tissue of the buttocks (Chow et al. 1978; Dabnichki et al. 1994; Todd et al. 1994; Brosh et al. 2000; Ragan et al. 2002; Oomens et al. 2003; Lin et al. 2004; Kuroda et al. 2005; Linder-Ganz et al. 2007; Linder-Ganz et al. 2009). Chow and Odell (Chow et al. 1978), using a simple axi-symmetric buttock FE model, found that the buttock tissue experienced small deformation when using a floating support design cushion. Dabnichki et al. (Dabnichki et al. 1994), employing contact elements in a simple 2D FE buttock model, found that the maximal compressive stress and displacement generated close to the bony prominences depended on the surface conditions, with the rough compliant surface generating the least compressive stress and tissue deformation. Oomens et al. (Oomens et al. 2003), using a simple 2D FE model, also studied the mechanical responses of the deep tissues related to cushion properties. They found that using a soft foam layer on the cushion provided a reduction in the normal stresses at the interface and in the deep tissue as the result of an increase in contact areas. These studies suggested an optimal method to reduce interface pressure and stress/strain on the skin and in the deep tissue of the buttock.

However, several intrinsic weaknesses are associated with the majority of the above models.

1.3.1 Limitations of Previous FE Models

Problems with geometry of current FE models

First, some of the models used simple geometrical objects to approximate the complicated anatomical structure (Brosh et al. 2000; Oomens et al. 2003; Kuroda et al. 2005). Second, for those models based on MRI recording, the MRI images were **not** performed in a setup close to real sitting, but in a supine posture (Protz et al. 1990; Todd et al. 1994) which differs intuitively from that in sitting posture in joint configuration and soft tissue stress-strain conditions.

It is well known that the accurate FE simulation of biomechanics needs accurate 3D geometry and realistic material models of a biological structure. Therefore, it is necessary to develop a comprehensive methodology for using 3D FE modeling technology in evaluating soft tissue response to external sitting load.

Problems with material models

Biological soft tissues have viscoelastic, anisotropic, and incompressible properties (Kosiak 1961). The description for such a complicated behavior of the soft tissues may be the bottle neck of applying the FE modeling to simulate clinical problems. Most currently reported models use material properties available in the literature and used the material properties in a way which is substantially different from the real scenario in the buttocks during sitting, in the following aspects.

In sitting, the soft tissue, especially the skeletal muscle, is subjected to ***compressive*** load ***transversely*** about the fiber orientation, while material properties given in literature are mostly obtained ***longitudinally*** along the fiber orientation, and mostly under ***tension***. Our laboratory data showed that, for the same tissue sample, material properties can be substantially different between testing in longitudinal (parallel to fiber direction) and transverse directions (Figure 1).

Those material properties used were from literature, mostly, obtained from fresh frozen cadaver or animal tissue without measuring tissue perfusion (Linder-Ganz et al. 2004), which inevitably introduced unknown discrepancies. Published values of Young's modulus of component tissue vary by up to an order of magnitude, presumably due to the method of measurement or estimation (Mason 1977; Agache et al. 1980; Larabee 1986; Larabee et al. 1986; Sarvazyan et al. 1994; Azar et al. 2000; Samani et al. 2001; Schnabel et al. 2001). Evidence shows the material property of biological tissue change substantially with time after dissection and with method of preservation (Metz et al. 1970). Literature has reported that material properties for postmortem soft tissues are significantly different from those of living tissues(Fung 1993). The material property of skeletal muscle is affected by tissue perfusion and muscle tone, i.e. a physiologically low level muscle contraction seen in skeletal muscle (Masi et al. 2008). Therefore, *in vitro* material test uses tissue samples which are excised from innervation and blood supply which hardly an appropriate "sample" for the living tissues.

Most earlier FE buttock models for sitting used linear material models to describe the non-linear biological tissue, which added more uncertainties on the FE model's output. More frequently, isotropic characteristic was assigned to the material models.

Another major weakness of current FE models for buttocks was that most of these models used a single material property for all soft tissue (Chow et al. 1978; Dabnichki et al. 1994; Todd et al. 1994) without differentiation among muscle, fat and skin. Also from Figure 1, it is evident that these tissues do not possess the same or even similar material properties.

No FE buttock model so far considered the specific material properties for the soft tissues in individuals with spinal cord injury (SCI) or pressure injury. Pressure or other mechanical loads cause alterations or adaptations of tissues (Sanders et al. 1995). Mechanical stress induced adaptations have been observed both in human

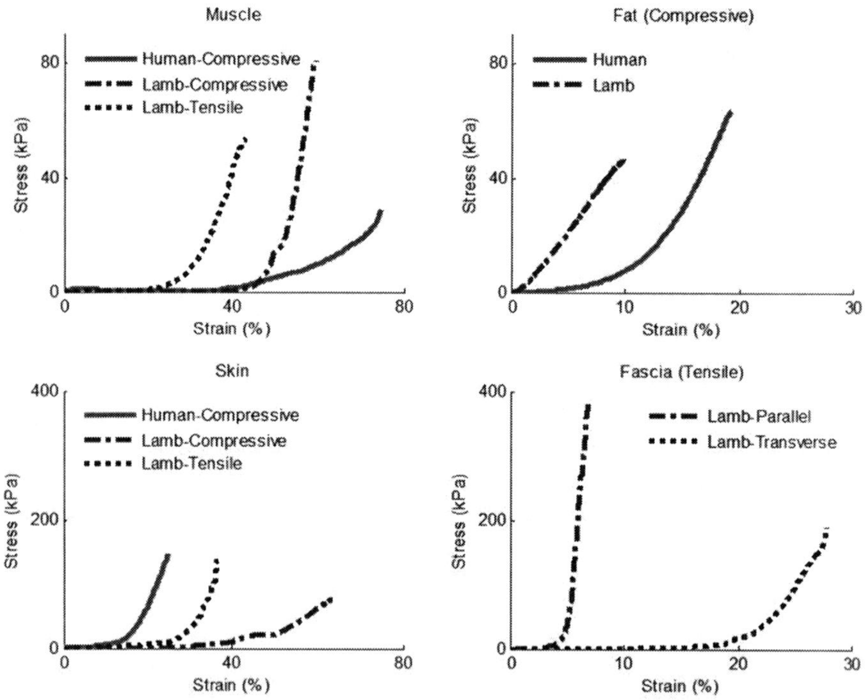

Fig. 1. Stress-strain results for muscle, fat, skin and fascia from tensile and compressive material test from human and animal tissue samples. Some of the samples were also tested in longitudinal (parallel to fiber orientation in figure) and transverse directions.

(Grimby et al. 1976; Stilwill et al. 1977; Scelsi et al. 1982; Lotta et al. 1991; Martin et al. 1992; Greve et al. 1993; Round et al. 1993; Edsberg et al. 1999; Edsberg 2001) and in animal models (Lieber et al. 1986a; Lieber et al. 1986b; Roy et al. 1991; Goldstein et al. 1998; Castro et al. 1999; Castro et al. 2000; Sanders et al. 2001; Wang et al. 2003; Landry et al. 2004) with and without SCI. Tissue at or adjacent to the PUs have been found to undergo significant adaptation or remodeling as a result of sustained pressure (Edsberg et al. 2000). Our preliminary data on rat tibialis anterior (TA) muscle stiffness also demonstrated such substantial difference between pressure injured and intact tissues. Figure 2 shows the plot of load-deformation curves for samples from intact TA and pressure-injured TA from the same rat. For obtaining these data, the TA on one side was kept intact while the other side was subjected to 200kPa compression for 4 hours 24 hours before the load-deformation testing.

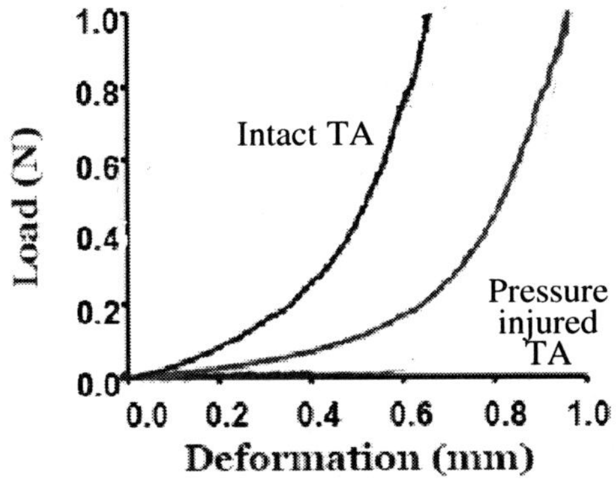

Fig. 2. Load-deformation data for *in-vivo* tests on rat tibialis anterior (TA) muscle. Data came from the pressure-injured TA and its contralateral intact TA on the same rat.

2 Methodology on Building an FE Model for Buttocks for a Seated Individual

2.1 The Importance of Building the Model in a Seated Configuration

As emphasized previously, the FE model for buttocks for a sitting posture should be built upon the geometry obtained from a sitting posture which is intuitively different in anatomical configuration (joint and tissues) from that of a supine or standing posture. In sitting, the hip and knee joints are both flexed to close to 90°, therefore, the loading relationship between segments is different between these 2 postures. With the flexion of the hip joint, the gluteus muscle slides laterally away from the ischial tuberosity (IT) (Kaplan 2003). Therefore, the tissue being compressed in sitting by the ITs is not that seen distally to IT in full hip extension as in standing or supine posture. Also because of this lateral sliding of the gluteus muscle, in sitting, the IT has less padding tissue to resist the pressure load, which can be mistakenly interpreted by taking the anatomical geometry in a supine posture. Our data of tissue thickness from 11 subjects at different hip flexion angle

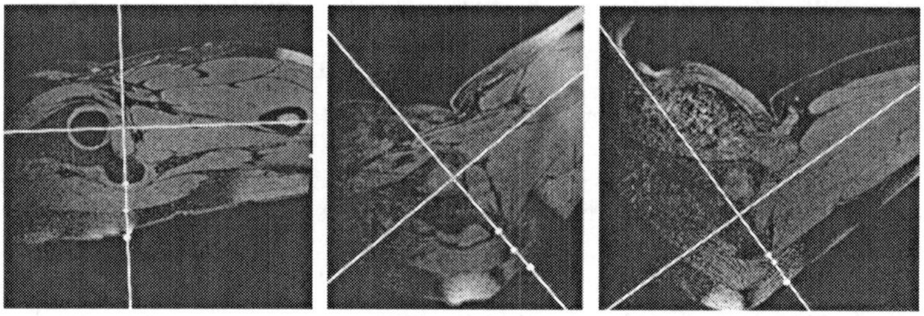

A. Buttock tissue thickness measurement on MRI images at different hip flexion angles

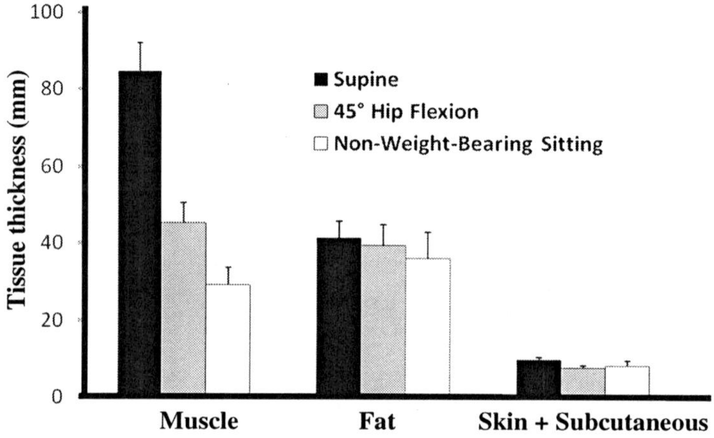

B. Data of measurement from A.

Fig. 3. Soft tissue thickness under the ischial tuberosity in three different hip flexion angles (0°, 45°, and 90°). Data were from MRI images of 11 able-bodied subjects.

measured from MRI images (Figure 3A) clearly showed this substantial difference in tissue composition and thickness between sitting and supine postures.

Therefore, the geometry of a FE model for buttocks for sitting related research should be acquired in a sitting posture or a posture resembling the hip and knee joint configuration of that of a sitting posture.

2.2 Building the Geometry of the Model

2.2.1 Building the Geometry in a Simulated Sitting Position

In a regular MRI scanner, an upright sitting posture cannot be achieved for a human adult because of the limited room. To overcome the restriction, we used a simulated sitting posture for acquiring the MRI images for buttock structure.

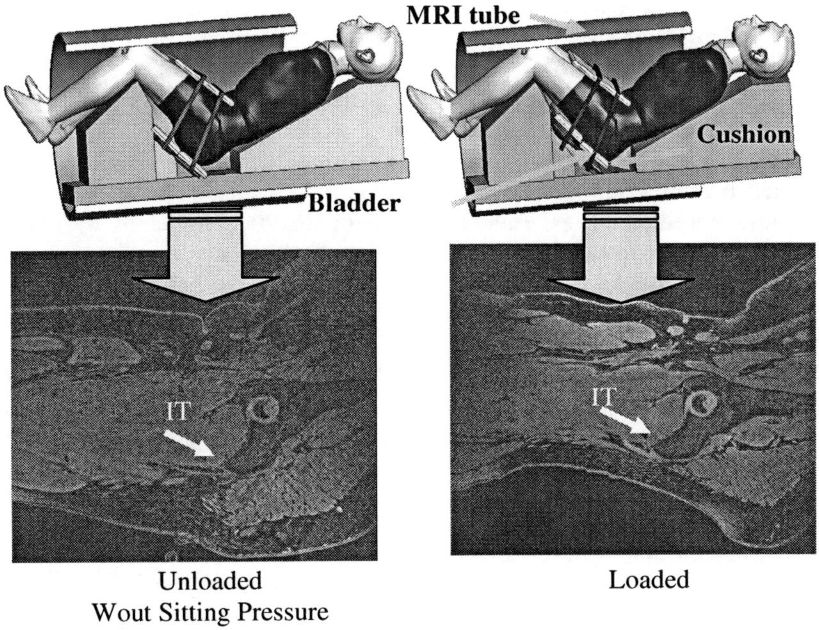

Fig. 4. Setup for recording MRI images from buttock-thigh area in a simulated sitting posture with simulated sitting load applied. Upper row: MRI setup to measure buttock-thigh structure under two loading configuration (Left: Unloaded: Without Sitting Pressure; Right: Loaded: With Sitting Pressure) for the simulated sitting posture. Two belts tied the cushion with the buttocks. An air bladder was placed in between of the 2 layers of the cushion. Inflation and deflation of the air bladder provided different loading levels. Lower row: Corresponding MRI images.

A custom built apparatus made of plastic foam and rubber was used to place the subject in a simulated sitting posture (Figure 4). Two loading conditions were used (Unloaded: without sitting pressure; Loaded: with sitting pressure) during the MRI recording. The sitting pressure was applied to the buttocks with an adjustable cushion placed under the buttock-thigh. The cushion was a sandwich structure consisting of two layers of stiff materials with a rectangular air bladder in between, while two belts tied the cushion tightly against the buttocks. Inflation and deflation of the air bladder provided different loading conditions as shown in Figure 4. The magnitude of the applied sitting pressure for the "With Sitting Pressure" configuration was selected from an actual average interface pressure under buttock (152.6±20.9mmHg, 20.3±2.8KPa) measured with a pressure mapping system, X2 (Xsensor Technology, Calgary, Canada) for this subject in an upright sitting posture. MRI images obtained from "Without Sitting Pressure" were used for development of the geometry of the FE model and those from "With Sitting Pressure" were used for validation of the FE analysis.

MRI images of buttock-thigh were obtained (Figure 4) for the above two loading conditions in one simulated sitting posture (80° flexion for hip and 90°

flexion for knee). The MRI parameters were 1.5 Tesla, 576×576 matrix, 35×35 cm DFOV, 0.6 mm inter-slice thickness.

2.2.2 Building the Geometry in a True Sitting Position

Although the above simulated sitting posture works well for acquiring the buttock geometry in a joint configuration similar to sitting, there are drawbacks. First, the simulated posture can only be achieved for individuals with relatively smaller body build, based on our experience, not taller than170cm in height. Second, since the subject is away from a load-bearing sitting posture, the sitting load can only be applied to the buttocks from the bottom, which is not the natural loading pattern during sitting. Therefore, using a setup which allows a true sitting posture be scanned provides an appropriate alternative.

In our study (Figure 5) and the work from Linder-Ganz (Linder-Ganz et al. 2009), an open MRI setup has been used for scanning the buttocks in a true sitting posture. In our study, a regular seat cushion without metal parts was used between the subject's buttocks and the seat. Additional cushioned material prevented the direct contact of the subject's body with the coil. MRI images were acquired in 2 conditions, "without buttock load" and "with buttock load" when the subject was in the upright sitting posture.

For "without buttock load" condition, an enhanced lumbar support was used at L4-L5 to support the spine. A cushion (buttock unloading cushion) was placed underneath the thighs to bear the sitting load while left the buttocks load free (Figure 5A, left). For the "with buttock load" condition, the subject remained in the above posture but the buttock unloading cushion was removed to allow the subject sit directly on the seat cushion with the upper body weight born by the buttocks. The buttock-cushion interface load was measured by using XSensor (XSensor Technology, Calgary, Canada) while the seating system was out of the MRI.

Volumetric MRI images of the right buttock-thigh including coccyx (256x256 matrix, 35cm FOV, 288 slices, 0.5mm thickness) were obtained using the 0.6T FONAR scanner. Several MRI opaque markers were placed on buttock skin for the purpose of registration of the images for later processing (Figure 5B).

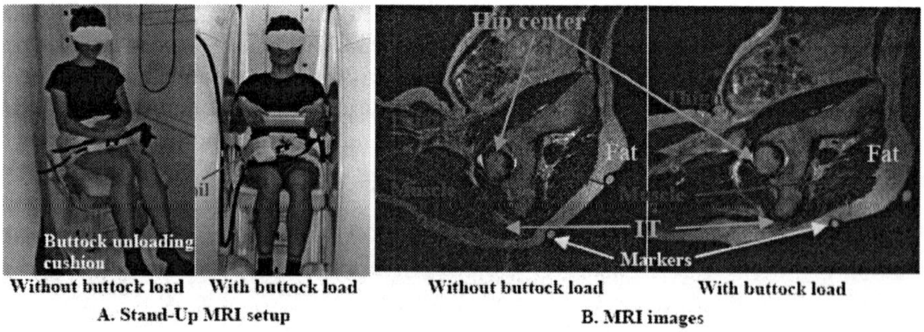

Fig. 5. Open (Stand-Up) MRI (Upright^{TM}, FONAR Corp., Melville, NY) setup to record buttock-thigh structure under sitting load. MRI images corresponding with and without buttock sitting load are shown.

2.3 Mesh Generation of the Final Geometry

After the imaging, the FE model geometry is usually created based on the 3D reconstruction of the MRI images under the non-weight bearing configuration. Contours of the femur, pelvis, skin, fat and muscle are identified and outlined from these images by using segmentation software such as WinSurf (SURFdriver, Kailua, HI).

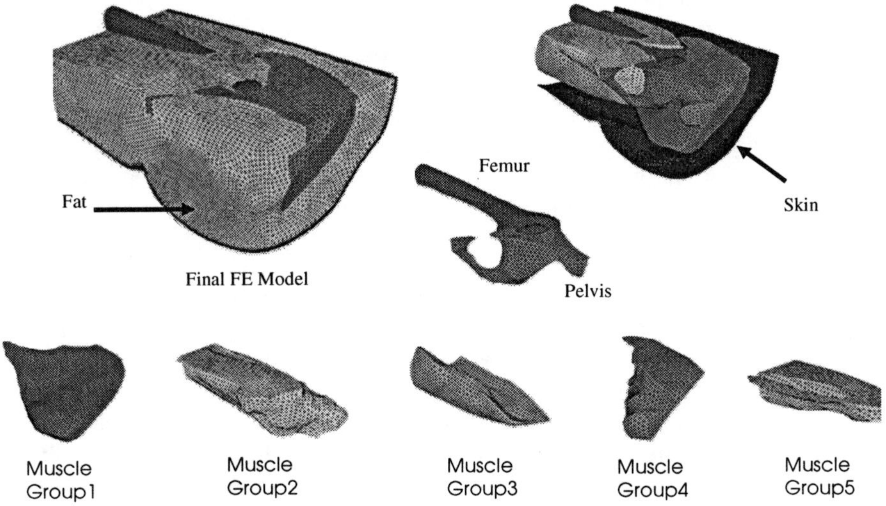

Fig. 6. Structural elements of the FE buttock model

Muscle is one of the primary tissues in buttock structure and is thought to be the most susceptible to pressure induced injury, especially in the vicinity of bony prominence such as ischial tuberosity where the pressure ulcers usually form. Therefore, the FE simulation for buttocks in sitting usually focuses on the stress/strain within the muscle tissue. The massive volume, separated locations, and various orientations of the buttock muscles bring issues for creating the muscle as one object in a buttock FE model. Our current model uses five groups to represent the various location and orientation of buttock muscles. Selection for muscles in each group depends on the location, and the line of action of each muscle. Considering the line of action for each muscle group is for applying initial muscle tone to each group, which will be described later. However, we believe that grouping muscles or not depends largely on particular application and the focus for outcome evaluation. The five muscle groups used in our model are:

Group1: *Minimus, Medius and Maximus Gluteus.*

Group2: *Adductor Longus, Adductor Brevis, Adductor Magnus, and Pectineus.*

Group3: *Biceps Femoris (long head), Semitendinosus, and Gracilis.*

Group4: *Piriformis, Superior and inferior Gemellus and Obturator Internus.*
Group5: *Vastus Lateralis, Vastus Intermedius, Vastus Medialis and Rectus Femoris.*

The segment information is then translated into HyperMesh (Altair Engineering, Inc., Troy, MI) to create the FE mesh. The final FE model consists of 453,502 four-node tetrahedral solid elements for the pelvis, femur, inner side of the skin, fat and five muscle groups, and 33,924 three-node triangle membrane elements for the outer side of the skin (Figure 6).

2.4 Selecting Material Model for Buttock FE Models

As pointed out previously, the biological tissue is non-linear, anisotropic, nearly incompressible material with viscoelasticity. Moreover, soft tissues in buttocks experience deformation as large as about 70% in practical sitting (Lin et al. 2004). Therefore, material model for FE simulation for buttocks in sitting should be selected accordingly. Adding to this complexity is the significantly altered material property for denervated tissues below the level of injury in SCI individuals. *In vivo* (blood perfusion preserved) compressive creep test on tibialis anterior (TA) muscle flaps from 10 neurologically intact and 8 chronic SCI rats in our laboratory revealed a 44.9% increase in SCI group in compressive deformation after 500s holding of a constant load (the constant holding load was chosen as 30% of the load needed to achieve 30% deformation for each TA muscle). This finding suggested that, in addition to the rapid muscle atrophy (Scelsi et al. 1982; Lotta et al. 1991), muscle tissue alters its fundamental properties post SCI and these alterations lower their pressure endurance threshold significantly (Daniel et al. 1981). This phenomenon calls for its proper representation in the material model chosen for simulation sitting for SCI population. Unfortunately, research on the changes of tissue material properties post SCI is scarce to none. Currently FE models, even when the focus is for evaluating sitting load in SCI wheelchair users, still employ material properties from the literature obtained from unrelated tissue samples.

In our developed FE model, static equilibrium equations with large deformation were assumed in the material model since pressure sores develop after prolonged loading with large deformation of the soft-tissue. The elastic *Moony-Rivlin* model (Dabnichki et al. 1994) (1st order polynomial model), which can account for large deformation behaviors of materials, was employed for skin, fat and muscle. The model was based on the following strain energy function W (Eq. 1):

$$W = C_{10}(\bar{I}_1 - 3) + C_{01}(\bar{I}_1 - 3) + \frac{1}{D_1}(J - 1)^2, \tag{1}$$

W is the strain energy per unit of reference volume; J is the total volume ratio; C_{10}, C_{01}, and D_1 are temperature-dependent material parameters; $\bar{I}_1$ and $\bar{I}_2$ are the first and second deviatoric strain invariants defined as following equations (Eqs. 2 and 3):

$$\bar{I}_1 = \bar{\lambda}_1^2 + \bar{\lambda}_2^2 + \bar{\lambda}_3^2, \quad (2)$$

$$\bar{I}_2 = \bar{\lambda}_1^{(-2)} + \bar{\lambda}_2^{(-2)} + \bar{\lambda}_3^{(-2)}, \quad (3)$$

The deviatoric stretches $\bar{\lambda}_i = J^{-1/3}\lambda_i$; Here, λ_i are the principal stretches.

The material parameters for skin, fat and muscle were determined based on values obtained from literatures (Dabnichki et al. 1994; Oomens et al. 2003). For all soft-tissues, D_1 was determined based on the method reported by Dabnichki et al, (Dabnichki et al. 1994) with an assumption that the materials were nearly incompressible (*Poisson ratio* $\cong$ *0.485*).

2.5 Assigning Boundary Conditions for Buttock FE Models

The boundary conditions for buttock FE models in sitting reflect the anatomical and practical scenarios in human sitting. In sitting, soft tissues are compressed by the sitting load coming from the seat-skin interface. The induced movement of the tissues is among the interests in sitting simulation. Movement does not usually occur in proximal-distal due to the anatomical restrain. Also the tissues on anterior side of the pelvic and femur do not usually move to anterior. To achieve efficient simulation, usually one half of the buttocks is used for FE simulation. The other half is assumed to experience the same loading condition with the same response.

Based on the above description, three boundary conditions (*BC*) are employed in our developed FE model (Figure 7). The femur and the pelvis are assumed as rigid bodies and constrained to obviate rigid body motion (BC1). The medial plane was constrained against the medial-lateral motions because of the symmetric

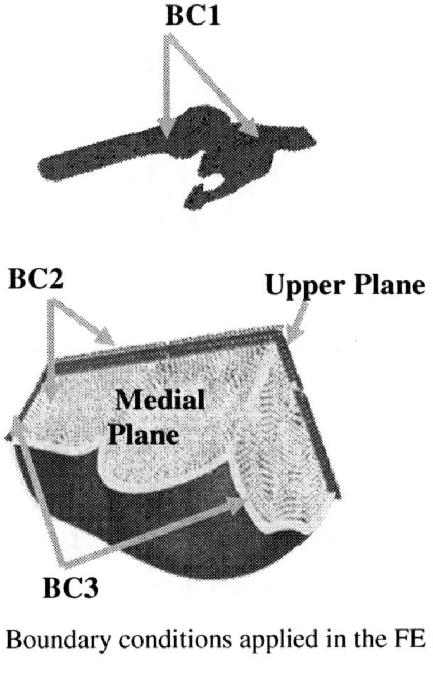

Fig. 7. Boundary conditions applied in the FE model

condition of the buttocks (BC2). Also for BC2 was that the upper plane was constrained from the anterior-posterior movement. The ends of tissues that connect to the rest of the body (distal end to the thigh, proximal end to lumbar region) were constrained against longitudinal motions (BC3).

2.6 Assigning General Loading Conditions for Buttock FE Models

When load from the sitting was considered as the specific loading applied as the input to the FE model, two general loading conditions (GLC in Figure 8) should also be included in the model, based on physiological facts.

First, in a sitting posture, buttock skin is not in a relaxed condition but with a low-level tension induced by hip flexion, which is represented by using an initial strain on the skin layer of the model (GLC1) (Figure 8). This initial strain is defined as the strain of the buttock skin when the subject moved from a full-hip extension posture to an upright sitting posture, when the hip and the knee both flex at about 90°. The value of this strain used in our current model was obtained from a single subject by calculating Green strain (Bonet et al. 1997) of painted dots on subject's buttock skin. Photographs were taken at 0° hip/knee flexion and again at a hip flexion angle of 80° and knee flexion of 90°. On these photographs, the array of dots painted on the buttock skin was digitized using custom codes on Matlab platform (Figure 9). Green strain was then calculated for an area of $15\text{x}15\text{mm}^2$ as 43.6±10.4% and 11.2±0.7% for the axial and transverse directions, respectively.

Secondly, skeletal muscles are consistently under a tension from a low level contraction, resting tone (Masi et al. 2008) or resting tension. To simulate this resting tone, we apply 1% of maximal muscle force estimated from $F_{max} = \sigma_{max} \times$ *Physiological Cross-Sectional Area (PCSA),* to each group of muscles along the line of action. In the above equation, σ_{max} is the maximal muscle tension (Morrey 1990) (GLC2 in Figure 8).

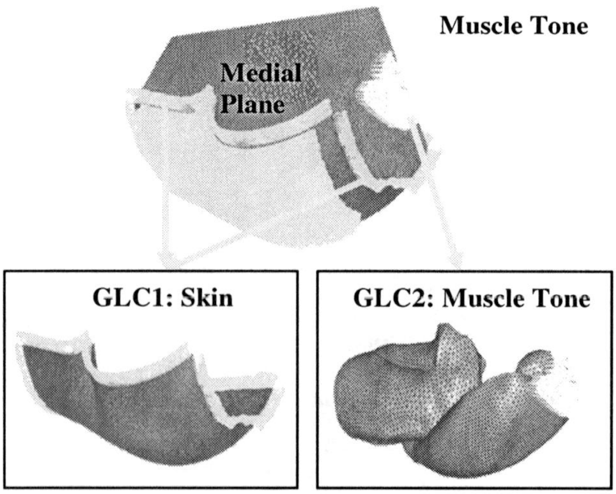

Fig. 8. General loading conditions applied in this FE model

3 Results

3.1 FE Model Solution

Upon establishment of the geometry and material model, the FE model is ready for a simulation. To solve the model simulation, the sitting interface pressure is used as the model load input. For this purpose, a sitting contact area is precisely identified from the corresponding MRI images and the sitting load is applied to this area.

Buttock contact pressure corresponding to the MRI scanning in loaded condition is used. In model development and validation stage, to be consistent with the loaded condition shown in MRI images, the FE model should be fed with the same input load to be able to perform the comparison in the validation process. Thus, we used a uniform contact pressure of 20.34kPa (152.60mmHg) because a constant pressure was actually applied to the buttocks when MRI images of loaded condition were taken. With this load input, the FE model was solved using ABAQUS software (ABAQUS 6.5, ABAQUS, Inc., Providence, RI).

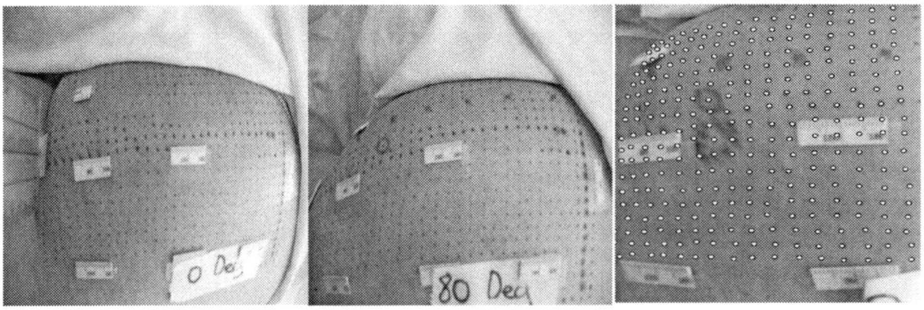

A) 0° hip flexion B) 80° hip flexion C) Digitized dot array

Fig. 9. Measurement of buttock skin initial tensile strain induced by a sitting posture relative to a standing posture

3.2 Outputs of the FE Model Simulation

I. Soft tissue deformation under sitting load

This prediction is performed for a sagittal piece of the entire buttock-thigh. This sagittal piece had a thickness of 12mm with its medial-lateral center at the tip of the ischial tuberosity. In this sagittal piece, the gross deformation of each layer was calculated using the changes in the coordinates of the FE nodes, 42 nodes per layer, in anterior-posterior direction when the buttock-thigh was loaded by 20.34 kPa.

II. Soft tissue internal stress induced by the sitting load

From the FE analysis, the internal pressure distribution and von-Mises stress distribution can be obtained for the entire or selected part of the structure. In our model, we calculated the internal stress for the muscle, fat and skin for the entire buttock-thigh, particularly in the area below the ischial tuberosity. Although it still is uncertain what mechanical parameters are the most relevant to tissue damage (Oomens et al. 2003), evidences supported that living cells are more vulnerable to deformation than to a high hydrostatic pressure (Bouten et al. 2003). Therefore, the von-Mises stress, which is related to the deformational energy stored in the material (Todd et al. 1994; Oomens et al. 2003; Linder-Ganz et al. 2004), was analyzed to identify the risk of possible tissue damage by sitting load.

3.3 Methodology on Validating an FE Buttock Model for a Sitting Posture

A FE model is a mathematic model which is built upon mathematical description of the structure and the materials. For a model to have the practical value in estimating the interested outcomes from successful simulation of the mechanical environment within the structure represented by the model, it should be validated before putting into practical application. An un-validated model, however complicated, can server at the best a qualitative tool. For a mathematical model simulating physiological process within a human body, it is always a challenge when it comes to the time for validation. Therefore, quite a few FE buttock models remained un-validated or were only validated using parameters which were used as the model inputs such as the interface pressure on the seat-buttock interface.

There are two ways to validate a FE model, through experiment on a physical model (Linder-Ganz et al. 2009), and through imaging measurement from a real sitting. Our developed model was validated using the imaging method.

Upon solved, the FE model was validated quantitatively via comparison of the output of the FE simulation with measurements from the MRI images. Two comparisons were performed: 1) the sitting induced gross displacement of soft tissues; 2) the sitting induced position shift of the muscle underneath the ischial tuberosity (Group1).

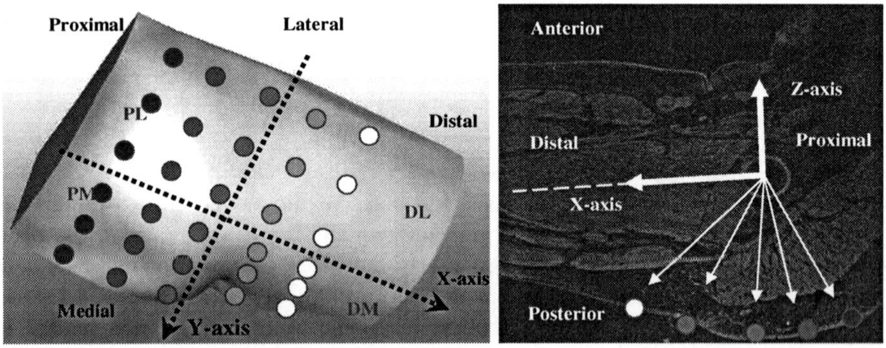

A) A posterior view of the coordinate system and the ROIs.

B) A sagittal MRI image showing the coordinate system and the ROIs.

Fig. 10. Validating FE model by comparing the imaging measurements and the simulation results of selected regions of interest (ROI)

I) Gross validation. *Sitting Induced Gross Displacement of Soft Tissues Computed by FE Analysis and that Measured from MRI Images:* It was assumed that the bony structure remained the same when loaded. A Cartesian coordinate system was defined based on the femur-pelvis bony structure. This coordinate system took the center of the femoral head as its origin. The X-axis was pointing to the distal along the femoral shaft. The Y- and Z- axes were pointing to the medial and the superior, respectively (Figure 10). Thirty regions of interest (ROIs) were identified over the skin of the sitting area (Figure 10). For measurement on MRI images, the morphological alteration of buttock soft tissues induced by sitting pressure of 20.34±2.79KPa was identified for these ROIs by comparing MRI images obtained in loaded condition to the matching image from unloaded condition. On each image, vectors pointing from the origin of the coordinate system to the specific locations on the skin were constructed to compute the gross displacement. In order to compare the differences in different portions of the buttock, 4 regions, the distal-medial (DM), distal-lateral (DL), proximal-medial (PM) and proximal-lateral (PL) regions, were defined (Figure. 5).

The same coordinate system mentioned above was established for the FE model. The 30 ROIs were also identified for the FE model for both the unloaded "Without Sitting Pressure" and loaded "With Sitting Pressure" configurations. The sitting induced changes of the coordinates of the FE nodes in these ROIs were taken as the FE predicted gross displacement. These displacements were then compared with the measurements from MRI images.

II) Selected validation for more details regarding the simulation focus. *Position Shift of the Muscle Group1 beneath the Ischial Tuberosity:* In the sitting condition, it is apparent that the tissue underneath the ischial tuberosity is loaded more than tissues at other locations. Therefore, the muscle here may be pushed away from its

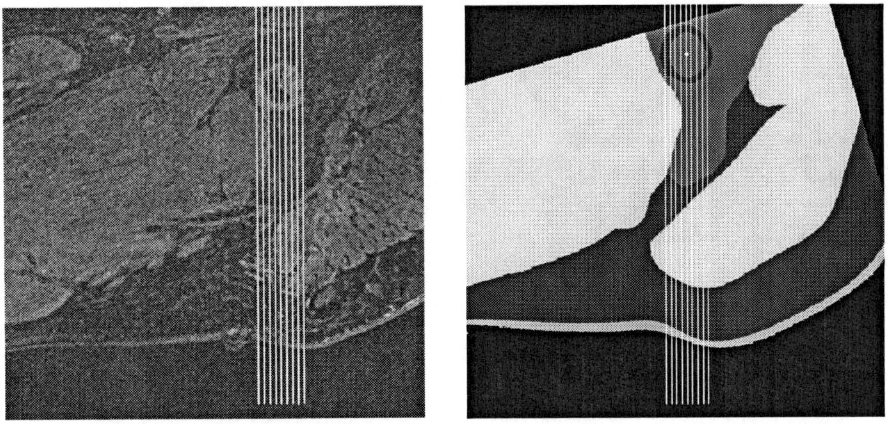

A) The selected area on MRI images for calculating muscle Group1 shift **B) The selected area on the FE model for calculating muscle Group1 shift**

Fig. 11. Validating FE model by comparing the imaging measurements and the simulation results of a selected region underneath ischial tuberosity original position. This shift of position of muscle Group1 can be used as a validation parameter. A small area right beneath the ischial tuberosity was selected to calculate the shift of the muscle Group1 induced by sitting load of 20.34 kPa. Taking the tip of the ischial tuberosity as the center, this area covered ±15mm in the proximal-distal direction and ±6mm in the medial-lateral direction (Figure 11). For this volume, in the medial-lateral direction, 12 sagittal slices were taken for the FE/MRI models. On each of these sagittal slices, 11 lines in anterior-posterior direction were defined with a distance of 3mm between neighboring lines. Therefore, altogether 132 anterior-posterior lines were determined in this volume. For MRI model, the proximal-distal shift of this volume under sitting load was measured as the average proximal-distal shift of these 132 lines. For the FE model, the proximal-distal shift of this volume was obtained by identifying the changes of the coordinates of the FE nodes within this volume.

3.4 Application of FE Models for Evaluating Buttock Loading in Sitting

Once validated, the FE model can be used to simulate various practical scenarios. Based on the purpose of the application, appropriate outputs can be selected. Most frequently, buttock FE models for sitting are used for evaluating different seat designs, including shape or material of the cushion.

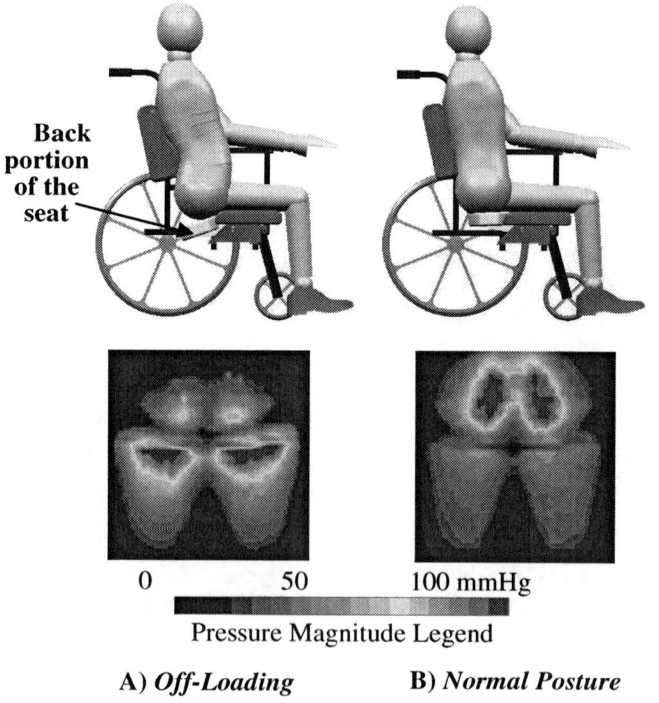

Fig. 12. Two sitting configurations of the studied seat

The above described FE model (Makhsous et al. 2007a) was used in an application to evaluate a novel seat cushion for its effectiveness on reducing the internal stress around the ischial tuberosity. The seat design features 2 configurations, *Normal* and *Off-Loading* (Makhsous et al. 2003; Makhsous et al. 2007b). A *Normal* posture was described as sitting upright on a flat seat and with a flat backrest, while an *Off-Loading* posture was sitting upright on a seat with partially removed ischial support, and with an enhanced lumber support (Figure 12). The Off-Loading configuration was intended to release the ischial load from sitting.

For evaluating this seat, the FE model input was the averaged interface pressure map measured from 35 able-bodied individuals (41.3±12.1years; 72.1±12.6kg; 168.0±8.5cm) in Normal and Off-Loading configurations (Figure 12). In the *Normal* configuration, the interface pressure was concentrated within the vicinity of ischial tuberosities, with the thighs taking substantially less interface pressure. In the *Off-Loading* posture, the concentrated interface pressure at ischial tuberosities was mostly shifted towards the thighs, and the middle part of the thighs took the most loads.

With the interface pressure as the load input to the FE model, simulation provided estimation of the internal stress and strain within buttock structure shown in Figure 13 and Figure 14.

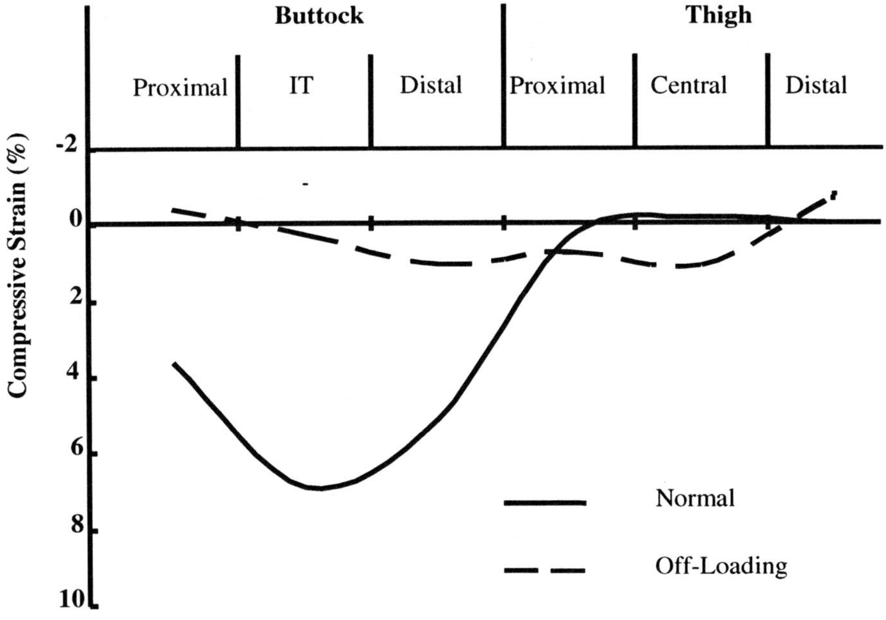

Fig. 13. Anterior-posterior compressive strain (%) of the muscle layer in sagittal plane in a portion covered 15mm thickness buttock-thigh from 7.5mm medial to the ischial tuberosity (IT) to 7.5mm lateral to IT

Compressive Strain on Muscles (Figure 13). The larger compressive strain in the *Normal* sitting (6.9%), which was seen underneath the ischial tuberosity, was shifted toward the thigh with a largely decreased value (0.3%) in the *Off-Loading* condition. On the other hand, the compressive strain on the center of the thigh region was increased from 0.2% in the *Normal* condition to 1.1±0.6% in *Off-Loading* condition.

Compressive Stress and von Mises Stress Distribution (Figure 14). The predicted compressive stress and von Mises stress distributions in the entire buttock-thigh structure and in muscles below the IT are shown in Figure 14. It was seen that, in *Normal* configuration, the high-pressure and stress region was at the location right beneath the bony prominence of ischial tuberosity. However, this high-stress region was relocated to the thighs with a reduction of magnitude in the *Off-Loading* posture (Figure 14).

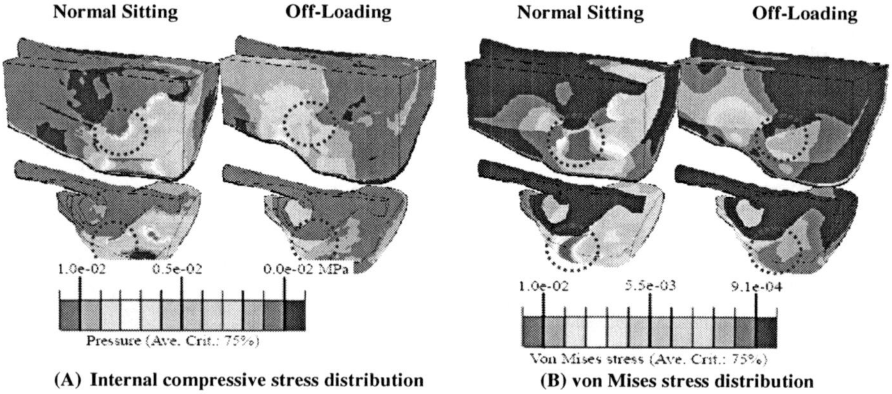

Fig. 14. FE simulation results on stress for a split seat in 2 of its configurations, *Normal* and *Off-Loading*. In each subset of the figures, the upper plots show results for the entire structure, while the bottom plots show data for the muscles below the ischial tuberosity.

This simulation confirms that the pressure relief provided by the *Off-Loading* posture has profound effect in decreasing the stress ***within*** the deep tissues of the buttocks. FE simulation shows that at the same time of releasing interface pressure, the *Off-Loading* posture significantly redistributed the internal compressive stress and von Mises stress within the deep tissues of the buttocks. In the *Off-Loading* posture, the internal compressive stress which was highly concentrated in the deep tissues at the location of the ischial tuberosity in *Normal* sitting posture were repositioned to the thigh with a substantially decreased magnitude.

The most common sites for sitting related PUs are areas close to a bony prominence. At these locations, such as the ischial tuberosity, tissue necrosis was frequently seen starting from deep in the tissues and progresses towards the superficial to become an open wound (Ankrom et al. 2005). Therefore, there is a need for a tool which can estimate the internal mechanical responses in the deep tissues, such as stress and strain, to external sitting load, to provide, from a mechanical point of view, a deep insight into the relationship between the formation of deep tissue damage and the interface pressure. With more of this understanding, it is possible to establish a criterion for evaluating designs of those new sitting systems which aim at achieving optimal pressure relief in sitting area for wheelchair users. The results in the current study proved that the *Off-Loading* posture is effective in reducing the risk of PUs by decreasing large compressive strain, high internal compressive stress in the deep tissue overlying ischial tuberosity.

It has been suggested that the muscular layer is the most *vulnerable* to pressure induced tissue damage due to its high metabolic rate and the dependence on tissue perfusion (Nola et al. 1980; Daniel et al. 1981). This vulnerability puts muscle tissue in a highly susceptible situation to localized compression, eventually leading to tissue degeneration in the form of a deep pressure injury. Therefore, our findings about the concentrated stress in the deep layer of the muscle tissue may be considered the evidence for this kind of deep tissue injury.

4 Future Direction for FE Models in Pressure Ulcer Research for SCI Population

FE simulation for sitting related topics has gained significant progress in recent years and proved itself as an extraordinary tool in understanding the comprehensive response to sitting loads in complex structure within buttocks. As demonstrated in this and other chapters, FE models can now be developed for the buttock anatomy with sufficient fidelity and acceptable accuracy and has been more and more used in the area investigating sitting phenomena, evaluating seating system design, and possible providing useful suggestion to various applications.

However, there are several important issues still hinder the complete acceptance of FE buttock models as a perfect tool for sitting related research and applications, especially when it comes to the practical scenario of wheelchair seating in SCI population. We have identified several critical issues for FE simulation in wheelchair sittings for this population, possible solutions and potential future directions.

4.1 Improving Material Properties

I: Material properties for buttock tissues in SCI samples. With the advances in medical imaging, image processing, and reverse engineering technology, the methodology and tools to obtain an accurate geometrical representation for any complex anatomical structure are being developed and become more and more sophisticated. The geometry of the FE models has progressed from an assembly of simplified 2D objects to faithful 3D structure with incredible resemblance to the real anatomy. Compare to this progress, material models for buttock tissue modeling, the other critical issue in FE sitting modeling, have gained little advance, especially not for the SCI samples. As known by probably every researcher in FE modeling area, a faithful description for each material involved in the FE model dictates the performance of the simulation, more importantly, the applicability of the model simulation. Therefore, the lack of a database for tissue material properties for SCI population is currently the bottle neck of applying FE simulation to simulate clinical problems. As evidenced by literature and our research, tissues, especially soft tissues, experience tremendous morphological and mechanical adaptation post SCI. The loss of proper muscle tone in muscles below the level of SCI due to denervation or disturbed neuromotor control makes additional contribution to post-SCI evolvement of tissue properties. Therefore, future researches should emphasize this need and a comprehensive yet careful investigation in researching such changes in the properties of the soft tissue below the level of the injury of the spinal cord is a must. Currently, this research is scarce to none. Several directions are worth to explore.

II: Material properties in SCI individuals. Data for material properties in SCI samples is in urgent needs. Collecting comprehensive material properties for SCI tissue samples can be a very tedious endeavor, including testing samples in various conditions within this population, such as tissue from acute and chronic stages of SCI for both genders and all age groups, and for people with SCI plus some other comorbidity which may also affect tissue properties. For example, the prevalence of diabetes in SCI is 15%~20%, which is higher than the 10.7% in general population. Diabetes is a medical condition which is known to affect soft tissue properties and increase the incidence of ulcers. Therefore, data for tissue properties from individuals with SCI and with diabetes should be obtained for better FE simulation application for these individuals.

III: Material properties in tissues post SCI from animal models. Collecting material property data from human tissues in SCI population will inevitably encounter several obstacles which prevent us from collecting data in a well controlled condition. For example, *in vivo* sample testing is almost impossible to perform on human subjects, which limits the data collection to excised tissue samples with yet highly limited availability. In well designed animal model, the condition for SCI and selected comorbidity can be strictly controlled and reproduced in multiple animals. At the same time, *in vivo* material testing on tissues with preserved blood perfusion and innervation is feasible and has been conducted in various setups including those from ours. By comparing material properties from SCI (and/or together with selected comorbidity) animals with those from the intact/healthy animals, the relative changes of tissue properties can be obtained. Then these relative properties may be used to calibrate human tissue properties for SCI individuals from that of the able-bodied individuals.

4.2 Subject-Specific Model

This direction has been initially explored by some groups of researchers (Lin et al. 2004; Wagnac et al. 2008; Linder-Ganz et al. 2009). However, being subject-specific usually means more resources and more work would be invested. Therefore, a valid methodology for scaling a prototype or primitive model to each individual is promising; however, no such work has been done yet. Research in this direction should be given more attention to help push forward the FE simulation to be a practical tool in both laboratory and clinics.

4.3 Using Combination of MRI/Open MRI

As described previously, there are two different methods to acquire the geometry of the buttocks for establishing a FE model, imaging in a simulated sitting posture in a high resolution MRI setup, or in a true sitting posture in an open MRI setup with less spatial resolution. Advantages and drawbacks are associated with each of the methods. To overcome the drawbacks and to take advantages from both of them, a possible solution is to combine the two methods to obtain the geometry both accurately using the simulated sitting posture in a high resolution MRI machine and with true sitting load using an open MRI setup.

Combination of these 2 types of MRI scans in constructing accurate and realistic FE model geometry has been reported for patellofemoral joint in researches related to patellofemoral joint loading simulation (Besier et al. 2008). The concept presented in these researches can be readily adapted to FE modeling for buttocks in sitting. Images can be obtained from regular MRI with higher resolution in the simulated sitting posture and then another scan for a loaded true sitting condition can be performed by using an open MRI machine. These 2 sets of MRI images can be registered to each other to produce a final series of MRI images which describes the buttock geometry in a sitting posture with soft tissue deformation induced by true sitting load. In this way, the most accurate yet the most realistic model geometry can be generated.

4.4 Apply More Realistic Loading to the Model

Currently, most of the FE buttock models use pressure reading on the buttock-seat interface as the input to drive the simulation. However, we know that this is not what actually happens during sitting. In a sitting posture, upper body weight is born by the buttocks with a load in the direction of gravity is applied through the spine to the sitting anatomy. Therefore, a more realistic way to simulate sitting would be to apply the upper body weight through the connecting bony structure, i.e. the sacrum which connecting the upper body to the pelvis. If the model geometry includes a sacrum, the upper body weight can be applied to the sacrum directly. However, many FE buttock models do not have the sacral segment then the upper body load should be applied to the surface of sacroiliac joint. The body weight of the buttocks which is below the sacrum level should be applied to the model as distributed load in the gravity direction.

4.5 Develop FE Models for Clinical Use

As described previously, a validated FE buttock model can serve as a powerful tool providing detailed and accurate information of internal loading in various sitting conditions. But currently, owing to its dependence on vast amount of complicated calculation, most of the FE models reside on powerful computers in research facility far from being used in practical application. Therefore, methodology of simplification should be researched to produce **fast, simple, and handy models** which can be used in real life, such as in wheelchair fitting clinic. However, such simplification can only be properly done when the knowledge of buttock tissue/structure behavior in sitting is fully understood without biased knowledge which brought forth by premature simplification. Ideally, with successful simplification, a deliverable turn-key package (or packages) should be produced to professionals in the seating field for practical application for various purposes, e.g. wheelchair fitting, office chair design and evaluation, vehicle seat design, etc.

References

1. Agache, P.G., et al.: Mechanical properties and Young's modulus of human skin in vivo. Arch. Dermatol. Res. 269, 221–232 (1980)
2. Ankrom, M.A., et al.: Pressure-related deep tissue injury under intact skin and the current pressure ulcer staging systems. Adv. Skin Wound Care 18(1), 35–42 (2005)
3. Azar, F.S., et al.: A finite model of the breast for predicting mechanical deformations during Biopsy procedureelement. In: IEEE workshop on Mathematical methods in biomedical image analysis, South Carolina, pp. 38–45 (2000)
4. Besier, T.F., et al.: The influence of femoral interanl and external rotation on cartilage stresses within the patellofemoral joint. J. Orthop. Res. 26(12), 1627–1635 (2008)
5. Bielser, D., et al.: Interactive cuts through 3-dimensional soft tissue. Eurographics 18(3) (1999)
6. Bonet, J., et al.: Nonlinear continuum mechanics for finite element analysis. Cambridge University Press, New York (1997)
7. Bosboom, E.M., et al.: Quantifying pressure sore-related muscle damage using high-resolution MRI. J. Appl. Physiol. 95(6), 2235–2240 (2003)
8. Bosboom, E.M., et al.: Passive transverse mechanical properties of skeletal muscle under in vivo compression. Journal of biomechanics 34(10), 1365–1368 (2001)
9. Bouten, C.V., et al.: The etiology of pressure ulcers: skin deep or muscle bound? Arch. Phys. Med. Rehabil. 84, 616–619 (2003)
10. Bradley, C.P., et al.: Geometric modeling of the human torso using cubic hermite elements. Ann. Biomed. Eng. 25(1), 96–111 (1997)
11. Bro-Nielsen, M., et al.: Real-time volumetric deformable models for surgery simulation using finite elements and condensation. Computer Graphics Forum 15(3), 57–66 (1996)
12. Brosh, T., et al.: Modeling the body/chair interaction - an integrative experimental-numerical approach. Clin. Biomech. (Bristol, Avon) 15(3), 217–219 (2000)
13. Castro, M.J., et al.: Influence of complete spinal cord injury on skeletal muscle cross-sectional area within the first 6 months of injury. Eur. J. Appl. Physiol. Occup. Physiol. 80(4), 373–378 (1999)
14. Castro, M.J., et al.: Influence of complete spinal cord injury on skeletal muscle mechanics within the first 6 months of injury. Eur. J. Appl. Physiol. 81(1-2), 128–131 (2000)
15. Chow, W., et al.: Deformation and stress in soft body tissue of a sitting person. J. Biomech. Eng. 100, 79–87 (1978)

16. Dabnichki, P., et al.: Deformation and stress analysis of supported buttock contact. Proc. Inst. Mech. Engrs. 208, 9–17 (1994)
17. Daniel, R.K., et al.: Etiologic factors in pressure sores: an experimental model. Arch. Phys. Med. Rehabil. 62, 492–498 (1981)
18. Dupont-Versteegden, E.E., et al.: Early changes in muscle fiber size and gene expression in response to spinal cord transection and excercise. Am. J. Physiol. 275, C1124–C1133 (1998)
19. Edsberg, L.E.: Microstructural evaluation of human skin subjected to static versus cyclic pressures. J. Rehabil. Res. Dev. 38(5), 477–486 (2001)
20. Edsberg, L.E., et al.: Microstructural and mechanical characterization of human tissue at and adjacent to pressure ulcers. J. Rehabil. Res. Dev. 37(4), 463–471 (2000)
21. Edsberg, L.E., et al.: Mechanical characteristics of human skin subjected to static versus cyclic normal pressures. J. Rehabil. Res. Dev. 36(2), 133–141 (1999)
22. Ferrant, M., et al.: 3D image matching using a finite element based elastic deformation model. In: Taylor, C., Colchester, A. (eds.) MICCAI 1999. LNCS, vol. 1679, pp. 202–209. Springer, Heidelberg (1999)
23. Fung, Y.C.: Biomechanics: Mechanical properties of living tissue, New York (1993)
24. Gefen, A., et al.: In vivo muscle stiffening under bone compression promotes deep pressure sores. J. Biomech. Eng. 127(3), 512–524 (2005)
25. Goldstein, B., et al.: Skin response to repetitive mechanical stress: a new experimental model in pig. Arch. Phys. Med. Rehabil. 79(3), 265–272 (1998)
26. Greve, J.M., et al.: Functional electrical stimulation (FES): muscle histochemical analysis. Paraplegia 31(12), 764–770 (1993)
27. Grimby, G., et al.: Muscle fiber composition in patients with traumatic cord lesion. Scand. J. Rehabil. Med. 8(1), 37–42 (1976)
28. Hagemann, A., et al.: Biomechanical modelling of the human head for physically based, nonrigid registration. IEEE Tran. Med. Imaging 18(10), 875–884 (1999)
29. Kaplan, R.L.: Physical Medicine & Rehabilitation Pearls of Wisdom. Boston Medical Publishing Corporation, Lincoln (2003)
30. Keyak, J.H., et al.: Prediction of femoral fracture load using automated finite element modeling. J. Biomech. 31(2), 125–133 (1998)
31. Kosiak, M.: Etiology of decubitus ulcers. Arch. Phys. Med. Rehabil. 42, 19–29 (1961)
32. Kuroda, S., et al.: Finite element analysis of undermining of pressure ulcer with a simple cylinder model. J. Nippon. Med. Sch. 72(3), 174–178 (2005)
33. Landry, E., et al.: Body weight, limb size, and muscular properties of early paraplegic mice. J. Neurotrauma. 21(8), 1008–1016 (2004)
34. Larabee, W.F.: A finite element model of skin deformation: I. Biomechanics of skin and soft tissue: a review. Laryngoscope 96, 399–405 (1986)
35. Larabee, W.F., et al.: A finite element model of skin deformation: III. The finite element model. Laryngoscope 96, 413–419 (1986)
36. Lieber, R.L., et al.: Long-term effects of spinal cord transection on fast and slow rat skeletal muscle. II. Morphometric properties. Exp. Neurol. 91(3), 435–448 (1986a)
37. Lieber, R.L., et al.: Long-term effects of spinal cord transection on fast and slow rat skeletal muscle. I. Contractile properties. Exp. Neurol. 91(3), 423–434 (1986b)
38. Lin, F., et al.: Finite Element (FE) Analysis for Evaluation of Pressure Ulcer on the Buttock: Part I: Development and validation, Summer Bioengineering Conference, Amelia Island, June 23 (2006)

39. Lin, F., et al.: FEM model for evaluating buttock tissue response under sitting load. In: 26th Annual Internation Conference of IEEE EMBS, San Francisco, CA. IEEE, Los Alamitos (2004)
40. Linder-Ganz, E., et al.: Mechanical compression-induced pressure sores in rat hindlimb: muscle stiffness, histology, and computational models. J. Appl. Physiol. 96(6), 2034–2049 (2004)
41. Linder-Ganz, E., et al.: Assessment of mechanical conditions in sub-dermal tissues during sitting: A combined experimental-MRI and finite element approach. J. Biomech. 40(7), 1443–1454 (2007)
42. Linder-Ganz, E., et al.: Real-time finite element monitoring of sub-dermal tissue stresses in individuals with spinal cord injury: toward prevention of pressure ulcers. Ann. Biomed. Eng. 37(2), 387–400 (2009)
43. Lotta, S., et al.: Morphometric and neurophysiological analysis of skeletal muscle in paraplegic patients with traumatic cord lesion. Paraplegia 29(4), 247–252 (1991)
44. Makhsous, M., et al.: Finite Element Analysis for Evaluation of Pressure Ulcer on the Buttock: Development and Validation. IEEE Trans. Neural Syst. Rehabil. Eng. 15(4), 517–525 (2007a)
45. Makhsous, M., et al.: Sitting with adjustable ischial and back supports: Biomechanical changes. Spine 28(11), 1113–1121 (2003)
46. Makhsous, M., et al.: Periodically Relieving Ischial Sitting Load to Decrease the Risk of Pressure Ulcers. Arch. Phys. Med. Rehabil. 88(7), 862–870 (2007b)
47. Martin, T.P., et al.: Influence of electrical stimulation on the morphological and metabolic properties of paralyzed muscle. J. Appl. Physiol. 72(4), 1401–1406 (1992)
48. Masi, A.T., et al.: Human resting muscle tone (HRMT): narrative introduction and modern concepts. J. Bodyw. Mov. Ther. 12(4), 320–332 (2008)
49. Mason, P.: Dynamic stiffness and crossbridge action in muscle. Biophys. Struct. Mech. 4(1), 15–25 (1977)
50. May-Newman, K., et al.: Homogenization modeling for the mechanics of perfused myocardium. Prog. Biophys. Mol. Biol. 69(2), 463–481 (1998)
51. Mazella, F., et al.: Auto acquisition of elastic properties of soft tissues for surgical simulation. Stanford-NASA biocomputation Center, Stanford (1999)
52. Metz, H., et al.: A comparison of the elasticity of live, dead, and fixed brain tissue. J. Biomech. 3(4), 453–458 (1970)
53. Miga, M.I., et al.: Model-updated image-guided neurosurgery using the finite element method: Incorporation of the Falx Cerebri. In: Taylor, C., Colchester, A. (eds.) MICCAI 1999. LNCS, vol. 1679, pp. 900–909. Springer, Heidelberg (1999)
54. Morrey, B.F., An, K.-N.: Biomechanics of the shoulder. In: Rockwood Jr., C.A., Matsen III, F.A. (eds.) The Shoulder, vol. 1, pp. 208–245 (1990)
55. Mow, V.C., et al.: Fluid transport and mechanical properties of articular cartilage. J. Biomech. 17, 377–394 (1984)
56. Nola, G.T., et al.: Differential response of skin and muscle in the experimental production of pressure sores. Plast. Reconstr. Surg. 66, 728–733 (1980)
57. Oomens, C., et al.: Can loaded interface characteristics influence strain distributions in muscle adjacent to bony prominences? Comput. Methods Biomech. Biomed. Engin. 6(3), 171–180 (2003)
58. Protz, P.R., et al.: Implementing magnetic resonanc imaging for the quantification of load-bearing buttocks tissues. In: The 13th Annual RESNA Conference, Washington, DC. RESNA Press (1990)

59. Ragan, R., et al.: Seat-interface pressures on various thicknesses of foam wheelchair cushions: a finite modeling approach. Arch. Phys. Med. Rehabil. 83(6), 872–875 (2002)
60. Round, J.M., et al.: Fibre areas and histochemical fibre types in the quadriceps muscle of paraplegic subjects. J. Neurol. Sci. 116(2), 207–211 (1993)
61. Roy, R.R., et al.: The plasticity of skeletal muscle: effects of neuromuscular activity. Exerc. Sport Sci. Rev. 19, 269–312 (1991)
62. Salcido, R., et al.: Histopathology of pressure ulcers as a result of sequential computer-controlled pressure sessions in a fuzzy rat model. Adv. Wound Care 7(5), 23–24, 26, 28 passim (1994)
63. Samani, A., et al.: Biomechanical 3D finite element modeling of the human breast using MRI data. IEEE Tran. on Medical Imaging 20(4), 271–279 (2001)
64. Sanders, J.E., et al.: Collagen fibril diameters increase and fibril densities decrease in skin subjected to repetitive compressive and shear stresses. J. Biomech. 34(12), 1581–1587 (2001)
65. Sanders, J.E., et al.: Skin response to mechanical stress: adaptation rather than breakdown–a review of the literature. J. Rehabil. Res. Dev. 32(3), 214–226 (1995)
66. Sarvazyan, D., et al.: Elastic imaging as a new modality of medical imaging for cancer detection. In: Proceedings of the international workshop on interaction of ultrasound with biological media, Valenciennes, France, pp. 69–81 (1994)
67. Scelsi, R., et al.: Muscle fiber type morphology and distribution in paraplegic patients with traumatic cord lesion. Histochemical and ultrastructural aspects of rectus femoris muscle. Acta. Neuropathol. (Berl) 57(4), 243–248 (1982)
68. Schnabel, J.A., et al.: Validation of non-rigid registration using finite element methods. In: Insana, M.F., Leahy, R.M. (eds.) IPMI 2001. LNCS, vol. 2082, pp. 344–357. Springer, Heidelberg (2001)
69. Sonka, M., et al.: Medical imaging. SPIE press, Washington(2000)
70. Stilwill, E.W., et al.: Histochemical and morphologic changes in skeletal muscle following cervical cord injury: a study of upper and lower motor neuron lesions. Arch. Phys. Med. Rehabil. 58(5), 201–206 (1977)
71. Sun, Q., et al.: Soft Tissue Stress in Buttock-Thigh of a Seated Individual Elucidated by a 3D FE Model. In: RESNA 28th Int. Conf., June 23-27 (2005)
72. Todd, B.A., et al.: Three-dimensional computer model of the human buttocks, in vivo. J. Rehabil. Res. Dev. 31(2), 111–119 (1994)
73. Wagnac, E.L., et al.: A new method to generate a patient-specific finite element model of the human buttocks. IEEE Trans. Biomed. Eng. 55(2), 774–783 (2008)
74. Wang, Y.N., et al.: How does skin adapt to repetitive mechanical stress to become load tolerant? Med. Hypotheses 61(1), 29–35 (2003)

Finite Element and Animal Studies of Scar Contractions Leading to Chronic Wounds

Cormac Flynn[1] and Brendan McCormack[2]

[1] Auckland Bioengineering Institute, Level 6, 70 Symond Street, Auckland 1010, New Zealand
c.flynn@auckland.ac.nz
[2] Institute of Technology, Sligo, Ireland
mccormack.brendan@itsligo.ie

Abstract. Abnormally healing or chronic wounds can result in excessive scar contraction. While beneficial in reducing the overall size of the wound, excessive scar contraction has numerous negative side-effects. It can deform the surrounding skin; inhibit motion where contraction occurs near joints; cause tissue necrosis and induce pain. This study presents an experimental and finite element model of scar contraction. The constitutive model of skin is represented by a validated orthotropic-viscoelastic law. The finite element model successfully simulated key characteristics observed in the experiment, in particular, the size and pattern of wrinkles that formed around the contracting scar. The orthotropic nature of skin significantly influences the orientation of the wrinkles – they form in a direction perpendicular to the Langer lines in skin. The size of the wrinkles is limited by the pre-stress inherent in skin. The wrinkle range decreased 67% and the wrinkle length decreased 83% when the skin tension was increased from 2.9 to 12.1 Nm-1. Also, a non-linear constitutive law better simulates the behavior of skin than a linear law. The proposed model will be useful in designing strategies to improve the healing of chronic wounds. These strategies would include the application of appropriate mechanical forces in the region of a wound to promote healing, minimize scarring and improve the quality of life of the patient.

1 Introduction

A healing wound undergoes a complex process consisting of three distinct phases – inflammation, fibroplasia and maturation (Stadelmann et al. 1998). During the initial inflammation phase, immediately after the occurrence of the wound, vascular vessels in the region dilate to reduce blood loss and platelets form to initiate clotting. Polymorphonuclear leukocytes (PMN), mononuclear leukocytes and wound macrophages form in succession to protect the wound from infection and remove wound debris. The fibroplasia phase starts within two to three days of the occurrence of the wound. Fibroblasts appear in the wound and there is an increase in collagen level for about three weeks, which correlates with an increase in the tensile strength of the wound. After about three weeks, the collagen rearranges into a more organized and structured lattice. This further increases the tensile strength of the maturing wound. The maturation phase of wound healing can last for more than a year, depending on the initial size of the wound (Stadelmann et al. 1998).

A. Gefen (Ed.): Bioengineering Research of Chronic Wounds, SMTEB 1, pp. 207–233.
springerlink.com

If a wound does not heal in a timely manner in the order of the phases described or does not acquire normal function, it is considered chronic (Robson 1997). One phenomenon that occurs in chronic wound healing is excessive contraction. Wound contraction is "the inward movement of the edges of an open wound, secondary to forces generated within that wound" (Rudolph 1980). Stadelmann et al. 1998 describes wound contraction as a persistent dominant force that exists even after the wound is healed. Contraction occurs as a result of a cellular mechanism within the wound involving contractile fibroblasts known as myofibroblasts. The myofibroblasts form intercellular attachments to one another and the wound edge. The entire wound bed shrinks and the wound contracts.

A certain degree of contraction can be beneficial in that it reduces the overall size of the wound. Excessive wound contraction, however, has many negative effects. The skin surrounding the contracted wound may wrinkle causing cosmetic deformity (Cerda 2005). In addition to being visually undesirable, the deformed and wrinkled skin may be an indicator of adverse stress fields, which can inhibit blood flow, leading to tissue necrosis and an unsightly scar (Bucalo and Iriondo 1995). Movement of the joints in the region of wound contraction can also be severely inhibited and painful. In some cases, joints can be dislocated (Rudolph 1980). Wound contraction can be problematic in breast reconstruction procedures. The contraction can compress around the breast implant resulting in significant pain (Rudolph 1980). Skin burns resulting in contracture usually requires surgery to restore normal motion and reduce pain (Gökrem et al. 2003).

Human skin is a complex material, which exhibits a non-linear orthotropic and viscoelastic stress-strain response (Wan Abas 1994; Daly 1982; Schneider 1982; Har-Shai et al. 1996; Silver et al. 2001). Skin *in vivo* is also in a state of anisotropic tension, which varies depending on body location, age and physical build. Due to its high visibility and importance, it has been studied in many areas of science and technology such as biomechanics, surgery, and cosmetology (Barbarino et al. 2009, Chabanas et al. 2003, Piérard et al. 2003). Much of this effort has been in the development of numerical models of human skin. The quality of these models has evolved significantly over the last few decades, such that several models can simulate accurately the mechanical characteristics of skin measured in various *in vivo* and *in vitro* experiments (Bischoff et al. 2004; Lanir 1983; Rubin et al. 1998).

Several finite element models have been developed to look at the stress fields around closed wounds of various shapes. Several of these models are two-dimensional and consider the skin as a single layer and ignore its non-linear stress-strain behavior (Lott-Crumpler and Chaudhry 2001; Retel et al. 2001). Kirby et al. 1998 developed a three-dimensional model but the skin was still assumed to be strictly linear elastic. The finite element analysis work of Yoshida et al. 2001 modeled the extrusion of skin when sutured using different methods. They used a simple linear isotropic constitutive law for skin. None of the models described above account for the *in vivo* tension in skin when simulating the closure of wounds. The skin tension would be an important factor in wound healing (Bucalo and Iriondo 1995). Cerda 2005 studied the effect of tension on the wrinkling around a contracting scar model but limited the analysis to isotropic polyethylene sheets.

There is a necessity for a model that can simulate the behavior of skin in the region of a healing wound. Wound healing models could be used to design strategies to aid the healing of chronic wounds. Such strategies could include determining appropriate mechanical forces to apply in the region of a wound (Cacou and Muir 1995) or what suitable pressure to apply on scars via garments, conformers or masks to promote healing and minimize scarring (Van den Kerckhove et al. 2001).

This chapter will examine the behavior of skin around a contracting scar. A simplified experimental model of a contracting scar in skin is presented. A finite element model of the experiment is also presented. An orthotropic-viscoelastic constitutive model represents the skin. The results of the finite element model are compared with those of the experiment. It will be shown that certain key characteristics in the constitutive model of skin are required in order to accurately simulate the behavior of skin around the contracting scar in the experiment.

2 Methods

2.1 Laboratory Experiment

Goat skin was used for the laboratory experiment, due to its availability. Its appearance and feel were also similar to human skin. The skin was cut into square samples of side 150 mm. Each sample appeared to be healthy and was free from any scars or blemishes. All hair was removed from the samples using a scissors and a razor. It was assumed that any possible micro-tearing of the outer skin layer by the razor was negligible. The orientation of each sample with respect to its orientation on the body was recorded. The orientation of each sample was such that the Langer or Borges lines were parallel to the side of the sample. Langer lines are obtained from cadavers, whereas Borges' lines are found by pinching the skin of a living person – the wrinkles are largest when the pinching is exerted at right angles to the relaxed skin tension lines (RSTL). According to Borges Albert 1960, Langer and Borges' lines correspond in most areas of the body apart from some facial areas, the elbows, fingers and lower abdomen. The differences are due to rigor mortis in the cadavers, which results in different lines of tension in the areas referred to above. They are the same for the skin used in this experiment. Once prepared, the skin was frozen 24 to 48 hours post mortem the samples until required for testing. Frozen storage has been reported as having a negligible effect on the mechanical properties of soft tissue (Foutz et al., cited in Wu et al. 2006a).

The skin sample was placed under equibiaxial tension on a square plate of side 300 mm (Fig. 1). The plate was supported by four legs of length 200 mm. There was a 10 mm radius hole in the centre of the plate. The edges of the hole and plate were raised 10 mm above the rest of the surface, such that the skin was only in contact with the edge of the hole. The edge of the hole had a 2 mm fillet to allow the skin to move smoothly over it. Holes were punched every 15 mm along the sides of the square sample. A hook was connected to each hole and fishing line ran from each hook over the rounded and lubricated edge of the plate. Weights were attached to the end of each line, thus applying an equibiaxial tension to the skin sample. The weights ranged from 0.06 to 0.3 N.

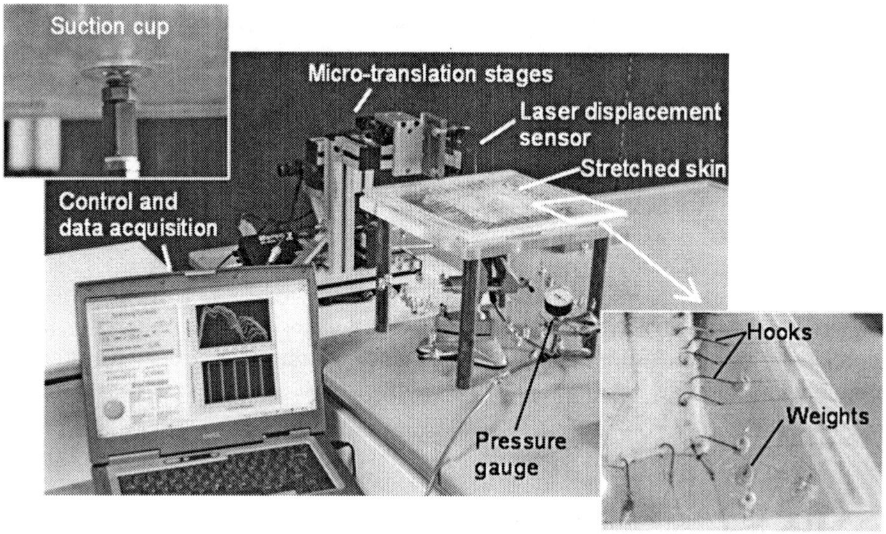

Fig. 1. Experimental rig with skin equibiaxially stretched on the supporting frame. Top-left inset shows the suction cup underneath the hole in the plate, which applies the vacuum pressure to the skin.

A square of side 100 mm was printed on the top surface of the skin prior to stretching it. It was noted that when the sample was under an equibiaxial tension, the sides of the square parallel to the assumed orientation of the Langer lines stretched less than the sides perpendicular to the assumed Langer line orientation. Skin is stiffer along the Langer lines. The assumed Langer line orientation in the sample was therefore correct. It was also noted that when the sample was under equibiaxial tension, the sides of the square remained orthogonal and parallel to each other, indicating there was no shear present in the sample.

A suction cup (Festo Type VAS-30-1/8-PUR) was located underneath the hole on the underside of the plate (Fig. 1). The cup was connected to a vacuum generator (Festo Type VAD-M5). The level of the vacuum was controlled by a flow-control valve and the pressure in the vacuum was measured using a vacuum gauge. As the pressure was reduced, the skin was drawn into the hole and wrinkles spread radially outwards. The sucking of the skin into the hole acted as a simplified model of a contracting circular scar. This is a similar approach to that taken by Cerda 2005, except that thin polyethylene sheets was used instead of animal skin.

The wrinkle profiles on the skin surface were measured using a laser profilometer system (Price 2005). It consisted of a charged coupled device (CCD) laser displacement sensor (LK-2000 Series, Keyence, Osaka, Japan) mounted on two micro-translation stages (M-111DG, Physik Instrumente Gmbh, Karlsruhe, Germany). With this arrangement, the laser-head could be moved in two orthogonal directions and a 15 x 15 mm square could be measured. The micro-translation stages were controlled by a DC-motor controller (Mercury II C-862, Physik Instrumente Gmbh, Karlsruhe, Germany). The controller was connected to the laptop computer through the series port. It was necessary to devise a system to measure a larger area than a 15 x 15 mm square. The profilometer was placed upon 5 mm grid paper. For each pressure and tension setting, the laser-head was located above the centre of the hole and a 15 x 15 mm square was measured and recorded. Upon completion, the profilometer was moved 15 mm on the grid paper and another 15 x 15 mm square was measured. The data from each 15 x 15 mm grid could then be joined together to give the skin profile over a larger area. The time taken to profile an area of dimension 30 x 30 mm was approximately 20 to 25 minutes.

The laser sensor was controlled by a laser controller unit (LK-2001, Keyence, Osaka, Japan), which was connected to a laptop computer (Dell Latitude D800 Pentium 1.6 GHz) via a data acquisition card (NI PCI-6036E, National Instruments, Austin, USA). A LabView code (National Instruments, Austin, USA) allowed the user to control the motion of the micro-translation stages and store the acquired data from it and the laser sensor.

Two different procedures were followed. In the first procedure, the vacuum pressure was kept constant at -0.5 Bar and the applied equibiaxial tension was varied. The variation in the equibiaxial tension represents the varying pre-tensions in *in vivo* human skin. *In vivo* pre-tensions vary according to location on the body and also according to the age of the person. The applied tensions were 18.4, 12.1, 9.2, 5.7 and 2.9 Nm^{-1}. In the second procedure, the skin sample was placed under a constant equibiaxial tension of 5.7 Nm^{-1} and the vacuum pressure underneath the hole reduced from 0.0 to -0.5 Bar in steps of 0.1 Bar. The reduction in the vacuum pressure represents the scar contracting with time. The thickness of the skin samples was measured using calipers. The average thickness was 1.4 mm. Based on this thickness, the levels of stress in the skin samples ranged from 2.07 to 13.14 kPa. Few values of *in vivo* pre-tension in human skin were reported in the literature. Alexander and Cook 1977 measured pre-tensions varying from 5 to 24 Nm^{-1} in the upper back of a male volunteer. Pre-tensions of 13 ±5 kPa have been measured in human volar forearm skin (Diridollou et al. 2000), which is equivalent to 12.35 ±4.75 Nm^{-1} using the 0.95 mm average skin thickness measured in that study. The applied tensions in the present study were chosen to cover the range of limited data in the literature.

In both procedures, at each vacuum pressure and equibiaxial tension, the wrinkle profiles were measured with the profilometer and the orientation of the wrinkles observed. The maximum wrinkle range and wrinkle length were calculated. The maximum wrinkle range was defined as the greatest vertical distance between a point on the crest of the wrinkle and a point at the bottom of an adjacent valley. Both points are at the same radial distance from the centre of the skin sample.

2.2 Constitutive Model of Skin

The skin is represented by an orthotropic-viscoelastic model similar to Bischoff et al. 2004. The skin is assumed to be a single homogeneous layer. It is assumed in this case that the mechanical properties of skin are dominated by the dermis and any contribution of the epidermis is ignored. A simplified rheological representation of the model is in Fig. 2. Network A represents the elastin fibers in the dermal layer that are in equilibrium. Network B represents the collagen fibers in the

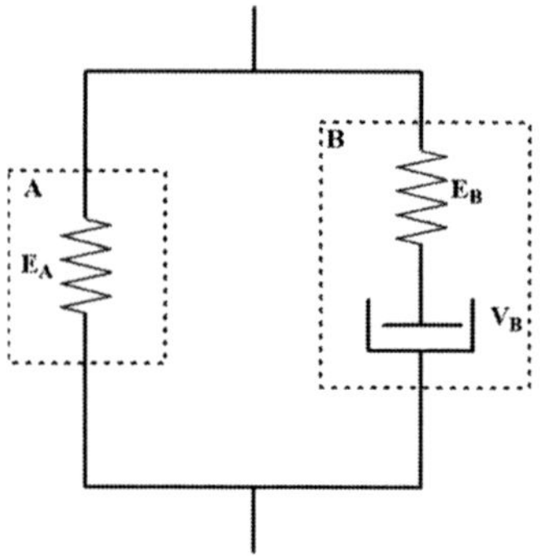

Fig. 2. Simplified rheological representation of skin with two parallel networks acting in unison. Network A represents collagen fibers in equilibrium, while B represents fibers moving with respect to the ground substance and other fibers in order to achieve a lower stress state.

dermis that are moving relative to the ground substance and other collagen fibers to achieve a lower stress state. The total Cauchy stress acting on the skin is the sum of the stresses in each network

$$\mathbf{T} = \mathbf{T_A} + \mathbf{T_B} \tag{1}$$

The total deformation gradient, **F**, is equal to the deformation gradients in each network

$$\mathbf{F} = \mathbf{F_A} = \mathbf{F_B} \tag{2}$$

The deformation gradient of Network B $\mathbf{F_B}$ can be further decomposed into an elastic and viscous component

$$\mathbf{F_B} = \mathbf{F_{Be}} \mathbf{F_{Bv}} \tag{3}$$

where $\mathbf{F_{Be}}$ is the deformation gradient for the elastic element and $\mathbf{F_{Bv}}$ is the deformation gradient for the viscous element in Network B.

Each network consists of an orthotropic hyperelastic element (Bischoff et al. 2002). The element is based on an assembly of eight fibers within an orthotropic unit cell as in Fig. 3(a). The strain energy potential for this cell is given by

$$W_X = W_{X0} + \frac{n_X k\theta}{4}\left[N_X \sum_i^4 \left(\frac{\rho_X^{(i)}}{N_X} \beta_{\rho_X}^{(i)} + \ln \frac{\beta_{\rho_X}^{(i)}}{\sinh \beta_{\rho_X}^{(i)}} \right) - \frac{\beta_{P_X}}{P_X} \ln\left(\lambda_{a_X}^{a_X^2} \lambda_{b_X}^{b_X^2} \lambda_{c_X}^{c_X^2} \right) \right] + B\{\cosh(J_X - 1) - 1\} \tag{4}$$

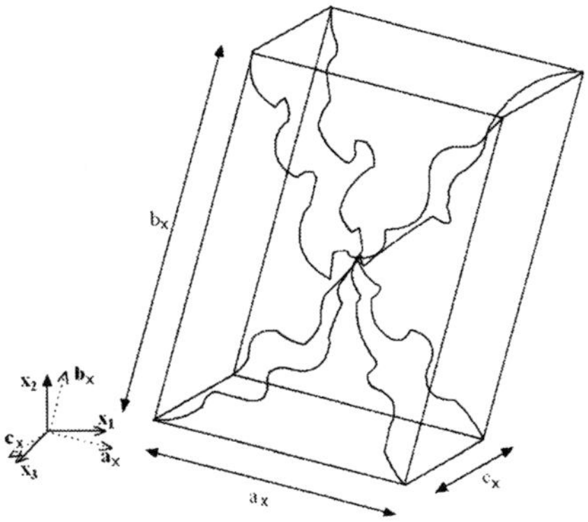

Fig. 3. 8-chain, three-dimensional orthotropic unit cell (adapted from Bischoff et al. 2002)

where the subscript X refers to Network A or B. The terms in the equation are explained in Table 1. The second Piola-Kirchhoff stress tensor in either network is given by

$$\mathbf{S_X} = \frac{dW_\mathbf{X}}{d\mathbf{E_X}} \tag{5}$$

where $\mathbf{E_X} = \left(\mathbf{F_X^T F_X} - \mathbf{I}\right)/2$ is the Lagrangian strain field for the elastic element in the network. The Cauchy stress tensor $\mathbf{T_X}$ can be calculated from the second Piola-Kirchhoff stress tensor as

$$\mathbf{T_X} = \mathbf{F_X S_X F_X^T} / J_\mathbf{X} \tag{6}$$

In order for $\mathbf{T}_\mathrm{B}$ to be evaluated for Network B, the deformation gradient $\mathbf{F}_\mathrm{Be}$ needs to be determined. The total velocity gradient of Network B is

$$\mathbf{L}_B = \dot{\mathbf{F}}_B \mathbf{F}_B^{-1} \tag{7}$$

where $\dot{\mathbf{F}}_B = \partial \mathbf{F}_B / \partial t$. This velocity gradient can be decomposed into elastic and viscous components

$$\mathbf{L}_B = \mathbf{L}_{Be} + \mathbf{L}_{Bv} \tag{8}$$

Substituting Equation 3 into Equation 7 and comparing with Equation 8 gives

$$\mathbf{L}_{Be} = \dot{\mathbf{F}}_{Be} \mathbf{F}_{Be}^{-1} \tag{9}$$

$$\mathbf{L}_{Bv} = \mathbf{F}_{Be} \dot{\mathbf{F}}_{Bv} \mathbf{F}_{Bv}^{-1} \mathbf{F}_{Be}^{-1} \tag{10}$$

Table 1. Definition of strain energy terms Equation 4. Subscript X refers to network A or B

W_{0X}	Strain energy of undeformed fibers
n_X	Number of fibers per unit volume of the network
k	Boltzman's constant, 1.38×10^{-23} JK^{-1}
	Absolute temperature
N_X	Number of freely jointed rigid links
$\beta_{\rho_X}^{(i)} = L^{-1}\left(\rho_X^{(i)} / N_X\right)$	Inverse Langevin function where $L(x) = \coth x - 1/x$
P_X	Undeformed fiber length; $P_X = \sqrt{a_X^2 + b_X^2 + c_X^2}/2$; a, b and c are the dimensions of the orthotropic cell in Fig. 3(a).
$\lambda_{a_X} = \sqrt{\mathbf{a}_X^{\mathbf{T}} \mathbf{C} \mathbf{a}_X}$, $\lambda_{b_X} = \sqrt{\mathbf{b}_X^{\mathbf{T}} \mathbf{C} \mathbf{b}_X}$, $\lambda_{c_X} = \sqrt{\mathbf{c}_X^{\mathbf{T}} \mathbf{C} \mathbf{c}_X}$	Principal stretches along the principal material axes for network. $\mathbf{C} = \mathbf{F}^{\mathbf{T}}\mathbf{F}$ is the right Cauchy stress tensor. $\mathbf{a}_X$, $\mathbf{b}_X$ and $\mathbf{c}_X$ form the axes that describe the orientation of the orthotropic cell in Fig. 3(a).
B	Bulk modulus
J_X	Volume ratio; $J = \det \mathbf{F}$

The velocity gradient can be decomposed into a symmetric part $\mathbf{D}_{Bv}$ and a skew-symmetric part $\mathbf{W}_{Bv}$ such that

$$\mathbf{L}_{Bv} = \mathbf{D}_{Bv} + \mathbf{W}_{Bv} \tag{11}$$

$\mathbf{D}_{Bv}$ is the rate of deformation tensor and $\mathbf{W}_{Bv}$ is the spin tensor of Network B. Following Boyce et al. 1989, $\mathbf{W}_{Bv}$ can be set to zero without loss in generality. Equating 10 and 11 and rearranging gives rate of change of the viscous deformation gradient

$$\dot{\mathbf{F}}_{Bv} = \mathbf{F}_{Be}^{-1}\mathbf{D}_{Bv}\mathbf{F}_{Be}\mathbf{F}_{Bv} \tag{12}$$

The rate of deformation tensor $\mathbf{D}_{Bv}$ is prescribed by (Bergström and Boyce 1998)

$$\mathbf{D}_{Bv} = \gamma_B \mathbf{N}_B \tag{13}$$

where $\mathbf{N}_B$ is the direction of the driving stress state and is given by

$$\mathbf{N}_B = \mathbf{T}_B' / \left(\sqrt{2}\tau_B\right) \tag{14}$$

$\mathbf{T}'_\mathbf{B} = \mathbf{T}_\mathbf{B} - trace[\mathbf{T}_\mathbf{B}]\mathbf{I}/3$ is the deviatoric part of the driving stress $\mathbf{T}_\mathbf{B}$ and $\tau_B = \left(trace[\mathbf{T}'_\mathbf{B}\mathbf{T}'_\mathbf{B}]/2\right)^{1/2}$ is the equivalent shear stress in Network B.

The viscoelasticity of skin is modelled using reptation dynamic theory, which was developed in the field of polymer science (de Gennes 1971), but has been used to model stress relaxation in cartilage (Fyhrie and Barone 2003). In reptation dynamics, the fibres are assumed to diffuse along their lengths in a snake-like manner. The theory is incorporated into the model through the expression for the effective creep rate of the collagen fibres in Network B

$$\gamma_B = C_1\left[\rho_{Bv}^{(i)} - 1\right]^{C_2}\tau_B^m \tag{15}$$

where $\rho_{Bv}^{(i)} = \left(trace\left(\mathbf{F}_{\mathbf{Bv}}\mathbf{F}_{\mathbf{Bv}}^T\right)/3\right)^{1/2}$ is the normalized deformed length of the viscous element in Network B. C_1, C_2, and m are material constants. Both C_1 and m are positive, while C_2 lies between 0 and -1.

Table 2. Orthotropic-viscoelastic model material parameters

n_A	Density of collagen fibers in Network A
n_B	Density of collagen fibers in Network B
B	Bulk modulus of skin
(a_A,b_A,c_A)	Dimensions of orthotropic unit cell of Network A
(a_B,b_B,c_B)	Dimensions of orthotropic unit cell of Network B
C_1, C_1, m	Viscoelastic material parameters

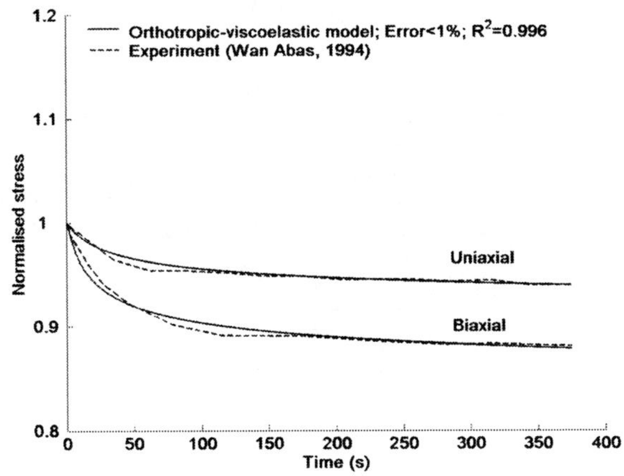

Fig. 4. Fitting orthotropic-viscoelastic model to *in vivo* uniaxial and biaxial stress relaxation tests of human forearm skin. Model parameters: $n_A = 1.641\text{x}10^{24}$; $n_B = 2.27\text{x}10^{22}$; $(a_A,b_A,c_A) = (2.265,1.378,1.894)$; $(a_B,b_B,c_B) = (1.481,1.419,0.694)$; B = 50 kPa; m=2.552; C_1 = 0.000034672 $s^{-1}(kPa)^{-m}$; C_2=-0.921. Model parameters are defined in Table 2.

The material parameters of the orthotropic-viscoelastic model (listed in Table 2) were optimized to fit experimental data from a series of *in vivo* uniaxial and biaxial stress relaxation tests on human skin (Wan Abas 1994). Both the model and experimental data are shown in Fig. 4, along with the model material parameters.

Further details on solving for the total Cauchy stress in the skin from Equations 1 to 15 can be found in (Flynn 2007; Bischoff et al. 2004, Bischoff et al. 2002).

2.3 Finite Element Study

ABAQUS/Explicit Version 6.5.1 (SIMULIA, Providence, RI) was used to develop a model of the laboratory experiment. The problem was assumed to be symmetrical so only one-quarter of the domain was modeled (Fig. 5). Symmetrical boundary conditions were placed along the appropriate edges of the model. The constitutive equation of skin developed in the previous section was implemented into the finite element model using the VUMAT user subroutine facility. The length and width of the skin section in the one-quarter model was 75 mm and it was 1.4 mm thick.

Approximately 30000 eight-noded brick elements (C3D8R) were used in the mesh. There were four elements across the thickness of the skin model. As most deformation occurred in the region of the central tube, the mesh density was much greater there. The aspect ratio of elements in this region ranged from 1.14 to 2.55. A mesh sensitivity analysis was carried out to ensure there was adequate convergence of the results. The density of the mesh was doubled until there was no significant change in the orientation, length and height of the resulting wrinkles.

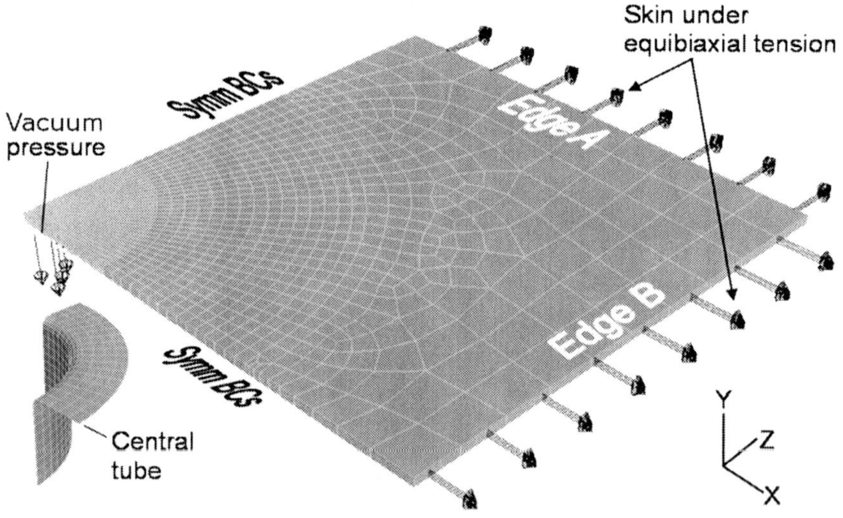

Fig. 5. Finite element model of skin: Skin meshed using approximately 30000 eight-noded brick elements; Biaxial stress applied to edges A and B and vacuum pressure applied smoothly over the inside area of the tube during second step. Skin raised above central tube in diagram for clarity only.

The central tube was modeled as a rigid body and was meshed using four-noded rigid elements.

It was assumed that goat skin was similar to most mammalian skin and was almost incompressible. Human skin has density of about 1100 kgm^{-3} (Leider and Buncke 1954) and it was assumed goat skin has a similar density.

The analysis consisted of two steps. For the first step, a biaxial tension was applied smoothly to the skin over a period of 5 seconds. The biaxial tensions applied in the model were 1.4, 2.1, 2.8, 5.7 and 9.1 Nm^{-1}. In the second step, a negative or vacuum pressure was applied over an area slightly smaller than the cross-sectional area of the tube. This had the effect of sucking the skin down into the central hole in the model. The time-period of the second step was 15 seconds.

As a result of the very fine mesh and the high non-linearity of skin, the maximum time step ABAQUS/Explicit could advance the solution in the current study was on the order of $1x10^{-6}$s resulting in very long analyses. A technique to reduce solution times, known as mass scaling, was employed. The analyses were assumed to be quasi-static in nature. Mass-scaling involved increasing the stability limit by artificially increasing the material density. For the analyses run in this study the density of the skin was increased by a factor of 100 in order to achieve favorable stability limits. To ensure the mass scaling was not having a dynamic inertial effect on the results it was checked that the kinetic energy of the system did not exceed 5% of the total energy over most of each analysis. This was recommended in the ABAQUS manual (Hibbitt et al. 2004).

The time period of the second step in the finite element analysis was 15 seconds and was a small fraction of the time period of the experiments, which were between one and two hours. The CPU time required to solve an analysis with a time period of two hours would be on the order of weeks, even if mass-scaling techniques were used.

The orientation, the mean of the maximum range and mean length of the wrinkles in the finite element analyses were recorded and compared with the experimental data. The stress fields around the wrinkles were also analyzed.

3 Results

3.1 Orientation and Distribution of Wrinkles

When the goat skin was under equibiaxial tension and the vacuum pressure was reduced, the skin was sucked into the central hole. Initially, the skin surface outside the hole was flat. However, beyond a certain critical vacuum pressure, wrinkles formed and extended out from the edge of the hole. It was observed in all the skin samples tested that the wrinkles tended to have one preferential orientation (Fig. 6(a)). This orientation was approximately aligned in a direction perpendicular to the assumed Langer lines in the sample.

This observation contrasted with the orientation of wrinkles formed around a contracting scar in an isotropic material, such as a polyethylene sheet, under equibiaxial tension. In this case, the wrinkles were distributed more evenly around the scar (Fig. 6(b)). There was not one preferential wrinkle orientation.

(a)

(b)

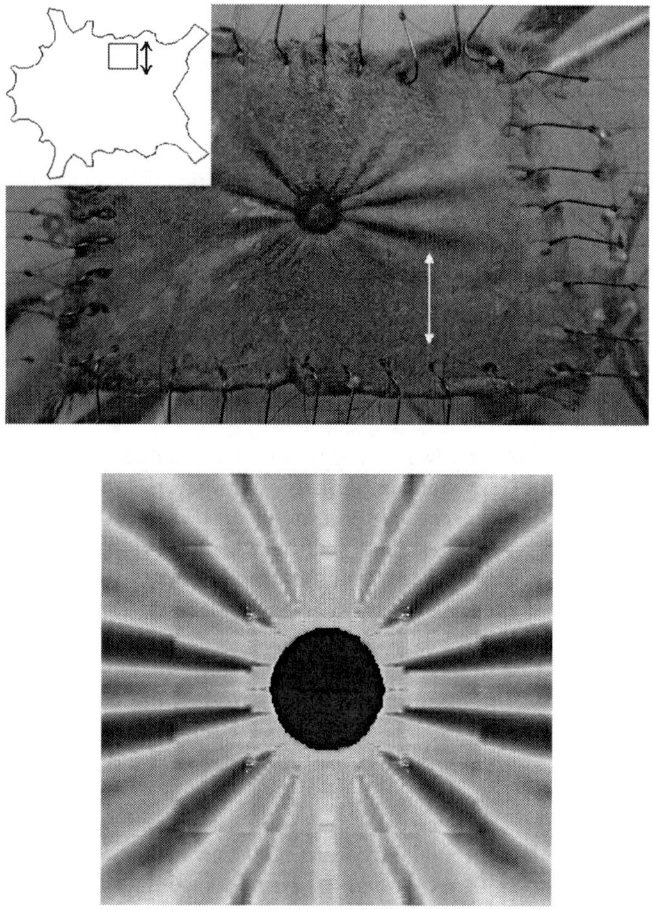

Fig. 6. Effect of orthotropy on the wrinkle distribution around contracting scars: (a) Wrinkles around a contracting scar in goat skin under an equibiaxial tension of 3.1 N m^{-1}; wrinkles have one preferred orientation perpendicular to the Langer lines (indicated by the arrows); inset indicates region of the hide where the skin sample was taken (Flynn and McCormack 2008b) (b) Height contours for wrinkles around contracting scars for isotropic polyethylene, under an equibiaxial tension of 2.8 Nm^{-1}; here, there is more than one wrinkle orientation.

This phenomenon was also observed in the finite element models. In the orthotropic skin model, the wrinkles formed around the scar when the vacuum pressure exceeded a critical value. The wrinkles were aligned in a direction perpendicular to the Langer lines in the model (Fig. 7(a)). The wrinkles formed around a contracting scar in an isotropic material model are shown in Fig. 7(b). The material parameters of this isotropic material were such that its stress-strain response was similar to the stress-strain response of the orthotropic material model in the stiffer direction. In this isotropic case, the wrinkles did not have a single preferred orientation.

(a)

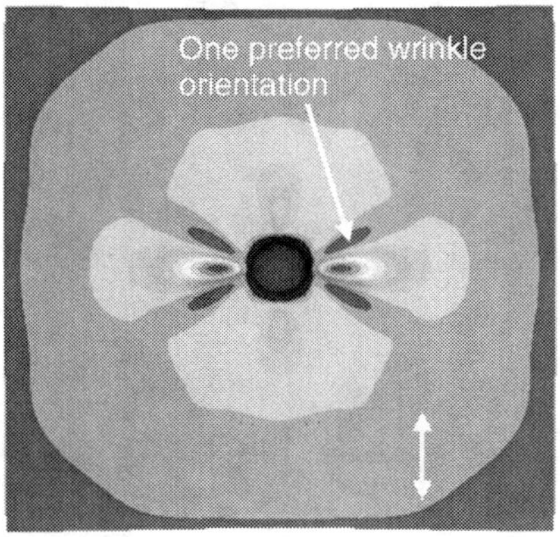

(b)

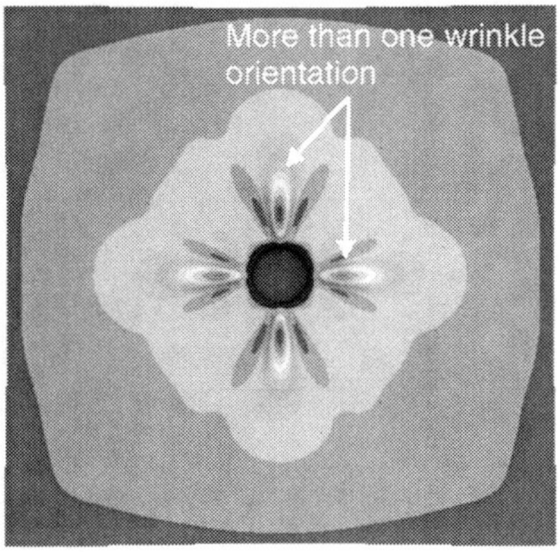

Fig. 7. Height contours for wrinkles around contracting scars for orthotropic and isotropic skin models, under an equibiaxial tension of 2.8 Nm^{-1}; For (a), the wrinkles have one preferred orientation, which is perpendicular to the Langer lines (indicated by the arrows); for an isotropic material (b) there is more than one dominant wrinkle orientation (Flynn and McCormack 2008b)

When the goat skin was under low-levels of tension, the wrinkles were more evenly distributed around the contracting scar. This is observed in Fig. 8(a), where the goat skin was subject to an equibiaxial tension of 2.9 Nm^{-1}. As the equibiaxial tension was increased, a preferential wrinkle orientation emerged, which corresponded to a direction perpendicular to the assumed Langer lines in the sample. Fig. 8(b) shows the single wrinkle that formed when the goat skin was under an equibiaxial tension of 12.1 Nm^{-1}.

This was also observed in the finite element models. When the orthotropic model had no tension applied, the wrinkles more uniformly distributed around the scar (Fig. 9(a)) than when there was an equibiaxial tension of 5.7 Nm^{-1} (Fig. 9(b)).

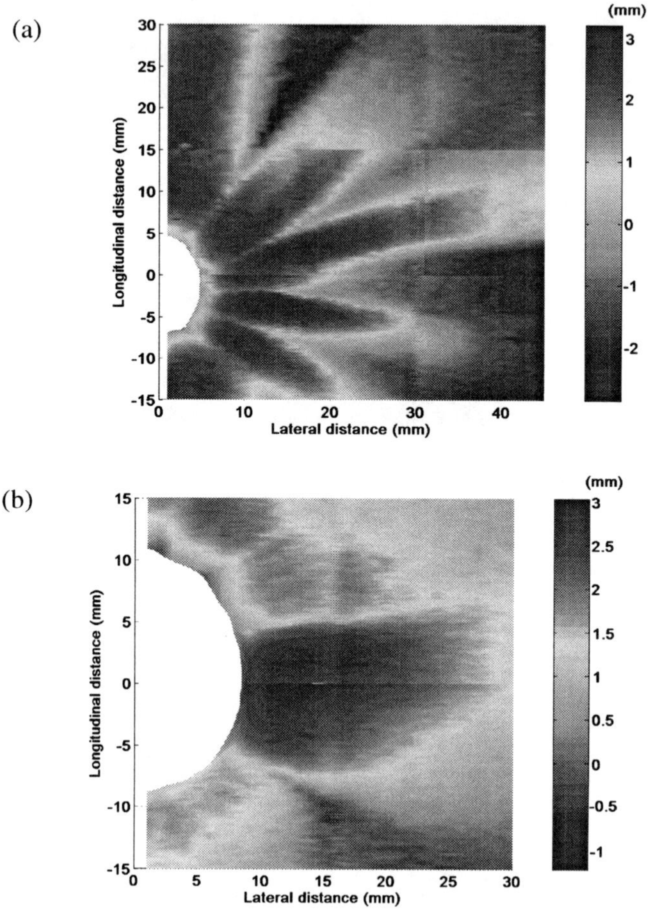

Fig. 8. Contour plots showing the heights of the wrinkles around the circular scar from a typical experiment. The pressure in the hole underneath the skin was -0.5 Bar. The Langer Lines lie approximately in the longitudinal direction. Equibiaxial tension in (a) was 2.9 Nm^{-1} and in (b) was 12.1 Nm^{-1}.

3.2 Dependence of Wrinkles on Skin Tension

When the vacuum pressure in the experiments was held constant at -0.5 Bar and the equibiaxial tension was increased, both the wrinkle range and length decreased non-linearly (Fig. 10(a) and (b)). The maximum range decreased from 2.86 to 0.85 mm when the tension increased from 2.9 to 18.4 Nm^{-1} (Fig. 10(a)). The length decreased from an average of 36 to 8 mm when the tension was increased from 2.9 to 18.4 Nm^{-1} (Fig. 10(b)).

For the models, the wrinkle range decreased non-linearly from 5.9 to 0.2 mm when the equibiaxial tension in the skin was increased from 1.4 to 9.1 Nm^{-1} as shown in Fig. 10(a). There was a corresponding non-linear decrease in the

(a)

(b)

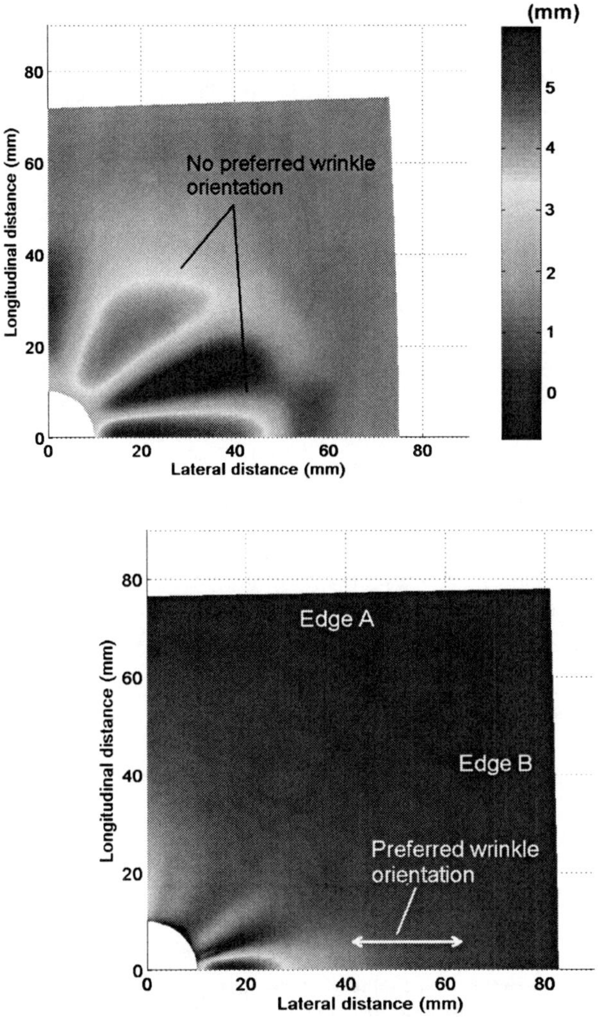

Fig. 9. Contour plots showing the heights of the wrinkles around the circular scar in the model. The pressure in the hole underneath the skin is -0.075 Bar. The equibiaxial tension in (a) was 0 Nm-1 and in (b) was 5.7 Nm-1. The Langer lines lie in the longitudinal direction.

predicted wrinkle length from 60 to 5 mm as shown in Fig. 10(b). The vacuum pressure in the hole underneath the skin models was much lower than the vacuum pressure in the experiment at -0.075 Bar.

3.3 Dependence of Wrinkles on Degree of Contraction

In the experiments, increasing the vacuum pressure modeled an increase in the level of scar contraction. The mean of the maximum ranges of the wrinkles

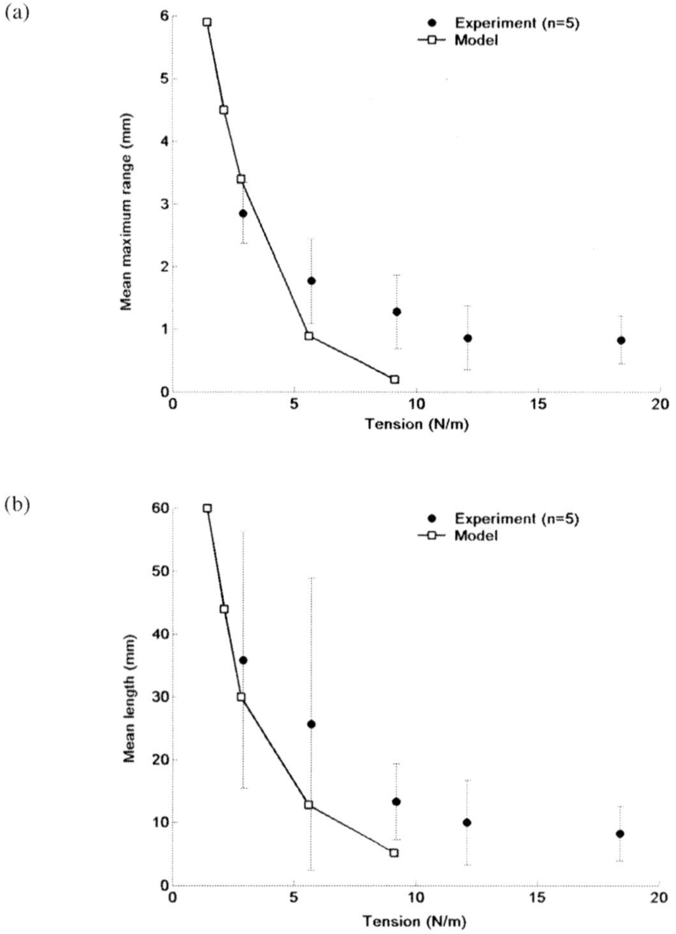

Fig. 10. The mean value and standard deviation of the (a) maximum range and (b) length of each measured wrinkle as a function of the applied equibiaxial tension. The experimental pressure in the hole underneath the goat skin is -0.5 Bar. In the models, the pressure in the hole underneath the skin is -0.075 Bar (data from Flynn and McCormack 2008b).

measured in the experiments decreased with decreasing scar contraction Fig. 11(a). The skin samples were under a biaxial tension of 5.7 Nm^{-1}. As the vacuum pressure decreased from -0.1 to -0.5 Bar, the maximum ranges of the wrinkles increased in an almost linear fashion from an average of 1 to 1.75 mm. The length of the wrinkles increased almost linearly from an average of 10 to 21 mm, as also shown in Fig. 11(a).

The maximum range and mean length predicted by the finite element models are shown in Fig. 11(b). The change of both the range and the length with pressure was almost linear as measured in the experiments. Both the magnitudes of the

(a)

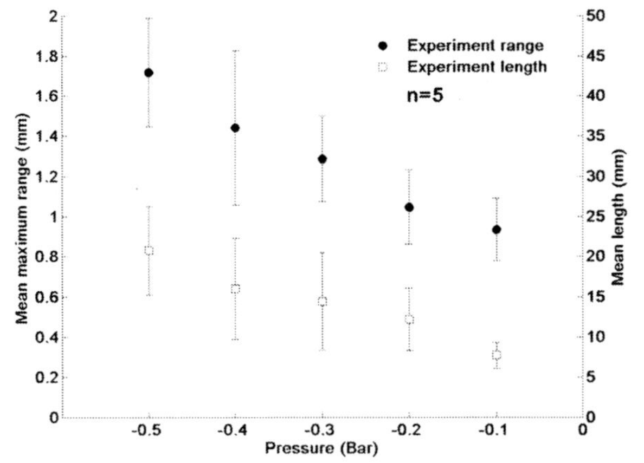

(b)

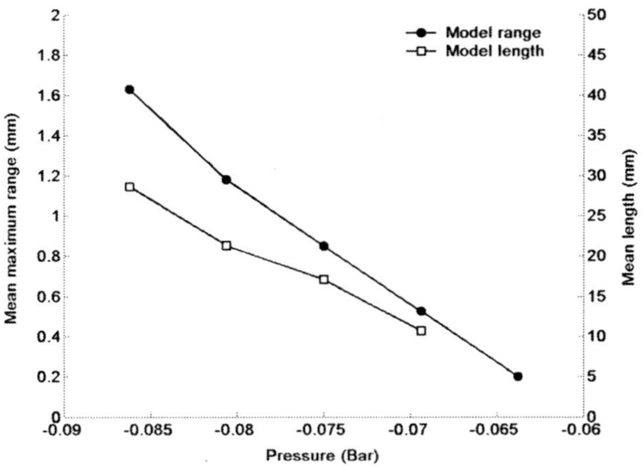

Fig. 11. (a) Experimental mean value and standard deviation of the maximum range and length of each wrinkle as a function of the vacuum pressure in the hole underneath the skin; (b) the corresponding quantities predicted by the finite element model. For the experiment and model, the applied equibiaxial tension was 5.7 Nm^{-1} (data from Flynn and McCormack 2008b)

wrinkle range and lengths were comparable with the experiment. However, the vacuum pressures in the model were much smaller than the pressures applied in the experiments. They ranged from -0.06 to -0.09 Bar in the model compared to -0.1 to -0.5 Bar in the experiments.

3.4 Stress Fields in Vicinity of Wrinkles

The Cauchy stress in the direction of the Langer lines, S_0, around the wrinkles of a contracting scar is shown in Fig. 12(a) and the Cauchy stress perpendicular to the

(a)

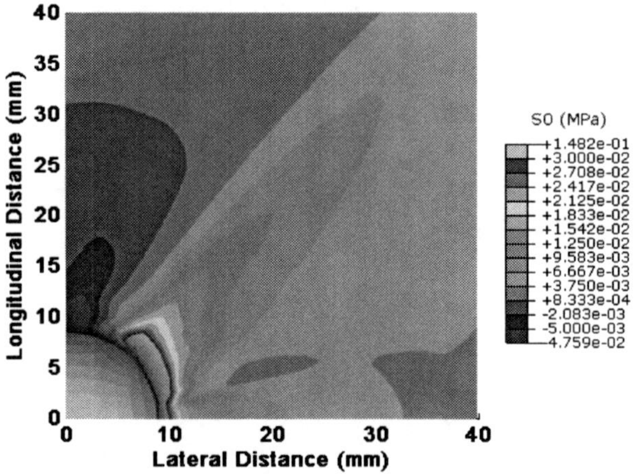

(b)

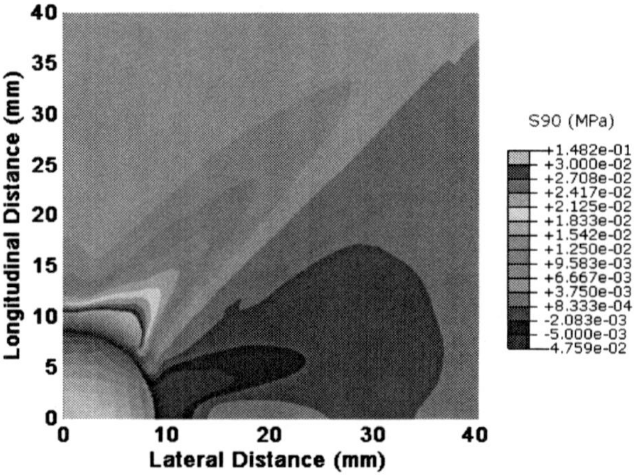

Fig. 12. Finite element analysis result showing (a) S0 and (b) S90 stress fields in the vicinity of wrinkles around a contracting scar. The stress fields to the wrinkle contours in Fig. (b). The pressure in the hole underneath the skin is -0.075 Bar. The equibiaxial tension was 5.7 Nm-1. The Langer lines lie in the longitudinal direction.

Langer lines, S_{90}, stress fields is shown in Fig. 12(b). These stress fields correspond to the wrinkle contours shown in Fig. 9(b). The pre-stress for this model was 5.7 Nm^{-1}, which is about 4 kPa given an undeformed skin thickness of 1.4 mm. Along the crest of the wrinkle, S_0 remains positive. Near the maximum range of the wrinkle, at about 18mm in the lateral direction, the stress is on the order of 10 kPa, which is significantly above the pre-stress of 4 kPa. There are more regions of elevated stresses of 10 to 20 kPa at an angle of 45° from the dominant wrinkle. The S_0 goes slightly negative 90° from the dominant wrinkle.

Along the dominant wrinkle direction, the S_{90} stress is compressive. In the region of this wrinkle, the S_{90} exceeds -5 kPa. Away from the wrinkle, the S_{90} significantly exceeds the 4 kPa pre-stress of the model. At an angle of 45° from the dominant wrinkle it ranges from 10 to 20 kPa.

3.5 Investigating the Assumption of a Homogeneous Stress Field

It was assumed that there was a homogeneous stress field in the skin before the formation of the contracting scar. In the experiment, the application of the tension

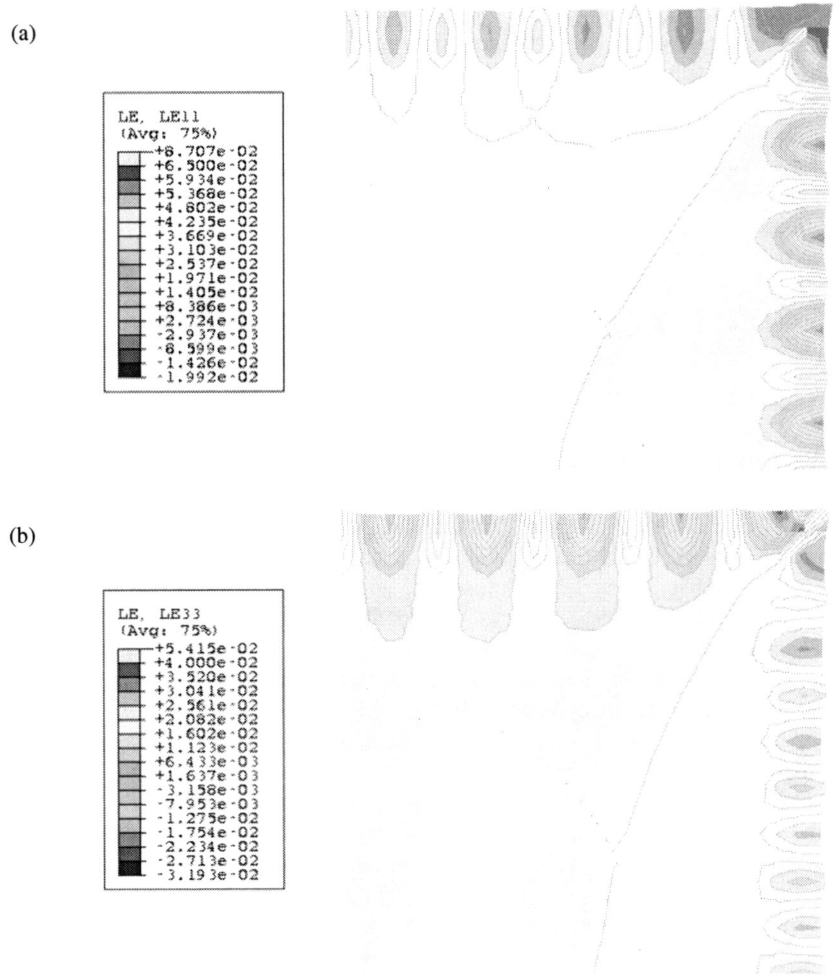

Fig. 13. (a) E_{11} and (b) E_{33} strain fields after skin model is equibiaxially stretched by applying loads at discrete points along edges. The strain field is mostly homogeneous except for the areas around each load point.

at discrete points along the edges of the sample would result in a heterogeneous stress field in the region of the attachment points. The biaxial tension in the finite element model was applied continuously along each edge of the skin. The extent of the heterogeneous stress field around the attachment points was investigated by simulating the equibiaxial stretching of the skin using discrete attachment points as per the experimental protocol. Point loads of 0.02625 N were applied at six points along each edge. This is equivalent to an equibiaxial tension 2.1 Nm^{-1}. The loads were applied smoothly over 5 seconds. Fig. 13(a) shows the E_{11} strain field and Fig. 13(b) shows the E_{22} strain field after the point loads were applied. Away from the attachment points, the strain fields are almost homogeneous. The area of maximum homogeneity (the bottom left section of the model) corresponds to the area of interest where the wrinkles form around the contracting scar. The stress field in this region was also found to be very homogeneous. In this region, the stress in the X direction was 1.4 kPa and in the Z direction was 1.3 kPa. These stresses are slightly lower than the 1.5 kPa stress field obtained when the loads are applied continuously along each edge of the skin. It can be concluded that although there are stress and strain inhomogeneities in the region of the discrete attachment points, the dimensions of the skin sample being modeled are large enough such that the area of interest lies in a sufficiently homogeneous stress and strain region. Therefore, simulating the laboratory experiment with continuously applied loads along the edges is a valid approach.

4 Discussion

The treatment of chronic wounds costs over US$1 billion in the United States and £3 billion in the UK annually (Walker et al. 2008; Posnett and Franks 2008). This financial cost does not account for the lost productivity, physical misery and impaired quality of life of the sufferer. As a result, chronic wound healing is a field subject to intense research. This research will lead to more effective treatment of chronic wounds.

The purpose of this study was to examine wrinkle formation around contracting wounds or scars. Samples of goat skin were equibiaxially stretched on a plate while the contracting scar was created by sucking the skin into a hole in the centre of the plate. A finite element model was also developed to simulate the laboratory experiment. The orientation and distribution of the wrinkles that formed as a result of the scar contraction were recorded. The dependence of the wrinkle range and length on the tension in the skin and the degree of scar contraction was also analyzed.

The physical and finite element models developed are a great simplification of an actual contracting scar. Wounds and scars are rarely circular in shape. There is no attempt in the current study to model the biological processes that lead to wound contraction. The time-scale of the scar contraction in the experiment was on the order of minutes, while in the finite element model, the scar contracted during a period of 15 seconds. Actual wound contraction processes last several weeks (Olsen et al. 1995). Despite these simplifications, the simplified model presented here identifies certain key characteristics of skin, which significantly influence the behavior of skin in the region of a chronically, contracting wound.

Comparing the results of the finite element model to the results of the experiments on real skin is an important development over the work of Cerda 2005 and Géminard et al. 2004, where isotropic latex and polyethylene membranes were used in place of real skin . Their models are unable to account for the influence of the orthotropy in real skin on the distribution of wrinkles around a contracting scar. The finite element model in the current study correctly predicted the orientation of the wrinkles around the contracting scar in the experiment on real skin. The reason for the single preferred wrinkle orientation observed in Fig. 6(a) and Fig. 7(a) may be attributable to the orthotropic characteristics of the material. In Fig. 7(a), the stiffness of the skin in the longitudinal-direction is greater than the stiffness in the lateral-direction. As the scar contracts, skin in the longitudinal-direction cannot stretch to the same extent as in the lateral-direction. Edge A in Fig. 5 displaces inwards to a greater extent than Edge B, causing the wrinkles to form predominantly in the lateral-direction. In an isotropic material, the stiffness is the same in all directions so wrinkles are evenly distributed around the contracting scar with no single preferred orientation (Fig. 6(b) and Fig. 7(b)). The results demonstrate that in order to make meaningful predictions of the deformation of skin in the region of chronic wounds, it is necessary to account for orthotropy in the skin model.

The finite element model showed that the stress field in the vicinity of the contracting scar was significantly different from the existing pre-stress in the skin (Fig. 12(a) and Fig. 12(b)). There were several regions with elevated stress levels of 5 to 15 kPa above the pre-stress in the skin. There were also regions of compressive stress. Stress has a very important role in wound healing. Regions of excessively high stress such as seen in Fig. 12(a) and Fig. 12(b) are likely to result in necrosis of the skin. The microcirculation in the region of a wound is reduced when the stress in the region is increased (Tønseth and Hokland 2004). Compressive stresses cannot be supported by the skin and so wrinkling occurs. The wrinkling and deformation of skin around a contracting wound are therefore strong indicators of abnormal physiological stresses, which may inhibit the complete healing of the wound.

The pre-stress in both the finite element and laboratory models significantly influenced the behavior of the wrinkles around the scar. It was observed that at lower levels of pre-stress, the wrinkles were more evenly distributed around the scar than at higher levels of pre-stress (Fig. 8 and Fig. 9). This phenomenon can be explained by the non-linear orthotropic stress-strain response of skin. At low levels of stress, the stiffness of the skin material is similar in the lateral and longitudinal directions (Fig. 14(a)). Therefore, at this stress-level skin behaves like an isotropic material and there is no single-preferred wrinkle orientation. As the pre-stress increases, the response of the skin becomes stiffer in the lateral direction than the longitudinal direction because it has entered the locking region. The skin is now behaving orthotropically and a single-preferred wrinkle orientation, perpendicular to the Langer lines, emerges. This is demonstrated by examining the distribution of wrinkles around a contracting scar in a hypothetical linear orthotropic material (Fig. 14(b)). For such a material, there is a single preferred wrinkle orientation at zero pre-stress in the direction perpendicular to the Langer lines. This contrasts

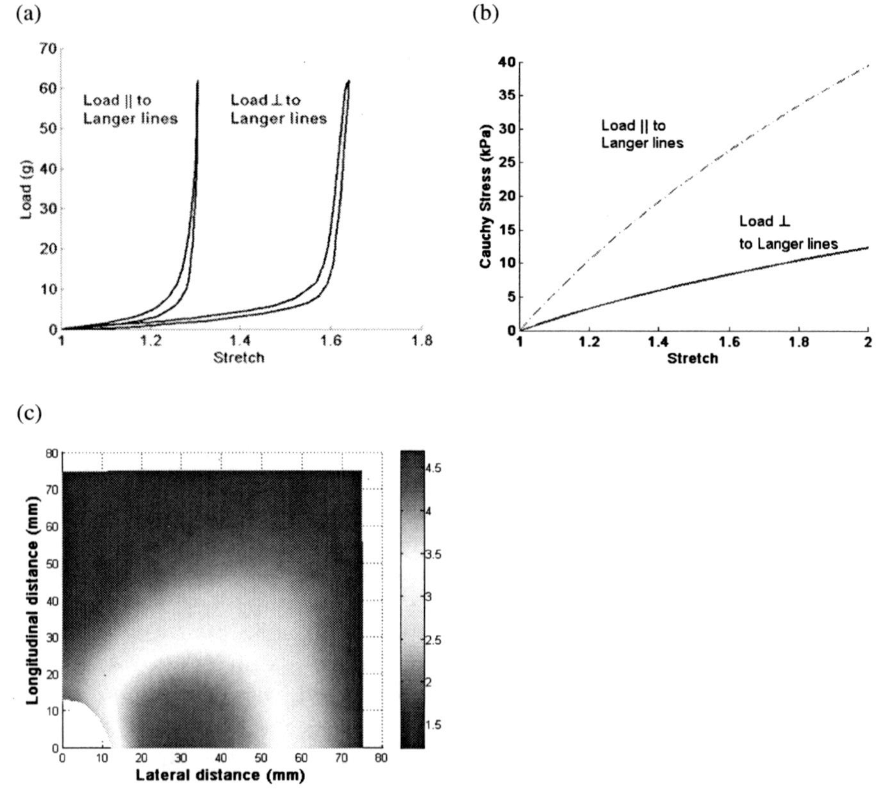

Fig. 14. Effect of non-linear orthotropy: (a) Uniaxial load-stretch response of rabbit skin (data from Lanir and Fung 1974). At low load or stretch levels, the stiffness of the skin is similar in both directions; (b) hypothetical linear orthotropic material; (c) wrinkles formed around a contracting scar in a linear orthotropic material with zero pre-stress. Wrinkle formed in direction perpendicular to Langer lines, which contrasts with the response of the non-linear orthotropic material in Fig. 4(a).

with the case of the non-linear orthotropic material response in Fig. 9(a). These results demonstrate the requirement for a non-linear orthotropic constitutive equation when predicting the behavior of skin around contracting scars.

The pre-stress in skin restricts the zone of wrinkling around the contracting scar. Both the finite element and laboratory models show a non-linear decrease in the length of the wrinkles as the equibiaxial tension in the skin is increased (Fig. 10(b)). While this result is expected, its demonstration in the context of a skin model is novel and provides evidence that inclusion of pre-stress or natural tension in a skin model is important when simulating wound contraction. Earlier wound closure models do not explicitly take into account the pre-stress in skin (Lott-Crumpler and Chaudhry 2001). With its inclusion of pre-stress, the proposed model in this study is an improvement over earlier skin models. Since stress plays an important role in wound healing and scar production (Cacou and Muir 1995;

Bucalo and Iriondo 1995), it is essential to include the pre-stress in a model of wound healing in order to make meaningful predictions.

In both the finite element model and laboratory experiment, the maximum range of the wrinkles non-linearly decreased as the tension in the skin increased (Fig. 10(a)). While both the length and range of the wrinkles measured in the experiment and predicted by the finite element model are qualitatively similar, the vacuum pressures for the experimental and model data were very different. The vacuum pressure for the experiment was -0.5 Bar, while it was -0.075 Bar for the finite element model.

Both the wrinkle range and length increased linearly with increasing vacuum pressure in the experiment (Fig. 11(a)) and the finite element model (Fig. 11(b)). There is reasonable qualitative agreement between the experiment and finite element mode. However, as in the case of varying tension, the range of vacuum pressure in the experiment (0 to -0.5 Bar) was much greater than the range in the finite element model (0 to -0.09 Bar).

One reason for the difference in vacuum pressures is that the mechanical properties of the goat skin used in the experiment and the human forearm skin used in the finite element model were different. A series of biaxial and compression tests should be performed on the skin samples used in the experiment. Material parameters for the orthotropic-viscoelastic model could then be obtained through non-linear optimization. Simulating the behavior of the goat skin under relatively simple homogenous biaxial and compression tests may improve the accuracy of more complex simulations such as the wrinkling of skin around a contracting scar.

Biological skin is a multi-layer material. The hypodermis was ignored in the model because it was not present in the skin samples used in the experiment. The epidermis was also ignored in the model, even though it was not removed from the skin samples used in the experiment. The living part of the epidermis is usually assumed to not make significant contributions to the mechanical properties of skin (Silver et al. 2001). However, the uppermost layer of the epidermis, the stratum corneum, is known to contribute significantly to the mechanical properties of skin (Batisse et al. 2002). Although about 50 times thinner than the dermis in many areas of the body, it can be up to 1000 times stiffer (Wu et al. 2006b; Yuan and Verma 2006). In the finite element model, there were fewer wrinkles than observed in the experiment (see Fig. 7 and Fig. 8). Flynn and McCormack 2008a have shown the presence of the stratum corneum in a multi-layer model of skin increases the number of wrinkles formed. It is reasonable to assume that if the stratum corneum was included in the skin model in this study more wrinkles would form around the contracting scar.

According to finite element analyses, the application of the equibiaxial tension at discrete points along the skin edges did not result in a heterogeneous stress field in the region of interest (Fig. 13). Other factors, however, could contribute to a non-homogeneous stress-state in the region of interest, such as a varying tension applied along each side of the sample. This variation may be due to friction between the fishing-line and the edge of the support plate, differences in the connection between the hooks and the skin or local differences in the skin thickness. Friction between the skin and the edge of the hole could also result in a heterogeneous stress field around the contracting scar. Friction was ignored in the finite element model. Steps were taken to minimize the effects of friction, such as lubricating the edge of the hole and milling a fillet radius on the edge of the hole.

The wrinkles formed around the contracting scars in this study had a single preferred orientation, which corresponded to a direction perpendicular to the Langer lines in the skin. When skin is pinched on the volar forearm, for example, the wrinkles are larger and longer when they are parallel to the Langer lines in the region (Borges 1989). The reason for the difference in the wrinkle orientation observed in this study and the orientation in *in vivo* skin is due to differences in the tension distribution. The goat skin in the experiment and the skin in the finite element model were in equibiaxial tension. *In vivo* skin is, in general, not in equibiaxial tension. It is greatest in the direction of the Langer or Borges' lines. The variation of the *in vivo* tension with direction and the material orthotropy of the skin influence the orientation of the wrinkles. While the equibiaxial tension used in this study is a limitation, it does allow us to study the effect of material orthotropy on wrinkle orientation in isolation from a varying tension field. The equibiaxial tension field did demonstrate the importance of pre-stress on the size of any wrinkles formed. Pre-stress was not explicitly considered in earlier wound closure and suture models.

The experiment presented in this study provides a means to validate a finite element model of scar contraction in skin. Key characteristics were observed in both the experimental and finite element model, including an increasing wrinkle range and length with increasing scar contraction and decreasing pre-stress. In particular, the non-linear orthotropic nature of skin has a significant influence on the wrinkle distribution around the contracting scar. Clearly, there are many improvements to be made to the finite element model. A more anatomically accurate model would include separate layers representing the epidermis, dermis and underlying hypodermis. A more realistic pre-stress distribution will also be required. The finite element model could eventually be coupled with mechanochemical models that simulate the biological processes within a wound that cause contraction (Olsen et al. 1995, Tranquillo and Murray 1992; Sherratt and Dallon 2002). Such a sophisticated model would provide a foundation to formulate better wound management strategies. A patient-specific model could be constructed to simulate the healing of a chronic wound. The model could determine whether wound contraction is likely to cause significant problems in that area of the body. If so, physical constraints could be applied in the model to see whether they improve the healing process. Certain chronic wound problems can therefore be anticipated and minimized. This would reduce the financial cost of chronic wound management and, most importantly, improve the quality of life of the patient.

Acknowledgments. We would like to acknowledge funding provided by the Council of Directors, Strand III Research program for this research. We would also like to thank Eamonn Price for technical assistance with the laser profilometer and Gordon Muir for support with the experimental set-up.

References

1. Albert, F.B.: Scar Prognosis of Wounds. Brit. J. Plast. Surg. 13, 47–54 (1960)
2. Alexander, H., Cook, T.H.: Accounting for Natural Tension in the Mechanical Testing of Human Skin. J. Investig. Dermatol. 69, 310–314 (1977)
3. Barbarino, G.G., Jabareen, M., Trzewik, J., Nkengne, A., Stamatas, G., Mazza, E.: Development and Validation of a Three-Dimensional Finite Element Model of the Face. J. Biomech. Eng. 131, 041006–041011 (2009)
4. Batisse, D., Bazin, R., Baldeweck, T., Querleux, B., Lévêque, J.-L.: Influence of Age on the Wrinkling Capacities of Skin. Skin Res. Technol. 8, 148–154 (2002)
5. Bergström, J.S., Boyce, M.C.: Constitutive Modeling of the Large Strain Time-Dependent Behavior of Elastomers. J. Mech. Phys. Solids 46, 931–954 (1998)
6. Bischoff, J., Arruda, E., Grosh, K.: A Rheological Network Model for the Continuum Anisotropic and Viscoelastic Behavior of Soft Tissue. Biomech. Model Mech. 3, 56–65 (2004)
7. Bischoff, J.E., Arruda, E.A., Grosh, K.: A Microstructurally Based Orthotropic Hyperelastic Constitutive Law. J. Appl. Mech. 69, 570–579 (2002)
8. Borges, A.F.: Relaxed Skin Tension Lines. Dermatol. Clin. 7, 169–177 (1989)
9. Boyce, M.C., Weber, G.G., Parks, D.M.: On the Kinematics of Finite Strain Plasticity. J. Mech. Phys. Solids 37, 647–665 (1989)
10. Bucalo, B.D., Iriondo, M.: Photoelastic Models of Wound Closure Stress. Dermatol. Surg. 21, 210–212 (1995)
11. Cacou, C., Muir, I.F.K.: Effects of Plane Mechanical Forces in Wound Healing in Humans. J. Roy. Coll. Surg. Edin. 40, 38–41 (1995)
12. Cerda, E.: Mechanics of Scars. J. Biomech. 38, 1598–1603 (2005)
13. Chabanas, M., Luboz, V., Payan, Y.: Patient Specific Finite Element Model of the Face Soft Tissues for Computer-Assisted Maxillofacial Surgery. Med. Image Anal. 7, 131–151 (2003)
14. Daly, C.H.: Biomechanical Properties of Dermis. J. Investig. Dermatol. 79, s17–s20 (1982)
15. de Gennes, P.G.: Reptation of a Polymer Chain in the Presence of Fixed Obstacles. J. Chem. Phys. 55, 572–579 (1971)
16. Diridollou, S., Patat, F., Gens, F., Vaillant, L., Black, D., Lagarde, J.M., Gall, Y., Berson, M.: In Vivo Model of the Mechanical Properties of the Human Skin under Suction. Skin Res. Technol. 6, 214–221 (2000)
17. Flynn, C.: The Design and Validation of a Multi-Layer Model of Human Skin. PhD Thesis. Institute of Technology, Sligo (2007)
18. Flynn, C., McCormack, B.A.O.: Finite Element Modelling of Forearm Skin Wrinkling. Skin Res. Technol. 14, 261–269 (2008a)
19. Flynn, C., McCormack, B.A.O.: A Simplified Model of Scar Contraction. J. Biomech. 41, 1582–1589 (2008b)
20. Fyhrie, D.P., Barone, J.R.: Polymer Dynamics as a Mechanistic Model for the Flow-Independent Viscoelasticity of Cartilage. J. Biomech. Eng. 125, 578–584 (2003)
21. Géminard, J.C., Bernai, R., Melo, F.: Wrinkle Formations in Axi-Symmetrically Stretched Membranes. Eur. Phys. J. E. 15, 117–126 (2004)
22. Gökrem, S., Özdemir, O., Demirseren, M.E., Katircioğlu, A., Can, Z., Sevin, K.: Correction of a Mild Breast Contracture with a New Technique: "V-Y-Z-Plasty". Eur. J. Plast. Surg. 26, 255–257 (2003)

23. Har-Shai, Y., Bodner, S.R., Egozy-Golan, D., Lindenbaum, E.S., Ben-Izhak, O., Mitz, V., Hirshowitz, B.: Mechanical Properties and Microstructure of the Superficial Musculoaponeurotic System. Plast. Reconstr. Surg. 98, 59–70 (1996)
24. Hibbitt, D., Karlsson, B., Sorenson, P.: Abaqus Analysis User's Manual Version 6.5 ABAQUS, Inc. Pawtucket, RI (2004)
25. Kirby, S.D., Wang, B., To, C.W.S., Lampe, H.B.: Nonlinear, Three-Dimensional Finite-Element Model of Skin Biomechanics. J. Otolaryngol. 27, 153–160 (1998)
26. Lanir, Y.: Constitutive Equations for Fibrous Connective Tissues. J. Biomech. 16, 1–12 (1983)
27. Lanir, Y., Fung, Y.C.: Two-Dimensional Mechanical Properties of Rabbit Skin–Ii. Experimental Results. J. Biomech. 7, 171–174 (1974)
28. Leider, M., Buncke, C.M.: Physical Dimensions of the Skin; Determination of the Specific Gravity of Skin, Hair, and Nail. A.M.A. Archives of Dermatology and Syphilology 69, 563–569 (1954)
29. Lott-Crumpler, D.A., Chaudhry, H.R.: Optimal Patterns for Suturing Wounds of Complex Shapes to Foster Healing. J. Biomech. 34, 51–58 (2001)
30. Olsen, L., Sherratt, J.A., Maini, P.K.: A Mechanochemical Model for Adult Dermal Wound Contraction and the Permanence of the Contracted Tissue Displacement Profile. J. Theor. Biol. 177, 113–128 (1995)
31. Piérard, G.E., Uhoda, I., Piérard-Franchimont, C.: From Skin Microrelief to Wrinkles. An Area Ripe for Investigation. J. Cosmet. Dermatol. 2, 21–28 (2003)
32. Posnett, J., Franks, P.J.: The Burden of Chronic Wounds in the UK. Nurs. Times 104, 44–45 (2008)
33. Price, E.: Design and Validation of a Device to Measure the Cutting Edge Profile of Osteotomes. Master Thesis. Institute of Technology, Sligo (2005)
34. Retel, V., Vescovo, P., Jacquet, E., Trivaudey, F., Varchon, D., Burtheret, A.: Nonlinear Model of Skin Mechanical Behaviour Analysis with Finite Element Method. Skin Res. Technol. 7, 152–158 (2001)
35. Robson, M.C.: Wound Infection: A Failure of Wound Healing Caused by an Imbalance of Bacteria. Surg. Clin. N. Amer. 77, 637–650 (1997)
36. Rubin, M.B., Bodner, S.R., Binur, N.S.: An Elastic-Viscoplastic Model for Excised Facial Tissues. J. Biomech. Eng. 120, 686–689 (1998)
37. Rudolph, R.: Contraction and the Control of Contraction. World. J. Surg. 4, 279–287 (1980)
38. Schneider, D.: Viscoelasticity and Tearing Strength of the Human Skin. PhD Thesis. University of California (1982)
39. Sherratt, J.A., Dallon, J.C.: Theoretical Models of Wound Healing: Past Successes and Future Challenges. C. R. Biol. 325, 557–564 (2002)
40. Silver, F.H., Freeman, J.W., DeVore, D.: Viscoelastic Properties of Human Skin and Processed Dermis. Skin Res. Technol. 7, 18–23 (2001)
41. Stadelmann, W.K., Digenis, A.G., Tobin, G.R.: Physiology and Healing Dynamics of Chronic Cutaneous Wounds. Am. J. Surg. 176, S26–S38 (1998)
42. Tønseth, K.A., Hokland, B.M.: Evaluation of Microcirculation and Wound-Closing Tension after Undermining the Skin. A Study in a Porcine Model Using Laser Doppler Perfusion Imaging. Eur. J. Plast. Surg. 27, 295–297 (2004)
43. Tranquillo, R.T., Murray, J.D.: Continuum Model of Fibroblast-Driven Wound Contraction: Inflammation-Mediation. J. Theor. Biol. 158, 135–172 (1992)

44. Van den Kerckhove, E., Stappaerts, K., Boeckx, W., Van den Hof, B., Monstrey, S., Van der Kelen, A., De Cubber, J.: Silicones in the Rehabilitation of Burns: A Review and Overview. Burns 27, 205–214 (2001)
45. Walker, M., Hadgraft, J., Lane, M.E.: Investigation of the Permeability Characteristics of Peri-Ulcer and Whole Ischaemic Skin Tissue. Int. J. Pharm. 357, 1–5 (2008)
46. Wan Abas, W.A.B.: Biaxial Tension Test of Human Skin in Vivo. Bio-Med. Mater. Eng. 4, 473–486 (1994)
47. Wu, J.Z., Cutlip, R.G., Welcome, D., Dong, R.G.: Estimation of the Viscous Properties of Skin and Subcutaneous Tissue in Uniaxial Stress Relaxation Tests. Bio-Med. Mater. Eng. 16, 53–66 (2006a)
48. Wu, K.S., van Osdol, W.W., Dauskardt, R.H.: Mechanical Properties of Human Stratum Corneum: Effects of Temperature, Hydration, and Chemical Treatment. Biomaterials 27, 785–795 (2006b)
49. Yoshida, H., Tsutsumi, S., Mizunuma, M., Yanai, A.: A Surgical Simulation System of Skin Sutures Using a Three-Dimensional Finite Element Method. Clin. Biomech. 16, 621–626 (2001)
50. Yuan, Y., Verma, R.: Measuring Microelastic Properties of Stratum Corneum. Colloid Surface B 48, 6–12 (2006)

Cellular Deformations under Compression in Cells Involved in Deep Tissue Injury

Noa Slomka, Shira Or-Tzadikario, and Amit Gefen

Department of Biomedical Engineering, Faculty of Engineering, Tel Aviv University,
Tel Aviv 69978, Israel
gefen@eng.tau.ac.il

Abstract. Deep tissue injury (DTI) is a serious lesion typically involving necrosis of skeletal muscle and fat tissues under intact skin. Currently, considerable research efforts are invested in understanding the mechanisms underlying the onset and progression of DTI. Recent studies indicated the involvement of deformation-related events at the cellular scale. Nevertheless, the specific processes at the cell level which ultimately lead to DTI are still unknown. We hypothesize that sustained deformations of soft tissues may lead to individual cell death, as a result of alteration in intracellular concentrations of cell metabolites that occur due to local plasma membrane stretches. A two-dimensional model of an adhered single generalized cell, three-dimensional models of adhered single myoblast and fibroblast, and a construct of cells embedded in ECM were developed. Finite-Element analyses of the compressed models were performed in order to study localized plasma membrane stretches. Models were compressed by a rigid plate, up to maximal global deformations of 65%, 35%, and 45%, respectively. Large deformation strain analysis was performed, and maximal local principal strains in the plasma membrane of the cells were obtained as function of the global deformation applied to the model. All models indicated that platen compression causes large tensional strains in segments of the plasma membrane. Three-Dimensional models of real cell geometry exhibited a maximal tensional strain of approximately 20%, at global cell deformation of 35%. These results support our above hypothesis, and may provide a new path in DTI research.

1 Introduction

Deep tissue injury (DTI) is a life-threatening pressure ulcer, involving the formation of a necrotic tissue mass (Agam and Gefen, 2007; Black et al., 2007; Bouten et al., 2003; Gefen, 2007; Stekelenburg et al., 2008). This injury initiates in internal muscular tissue that overlies bony prominences as a result of sustained soft tissue deformations. Subsequently, it progresses outwards until becoming visible once it emerges at the skin surface (Agam and Gefen, 2007; Black et al., 2007; Bouten et al., 2003; Gefen, 2007; Stekelenburg et al., 2008). Skeletal muscle, subcutaneous fat and skin tissues are all involved in the process of DTI formation. Currently, considerable research efforts are invested in understanding the mechanisms underlying the onset and progression of DTI. Several processes are thought to participate, including ischemic/hypoxic damage (Ceelen et al., 2008; Stekelenburg et al., 2007) ischemia-reperfusion injury (Peirce et al., 2000; Tsuji

A. Gefen (Ed.): Bioengineering Research of Chronic Wounds, SMTEB 1, pp. 235–248.
springerlink.com

et al., 2005), structural damage to cells (Breuls et al., 2003; Gawlitta et al., 2007a; Gefen et al., 2008), as well as impaired interstitial fluid flow and lymphatic drainage (Miller and Seale, 1981; Reddy and Cochran, 1981). Recently however, there are reports that indicate the involvement of deformation-related events at the cellular scale as being directly responsible for DTI (Ceelen et al., 2007; Gawlitta et al., 2007b). Nevertheless, the specific processes at the cell level which ultimately lead to DTI are still unknown.

Sustained loads in bony prominences leads to large deformations in the overlying soft tissues. These large deformations are likely to induce deformation and distortion of individual cells at the bone vicinity (Breuls et al., 2002). These deformations at the cellular level are further likely to induce large deformations of the plasma membrane of cells, as well as of membranes surrounding cell organelles. When localized tensional or compressive forces act, membranes are locally

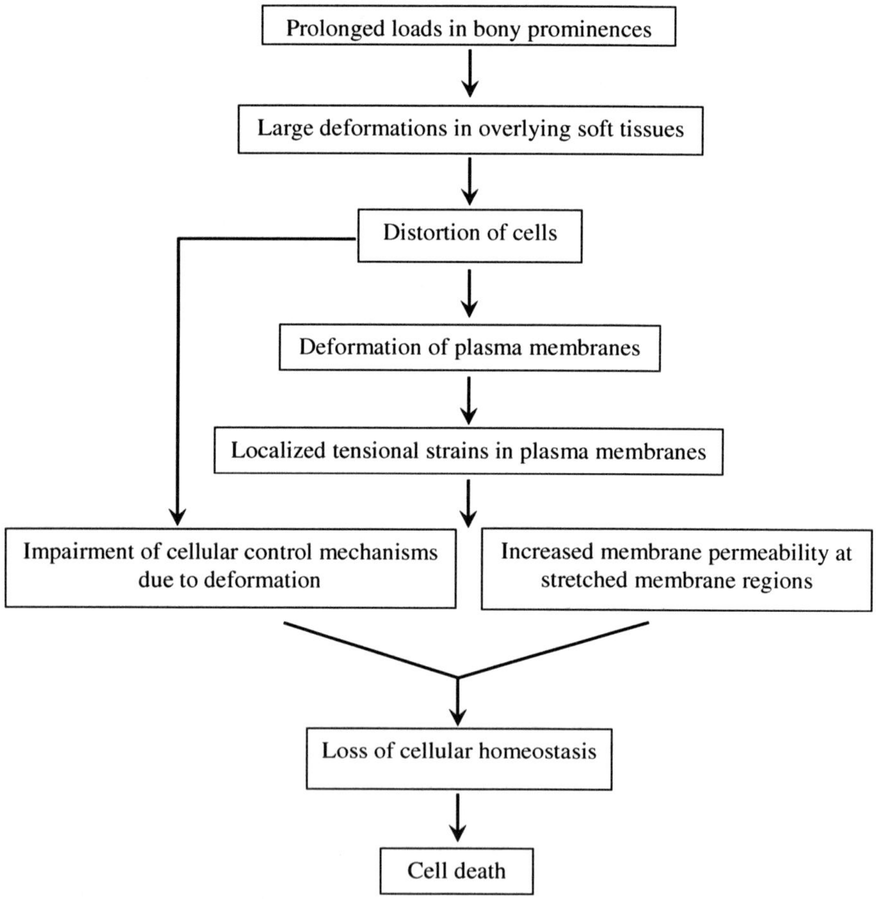

Fig. 1. The presently suggested mechanism of deformation-related cell death, caused by loss of cell homeostasis

stretched or folded, respectively. When membranes deform in tension, a local elevation of the plasma membrane permeability may occur, allowing increased diffusivity of ions, nutrients, and waste products into the cytosol, or outwards to the extracellular matrix (ECM). This stretch-induced increased permeability of the plasma membrane together with any deformation-related impairment of cellular control mechanisms may lead to loss of cellluar homeostasis, and eventually to cell death. This mechanism of deformation-related cell death is described in Figure 1.

Therefore, we hypothesize that sustained deformations of soft tissues may lead to cell death as a result of alteration in intracellular concentrations of cell metabolites that occur due to local plasma membrane stretches. Accordingly, the work described in this chapter included three main objectives: (i) demonstrate that compression of a schematic two-dimensional (2D) single cell, adhered to a culture dish, induces tensional strains in segments of its plasma membrane. (ii) Demonstrate that compression of three-dimensional (3D) cell models of a single myoblast and a single fibroblast, adhered to a culture dish, induces tensional strains in segments of plasma membranes. (iii) Demonstrate that compression of cells embedded in ECM, in a structure that represents a tissue-engineered construct, induces large compression deformations of individual cells, which indicates that cells are also subjected to localized tensional membrane strains when surrounded by ECM.

2 Two-Dimensional Finite Element Analyses of a Compressed Cell

A 2D finite element (FE) model of a compressed single cell in a cell culture was developed in order to study plasma membrane stretches during cell compression. The model was developed using commercial FE software (ABAQUS ver. 6.87-EF). The geometry of the model represents a cross-section through a single cell adhered to a culture dish, which is symmetrical around the central vertical axis (Fig. 2a). The model components are the cell cytoplasm enclosed in a plasma membrane. The geometrical features were a height of 12μm (Bouten et al., 2001), membrane thickness of 10nm (Becker et al., 2003; Chen and Moy, 2000), and curvature radii of the membrane at the peripheries of the cell base that are 4.68μm (Slomka et al., 2009) (Fig. 2a).

The cell membrane and cytoplasm were meshed with 92 elements and 1368 elements, respectively (of types CPS3 and CPS4 in ABAQUS). Consistent with previously reported cell models in the literature (Baaijens et al., 2005; Breuls et al., 2002; Peeters et al., 2005), the membrane and cytosol were assumed to behave as isotropic compressible materials with Neo-Hookean strain energy density (SED) functions (Slomka et al., 2009):

$$SED = \frac{1}{2}k(J_3^{\frac{1}{2}} - 1)^2 + \frac{1}{2}G(J_1 - 3J_3^{\frac{1}{3}}) \tag{1}$$

where $k = E/[3(1-2\upsilon)]$ is the bulk modulus, E is the instantaneous elastic modulus, υ is Poisson's ratio and $G = E/[2(1+\upsilon)]$ is the instantaneous shear

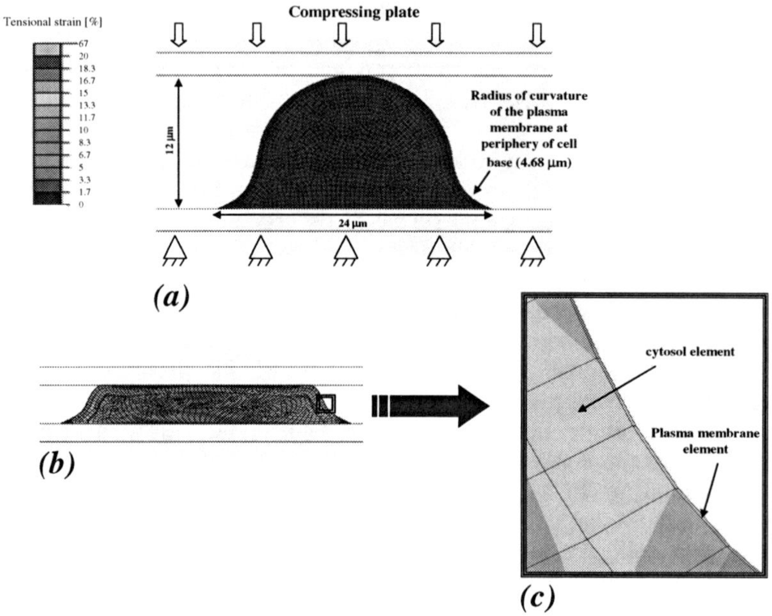

Fig. 2. (a) The initial undeformed single cell model: geometry, mesh and boundary conditions, and (b) the model- predicted strains for a maximal global deformation of 65%. A magnification of plasma membrane and adjacent cytosol elements during cell compression is provided in (c) to depict localized tensional strains which develop in the membrane for the maximal global deformation of 65%.

modulus. The parameters J_1=tr(B) and J_3=det(B) are the first and third invariants of the Finger tensor B, respectively. The shear modulus of the cytosol was set as 1kPa, and the shear modulus of the plasma membrane was set as 2.5kPa (Slomka et al., 2009). A Poisson's ratio of 0.45 was assigned for both the membrane and cytosol (Hochmuth et al., 1973). A "tie" condition was assumed at all membrane-cytosol interfaces. The cell model was compressed by a rigid cover glass to a maximum global deformation of 65% (Fig. 2a). A "no-slip" condition was set between the plasma membrane and the compressing glass. Frictional sliding (friction coefficient=0.2) was assumed between the base of the cell and the culture dish (Slomka et al., 2009).

The model was analyzed for maximal local principal strains in the plasma membrane as function of the global deformation applied to the cell. The peak tensional strain experienced by the plasma membrane, and the percentage of membrane perimeter subjected to tensional strain levels over P%, (P ranging from 1% to 11% strain, at intervals of 1% strain) were quantified.

Results showed that compression of a single cell causes localized tensional strains in segments of the plasma membrane (Fig. 2). Average tensional strains throughout the membrane increased nearly linearly with the global cell deformation until a global cell deformation of 50% (R=0.9981):

$$\text{Average membrane strain} = 0.0921 \cdot \%\ \text{global cell deformation} - 0.0996 \quad (2)$$

At higher (>50%) cell deformations, average membrane strains did not grow further significantly, i.e. they plateaued at approximately 4.5% membrane strain. Similarly, peak tensional strains in the membrane increased almost linearly with global cell deformation (R=0.9994):

$$\text{Peak membrane strain} = 0.1847 \cdot \%\ \text{global cell deformation} - 0.2678 \quad (3)$$

For a global cell deformation of 65%, a maximal membrane strain value of 11.5% was obtained. The predicted percentage of membrane perimeters subjected to tensional strain values exceeding a certain strain level (in the range of 1-11%, at intervals of 1% strain) are shown in Fig. 3 as a function of the global cell deformation. An asymmetrical-sigmoidal shape was exhibited for all datasets, and datasets were therefore fitted with the Richard's asymmetrical logistic function (Fig. 3), with correlation coefficients R>0.978 (Slomka et al., 2009). The results shown in this Figure (Fig. 3) revealed that the plasma membrane bears large tensional strains (i.e. over 5%) only for global cell deformations larger than approximately 25%. Sensitivity

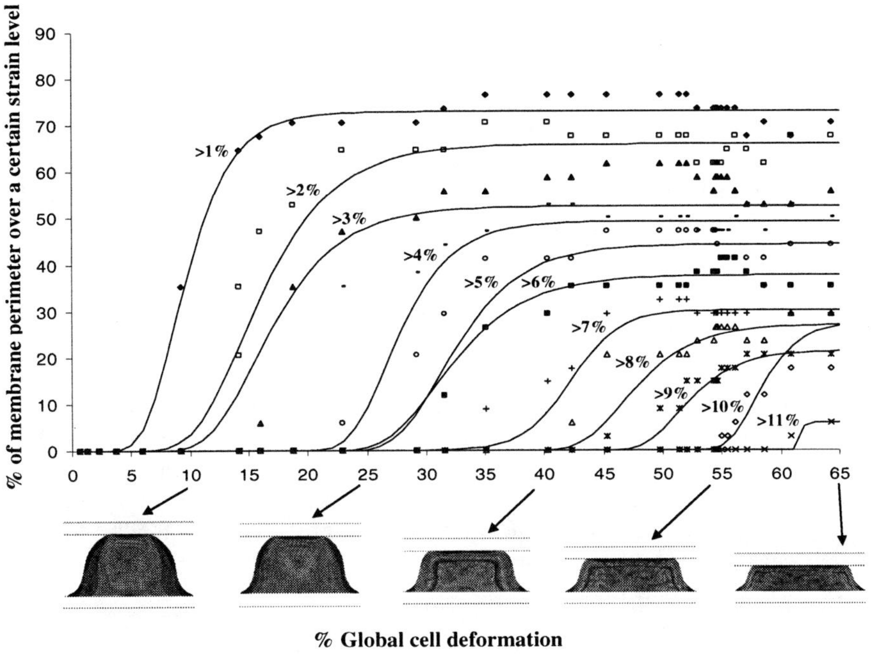

Fig. 3. Percentage of membrane perimeter subjected to tensional strain values exceeding a certain strain level (in the range of 1-11%, at 1%-strain-steps, as labeled on the individual curves), as function of the global cell deformation. Marks indicate data calculated from discrete finite element (FE) simulations, and solid lines indicate best fits of Richard's asymmetric logistic functions.

analyses were performed in order to evaluate the effects of mild variations in values of radii of curvature of the membrane at the peripheries of the cell base (Fig. 2), and in shear moduli of the plasma membrane and cytosol, on model predictions of membrane strains. These sensitivity analyses showed that changes in the radius of curvature of the membrane at the peripheries of the cell base had the greatest effect on model predictions of membrane strains (Slomka et al., 2009).

3 Three-Dimensional Finite Element Analyses of Compressed Myoblast and Fibroblast Cells

Three-dimensional FE models of a compressed skeletal muscle cell (from a murine C2C12 myoblast cell line) and a fibroblast cell (from a murine 3T3L1 fibroblast (pre-adipocyte) cell line) were further developed in order to study plasma membrane stretches in real 3D cell geometries. These cell types were chosen due to the involvement of muscle tissue and connective tissue in the development of pressure ulcers and DTI. The geometry of the cells was based on confocal microscopy images of actin-stained undifferentiated C2C12 and 3T3L1 cells, specifically stained with FITC-labeled phalloidin (Fig. 4). The z-stack images were imported into a solid modeling software (SolidWorks 2008, SolidWorks, MA, USA), and a 3D solid model of each of the cells, adhered to a culture dish were created, including cytosol and nucleus (Fig. 4b). The main geometrical characteristics of the C2C12 cell were a maximal height of 4.86μm and a maximal width at the cell base of 57.6μm. Similarly, the main geometrical characteristics of the 3T3L1 cell were a maximal height of 3.4μm and a maximal width at the cell base of 62.5μm. The solid models were then imported to an FE modeling software (ABAQUS v. 6.8, SIMULIA, RI, USA) for non-linear large deformation strain analyses. A plasma membrane with a thickness of 10nm was then configured in the FE model on the outer surface of the cytosol.

The C2C12 cell was meshed with 31,760 deformable elements (of types C3D10M and M3D6 in ABAQUS), of which 7820 represented membrane, 20,592 represented cytosol, and 3,348 represented the nucleus. The culture dish and the compressing plate were meshed with 9,558 and 9,383 elements (of types R3D3 and R3D4 in ABAQUS), respectively. The 3T3L1 cell was meshed with 3,993 deformable elements (of types C3D10M, C3D20R and M3D6 in ABAQUS), of which 1,432 represented membrane, 2,319 represented cytosol, and 242 represented the nucleus. The culture dish and the compressing plate for the fibroblast model were meshed with 669 and 274 elements (of types R3D4 and R3D3 in ABAQUS), respectively.

The SED function in Eq. 1 was used for the membrane, cytosol, and nucleus in all strain analyses. Mechanical properties assigned to the cytosol and membranes were as detailed in section 1.2. The shear modulus of the nucleus was assumed to be 1.7kPa (Jean et al., 2005). The cells were compressed in the FE simulations by a rigid plate, up to a maximal global deformation of 35%. A frictionless contact condition was set between the plasma membrane and the compressing plate. Frictional sliding, with a friction coefficient of 0.2, was assumed between the base of

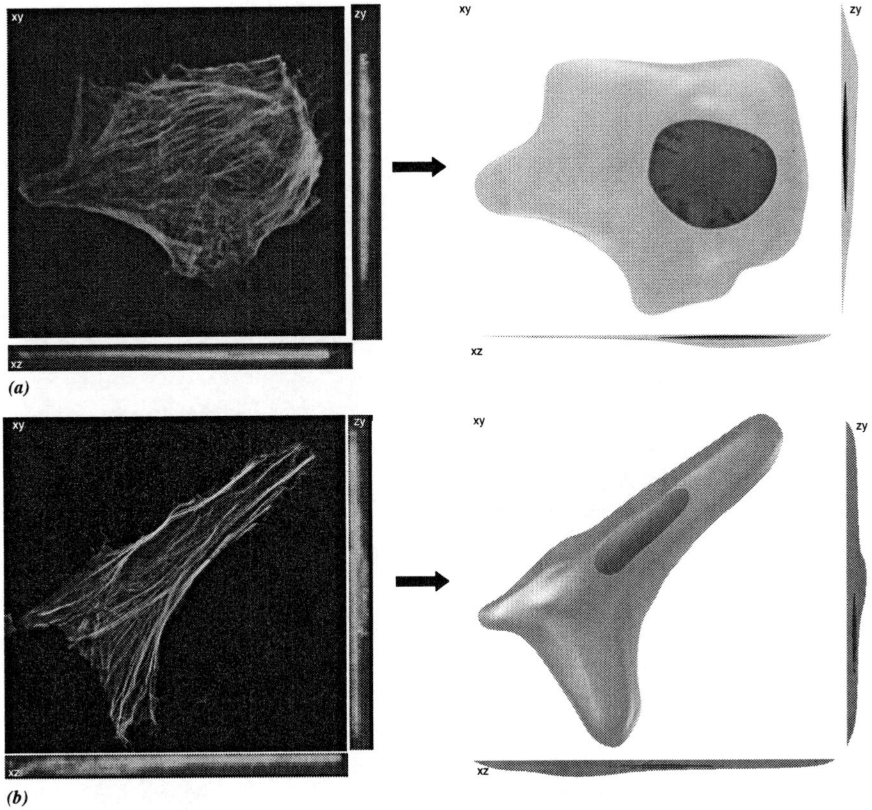

Fig. 4. Confocal microscope images of the cell base (on the left) and 3D reconstructions (on the right) of the (a) myoblast and (b) fibroblast, shown from three different points of view, xy, xz, and zy planes. This figure illustrates the similarity between the real and reconstructed cell geometries.

the cell and the culture dish. The culture dish was fixed for all translations and rotations, and the inferior surface of the cell was fixed for translation along the compression (vertical) axis. Large deformation strain analysis was performed, and maximal local principal strains in the plasma membrane were obtained as function of the global cell deformation.

The results of the FE analyses showed that platen compression induces substantial localized tensional strains in segments of the plasma membrane of both cell types (Fig. 5), which was consistent with the previously described 2D studies Tensional strains in the non-adhered membrane of the myoblast and fibroblast increased with the global cell deformation, until reaching a peak value of ~18% and ~15% at ~35% global cell deformation, respectively (Fig. 5). The percentage of membrane surface area subjected to tensional strains over 5% is shown in Fig. 6 for in both cell types. This plot indicates that the plasma membranes of the

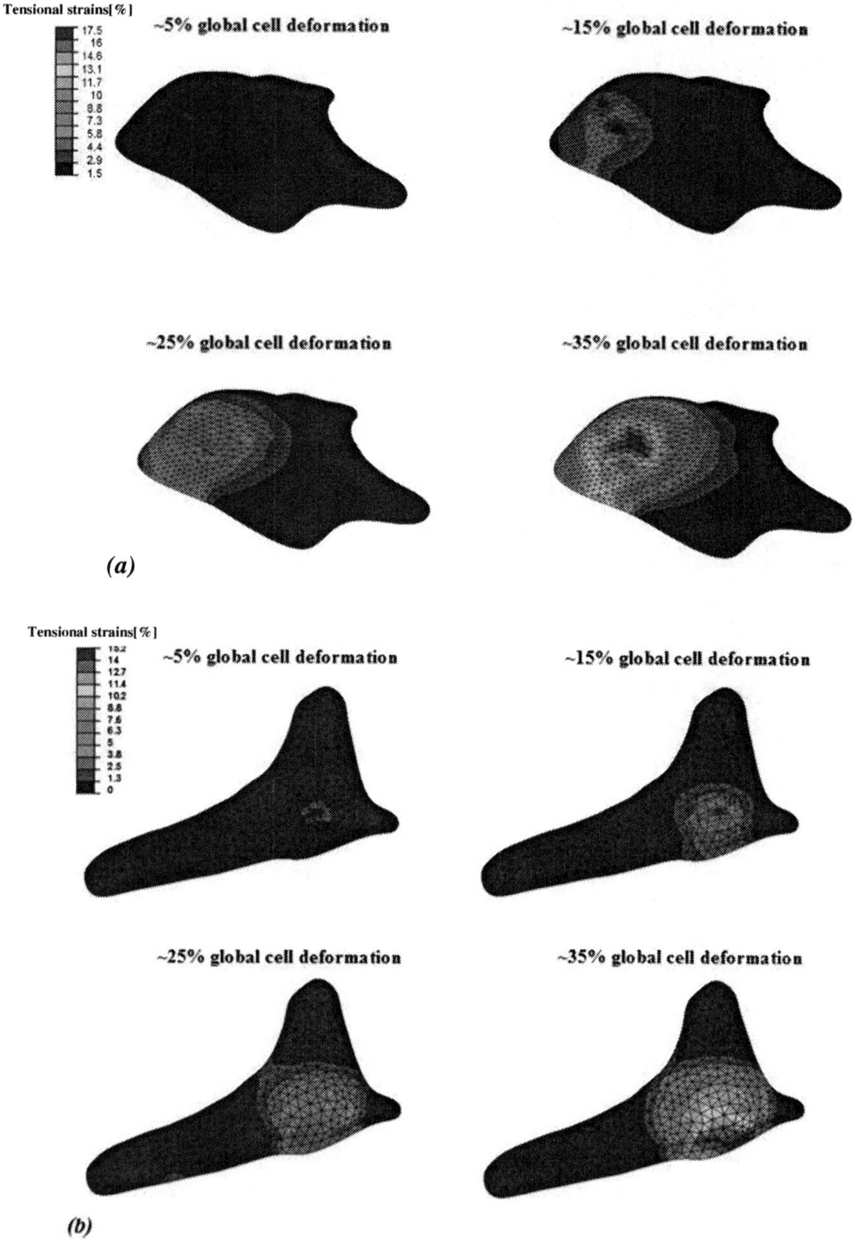

Fig. 5. Sequential steps of (a) myoblast and (b) fibroblast cell deformation under platen compression as predicted by the finite element models. Individual frames show tensional strains in the membrane at: ~5%, ~15%, ~25%, and ~35% global cell deformation.

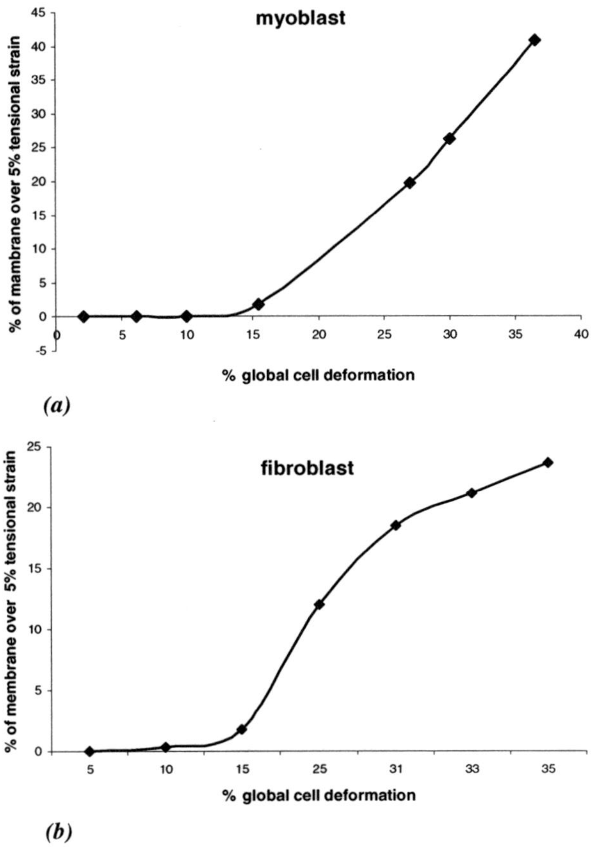

Fig. 6. Percentage of plasma membrane subjected to tensional strain values exceeding 5% for (a) myoblast and (b) fibroblast, as function of the global cell deformation. Marks indicate data calculated from the discrete finite element (FE) simulations.

modeled myoblast and fibroblast bear large tensional strains for global cell deformations exceeding ~15% and ~10%, respectively. The percentage of membrane surface area of the myoblast that experienced large deformations increased with the global cell deformation, until reaching a value of approximately 40% at 35% cell deformation. The same trend was exhibited for the membrane of the fibroblast, for which the surface area subjected to >5% tensile strains were approximately 25% at 35% cell deformation.

4 Finite Element Analysis of a Compressed Cell/Extracellular Matrix Construct

A 2D FE model of a compressed square construct of cells embedded in ECM was further developed using the same computational approach. The geometry of the model represents a microscopic 1mm × 1mm cross-section through a cell\ECM construct, which contains 53 cells in total. The geometry of each individual cell was idealized as being perfectly round at the undeformed state of the model, with cell diameter of 90μm. This cell diameter is representative of the magnitude of the diameter of pre-adipocyte cells (Sjostrom et al., 1971). The plasma membrane was not modeled at this scale of modeling because its thickness (~10μm) is negligible compared to the overall cell/ECM model dimensions. The cell/ECM model was meshed with 23,958 deformable elements (of types CPS3 and CPS4 in ABAQUS), of which 9,261 represented cells and 14,697 represented ECM (Fig. 7).

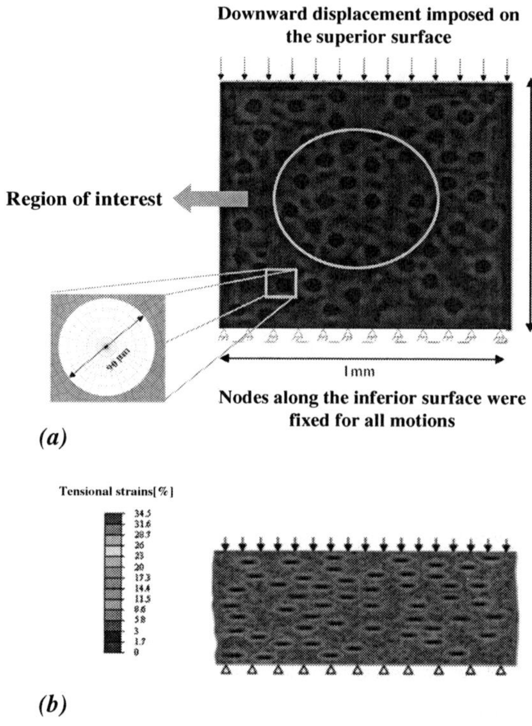

Fig. 7. The model of pre-adipocyte cells embedded in extracellular matrix (ECM). Model geometry, region of interest (ROI), mesh and boundary conditions are shown. *Note:* There are two cells in the bottom part that were meshed with a coarser grid because the contact they make with each other caused some convergence problems at large deformations. However, these cells are outside the ROI.

The SED function in Eq. 1 was again used for both cells and ECM in all strain analyses. For cells, mechanical properties were set same as for the individual cell models described previously. For the ECM, G was set as 10.1kPa with a Poisson's ratio of ν=0.38 (Jain et al., 1988). A "tie" condition was assumed at all cell-ECM interfaces. Compression of the model was obtained by enforcing a uniform downward displacement of its superior surface (Fig. 7). The inferior surface was fixed for all motions (Fig. 7). Global deformation was defined as the ratio of change in cell/ECM model height over the undeformed 1mm height. A maximal global deformation of 45% was achieved.

Consistent with the single cell models, maximal principal tensional strains at the perimeters of the cells, that is, where plasma membranes are located, were analyzed. To minimize boundary effects, as pointed out in St. Venant's principle, strain data were calculated only for the central region of interest (ROI), which was defined to be contained within a radius of 0.3mm (capturing 15 cells, Fig. 7). We determined the following outcome measures: (i) The mean of the values of average strains, calculated in the individual cells as a function of the global construct deformation, (ii) The mean of peak tensional strains across the individual cells, as a function of the global deformation applied to the cell/ECM construct, and (iii) The average deformation of cells, as a function of the global construct deformation. Deformation of an individual cell was defined as the ratio of change in its vertical dimension, over its undeformed 90μm diameter.

The cell/ECM construct compression simulations demonstrated again localized tensional strains in both the cells and ECM components, which increased with the deformation of the entire construct. Tensional strains experienced by the each cell typically occurred along the cell perimeter.

The mean of average tensional strains in cells increased linearly with the global construct deformation (R=0.997), until reaching a maximal value of 26.7% at a global construct deformation of 45.3%:

$$\text{Mean average strain} = 0.6083 \cdot \%\ \text{global construct deformation} - 1.6396 \quad (4)$$

The mean of peak tensional strains across cells also increased linearly with the deformation applied to the construct (R=0.9968), until reaching a maximal value of 27.3% for a maximal global construct deformation of 45.3%:

$$\text{Mean peak stain} = 0.6206 \cdot \%\ \text{global construct deformation} - 1.6516 \quad (5)$$

The mean of cell deformations consistently showed a linearly increasing trend with an increase in global construct deformation (R=0.9999), reaching a maximal value of 44.9% for a global construct deformation of 45.3%:

$$\text{Mean cell deformation} = 0.9936\ \%\ \text{global construct deformation} - 0.1855 \quad (6)$$

All of these parameters showed relatively small variability of values across individual cells (Fig. 7), with coefficients of variation lower than 20%. Variability of these parameters across cells maximized for the small construct deformations (~5%) and large deformations (>40%), but decreased at the midrange. Importantly, the data overall demonstrated that construct deformations of up to 45% induce similar deformations at the scale of individual cells. Hence, individual cell deformations can be approximated as the global construct deformations at construct deformation magnitudes lower than 45%.

5 Summary and Conclusions

The results of the computational studies of the 2D and 3D cell models and of the cell/ECM construct model were all supportive of our hypothesis that compressive deformations cause substantial tensional strains in plasma membranes of cells. These tensional strains are likely to affect the permeability of the plasma membrane and thereby lead to adverse events of permeation of molecules from the ECM into the cytosol at non-physiological concentrations. Likewise, such increased permeability may allow escape of essential molecules from the cytosol to the ECM. The threshold of compressive cell deformations that induce large tensional strains in the plasma membrane was ~25% for the idealized 2D cell geometry. The 3D cell models, describing realistic geometries of a muscle cell (myoblast) and a fibroblast (pre-adipocyte) exhibited lower thresholds, being ~15% and ~10%, respectively. This is likely because the 3D models included more complex membrane curvatures not accounted for in the 2D models. Different methodological approaches described in the literature, e.g. pipette aspiration, or cytoindentation, clearly show that the plasma membrane can structurally sustain these model-predicted large deformations without tearing (Slomka et al., 2009). The cell/ECM construct model, which more closely represents *in vivo* conditions, showed that construct deformations of up to 45% cause similar extents of compressive deformations of individual cells in the construct. Large tensional strains consequently develop at the plasma membranes of the embedded cells. In addition to the modeling results described in this chapter, we further characterized localized influx of calcium ions (Ca^{2+}) through a stretched membrane region, into the cytosol, using a mathematical model (Slomka et al., 2009). The work described herein, taken together with the results of the Ca^{2+} diffusion in Slomka et al. (2009), provides the proofs-of-concept for the above hypothesis, and can be utilized in order to plan future experiments and modeling, designed to study the membrane-stretch-induced cell death as related to pressure ulcers and particularly DTI, and to collect the pieces of information that are still missing.

References

1. Agam, L., Gefen, A.: Pressure ulcers and deep tissue injury: a bioengineering perspective. J. Wound Care 16, 336–342 (2007)
2. Baaijens, F.P., Trickey, W.R., Laursen, T.A., Guilak, F.: Large deformation finite element analysis of micropipette aspiration to determine the mechanical properties of the chondrocyte. Ann. Biomed. Eng. 33, 494–501 (2005)
3. Becker, W.M., Kleinsmith, L.J., Hardin, J., Becker, W.M.: The world of the cell, pp. 164–171. Benjamin/Cummings Pub. Co., San Francisco (2003)
4. Black, J., Baharestani, M., Cuddigan, J., Dorner, B., Edsberg, L., Langemo, D., Posthauer, M.E., Ratliff, C., Taler, G.: National Pressure Ulcer Advisory Panel's updated pressure ulcer staging system. Dermatol. Nurs. 19, 343–350 (2007)
5. Bouten, C.V., Knight, M.M., Lee, D.A., Bade, D.L.: Compressive deformation and damage of muscle cell subpopulations in a model system. Ann. Biomed. Eng. 29, 153–163 (2001)

6. Bouten, C.V., Oomens, C.W., Baaijens, F.P., Bader, D.L.: The etiology of pressure ulcers: skin deep or muscle bound? Arch. Phys. Med. Rehabil. 84, 616–619 (2003)
7. Breuls, R.G., Sengers, B.G., Oomens, C.W., Bouten, C.V., Baaijens, F.P.: Predicting local cell deformations in engineered tissue constructs: a multilevel finite element approach. J. Biomech. Eng. 124, 198–207 (2002)
8. Breuls, R.G., Bouten, C.V., Oomens, C.W., Bader, D.L., Baaijens, F.P.: Compression induced cell damage in engineered muscle tissue: an in vitro model to study pressure ulcer aetiology. Ann. Biomed. Eng. 31, 1357–1364 (2003)
9. Ceelen, K.K., Oomens, C.W., Stekelenburg, A., Bader, D.L., Baaijens, F.P.: Changes in intracellular calcium during compression of C2C12 myotubes. Exp. Mech. 49, 25–33 (2007)
10. Ceelen, K.K., Oomens, C.W., Baaijens, F.P.: Microstructural analysis of deformation-induced hypoxic damage in skeletal muscle. Biomech. Model Mechanobiol. 7, 277–284 (2008)
11. Chen, A., Moy, V.T.: Cross-linking of cell surface receptors enhances cooperativity of molecular adhesion. Biophys. J. 78, 2814–2820 (2000)
12. Gawlitta, D., Li, W., Oomens, C.W., Baaijens, F.P., Bader, D.L., Bouten, C.V.: The relative contributions of compression and hypoxia to development of muscle tissue damage: an in vitro study. Ann. Biomed. Eng. 35, 273–284 (2007a)
13. Gawlitta, D., Oomens, C.W., Bader, D.L., Baaijens, F.P., Bouten, C.V.: Temporal differences in the influence of ischemic factors and deformation on the metabolism of engineered skeletal muscle. J. Appl. Physiol. 103, 464–473 (2007b)
14. Gefen, A.: Risk factors for a pressure-related deep tissue injury: a theoretical model. Med. Biol. Eng. Comput. 45, 563–573 (2007)
15. Gefen, A., Cornelissen, L.H., Gawlitta, D., Bader, D.L., Oomens, C.W.: The free diffusion of macromolecules in tissue-engineered skeletal muscle subjected to large compression strains. J. Biomech. 41, 845–853 (2008)
16. Hochmuth, R.M., Mohandas, N., Blackshear Jr., P.L.: Measurement of the elastic modulus for red cell membrane using a fluid mechanical technique. Biophys. J. 13, 747–762 (1973)
17. Jain, M.K., Chernomorsky, A., Silver, F.H., Berg, R.A.: Material properties of living soft tissue composites. J. Biomed. Mater. Res. 22, 311–326 (1988)
18. Jean, R.P., Chen, C.S., Spector, A.A.: Finite-element analysis of the adhesion-cytoskeleton-nucleus mechanotransduction pathway during endothelial cell rounding: axisymmetric model. J. Biomech. Eng. 127, 594–600 (2005)
19. Miller, G.E., Seale, J.: Lymphatic clearance during compressive loading. Lymphology 14, 161–166 (1981)
20. Peeters, E.A., Oomens, C.W., Bouten, C.V., Bader, D.L., Baaijens, F.P.: Mechanical and failure properties of single attached cells under compression. J. Biomech. 38, 1685–1693 (2005)
21. Peirce, S.M., Skalak, T.C., Rodeheaver, G.T.: Ischemia-reperfusion injury in chronic pressure ulcer formation: a skin model in the rat. Wound Repair Regen. 8, 68–76 (2000)
22. Reddy, N.P., Cochran, G.V.: Interstitial fluid flow as a factor in decubitus ulcer formation. J. Biomech. 14, 879–881 (1981)
23. Sjostrom, L., Bjorntorp, P., Vrana, J.: Microscopic fat cell size measurements on frozen-cut adipose tissue in comparison with automatic determinations of osmium-fixed fat cells. J. Lipid Res. 12, 521–530 (1971)

24. Slomka, N., Or-Tzadikario, S., Sassun, D., Gefen, A.: Membrane-Stretch-Induced cell death in deep tissue injury: Computer model studies. Cellular and Molecular Bioengineering 2, 118–132 (2009)
25. Stekelenburg, A., Strijkers, G.J., Parusel, H., Bader, D.L., Nicolay, K., Oomens, C.W.: Role of ischemia and deformation in the onset of compression-induced deep tissue injury: MRI-based studies in a rat model. J. Appl. Physiol. 102, 2002–2011 (2007)
26. Stekelenburg, A., Gawlitta, D., Bader, D.L., Oomens, C.W.: Deep tissue injury: how deep is our understanding? Arch. Phys. Med. Rehabil. 89, 1410–1413 (2008)
27. Tsuji, S., Ichioka, S., Sekiya, N., Nakatsuka, T.: Analysis of ischemia-reperfusion injury in a microcirculatory model of pressure ulcers. Wound Repair Regen. 13, 209–215 (2005)

Tissue Engineered Models: A Valuable Tool in Pressure Ulcer Research

Cees Oomens[1] and Dan Bader[1,2]

[1] Biomedical Engineering Department, Eindhoven University of Technology, P/O Box 513
5600 MB Eindhoven, The Netherlands
c.w.j.oomens@tue.nl
[2] School of Engng. & Materials Science, Queen Mary University of London,
Mile End Road, London, E1 4NS, United Kingdom
d.l.bader@qmul.ac.uk

Abstract. Until the last two decades, research on the aetiology of pressure ulcers was primarily based on animal models, using rabbits, pigs, rats and dogs. Although these studies have been very valuable, there are ethical as well as conceptual reasons to look for alternative models. Sophisticated non-invasive techniques like Magnetic Resonance Imaging (MRI) and Ultrasound as well as advanced theoretical modelling have offered opportunities for studies with human volunteers. This chapter describes the activities in the laboratories of the authors in the last 15 years which have focused on the aetiology of pressure ulcers. This includes in vitro studies on single cells and on artificial skeletal muscle tissues. These investigations have provided considerable insight into aspects of deformation damage and the role of ischemia in pressure ulcer development. The results appeared consistent with animal studies and human studies.

1 Introduction

A pressure ulcer is a form of soft tissue degeneration caused by sustained mechanical loading, usually occurring in patients, who are bedridden or wheelchair bound. Pressure ulcers constitute a major problem in modern health care in hospitals, nursing homes, rehabilitation centres and in the community. Prevalence rates are high and are only likely to increase in association with the ever-increasing age of the population. Pressure ulcers can be very painful, for subjects with a normal sensation and are depressing for both sufferers and their carers. Additionally, they lengthen the hospital stay considerably and thus constitute a major burden to both the patients suffering from an ulcer as well as the overall costs of health care.

In the last 60 years, considerable research has been performed on the aetiology of pressure ulcers. Much of this research was based on the use of animal models (Kosiak, 1959,1961; Dinsdale, 1973, 1974; Daniel, 1981; Salcido, 1994; Bosboom, 2003; Stekelenburg, 2006, 2007). Although these studies have been very valuable, there are ethical as well as conceptual reasons to look for alternatives. Indeed in 2003 Bouten and colleagues proposed an hierarchical approach to study the aetiopathology of pressure ulcers. This approach which can be both experimental and computational in nature, examines the effects of loading in different, yet complementary, model systems with increasing complexity and length scales and incorporating one or more functional tissue units (see Figure 1).

A. Gefen (Ed.): Bioengineering Research of Chronic Wounds, SMTEB 1, pp. 249–262.
springerlink.com

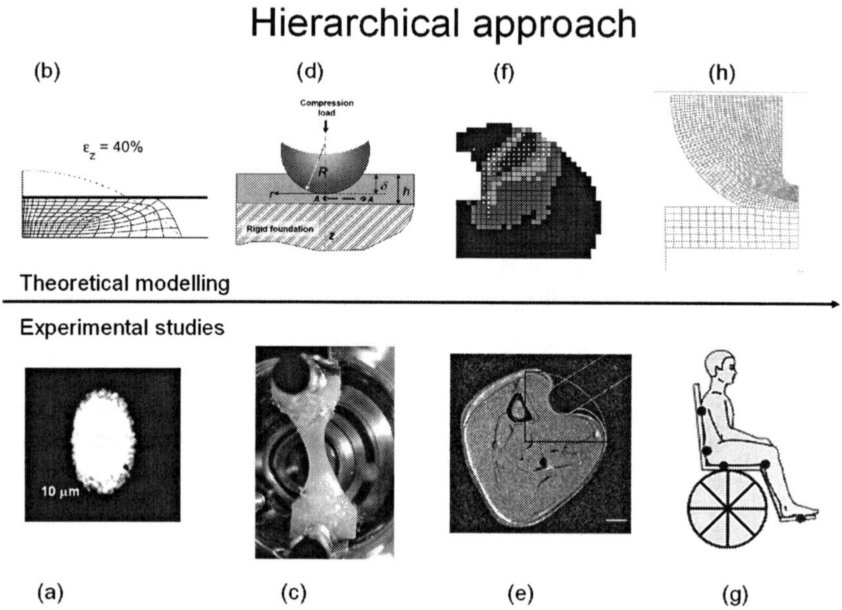

Fig. 1. Schematic of a hierarchical approach. (a) confocal image of a cell (Bouten et al., 2001), (b) finite element model of a loaded cell (Peeters, 2004), (c) image of a tissue engineered muscle (Gawlitta, 2007), (d) model of a muscle indented by a sphere (Gefen et al. 2008), (e) cross section of the lower leg of a rat while loaded with an indenter (Ceelen, 2008), (f) numerical comparison of strains and damage in the cross section of the leg of a rat (Ceelen, 2008), (g) A sitting person in a wheelchair (h) Model of deformation of the buttock while sitting in a chair (Oomens et al., 2003).

Thus, in-vitro models, ranging from the single cell (μm-scale) to cell-matrix constructs (mm-scale) and individual tissue layers (mm-cm-scale) might be used to study the relationship between cell deformation and cell damage, as well as the influence of the surrounding extracellular matrix and three-dimensional tissue architecture on this relationship. The advantage of such an approach is that at a smaller scale it is easier to separate different influencing factors of a certain phenomenon (for example to separate typical effects from ischaemia from those which are purely related to tissue deformation), which makes it easier to set up simple, relatively easy to interpret experiments with very limited ethical limitations. Extrapolation of these small scale experiments to human studies is problematic as the behaviour of cells and tissues differ significantly from in—vivo model systems. However by performing such small scale experiments the required number of in-vivo studies can be reduced considerably. In addition, focused hypotheses can be proposed in the in vivo studies and the design of the experimental protocols can be optimized accordingly.

A distinctive feature of the research in the laboratories of the authors involves the combination of computational and experimental studies. The latter are aimed at elucidating the relationships between mechanical loading, the pathophysiological response to loading and tissue breakdown in testing hypotheses related to the aetiopathogenesis of pressure ulcers. The computational models have been designed to predict the association between external and internal mechanical conditions within soft tissues and their functional units. The different length scales of these models can be coupled by multi-scale computer calculations that enable the prediction of the internal microscopic mechanical environment within a given model from global macroscopic modelling conditions, such as interface pressures.

The authors' research encompasses various levels of this multi-scale approach, ranging from studies on single cells, tissue engineered muscle and skin, animal models and studies with human volunteers. The present chapter is focussed on the in-vitro studies with cells and tissues to study the aetiology of pressure related deep tissue injury, i.e. pressure ulcers that are initiated in deeper tissue layers adjacent to bony prominences. This necessarily implies a focus on muscle cells and skeletal muscle tissue, each of which will be discussed in separate sections.

2 Single Cell Studies

Traditionally studies on the aetiology of pressure ulcers focused on the blood vessels.. The main hypothesis from these studies was that impaired perfusion and occlusion of blood and associated lymph vessels, resulted in mechanically-induced ischaemia leading to impaired transport of metabolites to and from the cell, leading to damage. However, this hypothesis can only partly explain the onset of tissue damage, as pressure ulcers can develop more rapidly than could be predicted by the establishment of ischaemia alone. Accordingly, Bouten et al. (2001) hypothesized that sustained deformation of the muscle cells in the tissue plays an important role. This deformation will inevitably trigger a number of events, such as local membrane stresses, volume changes, and remodelling of the cytoskeletal organization. In an in-vivo protocol it is impossible to deform the tissue without simultaneously provoking a degree of ischaemia. This motivated the design of an in-vitro protocol to study cells.

C2C12 mouse skeletal myoblasts were seeded in agarose constructs and cultured in growth medium for 4 days. Subsequently the cells were allowed to fuse into multinucleated myotubes for 8 days in differentiation medium, resulting in a population of spherical myoblasts, spherical myotubes and elongated myotubes (Bouten et al., 2001). These cells were seeded into three–dimensional, cylindrical agarose constructs which were subsequently compressed up to 40% strain. The cells were stained with calcein-AM, a vital stain that enabled a three-dimensional reconstruction of the morphology of the cells by means of a confocal microscope. The resulting cell deformations were estimated from confocal images.

Figure 2a shows a schematic of the experimental set-up that was used to compress individual constructs. Fig 2b gives an example of an unstrained round shaped myoblast. Figure 2c shows a myoblast in several stages of deformation. It was evident that the cells deformed to an oblate ellipsoid up to strains of about

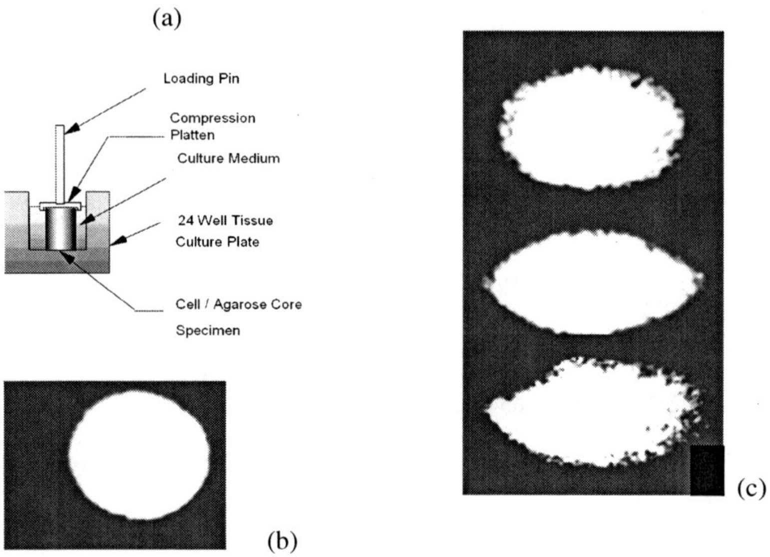

Fig. 2. (a) Schematic of set-up to study cell shape during loading (b) myoblast in an unstrained construct (c) myoblasts at different levels of applied strain. (adapted from Bouten et al., 2001).

20%. However, at higher strains most cells show localized buckling of their membranes, appearing as a pointed apex at the equators.

To study cell damage in response to sustained compression, cylindrical constructs were subjected to gross strains up to 20% (Bouten et al. 2001; Wang et al., 2005) for periods of 1, 2, 4, 12 and 24 hours. After that, they were fixed and histological slices were stained with haematoxylin and eosin and examined for cell damage. In this way a percentage of dead cells as a function of time could be constructed, as illustrated in Fig. 3.

Both studies indicted that within an unstrained construct, there is a certain percentage of dead cells which remain fairly constant with time. However, at an agarose compression of 20%, there was a monotonic increase in cell death with time. By contrast, at 10% strain, there was no such temporal increase in cell death, associated with a minimal cell deformation (Wang et al. 2005).

Peeters (2004, 2005) designed a loading device to compress single attached cells, either statically of dynamically, while measuring the resulting forces. The device, designed for use on the stage of a confocal microscope, incorporated a small incubation chamber to maintain optimum conditions during cell culture. Vital staining techniques were used to follow the three-dimensional cell deformations and the first signs of membrane damage and cell death. Viable cells were visualised by means of cell tracker green (CTG). The point of cell damage was visualised by the fluorescent stain propidium iodide (PI), which enters the cell when the nuclear membrane is damaged. In this way Peeters was able to examine

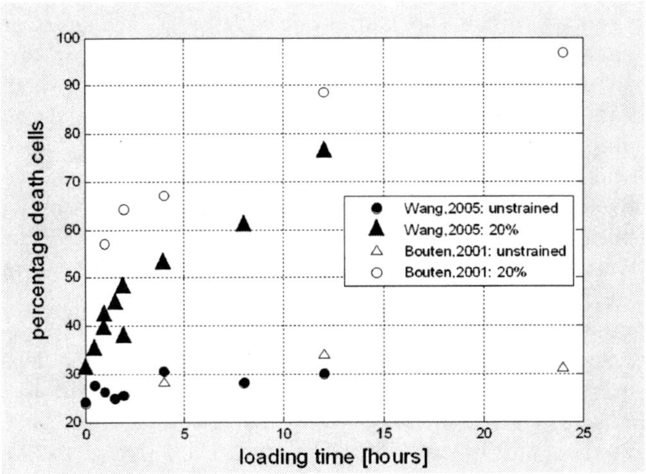

Fig. 3. Percentage of death cells as a function of time in unloaded state and in a situation with 20% strain. Data from: Bouten et al. 2001 and Wang et al. 2005.

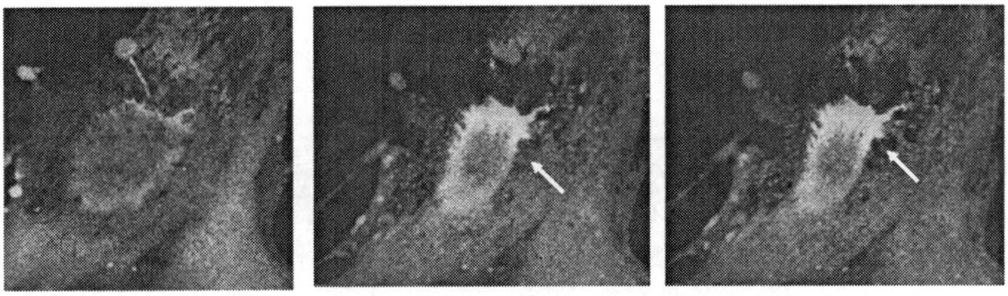

Fig. 4. Confocal images of a cell during relaxation after a time period of compression. The membrane of the nucleus is damaged and PI enters the cell (light gray signal near the white arrows), Adapted from Peeters, 2004.

the viscoelastic behaviour of individual cells (Fig.4). His experiments confirmed the findings of Bouten et al. (2001) that cells can be deformed to very high strains before membrane failure. This vital staining approach visualizing cells on a microscope stage could be adapted to be used with tissue models.

3 Tissue Engineered Skeletal Muscle Studies

From the single cell studies it could be concluded that sustained deformation of cells leads to damage, although the oxygen supply and removal of waste products were not disturbed, suggesting that cell shape alone were sufficient to cause damage. In the described studies, the cells were either dispersed in a culture medium and in direct contact with an indenter or embedded in a 3D agarose gel. In the latter case, as agarose does not provide anchor points for cell attachment, most cells adopted a spherical morphology. The question remained as to whether or not the behaviour of the cells would be similar in an environment which is more physiologically appropriate.

Breuls et al. (2003) adapted a method, originally published by Vandenburgh (1988), to develop tissue engineered skeletal muscle constructs by seeding C2C12 mouse myoblasts in a mixture of a collagen type I solution and matrigel. By allowing the cells and gels to attach to culture wells and changing growth medium into fusion medium the myoblasts differentiated into multinucleated, elongated fibres. These constructs were mechanically loaded using a compression device, designed to be used on a confocal microscope stage including an incubation chamber containing appropriate culture medium. CTG and PI were again used to visualise live and dead cells, respectively. Experimental constructs were loaded to mean compressive strains of 30% and 50%. The images in figure 5 show a top view of the indented side of the construct at different time points during loading. The white dots are nuclei of dead cells. The edge of the cylindrical indenter is clearly visible. It is clear that the number of dead cells is increasing with time and although there is a narrow band near the indenter edge with a high intensity it seems that under the indenter an equal distribution of dead cells was evident. This suggests that a reduced diffusion of oxygen is not likely to be the major determining factor for cell damage in this test protocol over an 8 hour time period.

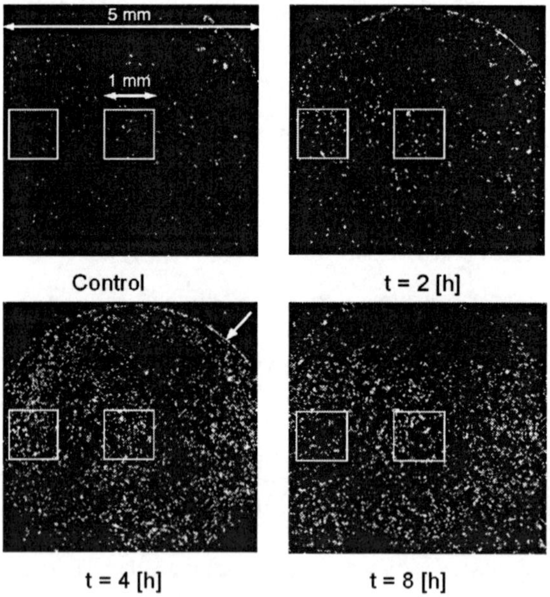

Fig. 5. Tile scan images covering an indented area (50%) of a muscle construct showing the distribution of dead cells at different time points. Adapted from Breuls, 2003.

The percentage of dead cells as a function of time showed a similar increase in time as in the experiments of Bouten (2001) and Wang (2005), but the change occured more rapidly at the higher strains employed by Breuls.

The work on bioartificial muscles (BAM's) was extended by optimizing the culture protocol (Gawlitta, 2008). In particular, test conditions were established to separate the effects of various mechanical and physicochemical factors on cell death. For example, the effects of hypoxia and deformation, both separately and in combination, were examined on a device and test chamber mounted on a confocal microscope. In this case Yo-PRO -1 (YP) was used to identify apoptotic cells and Propidium Iodide for necrotic cells. One control group and two experimental groups at either 20% or 40% strain were examined under both normal oxygen and hypoxic conditions. Fig. 6 shows the percentage of dead cells as a function of loading time for these six different test conditions. These data again confirmed that a strain leads to gradual increase of damage over time, which is higher than in a control situation and also that the increase is enhanced at higher strains. However, in this test protocol no statistically significant differences were observed between the hypoxic and the normoxic situations.

Gawlitta et al. (2007b) extended the tissue engineered muscle model to study the effect of glucose deprivation, pH change, lactic acid accumulation and deformation and combinations of these conditions. An air tight box was designed containing four 6-wells plates with BAM's (Fig. 7a). By either flushing with 95% N_2 and 5% CO_2 gas mixture or with mixture of 95% air and 5% CO_2 hypoxic or normoxic conditions were applied. Tissue compression was achieved by placing rectangular stainless steel weights on top of the constructs on day 0, with spacers of 350 μm at their four corners (see Fig. 7b). The influence of acidic pH values on sample viability was established by the addition of lactic acid to the culture medium. To assess the metabolic activity of the cells, glucose, lactate, pH and MTT assays were performed on the daily extracted medium samples. Additionally the viability of cultures was determined from the release of lactate dehydrogenase (LDH) in the medium.

It is beyond the scope of the present chapter to discuss the results extensively, but a typical result is depicted in Fig. 8. These graphs show results from experiments with high glucose medium and low glucose medium, to study the effect of a shortage of nutrients. The effect of glucose deprivation during a 5 day period in the absence of medium refreshment was significant from 24 hours onwards as shown, for example, with glucose utilization, where there was no glucose available to the cells in the low glucose medium (dotted line in figure 8A).

Accordingly metabolism was limited in the low glucose medium with lactate production reaching a maximum at day 1 and thereafter remaining constant (figure 8B). From the high glucose measurements it appeared that on average a ratio between lactate production and glucose consumption for normoxic condition was one. The control group associated with a high glucose medium revealed a lactate production up to day 3, thereafter remaining constant at approximately 20mM. As a result of glucose deprivation cell death was consistently higher in the low glucose group compared to the control group.

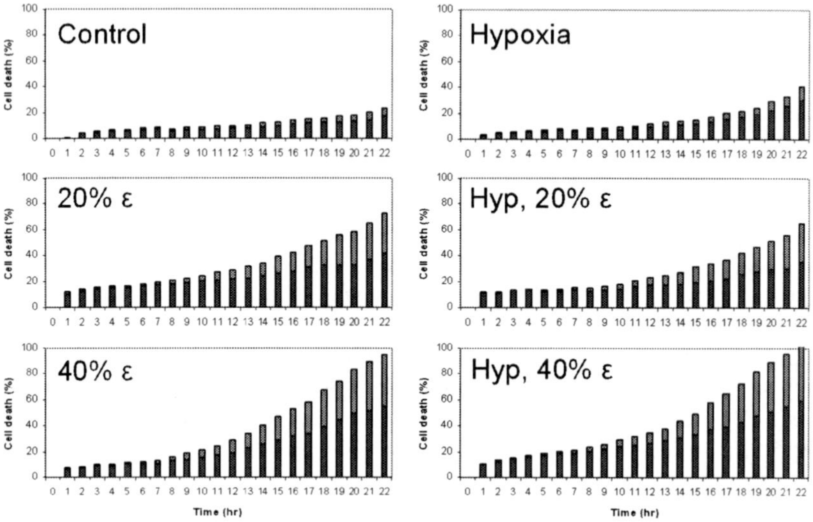

Fig. 6. Cell death as a function of time for control group and add 20 % strain and 40% strain. The left column show data in a situation with normal oxygen, the column on the right hand side show data under hypoxic condition. Light gray represents YP, dark gray represents PI. (Data extracted from: Gawlitta, 2007a).

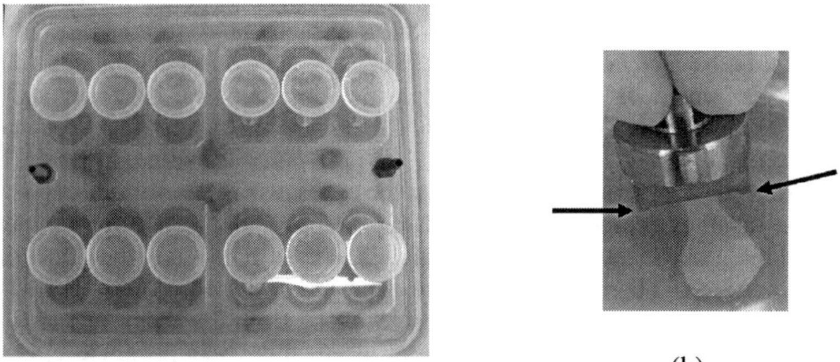

(a) (b)

Fig. 7. Photograph of the experimental set-up used to study ischaemic effects in skeletal muscle constructs. (a) top view of the air-tight box with four 6-well plates inside. (b) one of the stainless steel loads with arrows indicating 350 μm spacers.

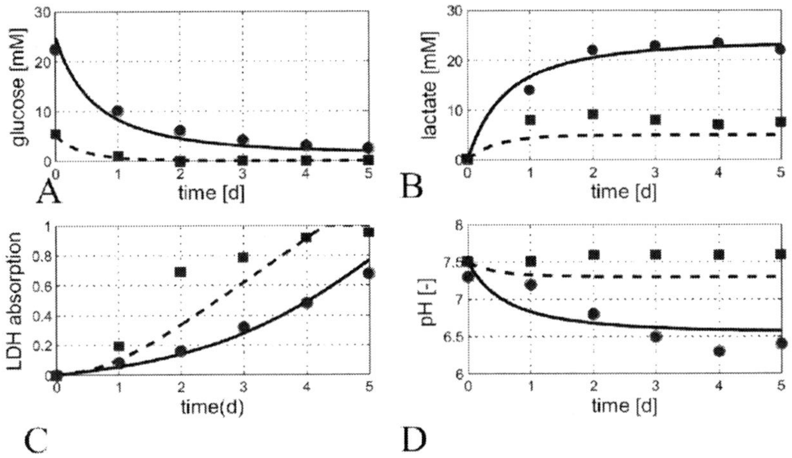

Fig. 8. Effects of glucose deprivation on temporal profiles for (A) glucose utiliziation (B) Lactate production (C) LDH absorption (D) medium pH. Lines are model results, markers are experimental results. Solid lines represent measurements in high glucose medium, dotted lines represent corresponding values in low glucose medium.

The main conclusion from the work of Gawlitta et al. (2007b) was that cells that are susceptible to damage when subjected to high deformation, involving shape change, which can occur fairly rapidly. By contrast, in the cell culture system, ischemic effects take a longer time before they lead to damage. When skeletal muscle cells become hypoxic they change from an aerobic to an anaerobic metabolism. This leads to a damaging situation for the cells when either nutrients (glucose) become short or when the concentration of lactic acid in the medium becomes too high. However, this represents a process of damage development which is slower than that associated with pure deformation.

Gefen et al. (2008b) continued the studies on BAMs in a set-up designed to establish a strain-time-cell-death threshold. The study was based on the concept of inducing a non-uniform concentric distributions of strains in the construct by indenting a BAM with a spherical indenter (see Fig. 9). Cell death was again monitored in the construct by using PI-staining.

Figure 10 shows a typical example of the increase in area of the PI-stained damage region over time that was observed in these experiments. For this specimen the damage started after 120 minutes of load application and the damaged area increased rapidly in a radial direction until about 200 minutes, thereafter remaining fairly constant. This suggests that at strains beneath the centre of the indenter the damage starts earlier than further away from the center where the strains are lower in magnitude.

Gefen et al. (2008b) were able to construct an average strain-time cell threshold curve (n=10), which was sigmoidal in form. A combination of loading time and true compressive strain resulting in values beneath the curve indicate that the cell

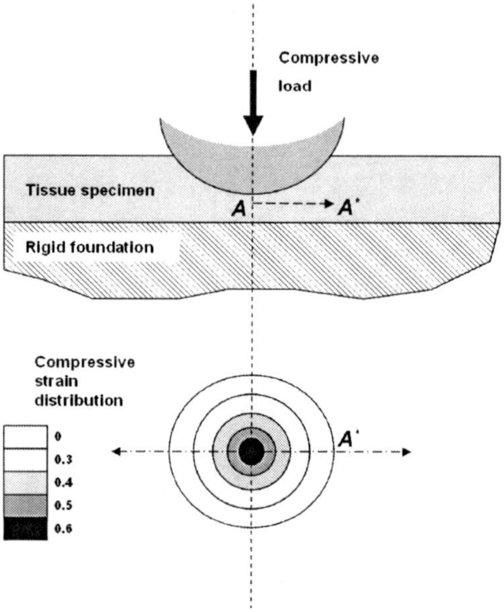

Fig. 9. A tissue-engineered construct loaded by a half-spherical indenter. This results in a non-uniform compressive strain distribution. The highest strain is found along the line of symmetry underneath the indenter. The strain decreases with distance from the line of symmetry line. Adapted from Gefen et al. (2008b).

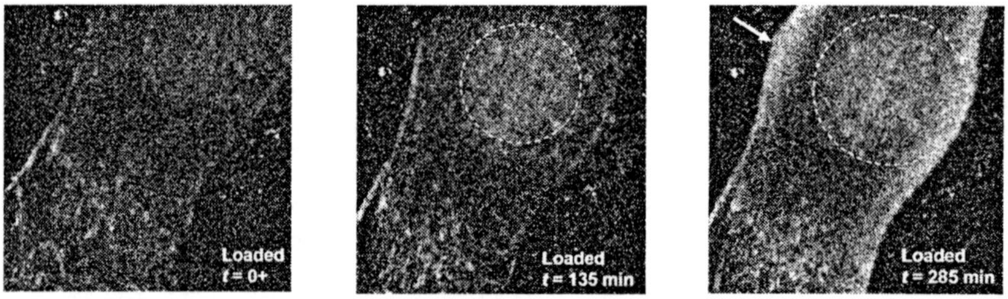

Fig. 10. Fluorescence images of the increase in area of the PI-stained damage region over time in a deformed bio-artificial muscle (BAM) (adapted from Gefen 2008b)

viability will be similar to unloaded controls. By contrast combinations above the curve will result in cell death. It is interesting to note that a similar sigmoidal curve for tissue damage was found in an animal model of deep tissue injury (Linder-Ganz et al., 2006).

Gefen et al. (2008a) also examined the diffusion of large molecules in BAMS subjected to compression. The motivation for this study was that impaired diffusion of critical metabolites in compressed muscle might contribute to both deep tissue injury and , play a role in the study of biochemical damage markers in blood

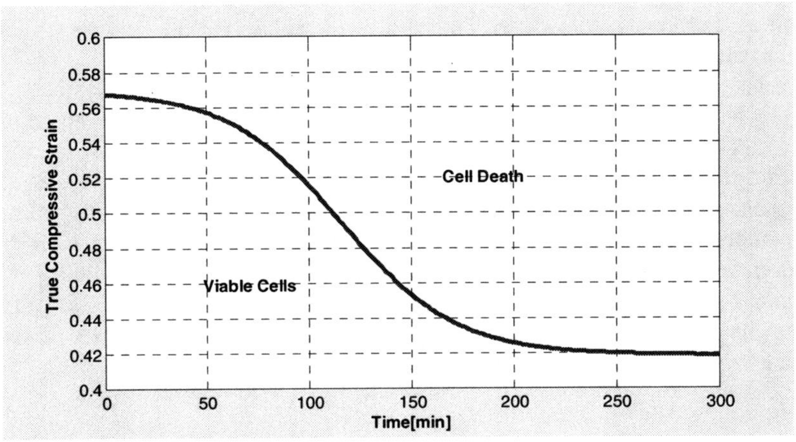

Fig. 11. The strain-time cell death threshold for bio-artificial muscle (BAM) specimens under large compressive strains. (based on data from Gefen et al., 2008b).

(in vivo) or in medium when using an in vitro model systems. Such a process would be particularly critical for high molecular weight molecules, whose diffusion coefficient would be reduced, thus hampering the transport of these molecules.

Diffusion coefficients were measured by means of fluorescence recovery after photo bleaching (FRAP). Prior to the measurements the BAMs were incubated overnight in medium containing fluorescein isothiocyanate conjugated dextrans of three different sizes, 10, 20 or 150 kDa, at a concentration of 0.75 μM. The authors assumed that uncharged molecules with similar sizes will have the same diffusion properties as the dextran model system. Measurements were performed on a confocal microscope. When the dextran is dispersed uniformly throughout the BAM, the fluorescence of the molecules will lead to images with a constant intensity. During a FRAP procedure a circular or rectangular area is photo bleached with a high power laser. Immediately after this procedure the photo bleached area will appear as a dark spot on the images. Because the fluorescent dextran diffuses to the dark area this spot will gradually disappear. From the transient behavior of the spot it is possible to derive diffusion coefficients for the different size dextran molecules. Gefen et al. (2008a)measured diffusion coefficients in the BAMs in the undeformed configuration and in the configurations deformed with compressive strains between 53 and 60%. They found an average decrease of the diffusion coefficients for all three molecular sizes of 47 %. Similar findings might be expected in skeletal muscle in vivo and this may have a negative influence on metabolic control.

4 Closing Remarks

In our laboratories animal experiments and in-vitro studies have run parallel for the last 15 years. The in-vitro studies have contributed enormously to our understanding of what is happening when soft biological materials are subjected to sustained compressive loading. The ability to isolate aspects of tissue ischemia from tissue deformation or to measure diffusion coefficients under certain well defined conditions was an essential feature of these in-vitro model systems. Qualitatively the observed phenomena were consistent (and sometimes quantitatively remarkably close) to the animal studies, even although isolated cells in culture or bio-artificial muscles are model systems with inherent limitations. We are also convinced that due to our increased understanding of pathophysiology from the in vitro studies, the number of animal studies necessary to fully understand damage development in skeletal muscle is considerably reduced. This provides a major advantage from an ethical point of view. However it should be clear from the above that besides ethical reasons there are very strong conceptual reasons to adopt a multi-scale approach with ex-vivo or in-vitro model systems.

References

1. Bouten, C.V.C., Knight, M.M., Lee, D.A., Bader, D.L.: Compressive Deformation and Damage of Muscle Cell Subpopulations in a Model System. Annals of Biomedical Engineering 29, 153–163 (2001)
2. Bouten, C.V.C., Breuls, R.G.M., Peeters, E.A.G., Oomens, C.W.J., Baaijens, F.P.T.: In-vitro models to study compressive strain-induced muscle cell damage. Biorheology 40(1-3), 383–388 (2003)
3. Bouten, C.V.C., Oomens, C.W.J., Baaijens, F.P.T., Bader, D.L.: The aetiology of pressure sores: skin deep or muscle bound? Arch. Phys. Med. Rehab. 84(4), 616–619 (2003)
4. Bosboom, E.M.H., Bouten, C.V.C., Oomens, C.W.J., Baaijens, F.P.T., Nicolay, K.: Quantifying pressure sore related muscle damage using high-resolution MRI. J. Appl. Physiol. 95, 2235–2240 (2003)
5. Breuls, R.G.M.: Experimental and theoretical analyses of compression induced muscle damage. PhD-thesis Eindhoven University of Technology (2003)
6. Breuls, R.G.M., Sengers, B.G., Oomens, C.W.J., Bouten, C.V.C., Baaijens, F.P.T.: Predicting local cell deformations in engineered tissue constructs: A multilevel finite element approach. J. Biomech. Eng. 124, 198–207 (2002)
7. Breuls, R.G.M., Bouten, C.V.C., Oomens, C.W.J., Bader, D.L., Baaijens, F.P.T.: Compression induced cell damage in engineered muscle tissue: An in vitro model to study pressure ulcer aetiology. Annals of Biomed. Eng. 31, 1357–1364 (2003a)
8. Breuls, R.G.M., Bouten, C.V.C., Oomens, C.W.J., Bader, D.L., Baaijens, F.P.T.: A theoretical analysis of damage evolution in skeletal muscle tis-sue with reference to pressure ulcer development. J. Biomech. Eng. 125, 902–909 (2003b)
9. Ceelen, K.K., Stekelenburg, A., Loerakker, S., Strijkers, G.J., Bader, D.L., Nicolaij, K., Baaijens, F.P.T., Oomens, C.W.J.: Compresion-induced damage and internal tissue strains are related. J. Biomech. 41(16), 3399–3404 (2008)
10. Daniel, R.K., Priest, D.L., Wheatley, D.C.: Etiologic factors in pressure sores: an experimental model. Arch. Phys. Med. Rehab. 62(10), 492–498 (1981)
11. Daniel, R.K., Wheatley, D.C., Priest, D.L.: Pressure sores and paraple-gia: an experimental model. Annals of Plastic Surgery 15(1), 41–49 (1981)

12. Dinsdale, S.M.: Decubitus Ulcers in Swine: Light and Electron Mi-croscopy Study of Pathogenesis. Arch. Phys. Med. Rehab. 54, 51–56 (1973)
13. Dinsdale, S.M.: Decubitus ulcers: role of pressure and friction in causation. Arch. Phys. Med. Rehab. 55, 147–152 (1974)
14. Gawlitta, D., Oomens, C.W.J., Baaijens, F.P.T., Bouten, C.V.C.: Evaluation of a continuous quantification method of apoptosis and necrosis in tissue cultures. Cytotechnology 46, 139–150 (2005)
15. Gawlitta, D., Li, W., Oomens, C.W.J., Baaijens, F.P.T., Bader, D.L., Bouten, C.V.C.: The relative contributions of compression and hypoxia to development of muscle tissue damage: an in vitro study. Ann. Biomed. Eng. 35(2), 273–284 (2007a)
16. Gawlitta, D., Oomens, C.W.J., Bader, D.L., Baaijens, F.P.T., Bouten, C.V.C.: Temporal differences in the influence of ischemic factors and deformation on the metabolism of engineered skeletal muscle. J. Appl. Physiol. 103(2), 464–473 (2007b)
17. Gawlitta, D., Boonen, K.J.M., Oomens, C.W.J., Baaijens, F.P.T., Bouten, C.V.C.: The influence of serum-free culture conditions on skeletal muscle differentiation in a tissue-engineered model. Tissue Engineering, Part(1), 161–171 (2008)
18. Gefen, A., Cornelissen, L.H., Gawlitta, D., Bader, D.L., Oomens, C.W.J.: The free diffusion of macromolecules in tissue-engineered skeletal mus-cle subjected to large compression strains. J. Biomech. 41, 845–853 (2008a)
19. Gefen, A.G., van Nierop, B.J., Bader, D.L., Oomens, C.W.J.: Strain-time cell death threshold for skeletal muscle in a tissue-engineered model sys-tem for deep tissue injury. J. Biomech. 41, 2003–2012 (2008b)
20. Kosiak, M.: Etiology and pathology of ischemic ulcers. Arch. Phys. Med. Rehab. 40, 62–69 (1959)
21. Kosiak, M.: The etiology of pressure sores. Arch. Phys. Med. Rehab. 42, 19–29 (1961)
22. Linder-Ganz, E., Engelberg, S., Scheinowitz, M., Gefen, A.: A Pressure-time cell death threshold for albino rat skeletal muscles as related to pressure sore biomechanics. J. Biomechanics 39, 2725–2732 (2006)
23. Oomens, C.W.J., Bressers, O.F.J.T., Bosboom, E.M.H., Bouten, C.V.C., Bader, D.L.: Can loaded interface characteristics influence strain distributions in muscle adjacent to bony prominences. Comp. Meth. Biomech. & Bio-med. Eng. 6(3), 171–180 (2003)
24. Peeters, E.A.G.: Biomechanics of single cells under compression, PhD-thesis Eindhoven University of Technology (2004)
25. Peeters, E.A.G., Oomens, C.W.J., Bouten, C.V.C., Bader, D.L., Baaijens, F.P.T.: Mechanical and failure properties of single attached cells under compression. J. Biomech. 38(8), 1685–1693 (2005)
26. Peeters, E.A.G., Oomens, C.W.J., Bouten, C.V.C., Bader, D.L., Baaijens, F.P.T.: Viscoelastic properties of single attached cells under compression. J. Biomech. Eng. 127(2), 237–243 (2005)
27. Salcido, R., Donofrio, J.C., Fisher, S.B., LeGrand, E.K., Dicky, K., Carney, J.M., Schosser, R., Liang, R.: Histopathology of pressure ulcers as a result of sequential computer controlled pressure sessions in a fuzzy rat model. Adv. Wound Care 7(5), 23–28 (1994)
28. Solis, L.R., Hallihan, D.P., Uwiera, R.R., Thompson, R.B., Pehowich, E.D., Mushahwar, V.K.: Prevention of pressure-induced deep tissue injury using intermittent electrical stimulation. J. Appl. Physiol. 102(5), 1992–2001 (2007)

29. Stekelenburg, A., Oomens, C.W.J., Bader, D.L.: Compression induced tissue damage; animal models. In: Bader, D.L., Bouten, C.V.C., Colin, D., Oomens, C.W.J. (eds.) Pressure Ulcer Research: Current and Future Perspectives. Springer, Heidelberg (2005)
30. Stekelenburg, A., Oomens, C.W.J., Strijkers, G.J., de Graaf, L.A.H.J., Bader, D.L., Nicolay, K.: A new MR-compatible loading device to study in-vivo muscle damage development in rats due to compressive loading. Med. Eng. Phys. 28, 331–338 (2006)
31. Stekelenburg, A., Oomens, C.W.J., Strijkers, G.J., Nicolay, K., Bader, D.L.: Compression-induced deep tissue injury examined with magnetic resonance imaging and histology. J. Appl. Physiol. 100, 1946–1954 (2006)
32. Stekelenburg, A., Strijkers, G.J., Parusel, H., Bader, D.L., Nicolay, K., Oomens, C.W.J.: The role of ischemia and deformation in the onset of compression induced muscle damage. J. Appl. Physiol. 102(5), 2002–2011 (2007)
33. Stekelenburg, A., Gawlitta, D., Bader, D.L., Oomens, C.W.J.: Deep tissue injury: how deep is our understanding? Arch. Phys. Med. Rehab. 89, 1410–1413 (2008)
34. Vandenburgh, H.H.: A computerized mechanical cell stimulator for tissue culture: effects on skeletal muscle organogenesis. In vitro cellular and developmental biology 24, 609–619 (1988)
35. Wang, Y.N., Bouten, C.V.C., Lee, D.A., Bader, D.L.: Compression-induced damage in a muscle cell model in vitro. J. Engineering in Medicine 219, 1–12 (2005)

Tissue-Engineered Models for the Study of Cutaneous Wound-Healing

M.W. Carlson, S. Dong, J.A. Garlick, and C. Egles*

Division of Cancer Biology and Tissue Engineering, Department of Oral and Maxillofacial Pathology, Tufts University, School of Dental medicine, 55 Kneeland Street, Boston, MA 02111
* Corresponding author
Phone: 617 636-2478
Fax: 617 636-2915
Christophe.egles@tufts.edu

Abstract. Understanding the molecular and cellular mechanisms underlying wound healing is pivotal for the advancement of new therapies designed to accelerate this process. In the last twenty years, new bioengineered human *in vitro*, three-dimensional (3D) tissue models, known as human skin equivalents, have been developed to study wound healing and test new types of dressings and drugs. In this article, we will discuss the advantages of these human tissue models compared to traditional two-dimensional (2D) assays. We also describe specific applications where human skin equivalents have been used to test new bioactive materials and compounds that may accelerate reepithelialization. We will conclude with a detailed protocol for the construction of wounded human skin equivalents.

1 Introduction

Skin functions to protect the body against toxins and microorganisms and to prevent dehydration. Because of its importance in maintaining homeostasis, the loss of skin integrity due to injury or disease is immediately followed by a regenerative process that involves multiple phases of repair that are spatially and temporally coordinated (Singer and Clark 1999). These events involve interactions between different cell types and extracellular matrix components that serve as a provisional matrix that will facilitate early wound organization that precedes restoration of functional tissue (Winter 1972; Mustoe 2004). Distinct stages have been identified during the wound healing process including inflammatory, proliferative, and remodeling phases. The inflammatory phase is characterized by hemostasis and inflammation that activate the clotting cascade and initiate an inflammatory response at the wound site. The proliferative phase is characterized by the reepithelialization of the wound by an epithelial tongue created by keratinocytes that proliferate just distal to the wound edge. At the end of this phase, the wound is physically closed and the remodeling phase ensues beneath the wound with generation of granulation tissue. It is within this evolving

A. Gefen (Ed.): Bioengineering Research of Chronic Wounds, SMTEB 1, pp. 263–280.
springerlink.com Springer-Verlag Berlin Heidelberg 2009

granulation tissue that dermal fibroblasts begin to organize this early stroma by producing non-collagenous and collagenous components of the extracellular matrix that lead to the organization and maturation of stroma under the wounded area. While simulating the complexity of these integrated events in an *in vitro* tissue is daunting, significant progress has been made in the last decade in the development of bioengineered HSEs that mimic the response of skin to wounding. In this review we will critically review the benefits of using 3D HSE models that have been adapted to study wound healing. We will present existing applications of these 3D tissue models that mimic cutaneous wounds in humans and will discuss their future use. Finally, we present a detailed protocol that will enable construction of 3D HSEs for the study of wound reepithelialization *in vitro*.

The mainstay of wound-healing assays in 2D monolayer tissue culture is the "scratch assay", that is designed to mimic directional cell migration and proliferative phases of wound re-epithelialization. The process (Figure 1) involves first removing cells in one area of a confluent tissue culture plate by "scratching" to create a "wound" that is subsequently filled by migrating cells from the edges of the scratch. The position of cells at the wound edge can be monitored at regular intervals until full closure of the wound and the migration rate of the cells can be calculated. Studies using these scratch wounds have allowed elucidation of key cellular and molecular factors involved in cell migration and proliferation following a wound-like injury that are linked to cell-matrix and cell-cell interactions. However, a major limitation to the use of this system is that it lacks the complexity of the wound healing microenvironment. It is therefore essential to move wound healing studies into experimental paradigms that provide skin-like tissues with 3D tissue architecture that are more predictive of *in vivo* wound responses. In recent years, our laboratory has adapted 3D Human Skin Equivalents (HSEs) to serve as "surrogate" tissues that provide more meaningful correlations between *in vitro* and *in vivo* wound healing responses. HSEs more closely mimic the *in vivo* environment than do 2D monolayer cultures, as the link that exists between the growth and differentiation in 3D tissues is optimal only when cells are spatially organized to display the morphological and biochemical features seen *in vivo*, which are lost in 2D culture systems (Bissell and Radisky 2001). HSEs consist of interactive dermal and epidermal layers that provide stromal-epidermal communication needed to spatially organize and support the growth of the overlying epidermis (Figure 2). This provides numerous advantages in the study of wound biology when compared to 2D monolayer culture systems (Carlson et al. 2007), as components of the model can be altered and tissue outcomes monitored in order to investigate specific morphological and molecular responses. The ability to manipulate either the cellular or tissue microenvironment provides opportunities to investigate wound healing in response to new types of therapeutics such as those aimed at accelerating the closure of chronic wounds (Figure 3). These tools are now playing a pioneering role by enabling the study of reepithelialization to advance our understanding of cellular and molecular mechanisms of wound healing by reproducing the intricate relationship that occurs between the different cell types and substrates involved in this process. These 3D models are now poised to advance the discovery and testing of agents targeted to improve clinical wound care.

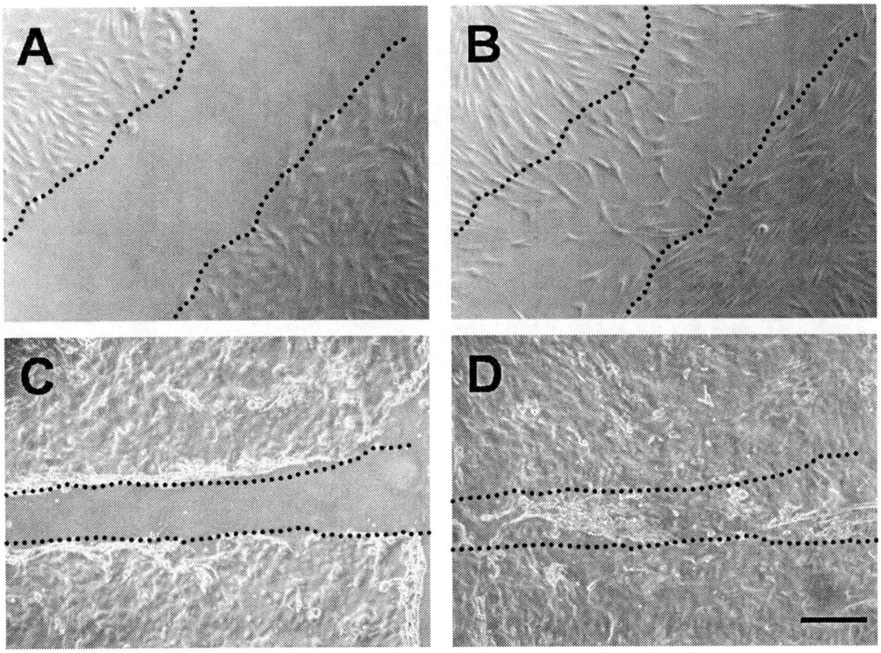

Fig. 1. Keratinocyte and fibroblast response to a scratch wound assay reveal different migration patterns.
Two-dimensional cultures of keratinocytes (C, D) and fibroblasts (A, B) were subjected to a scratch wound assay. Cells were scraped to create a "wound" (A, C) and the remaining cells migrated to fill the space after 48 hours (B, D). Keratinocytes migrate as a sheet of cells, always remaining in contact with each other. Fibroblasts migrate as individual cells and require longer time to "heal" the wound. Bar=50um.

2 Tailoring 3D Tissue Models for Specific Wound Healing Applications

Until now, *in vitro* studies of wound re-epithelialization have been helpful in studying keratinocyte migration in response to wounding, but have been of limited use in studying the complex nature of keratinocyte response during tissue repair in response to a wound, as these cultures do not provide the proper tissue architecture to study the *in vivo* wound response. We have previously found that HSEs adapted to study wound repair in human epithelium simulate the chronology of events that occur during re-epithelialization in human skin and have advanced the understanding of the healing of wounds in human skin and other stratified epithelia (Garlick and Taichman 1994; Garlick et al. 1996). This tissue model has allowed direct determination of the key response parameters of wounded epithelium including cell proliferation, migration, differentiation, growth-factor response, and protease expression.

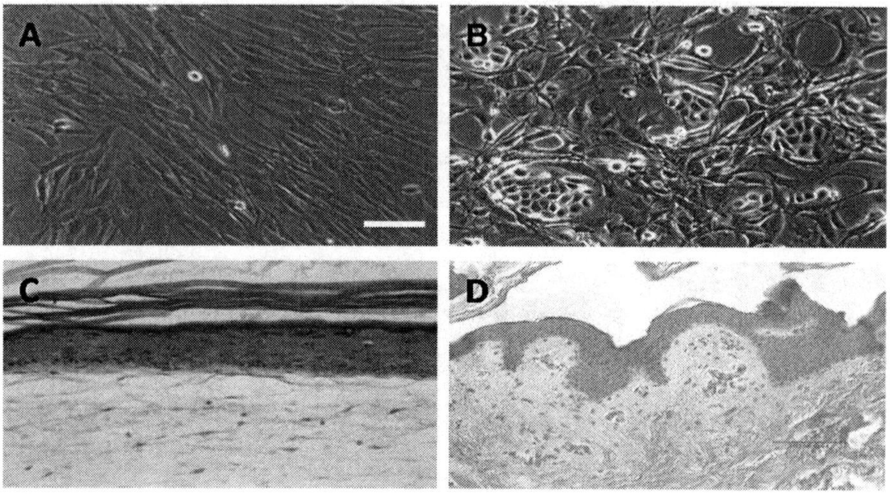

Fig. 2. The 3D culture technique.
Human dermal fibroblasts (A) and human keratinocytes (B) are cultured several days prior to the construction of an HSE. Fibroblasts are incorporated into collagen gels while they are highly proliferative. To achieve this, the fibroblasts are cultured to confluence and split at a 9:10 ratio and used the following day to construct the dermal layer of the collagen gels. To limit the number of differentiating keratinocytes and to increase the cellular growth fraction, the colonies should be kept small. After construction of an HSE (C), the keratinocytes exhibit all layers of differentiation, comparable to normal human skin (D).

A major advantage of studying wound healing in HSEs lies in the ability to control and modify the cellular environment in which these tissues are grown. It is possible to add soluble factors directly to these tissues to determine the phenotypic response to such modifications in environmental conditions. For example, growth conditions can be altered by modifying the phenotype of fibroblasts that are incorporated into the collagen gel to determine how epithelial-stromal cross talk can alter wound reepithelialzation (Andriani et al. 2003). Test agents or compounds can be added to the media to determine if they can accelerate tissue repair following wounding. A variety of ECM substrates can be incorporated into HSEs to answer specific questions on the impact of how adhesion may affect wound healing. For example, growing keratinocytes on a de-epidermalized dermis (Alloderm, LifeCell, Inc.) layered onto a contracted collagen gel provides an interface on which the assembly of basement membrane is optimized (Andriani et al. 2004). Alternatively, tissues grown on polycarbonate membranes coated with specific ECM proteins can be used to study how specific ECM or basement membrane proteins affect wound healing and tissue growth (Segal et al. 2008). Each of these environmental conditions can be tailored to answer specific questions relevant to cell proliferation, migration, and differentiation of keratinocytes in response to wounding.

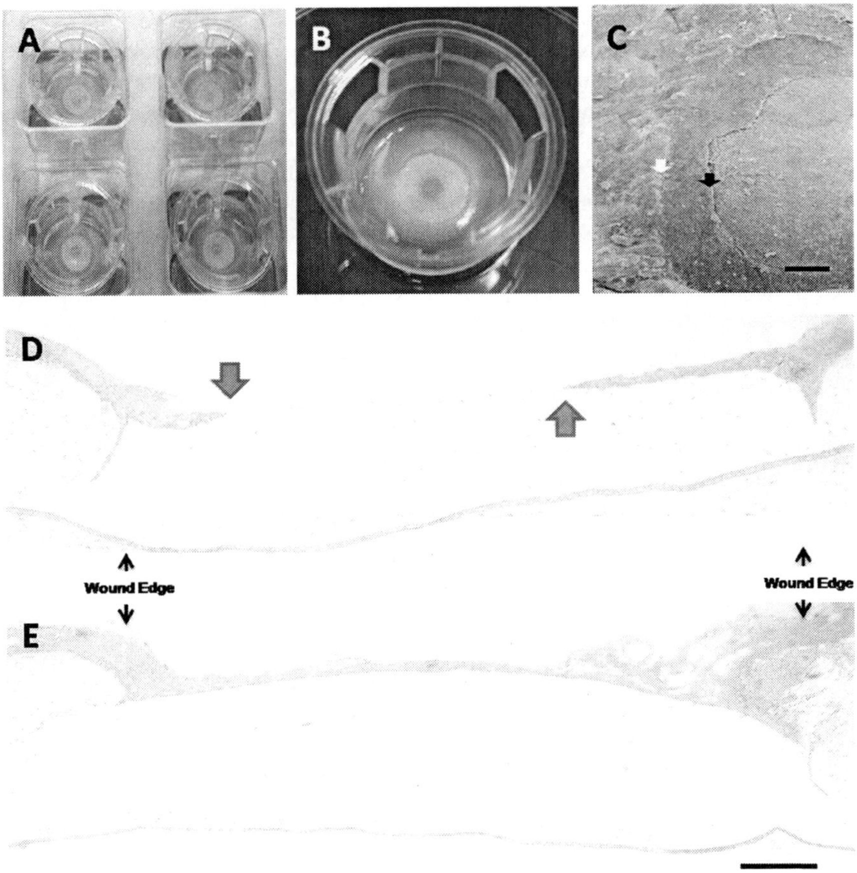

Fig. 3. Three-dimensional cultures mimic *in vivo* wound healing.
HSEs can be cultured in multi-well format to increase the number of conditions to be tested (A); magnified view of a single well (B). An excisional wound is created and the keratinocytes migrate to heal the wound (A, B). Reepithelialization begins with the migration of an epithelial tongue (D, arrows). SEM reveals the unwounded HSE (left of the white arrow) and the edge of the epithelial tongue (between white and black arrows) migrating across the collage gel (right of the black arrow) (C). After 48-72 hours, keratinocytes have completely covered the wound and reepithelialization is complete (E). Bar= 100um.

In light of the flexibility of HSEs to create wound microenvironments that enable the development and/or testing of novel wound therapies, we present three examples in which we have used them to study and screen novel approaches or materials to enable wound repair.

1. We have tested the potential of selected fibroblast subpopulations to enable cross-talk with the overlying epithelial cells that can stimulate wound

re-epithelialization (Segal et al. 2008). This has allowed us to further understand the role of epithelial-mesenchymal interactions in modulating skin repair by studying how the phenotype of human dermal fibroblasts was linked to the healing response of wounded HSEs. We compared fibroblasts extensively cultivated in 2D monolayer culture under 3 distinct conditions, on denatured collagen matrix, native collagen, or tissue culture plastic before incorporation into HSEs. Our results demonstrated that extended passage of dermal fibroblasts on denatured collagen was able to accelerate the rate of wound healing by 38 percent. This increased rate of wound closure was linked to the stimulation of keratinocyte proliferation seen when dermal fibroblasts expanded on denatured collagen were incorporated into the collagen gels on which wound reepithelialization occurred. Therefore, modification of cells that can mediate and direct interactions with their ECM or epithelial microenvironment, hold potential for novel cell therapy applications and for understanding mechanisms through which cross-talk between fibroblasts and keratinocytes may stimulate wound repair to reestablish tissue homeostasis. Use of such modified fibroblasts, that can be selected based on their interactions with specified extracellular matrix (ECM) components, may provide useful therapies to identify and select cell subpopulations that may best treat patients at elevated risk for the healing of chronic wounds.

2. We have established the utility of the 3D wound HSE model to test the capacity of novel agents designed to improve tissue repair by using these tissue platforms to measure the degree of reepithelialization when wounds are exposed to a soluble test agent. To accomplish this, multiple concentrations of the agent can be added directly to the culture media or provided topically to the surface of the tissue, and the degree of reepithelialization is compared to non-treated wounds. Reepithelialization can be directly measured and the tissue phenotype monitored for cell proliferation, differentiation, ECM production, and reestablishment of basement membrane. In this way, it is possible to screen large numbers of agents by testing their efficacy at various doses, thus paving the way for clinical testing of these new wound treatments.
3. We have used wounded HSEs as a platform to test novel nanotechnology-inspired biomaterials that act as a bioactive wound dressing, to topically-deliver compounds that may promote wound healing. As an example of this approach, we have tested two types of materials; a hydrogel made of self-assembling peptide (SAP) nanofiber that can physically be placed into the wound bed (Schneider et al. 2009) and silk mats, made of electrospun nanoscale silk fibers that can cover over the wound bed (Schneider et al. 2008). In both cases, a potential wound healing stimulant (EGF) was incorporated into these biomaterial dressings and determined that these dressings could augment wound reepithelialization by reducing the time of wound closure (Schneider et al. 2009, Schneider et al. 2008). These results proved that the wound HSE platform facilitated the testing of functional biomaterial devices on human tissues to enable the development of

biomaterials with properties that can be optimized for the new generation of biologically-active wound dressings. In addition, this platform has the flexibility to allow the simultaneous testing of synergistic growth factors by screening the efficacy of biomaterial-mediated delivery of multiple topically-acting growth factors. Other applications of this platform have been to utilize multi-functionalized biomaterials that incorporate antimicrobial molecules to deliver anti-bacterial factors that can provide wound protection that may stimulate wound closure.

3 Limitations and Future Directions of 3D Wound Healing Models

Despite its robust performance in mimicking *in vivo* skin, HSEs have a few disadvantages. First, while cultures can be kept at an air-liquid interface for up to 2-3 weeks, their surface cells do not desquamate and become excessively thickened. This timeframe is sufficient to study wound reepithelialization that occurs during the first week after wounding. However, late phases of wound healing such as reinervation, revascularization, or remodeling of the extracellular matrix in the wound bed require long term studies that cannot be realized using the *in vitro* model. The grafting of HSEs to the back of nude mice prior to wounding enables a more complete wound repair study. There are additional differences between the HSE model and *in vivo* human skin which may impact the translational nature of the wounding model. Integrin receptors not normally expressed in skin, such as the αv integrins, are constitutively expressed in keratinocytes grown in HSEs. In this light, keratinocytes grown in HSEs can be thought of as tissues undergoing *de novo* development such as newly re-epithelialized, healed wounds *in vivo*, wherein morphologic differentiation is complete but cells are still in a somewhat "activated" state. In addition, HSEs are somewhat deficient in barrier function.

While the HSE wound model described represents an important advance over simple, 2D monolayer culture systems, these tissues are continually being improved upon in several ways. 1) High-throughput platforms need to be developed for wounded HSEs that will further refine their utility as a screening tool for pre-clinical testing by enabling the simultaneous testing of a wider number of potential wound care products and therapies. 2) Incorporation of additional cell types, such as endothelial cells or inflammatory cells to HSEs would generate a more *in vivo*-like tissue microenvironment. For example, the use of microfluidic networks, capable of providing physiological fluid flow properties is currently under development for this purpose. 3) Integration of real-time imaging tools into 3D tissue platforms will allow dynamic monitoring of the sequential stages of wound healing responses. 4) Generation of wound HSEs that simulate the chronic wound environment, perhaps by modifying ECM components on which cells undergo reepithelialization to impede repair. 5) Incorporation of fibroblasts or keratinocytes from aged individuals to study the relationship between wound healing and aging. The design and development of interoperable, advanced HSE tissue platforms will impact directly on the science of wound healing and will create new synergies throughout the wound healing community that will economize and streamline the pharmaceutical design, screening and therapeutic application of lead candidate compounds and devices targeted to the treatment of human wounds.

4 Construction of 3D Human Skin

The 3D tissue model of normal skin consists of a stratified epithelium containing differentiated keratinocytes that are grown on a contracted Type I collagen gel populated with dermal fibroblasts. As seen in the scheme describing HSE assembly (Figure 4), normal tissues are constructed through the following sequential steps:

a) Acellular collagen layer – A thin, acellular layer of collagen is constructed first and acts as an attachment substrate for the cellular collagen that is fabricated above it. This thin layer prevents the cellular collagen from detaching from the insert membrane during contraction.
b) Cellular collagen layer – A Type I collagen gel with human dermal fibroblasts is constructed and allowed to contract for 7 days while submerged in media. Fibroblasts mixed within the collagen gel remodel the matrix upon production of extracellular matrix proteins (ECM) and cause contraction of the gel.
c) Epithelialization – Keratinocytes are added to the surface of the contracted gel and allowed to attach to generate a confluent cellular monolayer that will initiate tissue stratification.
d) Tissue stratification, organization, and differentiation – Tissues are raised to an air-liquid interface to enable complete stratification, as well as to provide conditions for full morphologic and biochemical differentiation.

5 Wound Healing Protocol

HSEs are constructed using skin keratinocytes and fibroblasts that are first grown in 2D, monolayer cultures in a way that optimizes their proliferative potential (Garlick et al. 1996; Kolodka et al. 1998). Following 3D tissue fabrication, basal keratinocytes in HSEs demonstrate a basal rate of proliferation that is similar to that seen in human skin as well as a well-differentiated, stratified epithelium (Garlick and Taichman 1994). HSEs are wounded with a scalpel to create an incisional wound or with a dermatological punch to create an excisional wound and transferred to an additional collagen gel that was prepared one week prior to wounding. Wound re-epithelialization then ensues in a very similar manner to that seen in full-thickness incisional wounds *in vivo* in the following discrete, temporal stages (Winter 1972): (a) Reepithelialization is initiated following a short delay (8 hours) as an epithelial tongue of keratinocytes begins to migrate onto the wound bed. (b) This is followed by a transient, proliferative burst in keratinocytes at the wound margin that occurs 24 hours after wounding. (c) Production of the matrix-degrading protease, MMP-1 is expressed in keratinocytes migrating onto the

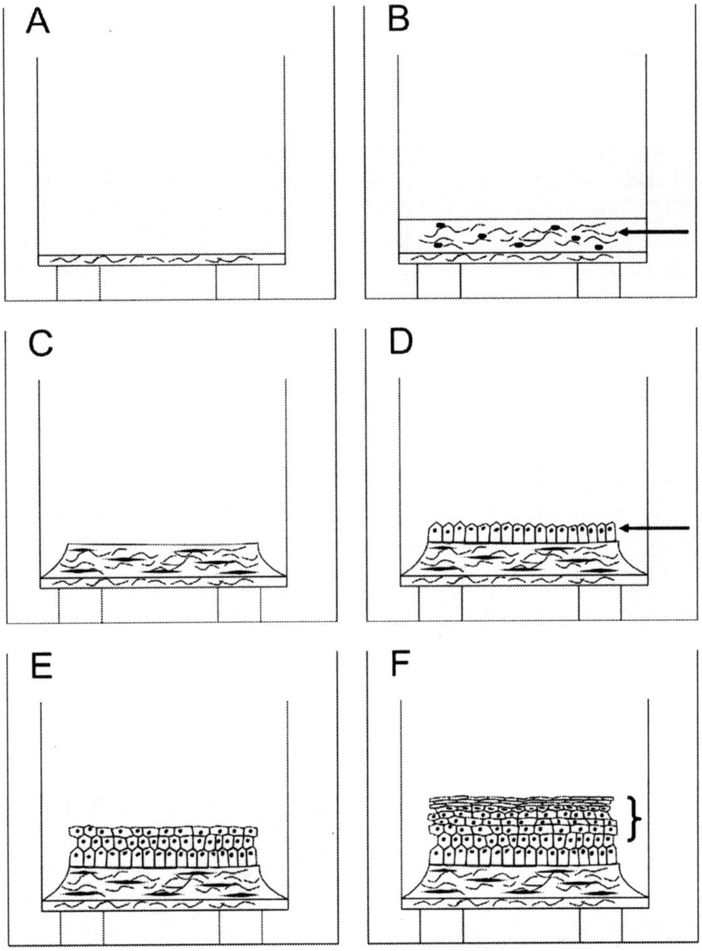

Fig. 4. Scheme of 3D tissue construction.
Panels A-F describes the construction of HSEs. Briefly, fibroblasts embedded in collagen is layered into the insert and allowed to contract over 7 days. Keratinocytes are added on top of the contracted collagen and differentiate over the course of 12 days at an air-liquid interface. A) A thin, acellular layer of collagen is first constructed and provides an attachment substrate for the cellular collagen. B) A collagen gel embedded with human dermal fibroblasts (arrow) is layered onto the acellular and contracts for 7 days. C) While submerged in media, dermal fibroblasts remodel the collagen matrix, causing it to contract away from the walls of the insert. The contracted collagen forms a plateau and keratinocytes are later added to the center of the plateau. D) Keratinocytes (arrow) are added and allowed to attach to the collagen (or intervening substrate such as Alloderm) to create a monolayer that will form the basal layer of the tissue. E) Tissues are raised to an air-liquid interface to initiate stratification. Keratinocytes stratify differentiate and form a suprabasal layer that mimics *in vivo* skin both morphologically and biochemically. F) Further exposure to the air-liquid interface and additional feedings with cornification media results in an increase in the thickness of the spinous and cornified layers of the tissue (bracket). This figure reprinted with permission of Current Protocols in Cell Biology.

wound bed resulting in the remodeling of ECM proteins at this interface, (d) Wound reepithelialization is complete as epithelial integrity is restored after 48-72 hours, (e) Basement membrane proteins are deposited to organize and reestablish the basement membrane interface. (f) Stratification of the restored tissue then occurs to regenerate a fully-differentiated, stratified squamous epithelium. Healing epithelial tissues are amenable to manipulation of medium composition (Carlson et al. 2007), modification of the ECM substrate interface on which keratinocytes are grown (Andriani et al. 2003; Segal et al. 2008) and alteration of its cellular constituents (Egles et al. 2008) to create microenvironments that mimic a variety of wound conditions in stratified epithelial tissues. For example, epithelial tissues can be constructed in the presence or absence of basement membrane proteins (Segal et al. 2008) or in the presence of various stromal cells that are incorporated into the contracted collagen gel.

5.1 Fabrication of Collagen Matrix with Dermal Fibroblasts

1. The day before incorporation, passage fibroblasts at a 9:10 split ratio so they will be mitotically active the next day when incorporated into the collagen gel. A 9:10 passage is performed by resuspending the trypsinized cells in 10 ml of medium, and adding 9 ml of the cell suspension to a new plate.

2. The following day, the collagen matrix is prepared by fabricating successive layers of acellular and cellular collagen onto the polycarbonate membrane. Prepare the acellular collagen as a mixture that is cooled on ice to prevent premature gelation (see note 3). To do this, pipettes should be chilled at -20°C for 15 minutes before use to prevent warming of collagen when it is mixed. Avoid air bubbles when mixing. Collagen should be a straw-yellow to light pink color to ensure optimal gelation. If the color is bright yellow, add a single drop of sodium bicarbonate and triturate until a straw-yellow color is seen.

3. Add 1 ml of acellular collagen to each insert. The insert is a single well in the 6-well tray that is shaped like a cup. The insert allows media to diffuse from only the bottom when the tissues are later raised to the air-liquid interface. Ensure that the matrix coats the entire bottom surface of the insert and allow it to gel at room temperature for 20 min. Do not move the tray while it is undergoing gelation. The color will turn pink when the collagen has fully gelled.

4. Trypsinize, count, and resuspend the fibroblasts to a final concentration of $3x10^5$ cells/ml. A total of $5x10^5$ fibroblasts will be used per 6-well tray.

5. Prepare the cellular collagen as a mixture that is cooled on ice (see note 3). Fibroblasts should be added last after collagen has been neutralized so that the cells will not be damaged by the alkaline pH that exists before neutralization. Resuspend the cell/collagen suspension by gentle trituration to evenly incorporate fibroblasts into the collagen gel.

6. Gently triturate the cellular matrix and add 3 ml into each insert on top of the gelled acellular collagen matrix. Gently transfer the mixture to the incubator for 30 min.

7. When the cellular matrix has turned pink and is completely gelled (usually less than 30 minutes), feed the gels with 12 ml of fibroblast medium by adding 10 ml of medium to the well around the insert and 2 ml of medium directly onto the insert.

8. Gels are then incubated for 5-7 days to allow complete gel contraction.

9. During the first few days, the sides of the gel contract and will form a plateau in the center. Gels are stable between 5 - 10 days after initial construction.

5.2 Addition of Keratinocytes to the Surface of Contracted Collagen Gels

1. Normal human keratinocytes are cultured on a feeder layer of mitotically inactivated mouse 3T3 fibroblasts. Keratinocytes should be grown to no more than 50% confluence to minimize the number of differentiated cells seeded onto the collagen gel. Alternatively, keratinocytes can be grown in monolayer culture in low calcium and serum-free medium (see note 4).

2. Remove the 3T3 feeder cells from the culture by incubating the plates in PBS/EDTA for 5 min. at 37°C. 3T3's can then be displaced by gentle pipetting so that keratinocytes will remain attached. It is important not to allow the cultures to incubate for an excessive time in PBS/EDTA, as the keratinocytes may detach from the plate as well. As soon as the 3T3's have begun to detach, replace the PBS/EDTA with PBS, gently rinse the plate 3 times with PBS until all 3T3's have been completely removed. PBS is then removed, leaving only keratinocyte colonies attached to the plate.

3. Trypsinize the keratinocytes with trypsin/EDTA (0.05%) for 5 minutes at 37°C to obtain a single cell suspension. Remove the detached cells into a 15ml tube containing keratinocyte medium (to neutralize the trypsin) and count them. The desired number of cells is dispensed into a 15ml tube from the tube with the trypsinized cells. Cells are then centrifuged at 2,000 x *g* for 5 min and resuspended in a volume so that a total of 5×10^5 keratinocytes can be used per insert.

4. Remove all fibroblast medium from the trays with the contracted collagen 20 minutes before seeding keratinocytes so that keratinocytes can be seeded onto a moist collagen gel. Keratinocytes should be seeded directly onto the contracted collagen gels in an aliquot of 50 µl containing 5×10^5 cells. To modify the nature of the substrate on which keratinocytes are seeded, de-epidermalized dermis or coated polycarbonate inserts can be applied directly on top of the contracted collagen gels at this point (see note 5).

5. Resuspend keratinocytes in a volume of Epidermalization I medium to a final concentration of 500,000 cells / 50 µl. Carefully add the 50 µl of the cell suspension to the center of the contracted collagen gel (or onto the center of the intervening substrate placed on the collagen gel). Do not move the tray for 15 minutes to allow the keratinocytes to attach. Constructs are then incubated at 37°C for 30-60 min. without any medium to allow the keratinocytes to fully adhere.

6. Add 12 ml of Epidermalization I medium to each insert by adding 10 ml to the bottom of the well and 2 ml gently into the insert on top of the keratinocytes. Incubate at 37°C.

7. Cultures are fed with medium (see section 3.1.5) every 2 days as follows:

   - Epidermalization I medium – 12 ml per well for the first 2 days.
   - Epidermalization II medium – 12 ml per well for the next 2 days.
   - Cornification medium – At this point, cultures are raised to the air-liquid interface by adding 7 ml per well to the bottom of the well so that the insert just contacts the medium. Aspirate medium from the inside of the insert so that tissues can be grown at the air-liquid interface. Additional feedings with Cornification medium are done every 2 days until termination of the experiment.

5.3 Fabrication of Three-Dimensional Wound Healing Model of Human Skin

1. HSEs to be wounded are first generated as described above. This protocol requires that an HSE with normal, primary keratinocytes and a second contracted collagen gel (onto which the wounded epithelium will be transferred) will first be simultaneously constructed.

2. Aspirate all medium from the HSE after 7-10 days of culture. Remove the insert from the tray and place it upside down in a sterile dish. Using a scalpel, cut away the insert membrane and place the culture in a sterile dish right-side up.

3. Trim the culture with the scalpel by cutting around the raised, mesa-like region to remove the part of the collagen gel not covered by keratinocytes. This will facilitate the removal and transfer of the wounded tissue from the membrane.

4. Cultures can be wounded with either an incisional or excisional wound. An incisional wound can be generated by incising tissues with a scalpel in a way that will allow the wound edges to be separated to generate an elliptical wound.

5. An excisional wound can be generated using an elliptical dermatological punch that completely penetrates the center of the tissue through the epidermis, collagen, and membrane. The excised tissue can be fixed and preserved for H&E staining in 10% formalin.

6. Use forceps to gently lift the edge of the wounded tissue by separating the collagen gel from the membrane. Drag the tissue onto a dental mirror while leaving the membrane behind. The transfer may be easier if the mirror is moistened with medium.

7. Unfold any wrinkles in the culture by gently moving the tissue back and forth on the mirror using the forceps. Once the culture is smooth, pull one side of the culture slightly over the edge of the mirror.

8. Carry the mirror directly over the second contracted collagen matrix so that the edge of the mirror and wounded tissue are in contact with the matrix. Slide the tissue onto the second collagen gel by teasing it gently with a closed forceps as the mirror is slowly pulled away, leaving the culture on the contracted collagen gel.

9. Using the forceps, tease apart the tissue wounded by incision to create an elliptical space that should be 2-3 mm at its greatest width. Smooth the tissue with the forceps to ensure that it is completely free of any folds or wrinkles.

10. Maintain the tissue at an air/liquid interface by adding 8 ml of Epidermalization II medium beneath the insert during reepithelialization, change the medium every 2 days until the end of the experiment.

5.4 Anticipated results

Several points regarding keratinocyte behavior in HSEs should be mentioned. The first concerns the length of time during which cultures can be maintained at the air-liquid interface. In our experience, cultures can be kept at this interface for up to 17 days. After this time, the surface layer of the epithelium becomes excessively thickened due to a failure to desquamate. As a result, lower layers of the epithelium become compressed and the longevity of cultures is limited. A second and related issue concerns the growth potential of keratinocytes in HSEs. While HSEs demonstrate a basal level of proliferation that is greater that of human skin, it has been shown that these cultures have a tremendous potential for cell growth and are very responsive to external growth stimuli.

It should be kept in mind that although keratinocytes grown in HSEs share many morphologic and biochemical features with *in vivo* skin, there are differences in tissue phenotype. For example, integrin receptors not normally expressed in skin may be constitutively expressed in keratinocytes grown in HSEs. This may be the result of HSEs being somewhat deficient in barrier function.

5.5 Time Considerations

Construction of HSEs require approximately 3-4 weeks from the time HFF are seeded in monolayer culture until 3D tissues are fully mature. HFF cells should be confluent two days before construction of the collagen gels. At that time, cells should be passaged at a 9:10 ratio to provide cells with a growth stimulus before incorporation into collagen gels. Complete contraction of the collagen gels requires 7-10 days, during which time the human keratinocytes cell cultures are initiated and expanded. The keratinocytes will need another week of culture before the HSEs are ready for wounding.

6 Notes

1. Milli-Q-water should be used for preparation of all medium and supplements, and all solutions should be filtered through a 0.22 µm filter for sterility.
2. Cholera toxin is very toxic. Use appropriate precautions when handling stock solutions.
3. Fabrication of the collagen gel requires that all components be kept on ice until the gel mixture is placed into the insert. This will ensure that collagen will not prematurely precipitate from these solutions. Plastic pipettes used for collagen should be chilled before use.

The amounts listed are for a single 6-well tray.

0.6 ml 10x MEM (Minimum essential medium with Earle's salts) (Cambrex, Walkersville, MD, cat. 12-684F)
54 µl 200mM L-Glutamine (Invitrogen, Carlsbad, CA, cat. 25030-081)
0.68 ml FBS (Hyclone, Logan, UT, cat. SH30071.01)
187 µl 71.2 mg/ml $NaHCO_3$ (Cambrex, Walkersville, MD, cat. 17-605E)
5 ml Bovine Type I Collagen (Organogenesis, Canton, MA, cat.# 200-055) Fresh and used immediately. It should not be stored once mixed.

- Keratinocyte proliferation and a high growth fraction are the most critical factors in the successful fabrication of HSEs. Most keratinocytes seeded onto HSE cultures will adhere to the connective substrate, but only replicating cells will grow after seeding. Keratinocytes that underwent a commitment to terminal differentiation while still in submerged culture will also attach to the substrate, but will not undergo further proliferation to form a well-stratified HSE. It is therefore important to grow keratinocytes so that a high growth fraction is present when monolayer cultures are seeded onto the contracted collagen gel of the HSEs. This can be accomplished by growing keratinocytes as small colonies at high clonal density in submerged cultures on 3T3 feeder layers, so that terminal differentiation will be minimized and the fraction of replicating cells will be maximized. Keratinocyte strains can be tested by screening them using a colony efficiency assay to determine those strains with the highest colony forming efficiency should provide optimal morphologic differentiation and tissue architecture of HSEs.

- We have found some variability in the degree to which fibroblast strains support keratinocyte growth after their incorporation into collagen gels. It appears that fibroblast support of HSE organization and growth is directly related to the degree to which fibroblasts are able to contract the collagen gel. In general, fibroblast strains demonstrating more shrinkage of the collagen gel before adding keratinocytes are better able to support keratinocyte growth. This parameter may be used to screen fibroblast strains for optimal growth-support when initiating HSEs.

7 Materials

7.1 Medium Components

0.5M EDTA, pH 8.0 (Invitrogen, Carlsbad, CA, cat.# 15575)

PBS (Invitrogen, Carlsbad, CA, cat.#14190)

0.25% Trypsin (Invitrogen, Carlsbad, CA, cat.# 15050)

5 mM EDTA ; add 5 ml of 0.5 M EDTA to 500 ml PBS

10% EDTA/PBS ; mix 50 ml of 5 mM EDTA with 450 ml PBS

0.1% Trypsin – mix 50 ml 0.25% trypsin with 75 ml PBS

50% Trypsin/EDTA – mix 50 ml of 0.1% trypsin with 50 ml of 5 mM EDTA

100x HEPES (800 mM) – dissolve 47.24 g in 250 ml ddH_2O, store at -20°C for up to one year (Sigma, St. Louis, MO, cat. H-4034)

100x Adenine (18 mM) – dissolve 0.972 g in 2.4 ml 4N NaOH, q.s. to 400 ml with ddH_2O, store at -20°C for up to one year (MP biomedicals, Solon, OH, cat. 100190)

500x Hydrocortisone (0.25 mg/ml) – dissolve 0.0538 g in 200 ml ddH_2O, store at -20°C for up to one year (Sigma, St. Louis, MO, cat. H-4881)

1,000x Cholera toxin (10^{-7} M) – dissolve 9 ng/ml in ddH_2O, store at -20°C for up to one year (Sigma, St. Louis, MO, cat. C-8052)

1,000x EGF (10 μg/ml) – dissolve 10 μg/ml in 0.1% BSA, store at -20°C for up to one year (Austral Biological, San Ramon, CA, cat. GF-010-9)

1,000x Insulin (5 mg/ml) – dissolve 50 mg in 10 ml of 0.005N HCl, store at -20°C for up to one year (Sigma, St. Louis, MO, cat.# I-2643)

10 nM triiodothyronine (T3) – add 1 ml T3 to 99 ml ddH2O for 500x stock (Sigma, cat # T-5516)

2 μM progesterone – dissolve 1 mg in 1 ml absolute ethanol, add 14.7 ml ddH_2O, dilute 1 ml in 100 ml DMEM for 1,000x stock (Sigma, St. Louis, MO, cat. P-8783)

Chelated BCS – Chelate serum by adding 10 g CHELEX 100 (Sigma, St. Louis, MO, cat. C-7901) to 100 ml serum and stirring for 3 hr. at 4°C, then filter through Whatman paper, then through a sterile filter.

Transferrin (5mg/ml) – (BioSource cat # 352-020, 200 ml)

500x PES – contains O-phosphorylethanolamine (0.01 mM final), Ethanolamine (10 μM final), and Selenium (10 μg/ml final) (BioSource, cat. P02-45-100)

7.1.1 O1O Medium

43 g DME powder (JRH Biosciences this is a special order medium base that is prepared in bulk), contains no glucose and no $CaCl_2$

5 L ddH_2O

0.5 g $MgSO_4$

18.5 g $NaHCO_3$

7.1.2 Keratinocyte Culture Medium

338 ml DME medium (Invitrogen, Carlsbad, CA, cat.# 11885)

112 ml F12 medium (Invitrogen, Carlsbad, CA, cat.# 11765)

25 ml FBS (Hyclone, Logan, UT, cat.# SH30071.03) (5% final)

5 ml 18mM Adenine (0.18 mM final)

3.4 ml 100x Penicillin/Streptomycin (Invitrogen, Carlsbad, CA, cat.# 15140-122)

5 ml 800 mM HEPES (8 mM final)

1ml 0.25 mg/ml hydrocortisone (0.5 μg/ml final)

0.5 ml 10^{-7} M cholera toxin (10^{-10} M final) (see note 2)

0.5 ml 10 μg/ml EGF (10 ng/ml final)

0.5 ml 5 mg/ml insulin (5 μg/ml final)

Store up to 2 weeks at 4°C.

7.1.3 Fibroblast Culture Medium

500 ml DME medium (Invitrogen, Carlsbad, CA, cat.#11885)

55.6 ml FBS (Hyclone, Logan, UT, cat.# SH30071.03) (10% final)

5.6 ml 800 mM HEPES (8 mM final)

3.4 ml 100x Penicillin/Streptomycin (Invitrogen, Carlsbad, CA, cat.#15140-122)

Store up to 2 weeks at 4°C.

7.1.4 3T3 Medium

500 ml DME medium (Invitrogen, Carlsbad, CA, cat.# 11885)

55.6 ml Bovine Calf Serum (Hyclone, Logan, UT, cat.# SH30072.03) (10% final)

3.4 ml 100x Penicillin/Streptomycin (Invitrogen, Carlsbad, CA, cat.#15140-122)

7.1.5 HSE Medium

	Stock Concentration	Final Concentration	Epidermalization I (ml)	Epidermalization II (ml)	Cornification (ml)
O1O			363	363	237
Ham's F12			120	120	237
L-glutamine	200 mM	4 mM	10	10	10
Adenine	18 mM	40 uM	1	1	1
Hydrocortisone	0.55 mM	1 uM	1	1	1
T3	500x	20 pM	1	1	1
Transferrin	5 mg/ml	10 ug/ml	1	1	1
Insulin	5 mg/ml	10 ug/ml	1	1	1
Progesterone	2 uM	2 nM	0.5	0.5	0
PES	500x	1x	1	1	1
Caclium Chloride	0.5 M	1.8 mM	0	1.8	1.8
Serum			0.5 (cBCS)	0.5 (FBS)	10 (FBS)

References

1. Andriani, F., Garfield, J., Fusenig, N.E., Garlick, J.A.: Basement membrane proteins promote progression of intraepithelial neoplasia in 3-dimensional models of human stratified epithelium. International journal of cancer 108(3), 348–357 (2004)
2. Andriani, F., Margulis, A., Lin, N., Griffey, S., Garlick, J.A.: Analysis of microenvironmental factors contributing to basement membrane assembly and normalized epidermal phenotype. The Journal of investigative dermatology 120(6), 923–931 (2003)
3. Bissell, M.J., Radisky, D.: Putting tumours in context. Nature reviews 1(1), 46–54 (2001)
4. Carlson, M.W., Iyer, V.R., Marcotte, E.M.: Quantitative gene expression identifies appropriate cell line models for individual cervical cancer pathways. BMC Genomics 8(1), 117 (2007)

5. Egles, C., Shamis, Y., Mauney, J.R., Volloch, V., Kaplan, D.L., et al.: Denatured collagen modulates the phenotype of normal and wounded human skin equivalents. The Journal of investigative dermatology 128(7), 1830–1837 (2008)
6. Garlick, J.A., Taichman, L.B.: Fate of human keratinocytes during reepithelialization in an organotypic culture model. Laboratory investigation; a journal of technical methods and pathology 70(6), 916–924 (1994)
7. Garlick, J.A., Parks, W.C., Welgus, H.G., Taichman, L.B.: Re-epithelialization of human oral keratinocytes in vitro. Journal of dental research 75(3), 912–918 (1996)
8. Kolodka, T.M., Garlick, J.A., Taichman, L.B.: Evidence for keratinocyte stem cells in vitro: long term engraftment and persistence of transgene expression from retrovirus-transduced keratinocytes. Proceedings of the National Academy of Sciences of the United States of America 95(8), 4356–4361 (1998)
9. Mustoe, T.: Understanding chronic wounds: a unifying hypothesis on their pathogenesis and implications for therapy. American journal of surgery 187(5A), S65–S70 (2004)
10. Segal, N., Andriani, F., Pfeiffer, L., Kamath, P., Lin, N., et al.: The basement membrane microenvironment directs the normalization and survival of bioengineered human skin equivalents. Matrix Biol. 27(3), 163–170 (2008)
11. Singer, A.J., Clark, R.A.: Cutaneous wound healing. The New England journal of medicine 341(10), 738–746 (1999)
12. Schneider, A., Wang, X.Y., Kaplan, D.L., Garlick, J.A., Egles, C.: Biofunctionalized electrospun silk mats as a topical bioactive dressing for accelerated wound healing. Acta Biomater 5(7), 2570–2578 (2009)
13. Schneider, A., Garlick, J.A., Egles, C.: Self-assembling peptide nanofiber scaffolds accelerate wound healing. PLoS One 3(1), e1410 (2008)

Tissue-Derived Materials for Adipose Regeneration

M.D. Ming-Huei Cheng[1], S. Uriel[2], and Eric M. Brey[2,3]

[1] Department of Plastic and Reconstructive Surgery, Chang Gung Memorial Hospital, College of Medicine, Chang Gung University, Taoyuan, Taiwan
minghueicheng@gmail.com
[2] Pritzker Institute of Biomedical Science and Engineering, Department of Biomedical Engineering, Illinois Institute of Technology, Chicago, IL USA
shiriwallach@gmail.com, brey@iit.edu
[3] Research Service, Hines Veterans Administration Hospital, Hines, IL USA

Abstract. Common chronic wounds, including pressure ulcers, diabetic foot ulcers, arterial and venous ulcers often involve damage to adipose tissue. In addition, defects or deficiencies in adipose tissue can occur due to lipoatrophies, trauma, and tumor resection. Clinical treatment options vary, but it is difficult to specifically regenerate or reconstruct adipose tissue. Materials that promote adipose regeneration have the potential to serve as alternatives or supplements to current treatments options. However, the majority of attempts to engineer adipose tissue have used standard synthetic or natural materials as scaffolds with little consideration of unique features of the adipose extracellular microenvironment. The extracellular matrix is different in composition and structure in every tissue and these differences play an important role in cell behavior and tissue function. An ideal material would account for these differences. Based on this knowledge a technique was developed where hydrogels can be extracted and assembled from any soft tissue. The structure, composition, and biological properties of these hydrogels vary depending on the tissue used for extraction and the mechanism of gelation. Hydrogels derived from adipose tissue using this technique promote greater preadipocyte differentiation *in vitro* and vascularized adipose formation *in vivo* than what has been observed with other materials in the absence of exogenous cells or growth factors. These complex, multi-component hydrogels hold great promise for soft tissue reconstruction.

1 Introduction

The treatment of chronic wounds, such as pressure ulcers, diabetic foot ulcers, arterial and venous ulcers, may require regeneration or replacement of damaged or missing adipose tissue. Congenital anomalies, traumatic or burn injuries, genetic and acquired lipodystrophies, and post-oncological resection can also lead to defects or damage in adipose tissue that can alter tissue function and form. The treatment of these defects varies depending on a number of biological, clinical, and practical considerations. Regardless of approach, it is difficult to regenerate or reconstruct functional adipose tissue. Tissue regions previously occupied by adipose are often reconstructed using another tissue type, supplemented with biomaterial bulking agents, or ignored altogether as more easily reconstructed tissues are

A. Gefen (Ed.): Bioengineering Research of Chronic Wounds, SMTEB 1, pp. 281–299.
springerlink.com

treated. Materials that promote formation of vascularized adipose tissue have the potential to serve as alternatives or supplements to existing treatment options, allowing for the regeneration of functional adipose in the appropriate tissue regions.

2 Etiology and Clinical Treatment Options

Chronic wounds can result in damaged or absent adipose tissue. Treatment options depend on the nature of the defect, patient health, and the background of the physician. In general, these approaches involve either enhancing the natural process of tissue regeneration, reconstruction of the defect with autologous adipose, reconstruction with other types of tissue, or filling the defect with biomaterial bulking agents (Figure 1). In many cases, the ideal option is the transfer of free adipose tissue to the defect. However, tissue transfer techniques, in general, are hindered by poor tissue function and perfusion in patients with chronic disease, and the free transfer of adipose is specifically difficult due to its high rate of absorption. Despite over a century of investigation, the absorption of transferred adipose can reach up to 70 percent. The most common congenital anomalies resulting in adipose deficiencies are hemifacial hypoplasia, Parry-Romberg syndrome and Poland syndrome. The best options for these patients would be reconstruction with donor adipose, but this is challenging due to the shortage of sufficient tissue required to achieve functional and cosmetic expectations. Free tissue transfers for soft tissue augmentation by combining various remote donor sites may be able to achieve improved results. However, these procedures may need to be further refined by inclusion of minor fat graft or injection of biomaterial fillers.

Soft tissue defects from chronic wounds, trauma, burn, or tumor resection often include regions of adipose tissue. The reconstruction of these defects depends on the size of the defect and health status of the patient. In a healthy adult, a wound with the largest dimension less than 5 centimeters can be healed by secondary intention. A vacuum assisted closure (VAC) device with a closed suction and dressing system can be used to continuously drain wound discharge resulting in enhanced granulation tissue formation within the wound bed. After the defect is filled with granulation tissue, the wound can be covered with a skin graft. The VAC helps to downgrade reconstruction ladder in which originally a local, regional or even free flap would be indicated in the absence of the granulation tissue formation. Patients with a similar size defect and concomitant vasculopathy can benefit from hyperbaric oxygen (HBO) treatments. This delivery of concentrated oxygen to local tissue enhances wound healing.

If a soft tissue defect is larger than 5 cm, local or free flap transfer is usually required for coverage of the defect. In a purely soft tissue defect with a healthy soft tissue or muscle covering the wound bed, the wound can be easily treated with a skin graft. However, if any vital structures are exposed, such as bone, tendon, nerve and artery, the exposed tissue needs to be covered with a tissue flap, or with a combination of soft tissue with a skin graft. The tissue flap can be local, pedicled, or free. Skin grafts have particular difficulty taking when there is exposed bone and tendon. An alternative is to use a tissue expander to expand the subcutaneous skin nearby for wound coverage; however, it is restricted to only

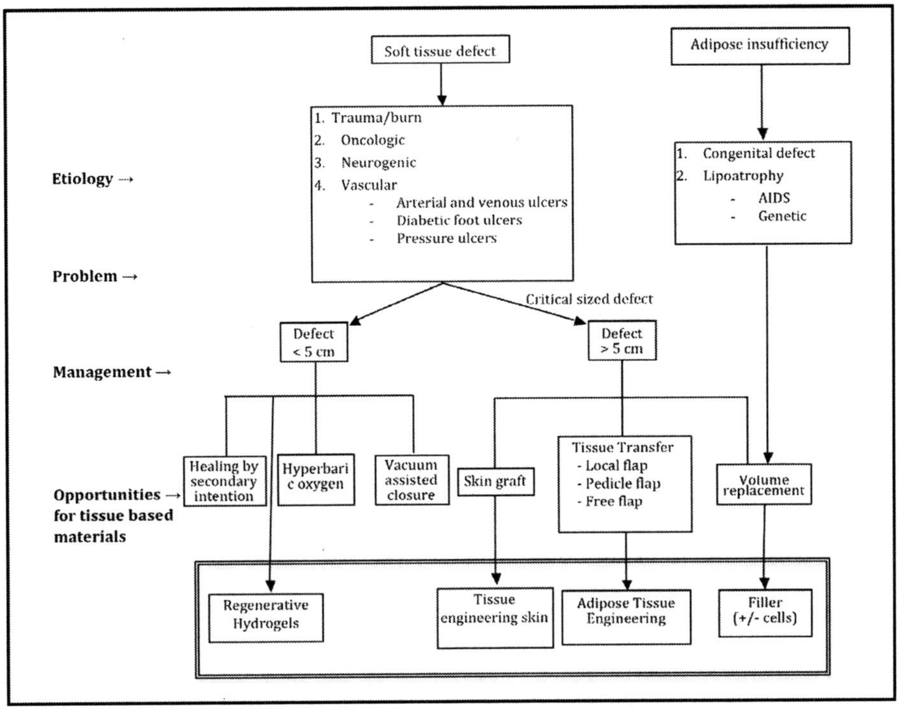

Fig. 1. Treatment options for adipose defects/deficiencies encountered clinically. Opportunities for improvement through the application of tissue-derived materials are shown in box enclosed by the double line.

cases of delayed reconstruction due to the time consuming process of expansion, requiring 2-3 months for completion.

These reconstructive procedures for large defects typically do not address insufficient or absent adipose tissue. This often leads to soft tissue volume insufficiency, a common problem encountered after reconstruction. This may be corrected with injection of autologous fat graft from liposuction. (Lam et al., 2008; Phulpin et al., 2009) This approach is also utilized for soft tissue augmentation in the sunken face (Guerrerosantos et al., 2007), wrinkling (Guyuron and Majzoub, 2007), and depression scar to improve aesthetic outcomes. Currently, the procedure can be achieved with fat graft injection from autologous liposuction or with hyaluronic acid injection. However, results are not consistent. Hyaluronic acid is absorbed over time with typical lifetimes of 1-2 years. In addition, hyaluronic acid only serves as a bulking agent without significant capacity for regenerating new adipose tissue. The viability of injected fat is extremely unpredictable. Due to an initial graft survival rate of 30% to 70% and gradual graft absorption, only 30% to 40% of patients can sustain long-term maintenance of the injected fat graft.

Adipose-reconstruction procedures have been used to treat genetic and acquired lipoatrophies where there can be significant loss of tissue from the face, arms, and legs. Genetic lipodystrophies are rare but can lead to significant alterations in tissue structure, resulting in either excessive loss (lipoatrophy) or buildup of subcutaneous adipose tissue. Acquired lipodystrophies are on the rise due to the prevalence of human immunodeficiency virus (HIV). Lipoatrophy may affect up to 50% of individuals with HIV treated with antiretroviral medications. Treatment of these patients with a small amount fat graft is feasible if the donor adipose is available for transfer. (Dollfus et al., 2009) Hyaluronic acid treatment encounters the same problems of short duration in these patients, so new, alternative materials are needed.

A number of approaches have been used to treat defects of absent or deficient adipose tissue. However, these methods are limited in their ability to provide a long-term solution that results in stable adipose tissue. Tissue transfer techniques have poor success when using adipose alone, so methods largely use other tissue types. Materials used to treat small adipose defects are typically bulking agents with little ability to regenerate new adipose. New biologically active materials that promote the formation of vascularized adipose could provide a new approach.

3 Biomaterials as Models of Disease

In addition to the potential for treatment of tissue defects, biomaterials that stimulate adipogenesis provide an excellent environment for the study of adipose regulated pathologies. Cell behavior is mediated by complex three-dimensional (3D) interactions between cells and the extracellular matrix (ECM). Cells cultured in 3D systems with ECM gels function more similar to *in vivo*. Materials that more closely recapitulate the natural adipose ECM environment could be used to gain significant insight into the normal and pathological function of adipose tissue and the contribution of the ECM to these processes. Obesity rates in adults and children are rising rapidly worldwide. Increased rates of obesity lead to increases in associated pathologies, including cardiovascular disease and diabetes. Greater insight into the mechanisms regulating adipose formation and growth could lead to new interventions for control of adipose-based pathologies.

4 Cell-ECM Interactions in Adipogenesis

Interactions with ECM play a critical role in regulating cell function. Synthetic or natural materials with chemical and/or structural similarity to native ECM have improved cellular interactions and biological performance. However, ECM composition and structure varies between tissues and at different stages of development in a single tissue. These differences are not well characterized. This has often resulted in generic biomaterial approaches to regenerative medicine, with a focus on the use of common, rather than tissue-specific, ECM proteins and peptides. Advances in biomaterials are needed that enable the design of tissue-specific microenvironments with the incorporation of proteins and proteoglycans similar to what is present in ECM *in vivo*.

Basement membranes (BM) are a subset of ECM that exist as thin layers surrounding various cell types, including endothelial cells and adipocytes, separating them from the tissue stroma. The BM helps regulate cell function. BMs are comprised of various proteins and polysaccharides, including various isoforms of laminin, collagen, nidogen/entactin, and proteoglycans.(Kleinman and Martin, 2005; Kleinman et al., 1986) Laminins are glycoproteins consisting of α, β, and γ chains assembled into a cross-shaped heterotrimer. The laminin isoforms present in BM vary between different tissues and with developmental stage. The currently accepted laminin nomenclature defines heterotrimers by sequential Arabic numerals of their α, β, and γ chains (e.g., laminin α1, β1, γ1 is denoted as laminin-111). (Aumailley et al., 2005)

Proteins in the BM appear to play an important role in adipogenesis and are essential for adipocyte maturation.(Kawaguchi et al., 1998) The deposition, synthesis and remodeling of BM molecules is enhanced during adipogenesis.(Chavey et al., 2003; Kawaguchi et al., 1999) As adipocytes begin to differentiate they first degrade the BM and then synthesize new BM components including collagen type IV, laminins and fibronectin. (Kubo et al., 2000) While the BM influences the maintenance and remodeling of adipose tissue, there is incomplete information regarding laminin isoforms present in mature adipose tissue. (Tzu and Marinkovich, 2008) Laminin-411 is the only laminin isoform known to be expressed by adipocytes (Niimi et al., 1997), but its role in adipogenesis is unknown. Previous studies have shown that preadipocytes adhere stronger to laminin-111 than BM proteins fibronectin and collagen IV (Patrick and Wu, 2003), but there is little information about mature or progenitor adipocyte interactions with other laminins. Increased understanding of the role that cell-ECM interactions play in regulating adipogenesis can be used to guide the design of new materials that support vascularized adipose formation.

5 Vascularization and Adipogenesis

Adipose tissue is highly vascularized. (Frye et al., 2005) In fact endothelial cells (ECs) and pericytes of the vasculature make up more than half of the cellular components in adipose tissue. (Kawaguchi et al., 1998) Vascular ECs provide a route for oxygen delivery, play a vital role in metabolism, and secrete soluble factors that regulate adipocyte survival. Adipose is one of the few adult tissues that can undergo both growth and regression throughout life (Dallabrida et al., 2003), and this process is tightly coupled to neovascularization. Inhibition of neovascularization using a number of different therapeutic agents has been shown to retard weight gain in rodent models of obesity. (Dallabrida et al., 2003)

The microvasculature consists of two primary cell types: ECs that line the vessel intima and mural cells/pericytes that exist outside the vessel connecting to the endothelium through the BM. Soluble factors either secreted by ECs or supplied by perfusion of the microcirculation have long been known to contribute to adipose growth. Until recently the mural component of the microvasculature was not thought to play a direct role in adipogenesis. These cells were considered to only to contribute to adipogenesis through regulation of vessel maturation and function. However, recent studies have shown that new adipose formation results from the differentiation and proliferation of progenitor cells present in the mural component of adipose vasculature. (Tang et al., 2008) These adipocyte progenitor cells are not present in the mural component of other tissue vasculature. These cells may be the source of the progenitor cells isolated from the stromal vascular fraction and commonly referred to as adipose derived stem/stromal cells (ADSCs). These findings identify an additional important contribution of the vasculature to adipogenesis, indicating that they are a source of mural cells that act as progenitor cells for continued adipose expansion.

The vasculature has been explored as a therapeutic target for improving adipose survival. Attempts to surgically transfer adipose tissue fail, in part, due to the low tolerance of adipocytes for hypoxia. Increased expression of angiogenic factors either through surgical manipulation (Zhong et al., 2009) or addition of therapeutic molecules (Yi et al., 2007)improves graft survival in rodent models. In addition, as described below, growth factors that stimulate neovascularization can induce adipogenesis in otherwise non-adipogenic materials depending on the surgical model used. (Kawaguchi et al., 1998; Rophael et al., 2007; Vashi et al., 2006) A successful biomaterial approach is likely to require both the ability to induce adipose differentiation and stimulate neovascularization.

6 Materials for Adipose Tissue Engineering

A number of materials, both natural and synthetic, have been investigated for adipose tissue regeneration. In general, these approaches have used common biomaterials tailored for application to adipose tissue by incorporating cells and/or soluble factors expected to induce adipogenesis and/or neovascularization. While cell-based therapies may ultimately be required for engineering vascularized adipose tissue these cells will likely require materials that provide an environment that maximizes adipogenesis. In this chapter we will focus on materials available for tissue engineering and their ability to induce adipogenesis.

6.1 Matrigel™

The material that has received the most attention for adipose regeneration is Matrigel™, a commercially available product consisting of select BM proteins at levels and ratios similar to those found *in vivo*. (Kawaguchi et al., 1998; Kawaguchi et al., 1999; Rophael et al., 2007; Vashi et al., 2006) Matrigel™ is extracted from a spontaneously occurring mouse tumor, the Engelbreth-Holm-Swarm (EHS) sarcoma, (Kleinman and Martin, 2005; Kleinman et al., 1986) and assembles into a fibrous gel network by incubation at 37°C. The popularity of Matrigel™ for adipose tissue engineering derives from landmark studies showing that the gels induce neovascularization and adipose formation when supplemented with more than 1 ng/ml of FGF-2 and injected subcutaneously into rodents. (Kawaguchi et al., 1998) In addition, Matrigel™ with FGF-2 has been shown to induce adipogenesis in silicone chambers implanted subcutaneously around the epigastric artery and vein of rodents.(Cronin et al., 2004; Cronin et al., 2007; Knight et al., 2006) This model of vascularized tissue formation has been used extensively to show that supplementing Matrigel^{TM} with many factors known to stimulate neovascularization increases adipose formation. Matrigel^{TM} has also been shown to improve the longevity and volume maintenance of adipose grafts when mixed with adipocytes. (Piasecki et al., 2007; Piasecki et al., 2008)

Until recently, more extensive adipose tissue formation had been observed in Matrigel^{TM} than other natural and synthetic scaffolds(Cronin et al., 2007), but its application is limited by a number of concerns. First, it is a tumor-derived xenogeneic material that has limited potential for any direct clinical use. Second, Matrigel^{TM} induces little adipose formation in the absence of growth factors or adipose grafts, suggesting that Matrigel^{TM} is conductive rather than inductive for adipogenesis. Interestingly, while Matrigel^{TM} has been shown to stimulate extensive adipogenesis *in vivo* it only moderately stimulates preadipocyte differentiation into mature adipocytes *in vitro*.(Cheng et al., 2009) This may be because the composition of Matrigel^{TM} is not reflective of adipose ECM. While sometimes referred to as "laminin-rich ECM", the only laminin isoform present in Matrigel^{TM} is laminin-111. This isoform plays an important role in early embryogenesis, but is not expressed in most mature organs, and there is no evidence that it is present in the BM surrounding adipocytes. Alternatives to Matrigel^{TM} are needed to translate these results into clinical application.

A material developed from skeletal muscle of various animal species (rat, pig, human) has been developed as a potential alternative to Matrigel^{TM}. (Abberton et al., 2008) The resultant material, known as Myogel, contains laminin isoforms distinct from Matrigel^{TM} and induces similar levels of preadipocyte differentiation *in vitro* and vascularized adipose formation *in vivo*. The results *in vivo* occurred without the addition of exogenous growth factors, possibly due to high levels of FGF-2 in the material. (Abberton et al., 2008) Skeletal muscle was selected for these investigations due to its relative abundance, which could allow scale-up for clinical application. However, it remains unclear why muscle-based materials would provide an improved environment for adipogenesis over materials from other tissue sources. Studies into the mechanism for this affect may provide additional insight into the role ECM proteins play in adipocyte differentiation.

6.2 Collagen

Type I collagen gels are commonly investigated for tissue engineering applications due to their ready availability and well-established biocompatibility. Alone, collagen gels are not adipogenic, but when supplemented with FGF-2 they stimulate adipogenesis in fat pad defects (Hiraoka et al., 2006) and the rodent pedicle model. (Vashi et al., 2006) These results indicate that, like Matrigel^{TM}, the *in vivo* adipogenic properties of collagen appear to depend on the ability to induce neovascularization. The local stimulation of neovascularization accelerates adipogenesis into materials that are otherwise not adipogenic. This approach is not likely to be successful clinically where neovascularization alone is not sufficient to reconstruct adipose.

6.3 Fibrin

Fibrin promotes cell migration and neovascularization and has been explored for a number of applications in regenerative medicine. Fibrin gels have been investigated as materials for adipose tissue engineering. These gels can be used as a carrier for injected adipocytes that induce adipose tissue regeneration when implanted in rodent models. (Cho et al., 2007; Torio-Padron et al., 2007) However, the adipogenic properties of the injectable system appears to result entirely from the injected adipoctyes as fibrin alone does not appear to have adipogenic properties. (Torio-Padron et al., 2007)

6.4 Poly(ethylene glycol)

Poly(ethylene glycol) (PEG)-based hydrogels have received great attention for tissue engineering applications. PEG is hydrophilic and resistant to protein adsorption and cellular adhesion, effectively providing a "blank slate" upon which proteins and peptides can be incorporated to introduce selective biologic properties. The excellent biocompatibility of PEG and the ease in which biological molecules can be incorporated into its structure make it an excellent material for regenerative medicine applications. However, the design challenge for PEG hydrogels is in the selection of the appropriate biological signals and polymerization conditions for controlled and specific tissue regeneration.

Incorporation of PEG with laminin-111 (YIGSR) and peptide sequences degraded by cellular enzymes results in an environment permissive for the proliferation of preadipocytes. (Patel et al., 2005) However, no evidence is presented to suggest that these gels promote cell differentiation. Three-dimensional culture of embryonic germ cells in PEG improves adipose formation in response to soluble factors over traditional 2D culture. (Hillel et al., 2008) Interestingly this appears to occur in the absence of any cell adhesion sequences present in the PEG hydrogels. It may be that the advantage of PEG in this case is in the facilitation of cell-cell contacts.

Porous PEG hydrogels containing FGF-2 are not adipogenic when implanted subcutaneously, but can improve adipose formation when combined with adipocytes or mesenchymal stem cells cultured in adipogenic conditions prior to implantation. (Stosich et al., 2007) We have recently found that, unlike collagen and Matrigel^{TM}, supplementing PEG with growth factors that stimulate neovascularization (FGF-1 and VEGF) does not result in significant adipogenesis in a vascularized pedicle model (unpublished data). When combined with other studies these results suggest that PEG-based hydrogels that stimulate neovascularization alone will not be adipogenic. In order to design an appropriate scaffold for vascularized adipose formation additional, adipose-specific signals need to be incorporated into these synthetic hydrogels.

6.5 Other Materials

Many other common biomaterials have been investigated for vascularized adipose formation. In the rodent epigastric pedicle model poly(D,L-lactic-co-glycolic acid)

(PLGA) not only did not induce adipogenesis(Cronin et al., 2004), but it inhibited the natural adipogenic process.(Dolderer et al., 2007) However, porous PLGA scaffolds have been shown to support the growth and adipogenic differentiation of ADSC. Surface patterning of another polyester, poly(lactic acid) (PLLA), induces lipid accumulation in preadipocytes *in vitro*(Chaubey et al., 2008). These results suggest that both the chemical and physical properties of synthetic materials contribute to their adipogenic properties. While many biomaterials have been investigated for adipose tissue regeneration they are, at best, conductive for adipogenesis. There has been little success in identifying materials that actively promote adipogenesis.

7 Tissue-Derived Hydrogels

The majority of attempts to engineer vascularized adipose tissue have used standard synthetic and natural materials as scaffolds with little consideration of unique features of adipose ECM. The insoluble microenvironment of the cell is different in every tissue and these differences play an important role in cell behavior and tissue function. An optimal tissue engineering therapy would likely have to account for these differences in biomaterial scaffold design. Based on this knowledge a technique was developed where hydrogels can be extracted and assembled from any soft tissue. (Uriel et al., 2009) The structure, composition, and biological properties of these hydrogels vary depending on the tissue used for extraction. These differences reflect known variations in ECM composition between the tissue sources.

7.1 Extraction and Gelation of Tissue-Derived Hydrogels

The technique for the isolation and gelation of hydrogels from soft tissues has been described previously (Uriel et al., 2009) and is summarized in Figure 2. It is a modification of the technique used to isolate from Matrigel™ from EHS tumors. (Kleinman et al., 1986) The procedure is successful due to the urea solubility of BM proteins and proteoglycans linked by secondary bonds. Molecules with a high number of covalent crosslinks, such as stromal collagens, are present in only small amounts in the resultant gels. (Uriel, 2009)

The concentration of intracellular components can be high in tissue-derived materials. This is a significant concern when developing materials for clinical application. For this reason, tissue samples are first exposed to a brief enzymatic and mechanical process to remove cells. Dispase, a mild metalloprotease, is used to loosen cells from their underlying BM. Tissues samples are cut into small sections (1–2mm in thickness), and suspended in dispase solution at 2mL/g of tissue (5000 total caseinolyptic units of dispase per 100 ml) and incubated for 15 min at 4°C. The sections are rubbed over a cell sieve to separate cells from the remaining tissue. Following treatment there is a drastic reduction in levels of intracellular proteins and DNA (462±15 to 184±10 ng/mL) in the extracts. (Uriel et al., 2009)

The decellularized tissues are homogenized in a high salt buffer solution containing protease inhibitors. The homogenized mixture is centrifuged and the pellet

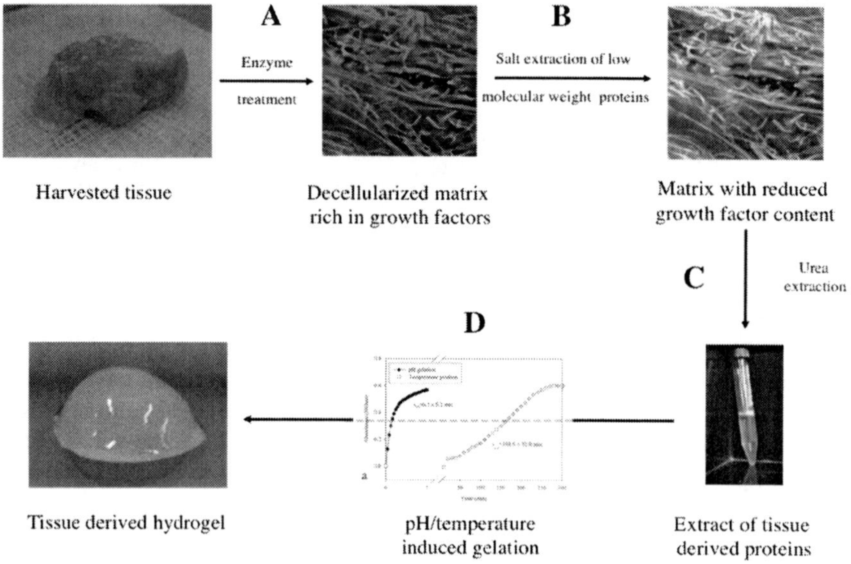

Fig. 2. Procedure for the development of tissue-derived hydrogels. A tissue sample is harvested and (A) exposed to an enzymatic procedure to remove cells. (B) The decellularized tissue is then homogenized in a salt solution to increase surface area and remove low molecular weight proteins (e.g. growth factors) bound to the tissue. (C) Matrix proteins are extracted in a urea solution and then (D) induced to assemble into gels by incubation at 37° C or in low pH.

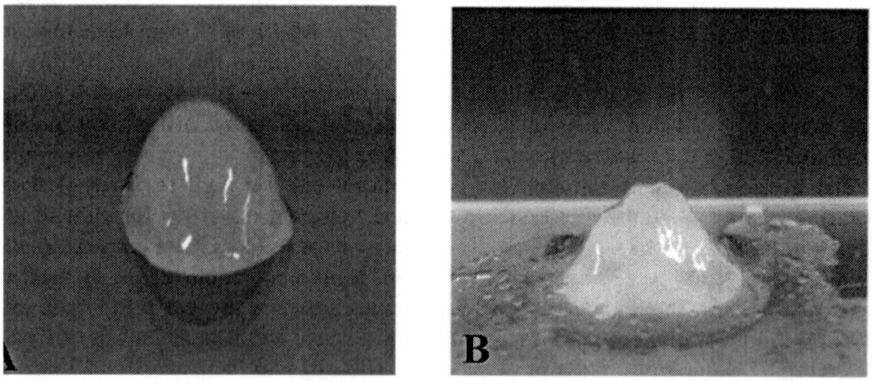

Fig. 3. Images of gelled formed by incubation (A) in low pH and (B) at 37° C

resuspended in a 2M urea buffer at 1mL of buffer/g of tissue and stirred overnight at 4°C. The mixture is then centrifuged and the supernatant stored, while the pellet is resuspended in the urea buffer. The pellet is homogenized and centrifuged again. The supernatants from both centrifugation steps are combined and used to form hydrogels.

This technique has been successfully applied to a number of tissues and organs, including subcutaneous adipose, dermis, multiple tumor xenografts, aorta, and pancreas. (Cheng et al., 2009; Uriel, 2009; Uriel et al., 2008; Uriel et al., 2009) All extracts derived from this procedure rapidly assemble into gels with a reduction in pH to 4.0. The gels formed at low pH are stable even after equilibration into buffer at physiologic pH. Extracts from some tissues (adipose and dermis) also assemble into gels by incubation 37°C (Figure 3). The process of temperature gelation is slower than pH-induced gelation.

7.2 Composition of Tissue-Derived Hydrogels

The extracts consist of a complex mix of BM, ECM, and growth factor proteins and polysaccharides. The composition varies depending on the tissue used for extraction (Table 1). These compositional differences reflect known differences in the ECM of the tissue sources, with only extracts from dermis containing collagen VII and laminin β3. Materials derived from subcutaneous adipose contained the α4 chain of laminin. As discussed in previous sections this isoform appears to

Table 1. Comparison of Properties of Tissue Extracts From Subcutaneous Adipose and Dermis

	Adipose	**Dermis**
Protein Concentration (mg/ml)	1.8-5.2*	10-17
Collagen (%)	59.3 ± 7.2	59.0 ± 5.6
sGAG (%)	2.80 ± 0.36*	0.40 ± 0.08
Glycoprotein (%)	25.4 ± 3.8	26.8 ± 0.4
BM proteins	Collagen IV Laminin α4* Nidogen Fibronectin	Collagen VII* Collagen IV Laminin β3* Nidogen Fibronectin
Gelation time (min)	Temp: 3.6 ± 2.3* pH: 0.6 ± 0.1	Temp: 168.6 ± 52.0 pH: 0.5 ± 0.2
Growth Factors (ng/ml)	FGF-1: 6.04 ± 0.08* FGF-2: 3.65 ±0.01	FGF-1: 11.0 ± 4.0 FGF-2: 3.2 ± 0.3

Data are from Uriel *et al.*(Uriel et al., 2008; Uriel et al., 2009), and Cheng *et al. (Cheng et al., 2009)* * indicates differences between the two hydrogels play an important role in adipocyte differentiation. Interestingly, adipose derived gels have significantly higher levels of sulfated glycosaminoglycans (sGAG) than dermis derived gels or Matrigel™. Tissue-derived hydrogels support integrin-mediated interactions with cells. The specific integrin mediating this interaction varies with tissue source, consistent with differences in hydrogel composition. (Cheng et al., 2009; Uriel et al., 2009)

7.3 Physical and Mechanical Properties of Tissue Derived Hydrogels

The physical and mechanical properties of materials play an important role in their biological function. Tissue extracts from many tissues gel with a reduction in pH or incubation at 37ºC. While hydrogels from a single tissue formed by pH and temperature mechanisms have the same composition, their physical and mechanical properties vary significantly (Table 2). The pH mechanism results in more rapid gel assembly than temperature, but both mechanisms result in gels consisting of a 3D network of nanofibers with structure similar to ECM *in vivo* (Figure 4). Interestingly, the temperature gels appear to have two distinct fiber sizes, specifically a longer fiber not present in pH gels. On average the pore and fiber sizes are significantly smaller in the pH gels. This more compact structure is consistent with a compressive modulus that is nearly two orders of magnitude larger for pH than temperature gels. The differences in mechanical and physical properties of the gels may result in different biological properties and should be further investigated to identify the conditions optimal for adipose regeneration.

The assembly of BM *in vivo* has largely been considered a result of cells achieving a threshold concentration at physiologic temperature, but recent studies suggest that assembly may be triggered by a local acidic pH at the cell surface of cell membranes. (Barroso et al., 2008; Freire and Coelho-Sampaio, 2000) These results may explain why concentrations of BM observed *in vivo* are lower than minimal thresholds required for temperature assembly *in vitro*. These tissue-derived materials may provide method for studying the mechanisms of BM assembly.

Table 2. Comparison of Physical and Mechanical Properties of Hydrogels Formed By pH and Temperature Gelation

	pH	Temperature
Gelation time (min)	0.5 ± 0.2*	168.6 ± 52.0
Fiber size (nm)	32 ± 15*	55 ± 14
Pore Size (nm)	160 ± 64*	359 ± 143
Young's Modulus (Pa)	94.9±24.1*	1.2±0.8
Volume fraction (unitless)	0.017±0.002	0.015±0.003

Data are from Uriel *et al.*(Uriel, 2009; Uriel et al., 2008; Uriel et al., 2009)
* indicates significant differences between the two hydrogels

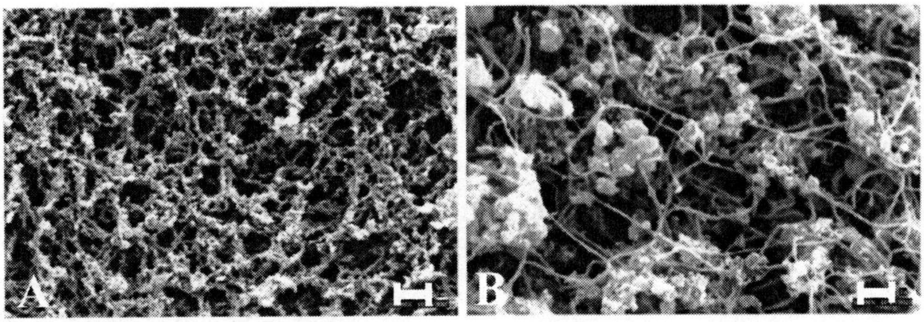

Fig. 4. Tissue derived gels have a fibrous nanostructure similar to *in vivo* matrix regardless of gelation mechanism. SEMs of gels formed by (A) pH- and (B) temperature-induced gelation. Notice the presence of two distinct fibril structures in the temperature gels when compared to the acid gels. Scale bars on each image are 1 μm.

8 Cell Differentiation on Tissue-Derived Hydrogels

The adipogenic properties of the tissue-derived materials were first explored through the culture of preadipocytes on materials from different sources. Materials were isolated from subcutaneous adipose and dermis tissues. Preadipocytes were harvested from rat epididymal fat. In some studies, these cells have been referred to as ADSCs. (Xu et al., 2008) The adipose, dermis and Matrigel™ hydrogels were assembled by incubation at 37°C for 90 minutes. Preadipocytes (P1) were trypsinized and seeded at 10,000 cells per well on the gels. Oil-Red O staining was utilized to determine differentiation via lipid-loading. The stained cells were imaged under fluorescence and brightfield microscopy. Interestingly, preadipocytes cultured on gels derived from subcutaneous adipose aggregated and rapidly differentiated, forming large lipid-loaded colonies (Figure 5). While dermis-derived gels and Matrigel™ accelerated differentiation over tissue culture plastic, very few aggregates were observed and they were much smaller than those observed on adipose derived gels. It appears that tissue-derived gels in general promote more rapid preadipocyte differentiation than culture on tissue culture plastic. However, only adipose-derived gels induced formation of large adipose tissue aggregates.

9 Tissue-Derived Hydrogels Promote Vascularized Adipose Formation *in vivo*

The potential of adipose derived matrices to induce adipogenesis has also been investigated *in vivo*. (Uriel et al., 2008) Adipose hydrogels formed by both pH and temperature mechanisms were implanted into the rat epigastric pedicle bundle *in vivo*. Adipose levels in tissue-derived gels formed by both pH and temperature mechanisms were significantly greater than Matrigel™ (Figure 6). The persistence of the adipose formed depended on mechanism of gelation with acid gels having more adipose at 6 weeks than temperature. These results indicate that hydrogels

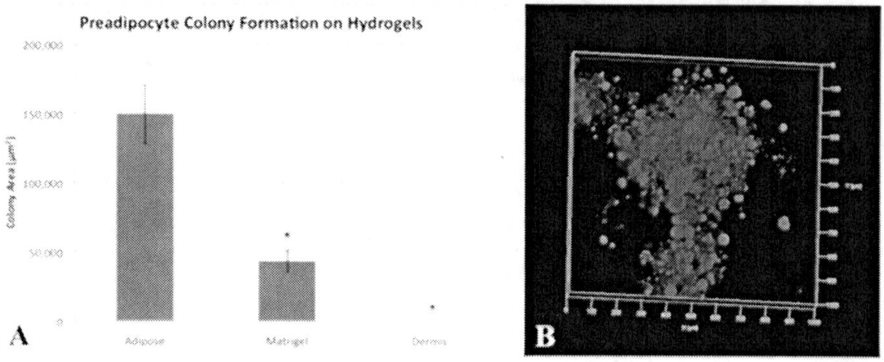

Fig. 5. Gels derived from subcutaneous adipose induce preadipocyte aggregation and lipid loading, resulting in the formation of adipocyte colonies. (A) The size of colonies formed is much larger with preadipocytes cultured on adipose derived gels then Matrigel or dermis derived gels. * indicates statistical difference ($p < 0.05$) in comparison to preadipocytes cultured on adipose derived gels. (B) Three-dimensional confocal image of lipids in adipocyte aggregates formed on adipose derived gels and stained with Oil Red O.

derived from subcutaneous adipose tissue promote rapid preadipocyte differentiation *in vitro* and vascularized adipose formation *in vivo*.

It was not clear from these studies if the adipogenic properties of the materials were due to the adipose source of the extracts or if the technique preferentially isolates adipogenic factors from soft tissues. In order to investigate this question similar studies were performed using hydrogels extracted from dermis. (Cheng et al., 2009) Interestingly, dermis-derived hydrogels induced greater vascularized adipose formation than Matrigel in the *in vivo* pedicle model (Figure 6). There were no differences between adipose and dermis derived hydrogels in their adipogenic properties in the vascular pedicle model. However, only the adipose-derived hydrogels induce significant preadipocyte differentiation. These results suggest that there are specific signals in adipose-derived hydrogels that directly induce differentiation of preadipocytes.

The similar effects of both dermis and adipose hydrogels in the *in vivo* model may be due to the ability of both hydrogels to induce significant neovascularization. Immunohistochemical staining for a sensitive EC marker (CD31) indicated the presence of a high density of vessels in the resultant vessels (Figure 6C). Both adipose and dermis derived hydrogels have biologically active levels of FGF-1 and FGF-2 (Table 2) and they induce vessel assembly *in vitro*.(Uriel et al., 2009) Hematoxylin and eosin stains were scored for vascular density. Both dermis and adipose derived hydrogels had greater vascular density based on histological scoring than Matrigel™. In addition, samples that had high vascular scores had greater adipose area on average than those with low vascular scores (% adipose area = 18.0±2.2% vs. 7.4±0.9%, $p < 0.001$). These findings suggest the importance of neovascularization in the stimulation of adipogenesis.

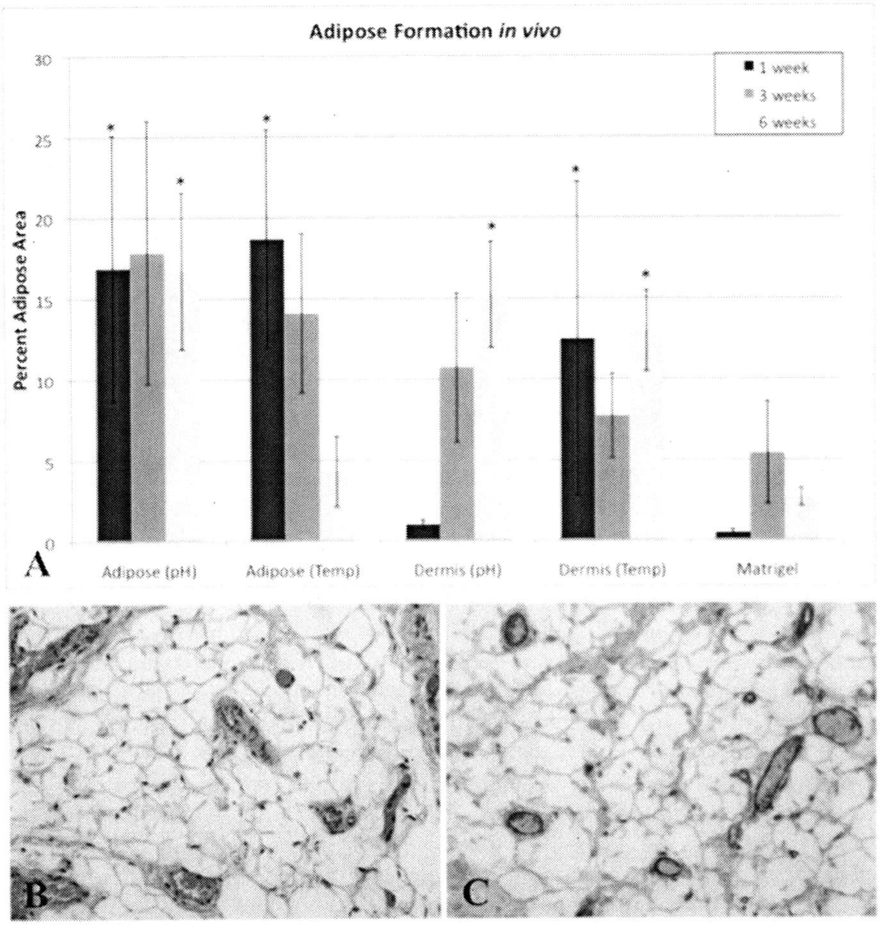

Fig. 6. Tissue derived hydrogels stimulate the formation of vascularized adipose *in vivo.* (A) Comparison of % area of adipose tissue following implantation of tissue derived hydrogels from adipose and dermis and Matrigel™ implanted in an epigastric pedicle model for 1, 3, and 6 weeks. Tissue derived gels were induced to form by direct injection (temperature) or first exposure to acidic conditions (pH). * indicates statistical significance ($p<0.05$) over Matrigel™ at the same time point. (B) Hematoxylin and eosin stain of vascularized adipose formed by implantation of adipose derived gels (pH) for 6 weeks. (C) Immunohistochemical stain for CD31 stain, a sensitive marker for endothelial cells, showing extensive vasculature (brown area) formed in the adipose tissue.

10 Synthetic Mimics of Tissue Derived Materials

While tissue derived hydrogels have shown great potential, natural materials are often hindered by a number of issues, such as variability in the product. For this reason, the development of synthetic "mimics" of adipose-derived hydrogels may allow for large-scale production of clinically-applicable products with consistent properties. Studies described in the previous section resulted in the findings that 1) adipose levels in the implanted hydrogels appeared to correlate with increased vascularization, 2) both dermis and adipose derived hydrogels contain biologically active levels of angiogenic growth factors, and 3) previous studies have shown that factors that promote neovascularization can stimulate adipogenesis in otherwise non-adipogenic materials. These results suggest that the addition of angiogenic growth factors into synthetic hydrogels may create a material that stimulates adipose formation *in vivo*.

Our first pass at creating a synthetic mimic of the tissue-derived hydrogels involved introduction of VEGF and FGF-1 into degradable PEG hydrogels. PEG hydrogels do not readily degrade under physiologic conditions, so copolymers were synthesized with poly (L-lactide) (PLLA), a hydrophobic and biodegradable polymer that has been studied extensively for tissue engineering applications. Hydrogels formed from acrylated PEG-PLLA (PEG-PLLA-DA) copolymers degrade by hydrolysis under physiologic conditions due to the presence of PLLA units into the polymer backbone.(Chiu et al., 2009) These degradable hydrogels were loaded with varying concentrations of growth factor from 10-50 ng/ml and implanted in the epigastric pedicle model. The level of 10 ng/ml FGF-1 is at the same order of magniture of the levels in adipose and dermis gels. While significant neovascularization occurred in the hydrogels, little adipogenesis was observed. In fact, the growth factors did not increase adipose levels above controls (hydrogels without growth factors), and the levels were much lower than adipose hydrogels. These results indicate that materials that stimulate neovascularization alone are not sufficient to promote adipogenesis in the epigastric pedicle model. When combined with literature on the contribution of laminin to adipogenesis, these results suggest that in addition to factors that stimulate neovascularization other, tissue-specific BM signals may be required for creation of a successful mimic of adipose-derived hydrogels.

11 Summary and Conclusions

Materials that stimulate vascularized adipose formation could be used to treat defects resulting from chronic wounds, congenital defects, acquired lipodystrophies, trauma, and tumor resection. To this point, standard natural and synthetic materials have been investigated for engineering vascularized with little consideration of unique features of the adipose extracellular microenvironment. Tissue derived hydrogels have unique composition from any currently available materials. The materials consist of a mixture of ECM proteins and growth factors unique to the tissue source. Hydrogels derived from adipose tissue using this technique promote greater adipose differentiation and formation than what has been observed with other materials in the absence of exogenous cells or growth factors. These materials can be used to study mechanisms of BM assembly, adipose-mediated diseases, and to enhance regeneration in adipose defects and deficiencies.

Acknowledgements. The research described in this chapter has been supported trough funding from the Veterans Administration, the National Science Foundation (0552896, 0731201), the Taiwanese National Science Council (NSC96-2314-B-182A-075-MY2), and the Illinois Institute of Technology Education and Research Initiative Fund. The authors would like to thank Kathan Amin, Yu-Chieh Chiu, Megan Francis-Sedlak, Jung-Ju Huang, Bin Jiang, Shu-Wei Kao, Jeffery Larson, Stephanie Lucas, Jessica Martinez, Krithika Mittakanti, Monica Moya, Sophia Pilipchuk, and Rui Wang for contributions to this work.

References

1. Abberton, K.M., et al.: Myogel, a novel, basement membranerich, extracellular matrix derived from skeletal muscle, is highly adipogenic in vivo and in vitro. Cells Tissues Organs. 188, 347–358 (2008)
2. Aumailley, M., et al.: A simplified laminin nomenclature. Matrix Biol. 24, 326–332 (2005)
3. Barroso, M.M., et al.: Artificial laminin polymers assembled in acidic pH mimic basement membrane organization. J. Biol. Chem. 283, 11714–11720 (2008)
4. Chaubey, A., et al.: Surface patterning: tool to modulate stem cell differentiation in an adipose system. J. Biomed. Mater. Res. B. Appl. Biomater. 84, 70–78 (2008)
5. Chavey, C., et al.: Matrix metalloproteinases are differentially expressed in adipose tissue during obesity and modulate adipocyte differentiation. J. Biol. Chem. 278, 11888–11896 (2003)
6. Cheng, M.H., et al.: Dermis-derived hydrogels support adipogenesis in vivo. Journal of Biomedical Materials Research (in Press, 2009)
7. Chiu, Y.C., et al.: Formation of Microchannels in Poly(ethylene glycol) Hydrogels by Selective Degradation of Patterned Microstructures. Chemistry of Materials (in Press, 2009)
8. Cho, S.W., et al.: Engineered adipose tissue formation enhanced by basic fibroblast growth factor and a mechanically stable environment. Cell Transplant. 16, 421–434 (2007)
9. Cronin, K.J., et al.: New murine model of spontaneous autologous tissue engineering, combining an arteriovenous pedicle with matrix materials. Plast. Reconstr. Surg. 113, 260–269 (2004)
10. Cronin, K.J., et al.: The role of biological extracellular matrix scaffolds in vascularized three-dimensional tissue growth in vivo. J. Biomed. Mater. Res. B. Appl. Biomater. 82, 122–128 (2007)
11. Dallabrida, S.M., et al.: Adipose tissue growth and regression are regulated by angiopoietin-1. Biochem. Biophys. Res. Commun. 311, 563–571 (2003)
12. Dolderer, J.H., et al.: Spontaneous large volume adipose tissue generation from a vascularized pedicled fat flap inside a chamber space. Tissue Eng. 13, 673–681 (2007)

13. Dollfus, C., et al.: Correction of facial lipoatrophy using autologous fat transplants in HIV-infected adolescents. HIV Med. 10, 263–268 (2009)
14. Freire, E., Coelho-Sampaio, T.: Self-assembly of laminin induced by acidic pH. J. Biol. Chem. 275, 817–822 (2000)
15. Frye, C.A., et al.: Microvascular endothelial cells sustain preadipocyte viability under hypoxic conditions. In Vitro Cell Dev. Biol. Anim. 41, 160–164 (2005)
16. Guerrerosantos, J., et al.: Classification and treatment of facial tissue atrophy in Parry-Romberg disease. Aesthetic. Plast. Surg. 31, 424–434 (2007)
17. Guyuron, B., Majzoub, R.K.: Facial augmentation with core fat graft: a preliminary report. Plast. Reconstr. Surg. 120, 295–302 (2007)
18. Hillel, A.T., et al.: Embryonic Germ Cells Are Capable of Adipogenic Differentiation. In Vitro and In Vivo. Tissue Eng. Part A (2008)
19. Hiraoka, Y., et al.: In situ regeneration of adipose tissue in rat fat pad by combining a collagen scaffold with gelatin microspheres containing basic fibroblast growth factor. Tissue Eng. 12, 1475–1487 (2006)
20. Kawaguchi, N., et al.: De novo adipogenesis in mice at the site of injection of basement membrane and basic fibroblast growth factor. Proc. Natl. Acad. Sci. USA 95, 1062–1066 (1998)
21. Kawaguchi, N., et al.: Reconstituted basement membrane potentiates in vivo adipogenesis of 3T3-F442A cells. Cytotechnology 31, 215–220 (1999)
22. Kleinman, H.K., Martin, G.R.: Matrigel: basement membrane matrix with biological activity. Semin. Cancer Biol. 15, 378–386 (2005)
23. Kleinman, H.K., et al.: Basement membrane complexes with biological activity. Biochemistry 25, 312–318 (1986)
24. Knight, K.R., et al.: Vascularized tissue-engineered chambers promote survival and function of transplanted islets and improve glycemic control. FASEB J. 20, 565–567 (2006)
25. Kubo, Y., et al.: Organization of extracellular matrix components during differentiation of adipocytes in long-term culture. In Vitro Cell Dev. Biol. Anim. 36, 38–44 (2000)
26. Lam, S.M., et al.: Limitations, complications, and long-term sequelae of fat transfer. Facial Plast. Surg. Clin. North Am. 16, 391–399 (2008)
27. Niimi, T., et al.: Differentiation-dependent expression of laminin-8 (alpha 4 beta 1 gamma 1) mRNAs in mouse 3T3-L1 adipocytes. Matrix Biol. 16, 223–230 (1997)
28. Patel, P.N., et al.: Poly(ethylene glycol) hydrogel system supports preadipocyte viability, adhesion, and proliferation. Tissue Eng. 11, 1498–1505 (2005)
29. Patrick Jr., C.W., Wu, X.: Integrin-mediated preadipocyte adhesion and migration on laminin-1. Ann. Biomed. Eng. 31, 505–514 (2003)
30. Phulpin, B., et al.: Rehabilitation of irradiated head and neck tissues by autologous fat transplantation. Plast. Reconstr. Surg. 123, 1187–1197 (2009)
31. Piasecki, J.H., et al.: An experimental model for improving fat graft viability and purity. Plast. Reconstr. Surg. 119, 1571–1583 (2007)
32. Piasecki, J.H., et al.: Purified viable fat suspended in matrigel improves volume longevity. Aesthet. Surg. J. 28, 24–32 (2008)
33. Pu, L.L., et al.: Autologous fat grafts harvested and refined by the Coleman technique: a comparative study. Plast. Reconstr. Surg. 122, 932–937 (2008)
34. Rophael, J.A., et al.: Angiogenic growth factor synergism in a murine tissue engineering model of angiogenesis and adipogenesis. Am. J. Pathol. 171, 2048–2057 (2007)

35. Stosich, M.S., et al.: Vascularized adipose tissue grafts from human mesenchymal stem cells with bioactive cues and microchannel conduits. Tissue Eng. 13, 2881–2890 (2007)
36. Tang, W., et al.: White fat progenitor cells reside in the adipose vasculature. Science 322, 583–586 (2008)
37. Torio-Padron, N., et al.: Engineering of adipose tissue by injection of human preadipocytes in fibrin. Aesthetic. Plast. Surg. 31, 285–293 (2007)
38. Tzu, J., Marinkovich, M.P.: Bridging structure with function: structural, regulatory, and developmental role of laminins. Int. J. Biochem. Cell. Biol. 40, 199–214 (2008)
39. Uriel, S.: Isolation and Characterization of Tissue Specific Basement Membrane Extracts. Biomedical Engineering. Illinois Institute of Technology, Chicago (2009) vol. Ph.D
40. Uriel, S., et al.: The role of adipose protein derived hydrogels in adipogenesis. Biomaterials 29, 3712–3719 (2008)
41. Uriel, S., et al.: Extraction and Assembly of Tissue-Derived Gels for Cell Culture and Tissue Engineering. Tissue Eng. Part C Methods (2009)
42. Vashi, A.V., et al.: Adipose tissue engineering based on the controlled release of fibroblast growth factor-2 in a collagen matrix. Tissue Eng. 12, 3035–3043 (2006)
43. Xu, Y., et al.: Neurospheres from rat adipose-derived stem cells could be induced into functional Schwann cell-like cells in vitro. BMC Neurosci. 9, 21 (2008)
44. Yi, C.G., et al.: VEGF gene therapy for the survival of transplanted fat tissue in nude mice. J. Plast. Reconstr. Aesthet. Surg. 60, 272–278 (2007)
45. Zhong, X., et al.: Improved fat graft viability by delayed fat flap with ischaemic pretreatment. J. Plast. Reconstr. Aesthet. Surg. 62, 526–531 (2009)

Clinical and Molecular Perspectives of Deep Tissue Injury: Changes in Molecular Markers in a Rat Model

Takashi Nagase[1], Hiromi Sanada[1], Gojiro Nakagami[1], Yunita Sari[1], Takeo Minematsu[2], and Junko Sugama[2,3]

[1] Department of Gerontological Nursing and Wound Care Management, Division of Health Science and Nursing, University of Tokyo Graduate School of Medicine, Tokyo, Japan
[2] Department of Advanced Skin Care (Miss-Paris), Division of Health Science and Nursing, University of Tokyo Graduate School of Medicine, Tokyo, Japan
[3] Department of Clinical Nursing, School of Health Sciences, College of Medical, Pharmaceutical and Health Sciences, Kanazawa University, Kanazawa, Japan

Abstract. Deep tissue injury (DTI) is a pressure-related injury to subcutaneous tissues under intact skin. DTI has recently been a focus of enthusiastic debates among wound care specialists, particularly regarding how DTI should be categorized into the conventional pressure ulcer (PU) classification. Pathophysiologically, DTI is regarded as an antithesis (bottom-up) of the conventional understanding of PU formation (top-down), and some researchers have suggested that all deep PUs are derived from DTI. On the other hand, the concept DTI can be applied to several specific conditions characterized by subcutaneous tissue damage, some of which were not originally recognized as PUs. In the first part of this chapter, we discuss the novelty and conceptual confusion of DTI from the clinical standpoint, and briefly review several specific types of DTI, as found in immobilized patients, in Asian patients with excess bony prominences, and in patients with spinal cord injury. In the second part of this chapter, we review the molecular aspects of DTI pathophysiology. Damage to subcutaneous tissues such as muscle and adipose tissue is predominantly repaired by the conventional wound healing process with granulation and scar formation. However, there may also be endogenous regenerative reactions in these tissues, such as myogenesis, possibly adipogenesis, and angiogenesis, which will be reviewed at the molecular level later in this chapter. Furthermore, we have established a rat model of DTI. We will present our own experimental results of muscle damage in the DTI model, and the elevation of biochemical and molecular markers indicating muscle damage and regeneration, including creatine phosphokinase (CPK), hypoxia-inducible factor-1 (HIF-1), and Hedgehog (Hh) signaling molecules.

1 Deep Tissue Injury (DTI)—Clinical Spectrum and Controversies

1.1 Novel Concept of DTI and Its Classification

The surface of the human body is covered by skin and subcutaneous tissues. The superficial and deeper layers of the skin are the epidermis and the dermis,

A. Gefen (Ed.): Bioengineering Research of Chronic Wounds, SMTEB 1, pp. 301–341.
springerlink.com

respectively. Subcutaneous tissues include the adipose and skeletal muscle tissues. The adipose tissue lies just beneath the dermis, and the skeletal muscles are located beneath the adipose layer and are bordered by the deep fascia. The bony skeleton, joints or abdominal cavity are covered by these skin/subcutaneous tissue layers. Here, the term deep tissue injury (DTI) refers to the deep tissues, namely the subcutaneous tissue (adipose tissue and skeletal muscles).

DTI is defined as "a pressure related injury to subcutaneous tissues under intact skin." This is based on the draft definition developed by the United States National Pressure Ulcer Advisory Panel (NPUAP) in 2001, and is followed by the sentence "Initially these lesions have the appearance of a deep bruise, and they may herald the development of subsequent development of a Stage III–IV pressure ulcer (PU) even with optimal treatment" [18]. DTI is now a hot topic among wound care specialists, because the concept of DTI is quite novel and somewhat provocative, considering the traditionally accepted "common-sense" view of the pathophysiology of PUs [2, 39, 40, 156].

Conventionally, it has been considered that PUs develop from the superficial skin layer and progress toward the deep tissues such as the fat and the muscle (top-down), as in the classification of burns. Since Shea's classical paper [131], the staging/grading classification of PUs has been dependent on the damaged tissue layer from the superficial layer to the deep layer, reflecting this top-down hypothesis. By contrast, DTI offers an explanation for the development of PU in a "bottom-up" manner. A considerable body of evidence supports the bottom-up mechanism, at least in part. For example, many specialized clinicians in wound care are likely to have experienced cases with initial symptoms that were categorized as Stage/Grade I–II, which rapidly deteriorated into deep Stage/Grade III–IV PUs. It has already been shown in animal studies that deep tissue is more susceptible to pressure and mechanical forces than the skin [35, 104]. DTI-like tissue damage was described in some classical papers by Paget [108] and Shea [131]. We believe that Shea's description of closed pressure ulcers is still very important, and we will describe it later in this review. Recent biomechanical analyses and computer simulations also showed that deep tissue close to bony prominences is the focus of the mechanical stress and tissue distortion when pressure is applied to the skin [9, 21, 54, 55, 81].

The conventional frameworks for PU risk factors are now challenged by the DTI concept. The Braden Scale [22] is one of the most well-known systems for PU risk assessment. This scale consists of six categories of risks: sensory perception, moisture, activity, mobility, nutrition and friction & shear. These factors were assumed to contribute equally to both superficial and deep PUs. However, the emergence of the DTI concept has proposed that the superficial PUs are predominantly caused by shear forces, whereas deep PUs are mainly caused by sustained compression of the tissue [9, 14, 21]. There is also an argument that, because the superficial PU-like skin injury can be induced by maceration or friction, even without pressure, these lesions should be excluded from the classification of PUs [14]. Differentiating between the possible risk factors for superficial PU, deep PU and DTI is now confusing.

The concept of DTI has also dramatically affected the classification of PUs. One of the most striking examples is the 2007 revision of PU classification in the NPUAP [18, 102] In addition to the previous four stages, this revision included "suspected DTI" and "unstageable" as new classification subgroups. In their classification, DTI is re-defined in a more descriptive manner as a "Purple or maroon localized area of discolored intact skin or blood-filled blister due to damage of underlying soft tissue from pressure and/or shear. The area may be preceded by tissue that is painful, firm, mushy, boggy, warmer or cooler as compared to adjacent tissue." Although this definition vividly describes the possible observational symptoms among Caucasian patients, this definition is not based on possible pathophysiological mechanisms of DTI, and is not applicable for patients with dark pigmented skin, as was also noted by the NPUAP. By contrast, DTI can be categorized as Grade IV according to the classification system developed by the European Pressure Ulcer Advisory Panel (EPUAP) [46], because their Grade IV includes both with and without full-thickness skin loss, and DTI can be classified as Grade IV without full-thickness skin loss. Therefore, the EPUAP has not reached a consensus that DTI is a novel, specific entity, despite classical grading systems.

In February 2009, a draft version of the NPUAP/EPUAP pressure ulcer guideline [47] was presented at the 11th National NPUAP Biennial Conference. This represents a landmark and informative collaboration between these advisory panels. Together, they have proposed a new classification term "Category" as a substitute for Stage in NPUAP and Grade in EPUAP. However, the two categories unstageable and suspected DTI, included in the 2007 version of NPUAP classification, were not officially included in the new NPUAP/EPUAP International Classification System. In this draft guideline, unstageable and suspected DTI were included as "Additional Categories for the USA"! This discrepancy between the two leading expert groups reflects the obstacles faced by clinicians on how to understand DTI within the conventional conceptual frameworks.

1.2 Clinical Confusion and Controversies of DTI: Is This a New Way of Understanding the Pathophysiology of Deep PUs, Or a Novel Clinical Entity?

The recent interest in DTI is partly due to the NPUAP statement for DTI, which was prepared in 2001, and published in 2003 by Black and Black [17] in the form of a case report. This case involved 67-year old woman with a sacral lesion who had undergone bowel resection surgery. Although it initially seemed like a Stage I PU, it showed rapid deterioration into a deep PU. Her symptoms of the lesion were initially misdiagnosed as an infarct, perirectal abscess, cellulitis, fracture of the sacrum, pilonidal disease, cautery burn and prep solution burn, which were finally ruled out. This case represented the, then novel, clinical entity of DTI, because the skin changes observed in this patient were not recognized as PUs before the concept of DTI.

However, at the same time, the mechanisms of DTI formation should have been recognized as a common phenomenon among deep PUs, because the above-mentioned "bottom-up" directed progression of DTI frequently occurs in deep PUs. Indeed, Berlowicz and Brienza [14] suggested that all deep PUs can be derived from DTIs. This concept is quite attractive and reasonable because the four mechanisms of pressure-related tissue damage (ischemia, lymphatic or interstitial deficiency, reperfusion and deformation; as described later) are applicable to both the superficial skin and the deep tissue. Therefore, the question that needs to be answered is: is DTI a specific condition of tissue damage or is it a common phenomenon among deep PUs?

In fact, we suggest that both factors are involved. DTI may be a common mechanism among the majority of deep PUs. Meanwhile, the concept of DTI enhanced our understanding that skin/tissue damage is a variation of DTI, and was not originally regarded as a PU in the previous diagnostic system. These specific variations of DTI include following three conditions, at least: 1) Relatively obese patients with prolonged immobilization; 2) Extended undermining commonly seen in thin Asian patients; and 3) Ischial seroma seen in patients with spinal cord injury and which were not ruptured. We will describe and discuss these conditions in the following sections.

1.2.1 DTI Due to Prolonged Immobilization

The initial case reported by Black and Black [17] represents a novel clinical entity of DTI, namely sacral injury after a long period of immobilization due to long surgery time. We believe that this type of DTI is often associated with the following characteristics: 1) Long immobilization time, such as long surgery time or protracted unconsciousness due to excess drug intake; 2) Most commonly observed in relatively obese patients; 3) Are found in an area with thick soft tissue without remarkable bony prominences; and 4) Feature painful subdermal induration. Similar cases were previously reported by Aronovitch [4, 5] as "intraoperatively acquired PU". The author claimed that "many nurses are not familiar with the differences between 'traditional' and intraoperatively acquired PUs. Intraoperatively acquired PUs are initially described as unexplained, burnlike lesions observed after a lengthy operative procedure. Such lesions may appear up to 3 days after surgery, although they usually observed as reddened areas that appear 24–48 hours after a surgical procedure. The reddened area then undergoes ecchymotic changes (bruising) and may blister. Within 2 to 6 days necrosis is present, which may progress to a full-thickness wounds" [4]. Her report included two important points: 1) Her description of intraoperative PUs shares some characteristics with some common DTI symptoms; and 2) Intraoperatively acquired PUs (which includes at least some DTIs) should be distinguished from the "traditional" PUs by the nurses or caregivers.

We previously reported another case in a similar situation [96]: a 39-year-old patient who was obese (more than 100 kg body weight) and who underwent liver transplantation, an operation that lasted over 16 hours (Fig. 1a,b). In this patient, ultrasonographic findings revealed severe deep tissue damage beneath relatively minor dermal erosion. In this case, tissue damage was not consistent with the location of bony prominences; induration was observed within the buttock lesion without a prominent bony mass. Interestingly, this finding was associated with a sharp pain and a hard subcutaneous tissue mass. This lesion resolved within several weeks with intensive pressure relief. This case proved to be very informative for us: focusing on reducing pressure can sometimes avoid DTIs that can often "unavoidably" progress into deep PUs.

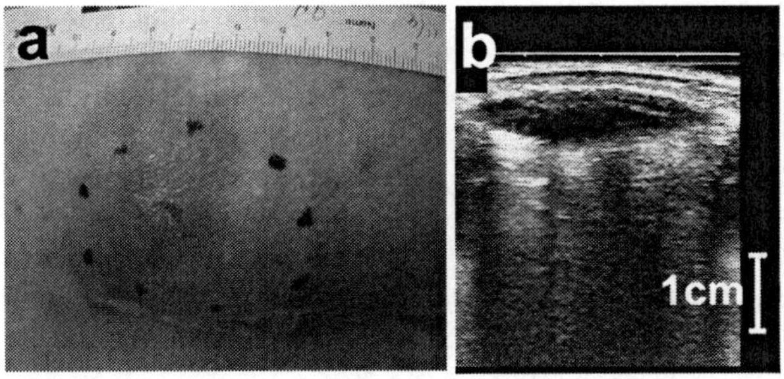

Fig. 1. A case with DTI due to prolonged immobilization after surgery. (a) The dotted area indicates the hard DTI mass with very slight superficial damage. (b) Ultrasonographic finding of the lesion. Reproduced from ref. [96] with permission from the publisher.

In these relatively obese cases with prolonged immobilization, ultrasonography may provide valuable information. Our case described above exhibited a very clear low-echoic lesion with clear demarcation (Fig. 1b). We considered that this lesion was a subcutaneous seroma or a fluid-filled cyst, but we obtained no fluid by puncture and, mysteriously, the ultrasonographic findings disappeared with resolution of the induration. Therefore, we believe that this case represents an atypical case of DTI. A recent report by Aoi and our group [3] described abnormal ultrasonographic findings in more typical cases of these types. Together with Aoi, we classified four ultrasonographic findings that can specifically predict DTI: unclear layered structure, hypoechoic area, discontinuous fascia and heterogenous hypoechoic area. These ultrasonographic findings may be an important clue to help predict whether the deep tissue is damaged, which necessitates prompt intervention.

As stated earlier, this category of DTI is sometimes misdiagnosed as other skin problems such as contact dermatitis due to antimicrobial fluid that has dripped and gathered around the perianal region during an operation, or a burn due to an electrical current between the dripped perianal fluid and an electric scalpel. However, at least some of these cases should be considered as DTI, particularly in cases with longer operation times. If the staff in the operating theater can identify the risks for DTI, a considerable number of DTIs can be prevented.

1.2.2. Extensive Undermining in Asian Patients

If the wound undermining extends beneath the healthy skin, the undermining can be categorized within a subtype of DTI. Undermining is included as a deficiency of the wound edge, an important components in the modern TIME (an acronym for tissue, infection, moisture, and edge) concept of wound bed preparation [125, 126]. Undermining has been described as a symptom of severe, deeply progressive PUs in classical reports. For example, Campbell [27] recommended exploration of the undermining in all directions by exerting gentle upward pressure of a curved clamp, causing dimpling of the skin. These points are dotted by ink, and joined by a continuous line. He described that these maneuvers should be done at the time of surgical debridement and flap surgery rather than as part of daily bedside assessment. Similarly, in 1975, Shea [131] described that "the deep fascia, being relatively avascular and physically resistant, limits the depth of the penetration of the ischemic necrotic process encouraging peripheral spread and undermining" in his Grade III classification, suggesting that undermining is likely to extend along the layer of the deep fascia. Of note, Grade III represents PUs to the depth of the deep fascia in the Shea's original classification, and is not the same as the present classification of Stage/Grade/Category III PUs with just full-thickness skin loss.

Undermining is also included as a parameter in the assessment tool Pressure Sore Status Tool (PSST), now called the Bates-Jensen Wound Assessment Tool (BWAT), which are used in all types of chronic ulcers [11]. However, we now believe that the problem of undermining is relatively underestimated in Western countries. It seems to be only one of the problems belonging to "E", at least from our understanding of recent papers, such as a review by Schulz *et al.* [126]. In that review, the undermining in the patient, as shown in a photograph, appeared to be very slight, and only a few centimeters in length. Furthermore, we have been unable to find a description of or recommendations for the treatment of undermining in the evidence-based guidelines published in Western countries.

The situation is quite different in Japan, and possibly other Asian countries. Several assessment tools describing status of the PUs, such as the Pressure Ulcer Scale for Healing (PUSH) tool developed by NPUAP [146] and PSST have been introduced in Japan. In addition, the Japanese Society of Pressure Ulcers (JSPU) originally established the DESIGN tool in 2002 (Table 1) based on the situation in Japan, and our group published it in English literature in 2004 [121]. The DESIGN tool is now frequently used by Japanese clinicians nationwide. DESIGN is an acronym for each parameter included in the tool, where D=depth, E=exudate, S=size, I=infection (and inflammation), G=granulation tissue, and N=necrotic tissue. In addition, as presented in Table 1, a parameter, P, is included in this tool. What is P?

P represents the undermining of the skin edge, because most Japanese clinicians refer to the undermining as a "pocket" as medical jargon. Why is it called a "pocket"? The reason seems to be that the typical undermining in Japanese PU patients is so extensive and wide that it looks like a pocket in clothes. In Table 1, the classification of P is based on the total PU size (the total area of the exposed wound plus undermining is estimated as the length [cm] × width [cm]) minus the size of the exposed part (estimated as the length [cm] × width [cm]). We believe that many Western readers will be surprised by the large value for P. Indeed, we often encounter extensive undermining, despite a very small wound opening (Fig. 2).

The fact that undermining is a very serious problem in the Japanese population is reflected in the recent revision of DESIGN by the JSPU. Although the original version of DESIGN equally scored each parameter, it lacked a statistical rating for differences among these parameters. After reevaluation of DESIGN using data from more than 1000 patients, JSPU introduced a revised version, DESIGN-R [141], where R means Rating (Table 1). Scores for each parameter were differentially estimated by multivariate analysis, reflecting their relative clinical importance. It is noteworthy that the coefficient for P is very high compared with the other parameters.

Table 1. The DESIGN tool showing original and revised ratings [121, 141]
The scores for the revised version (DESIGN-R) are indicated in bold text. The total sum of the scores (excluding scores for depth) represents severity, and is better in the revised version. Please note the high scores for Pocket (=Undermining), which indicates that undermining is a serious problem among Japanese patients.

Depth: this should be measured at the deepest point of the wound. If the wound becomes shallower, the decreased depth should be reflected in the assessment.							
d	0	0	No particular skin lesion and no redness	D	3	3	Lesion extends into the subcutaneous tissue
	1	1	Persistent redness		4	4	Lesion extends to the muscle, tendon and bone
	2	2	Lesion extends into dermis		5	5	Lesion extends into the articular or body cavity
						U	It is impossible to measure the depth
Exudate: amount							
e	0	**0**	None	E	3	**6**	Heavy: requires dressing changes more than twice a day
	1	**1**	Slight: does not require daily dressing change				
	2	**3**	Moderate: requires daily dressing change				
Size: the area of a skin injury (Length × width). Longest measurement in the wound is length: width is longest measurement perpendicular to that axis.							
s	0	**0**	None	S	6	**15**	100 cm² or larger
	1	**3**	Smaller than 4 cm²				
	2	**6**	4 cm² or larger, but smaller than 16 cm²				
	3	**8**	16 cm² or larger, but smaller than 36 cm²				
	4	**9**	36 cm² or larger, but smaller than 64 cm²				
	5	**12**	64 cm² or larger, but smaller than 100 cm²				
Infection:							
i	0	**0**	None	I	2	**3**	Clear signs of local infection (e.g., inflammation, pus and foul smell)
	1	**1**	Signs of inflammation (fever, redness, swelling, and pain around the wound)		3	**9**	Systemic impact, such as fever
Granulation tissue: percentage of healthy granulation.							
g	0	**0**	Granulation cannot be assessed because the wound is healed or too shallow	G	3	**4**	Healthy granulation tissue occupies 10% or more, but less than 50%
	1	**1**	Healthy granulation tissue occupies 90% or more		4	**5**	Healthy granulation tissue occupies less than 10%
	2	**3**	Healthy granulation tissue occupies 50% or more, but less than 90%		5	**6**	No healthy granulation tissue exists
Necrotic tissue: when necrotic and non-necrotic tissues are mixed, the dominating condition should be used for assessment.							
n	0	**0**	None	N	1	**3**	Soft necrotic tissue exists
					2	**6**	Hard and thick necrotic tissue is attached to the wound
Pocket: the area obtained by subtracting the ulcer from the entire affected area, including the pocket							
p	-	**0**	None	P	1	**6**	Smaller than 4 cm²
					2	**9**	4 cm² or larger, but smaller than 16 cm²
					3	**12**	16 cm² or larger, but smaller than 36 cm²
					4	**24**	36 cm² or larger

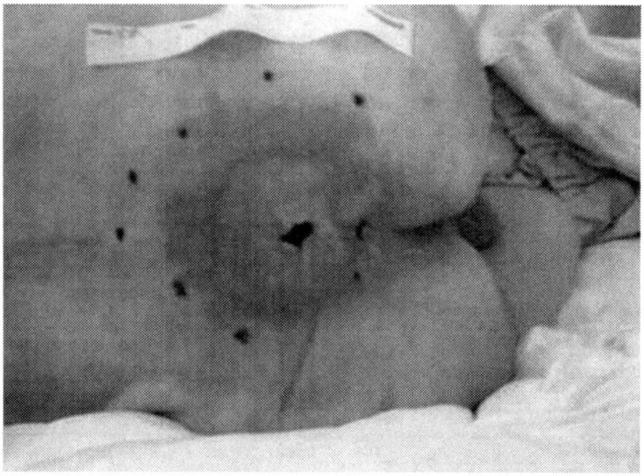

Fig. 2. A case with a sacral pressure ulcer with extensive undermining. The dotted area indicates the wide area of the undermining. This type of undermining is frequently seen in Japanese patients.

What is the mechanism responsible for the development of such extensive undermining in Japanese people? One possible reason is that the bony prominences in Japanese patients protrude more extensively compared with the surrounding soft tissues, which are often atrophic and thin in many elderly bedridden patients in Japan [105]. Ohura and Ohura [106] discussed the formation of undermining in relation to the DTI concept. They classified the mechanisms of undermining into two categories: the discharge type and the external force type. Discharge undermining is found in the initial phase of PU, and is due to the finding that the deep tissue shows more extensive liquefied necrosis adjacent to the bony prominence than the superficial skin. This notion clearly reflects the concept of DTI. On the other hand, external force undermining is found in the later phase of PU, and is mainly due to the application of a shear force to the skin and soft tissue surrounding the bony prominence. Morphologically, the latter type is characterized by asymmetric undermining in a direction towards the bony prominence. Both types of undermining are typically seen in a population with very thin soft tissue and exaggerated prominent bone protrusion, and are quite different from the DTI observed in obese patients, and feasibly in many Western countries.

Taken together, we stress the significance of undermining in Asian patients as an important subtype of DTI.

1.2.3 Ischial DTI Among Patients with Spinal Cord Injury: Ischial Bursitis?

Spinal cord injury (SCI) is one of the greatest risk factors for ischial PUs. Ischial PUs in SCI patients are mainly due to an inappropriate sitting position or cushioning in the wheelchair during their daily life The recurrence rate of PUs in SCI patients is unacceptably high [12, 60] and are occasionally life-threatening. For instance, Imai *et al.* [64] analyzed 960 patients with SCI over a 30-year period and found that the mortality ratio of septicemia, mostly from infected PUs, was 68-fold greater than in the general population. Indeed, Christopher Reeve, an actor in the Superman series of films, suffered from SCI and died at only 52 years of age owing to PU-derived sepsis.

One of the remarkable characteristics of ischial PUs in SCI patients is the frequent undermining and subdermal wide spaces [12]. Discharge is either purulent or un-infectious, with a serous or bloody-serous appearance in the latter cases. The wound may be as small as a pinhole in extreme cases, without inflammatory redness, and the shape of the surrounding subdermal cavity looks like a subdermal seroma or any type of a cyst. This strongly suggests that at least some cases of ischial PUs in patients with SCI originate from unruptured cysts or DTI. However, the probability of clinicians observing such unruptured cysts are likely to very low, because most SCI patients noticed their ischial lesion after rupture due to sensory disturbance.

Is there latent DTI in SCI patients? A recent paper by Kanno *et al.* [71] demonstrated low-echoic lesions beneath the skin in SCI patients by ultrasonography. This report included 129 areas (sacral and bilateral ischial regions) in 43 SCI subjects. The eight areas exhibited subdermal fluid collection by inspection and palpation, and these areas were detected by ultrasonography. Interestingly, a further nine areas, although seemingly healthy, had low-echoic lesions. This report indicates that latent DTI is found among some SCI patients, and ultrasonography is useful for its detection. However, the most critical timing is when these latent DTIs develop into a large cyst-like lesion. Some patients who suffered from this type of cyst-like lesion said that the formation and rupture of the lesion are quite rapid, often within a few days. The details of this type of DTI cannot be examined unless the clinician can detect the narrow timeframe of the formation of cyst-like DTI *just before rupturing*.

This notion prompted us to identify SCI patients with this rare condition. It required intensive patient education for early detection of the lesion, and rapid communication from the patients using E-mail. We finally obtained a chance to observe a SCI patient, a 58-year-old male, with an unruptured cyst-like DTI. The cyst was a few centimeters in diameter, and fluctuated on palpation. Although the patient was immediately hospitalized and vigorous pressure-relief care was given, the lesion enlarged rapidly over 1 week, reaching a diameter of nearly 10 cm (Fig. 3a). Interestingly, when the cyst was punctured, there was about 30 ml of fluid lacking debris with a serous-bloody appearance (Fig. 3b), and bacterial culture was negative. Ultrasonography (Fig. 3c) and MRI imaging (Fig. 3d) revealed a multicellular cystic lesion with septa and smooth surfaces. These findings strongly suggest that the DTI in this case is quite uncommon, without the massive deep tissue damage associated with infection or necrosis. Therefore, what is the diagnosis for this case?

Based on the previous reports by Rubayi and Montgomerie [120] and Tajima *et al.* [142], we suspected that this lesion was ischial (or ischiogluteal) bursitis. Bursitis itself is a relatively common disease, in which synovial fluid accumulates within the bursa around the joint capsule due to frequent repeated mechanical stimuli, and is most commonly observed in the shoulder, knee and elbow joints [16]. Treatment for bursitis includes immobilization, puncture, injection of steroid

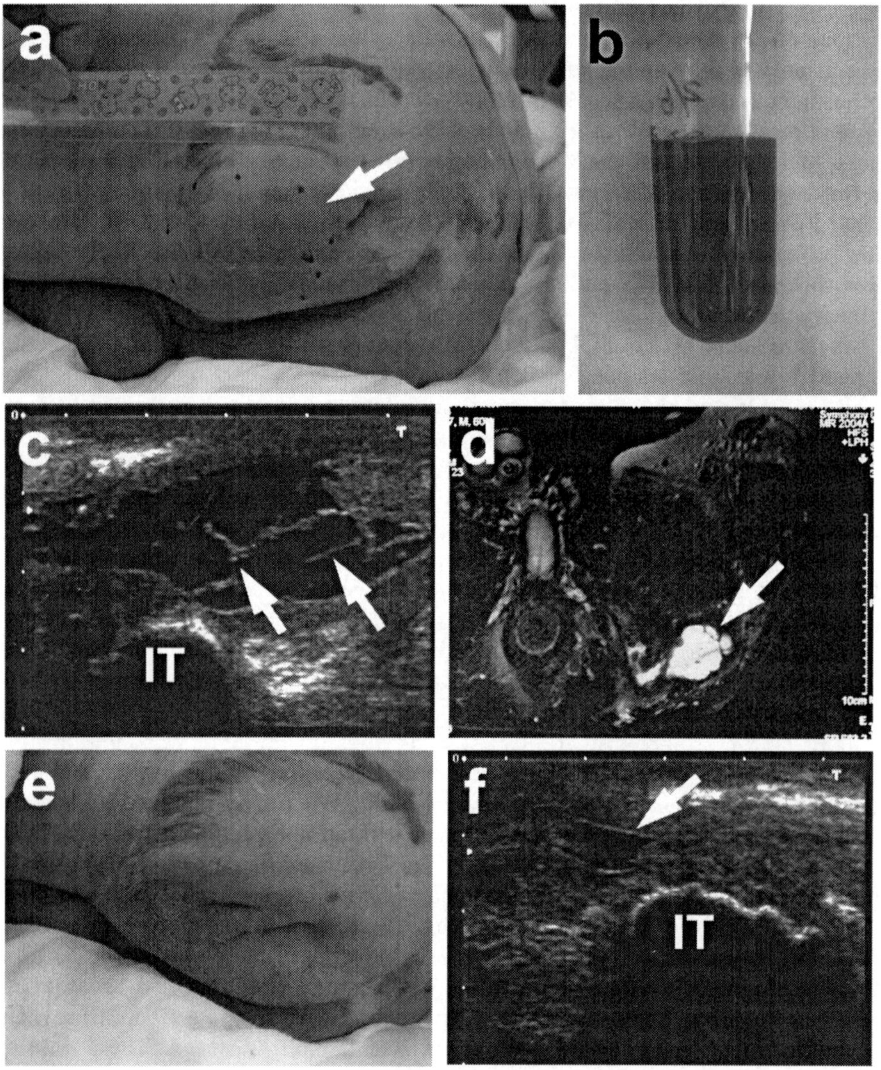

Fig. 3. A paraplegic patient with DTI in the ischial region. (a) The dotted area indicates floating induration due to bursa-like DTI (arrow). (b) Non-purulent serous-bloody appearance of the punctured sample. (c) Ultrasonography reveals a low-echoic lesion with multiple septa (arrows). (d) MRI T2-weighted image also shows a multilobular cystic lesion. These findings resemble those in ischial bursitis. (e) Appearance of the lesion 5 days after steroid injection. (f) Ultrasonography shows a very small low echoic lesion (arrow) 6 days after the injection. IT: ischial tuberosity.

within the bursa and surgical excision. However, bursitis frequently recurs. Bursitis in the ischial region is relatively rare [1] but is occasionally seen in people with frequent sitting position (Weaver's or Tailor's bottom). Ischial bursitis causes pain in patients without sensory disturbance. Nevertheless, our ultrasonographic and MRI findings were very similar to the previous cases reported by Kim *et al.* [73], and Cho *et al.* [30]. Definitive diagnosis requires tissue biopsy, which was not available in our case.

If this lesion is a bursitis, steroid injection is a treatment option [16]. Indeed, steroid injection (20 mg triamcinolone acetonide) had a surprisingly profound effect in this DTI case. Skin dimples were observed on the skin surface, and the fluid volume was dramatically reduced on ultrasonography within 5 or 6 days after injection (Figs. 3e,f). Although this patient was discharged from the hospital, his ischial DTI was observed twice at intervals of several months, and finally ruptured.

Since then, we have encountered several similar unruptured DTIs in the ischial region of SCI patients. They were sometimes infected (from an undefined origin) with symptoms of local and/or general infection, and oral or intravenous antibiotics were necessary. We administered steroid injections in these cases after confirming that the puncture fluid was negative for bacterial culture. Although steroids cannot fully prevent recurrence, steroid injection should be considered as one of the options for inhibiting rupture of ischial DTIs in SCI patients, because ischial PUs are very difficult to treat once they have progressed into an open and deeply undermined PU.

1.2.4 Unsolved Problems of DTI: Shea's "Closed Pressure Sore" Revisited

Here, we have described three subtypes of DTI based on our experience of the Japanese population, and consider that the concept of DTI can be applied to broader situations than described in the 2007 NPUAP classification [18, 102]. DTI is a novel bottom-up concept for the development of many, if not most, deep PUs. Moreover, DTI includes several specific subtypes. In other words, the DTI concept is a mixture of both the general and specific conditions. This certainly led to the confusion surrounding DTI, as featured in the NPUAP/EPUAP joint classification, as described previously [47]. To resolve this confusion, more detailed etiopathologic information is needed. Such information will serve as a sound basis for understanding and differentiating popularized and specified patterns within the DTI spectrums.

The pioneering work by Shea [131] of the classification of PUs offers some important insight in this direction. It is noteworthy that Shea tried to unify the intensive clinical observation, pathophysiological understanding based on histological analysis available at his time, and therapeutic interventions. However, his classification is somewhat different from the modern NPUAP or EPUAP classification systems. In his paper, Grade I and II correspond to the Stage/Grade/Category II and III (superficial) in the NPUAP, EPUAP and joint classifications. In Shea's paper, Grade III is a full-thickness PU that extends as depth as the fascia, as stated above. In his framework, tissue damage to the deep fascia is a sign or criterion for severe PUs, because necrosis or undermining can easily spread along the deep fascia. He proposed that Grade I–II PUs can be cured by conservative treatment, whereas

Grade III–IV PUs require vigorous surgical repair. Although his treatment recommendations have dramatically changed with more than 30 years of progress in this field, his fundamental concepts are still very important; the classification should be based on the pathophysiological understanding of PUs, not on mere observational descriptions, as seen in modern classifications such as those devised by the NPUAP or EPUAP.

Interestingly, aside from the Grade I–IV classification, Shea [131] described an additional DTI-like category, which he referred to as closed pressure sores. He reported that closed pressure sores "commonly develop about the pelvis overlying bony prominences such as the ischial tuberosity and the greater trochanter in a relatively healthy trained wheelchair bound paraplegic". This description shows surprising similarity to our cases of ischial DTI, as described above. His pathophysiological interpretation was as follows: "In some situations repeated prolonged pressure insults combined with sheer stress causes an ischemic necrosis in the subcutaneous fat without skin ulceration leading to the development of a bursa-like cavity filled with necrotic debris," which is strikingly similar to the modern concept of DTI. He continued that "the pigmented, thickened, and fibrotic overlying skin eventually ruptures, creating a small skin defect draining a large base which becomes contaminated by locally resident bacteria," which suggests that he had already encountered rupture of cyst-like DTI in SCI patients. Our opinion toward Shea's description is that: 1) his keen insight gained over 30 years ago should be regarded as a pioneering conceptualization of DTI; but 2) his consideration of the pathophysiological mechanism might not be applicable to closed pressure sores in the ischial region of SCI patients, because it may actually be a bursitis, based on our experience.

Over the 30 years since Shea's original publication [131], there have been enormous changes and progresses in basic sciences, particularly in the field of molecular biology and stem cell biology. To overcome the confusion surrounding the understanding of DTI, as stated above, we need to unite the clinical observations and molecular/cellular perspectives to create a new conceptual framework of PUs and DTI, as Shea tried to do 30 years ago. In the next chapter, we briefly mention the cellular and molecular mechanisms involved in the breakdown and autonomous repair/regeneration of deep tissues.

2 Pathways Involved in Deep Tissue Breakdown in Response to Mechanical Stress

Scientific research for the mechanisms of PUs have made great progress, and is based on evidence-based clinical studies, biomechanics and computer simulation analyses [9]. In contrast, the pathways linking mechanical stress to injury at the tissue and cellular level are not fully elucidated yet, and have become a target for many scientists worldwide. Based on the previous reports [9, 14, 21, 136], a consensus seems to have been established that there are four pathways that seem to be involved in pressure-related tissue breakdown: 1) ischemia, 2) impaired interstitial and lymphatic flow, 3) reperfusion, and 4) deformation. Here, we provide a brief explanation for each of these pathways.

Ischemia can be caused by defective of blood circulation owing to compression of the vessels or thrombosis. Oxygen and nutrient supply become insufficient, and anaerobic metabolism is promoted, causing an increase in the lactate level and a reduction in pH in the ischemic/hypoxic tissues. These factors may promote cellular and/or tissue necrosis [136]. Impaired interstitial and lymphatic flow occur as a result of tissue compression, resulting in localized deposition of waste or toxic substances [86, 89]. Localized edema may also exert an internal compression effect, as in compartment syndrome, forming a vicious cycle of local ischemia [3]. Reperfusion injury does not occur while the blood supply is impaired, but only after the compression is released and the circulation recovers. Toxic substances such as reaction oxygen species, produced during the ischemic period, are widely distributed after reperfusion, further worsening the tissue damage over and above that directly caused by ischemia [145].

Deformation is also a major focus of recent research, representing an important pathway that causes damage in response to mechanical stimuli. Numerous studies, mainly *in vitro*, have shown that morphological deformation of cells due to external mechanical stress directly affects the viability of cells, irrelevant of blood supply or hypoxia. For instance, Gawlitta *et al.* [52, 53] reported *in vitro* studies of cultured cells and engineered muscle tissues, and showed that deformation can cause cell death, and that cellular viability can be modified by both hypoxia and deformation. Their group, which includes an editor of this book, Prof. Gefen, has been involved in very interesting research projects, which may be discussed in other chapters in this book. Meanwhile, Edsberg [42] described histological changes of skin tissue cultured *in vitro*, to which a mechanical load was applied, and found that the direction of collagen and other extracellular matrix fibers in the dermis changed from a multidirectional structure to a horizontal/parallel alignment. Such deformation may affect the vulnerability of the skin.

All four pathways are commonly observed in superficial and in deep tissues. However, the deep tissue is more prone to damage as a result of these pathways, because the deep tissue, which includes muscle, is metabolically more active than the skin [21].

3 Wound Healing Versus Regeneration in the Deep Tissues: The Role of Somatic Stem/Progenitor Cells

The wound healing process is activated after the skin or other tissues are damaged. The mechanisms involved in wound healing have been extensively investigated [43]. Bleeding is inhibited by coagulation factors and platelets in the hemostasis phase. Devitalized necrotic tissue and microorganisms are cleared in the inflammatory phase, and granulation tissue, which is rich in vascularity and fibroblasts, fills the skin and tissue defects in the tissue formation phase. Finally, the granulation tissue is contracted into fibrous scar tissue, which is re-epithelialized by migrating keratinocytes from the surrounding skin in the tissue remodeling phase. In this canonical wound healing framework, tissue defects ultimately result in fibrosis, except for the epidermis. Of note, the epidermis can regenerate, at least partly, because epidermal stem cells and progenitor (transient amplifying) cells are highly localized in the basal layer and in the hair follicle bulge regions [107]. Although several reports have shown that the dermis also contains small numbers of multipotent stem cells [49, 147], the availability of stem cells is very limited and the dermis has no regenerative potential. Therefore, defects in the dermis are only repaired by the fibrotic scar tissue. But, what about the deep tissues?

In the following sections, we discuss whether the deep tissues possess their own stem/progenitor cells and to what extent the deep tissues can regenerate in some details. This point is very important, because tissue repair is generally determined according to the shifting equilibrium between regeneration and scar formation, and the endogenous regenerative potential could be exploited for the treatment of DTI in the future.

3.1 Regeneration of Muscle Tissue: Satellite Cells

It is well established that skeletal muscle possesses regenerative potential [63]. This is commonly experienced in our daily life; for example, muscular hypertrophy seen in athletes is due to the endogenous regenerative mechanisms within the adult skeletal muscles. Skeletal muscle is composed of bundles of huge numbers of muscle fibers, longitudinal multinucleated cells, which are derived from cell fusion of multiple myoblasts of embryonic origin. Nuclei in the mature muscle fibers are located in the periphery of the fibers, just beneath the cytoplasmic membrane called sarcolemma [25]. In addition, there are at least several subtypes of muscle stem/progenitor cells within the adult skeletal muscles [57], which contribute to regeneration. Satellite cells are a representative population of the muscle progenitor cells [75] (Fig. 4).

Satellite cells are located alongside the muscle fibers, in the narrow space between the sarcolemma and the basal membrane that wraps the muscle fiber along its whole length. Satellite cells are dormant in normal conditions, possibly mediating homeostasis of the muscle. When the muscle tissue is damaged or injured (including the four pathways of breakdown described above), the muscle fibers are degenerated and necrotized by endogenous autodigestive enzymes and by the infiltrated inflammatory cells. At the same time, satellite cells are released from the disrupted basal lamina and plasma membrane, and activated by cytokines produced by the surrounding inflammatory cells. The cytokines released at the injured site include insulin-like growth factor-1 (IGF-1), hepatocyte growth factor (HGF), epidermal growth factor (EGF), transforming growth factors (TGF-α and TGF-β) and platelet derived growth factors (PDGF-AA and PDGF-BB) [25, 63]. The activated satellite cells proliferate and become myoblasts. These myoblasts fuse to each other to form multinucleated immature myotubes. The nuclei are centrally localized in the immature, regenerating myotubes. When the myotubes mature into new muscle fibers, the nuclei migrate to the periphery. Therefore, the presence of centralized nuclei within normal-looking muscle fibers in histological sections is indicative of muscle regeneration [63].

Notably, the regenerative potential of adult tissue is often considered to be a recapitulation of the embryonic morphogenetic process. This is also true in muscle regeneration. Skeletal muscle is embryologically derived from the paraxial mesoderm (mainly the myotome in somites), and the myoblasts migrate from the mesoderm and fuse to form immature myotubes and eventually mature muscle fibers in

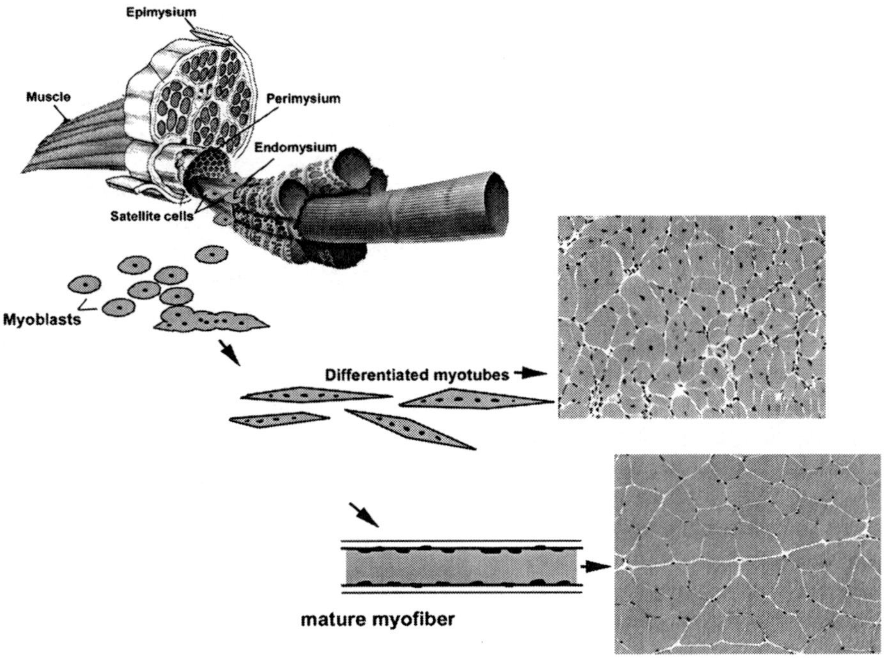

Fig. 4. Schematic drawing of the structure of skeletal muscle and the localization of satellite cells. The endomysium is the connective tissue layer that surrounds individual myofibers. The perimysium surrounds fascicles or bundles of myofibers. The epimysium is the superficial connective tissue layer that surrounds the skeletal muscle. On muscle injury, the satellite cells are released and are activated to become myoblasts, which eventually differentiate into immature (myotubes) and mature muscle fibers. The nuclei are located in the central portion of the myotubes (immature myofibers), but they eventually migrate to the periphery of the myofiber when the muscle fibers mature. Reproduced from ref. [63] with permission from the publisher.

normal development [57], in a similar manner to muscle regeneration by satellite cells. This analogy is also applicable at the molecular level. Gene expression of paired-box transcription factor *Pax7* and so-called muscle regulatory factors (MRFs), such as *Myf5* and *MyoD*, is common for both the embryonic myotome and the adult muscle satellite cells [57, 111, 127]. MRFs are a superfamily of basic helix-loop-helix transcription factors, and Myf5 and MyoD act as "master control genes" for muscle fate determination in normal development. Other members of MRFs, myogenin and MRF4, are expressed in the later stage of development, primarily at the stage of myoblast differentiation into myotubes [111]. A more recent review suggested heterogeneity of satellite cells according to the combination of Pax7 and MRF expression [75]. Taken together, satellite cells, like other types of somatic stem cells, may be regarded as residual embryonic cells within the adult muscle tissues.

Therefore, how active is the endogenous regenerative potential of the muscle, particularly in the clinical settings? Many readers are likely to have experienced that large PUs that progress as deep as the muscle layer are repaired, but with massive fibrous scar formation. As stated above, the tissue repair mechanism represents a "battle" between regeneration and scar formation. Muscular hypertrophy in athletes is an extreme example; their muscle training is regarded as very slight damage that triggers over-regeneration by activated satellite cells, without any scar formation. By contrast, more severe damage does not activate but instead destroys satellite cells, resulting in healing that is solely dependent on fibrosis. Considerable fibrosis may be inevitable in many cases of muscle injury, between these two extremes. For example, more muscle fiber regeneration and less scar formation were observed when a lacerated muscle was sutured than when it was left open [85]. Huard *et al.* [63] described in their review that muscle degeneration and inflammation occur in the first few days, and the regeneration process starts within 1 week after injury, peaks at 2 weeks and decreases at 3–4 weeks post-injury. They also reported that fibrosis begins at 2–3 weeks post-injury, and increases over time. The level of regenerative potential is dependent on animal species and on the severity of damage [57]. In terms of animal studies with rats, Gonda *et al.* [59] reported that even complete ligation of the nutrient vessels of the soleus muscle (but keeping the nerve intact) resulted in a remarkable increase in putative satellite cells after 7 days without massive muscle necrosis.

How can we enhance regeneration and reduce fibrosis after muscle injury? Although this is a very important question from the clinical point of view as an intervention for DTI, this may be beyond the scope of this chapter. Instead, we will comment on the following. 1) Although satellite cells are a candidate for stem cell therapy [75], we believe that satellite cells are not ideal or practical, because harvesting is invasive and culturing may be relatively difficult compared with adipose tissue-derived stromal cells (as described in the next section). 2) Administration of cytokines (either protein or gene transfer) to activate satellite cells needs careful consideration, because cytokines such as IGF-1 may also stimulate fibrosis [63]. 3) Other factors such as angiogenesis should be taken into account when selecting clinical interventions.

3.2 Regeneration of Adipose Tissue?: Adipose-Derived Stromal Cells

Adipose tissue is somewhat mysterious compared with skeletal muscle. Although the number of adipose cells can increase throughout life [116], it is unclear whether adipose tissue has a regenerative potential. In animal studies, Pajvani *et al.* [109] selectively induced massive apoptosis in fatty tissue in genetically engineered mice with drug-induced *Caspase 8* activation. Within a few weeks after cessation of drug administration, the adipose tissue showed regrowth almost to the normal level. Another study using rats also showed regrowth of the fatty tissue after lipectomy [62]. Interestingly, regrowth of the fatty tissue seems to be compensative, that is, adipose tissue hypertrophy of non-lipectomized sites was observed. Similarly, a classical study by Reyne *et al.* [117] using rabbits showed that lipectomy of the dorsoscapular, omental and inguinal fat did not lead to regeneration, whereas lipectomy of the perirenal fat pad showed remarkable regeneration. These data suggest that the regenerative potential of missing adipose tissue may be site specific. In humans, liposuction, a form of esthetic plastic surgery, has been performed in a large number of patients, and should provide an opportunity to estimate the regenerative potential of localized fat loss (artificial DTI). Long-term assessment of 209 patients after liposuction, as reported by Rohrich *et al.* [119] revealed weight-gain in 43% of the cases; however, it was not described whether the regions involved showed local regeneration of fat tissue. Taken together, we suggest that adipose tissue has a weak regenerative potential, but it should be regarded as a part of total body fat homeostasis, not a part of the wound healing process. This interpretation is quite reasonable, because recent studies clearly showed that adipose tissue is an endocrine organ that secretes several adipocytokines such as leptin and adiponectin [84]. Accordingly, endogenous regeneration of the fatty tissue under DTI is undesirable. It is reasonable to suggest that fatty tissue damage in DTI is mostly repaired by the fibrous scar.

It is also unclear whether such endogenous regeneration and the regrowth potential of the fat is a refrain of the embryonic process of adipogenesis. Embryonic adipogenesis commences at 14–15 weeks of gestation in humans [113]. Adipose tissue is commonly regarded as a derivative of the mesoderm. However, information of the precise embryonic origin of the adipose tissue is surprisingly limited [15, 56]. Atit *et al.* [8] showed that interscapular brown fat is a derivative of the paraxial mesoderm using mice with genetic labeling of the *Engrailed-1* gene. Notably, the adipose tissue, bone and dermis in the cranial region are not derived from the mesoderm, but from ectodermal neural crest cells, which are a subpopulation of cells that migrate from the dorsal-most region of the neural tube into the mesenchyme of the embryo [79]. The conventional view that adipose tissue are of mesodermal origin was mainly based on the *in vitro* finding that mesenchymal stem cells (MSCs) derived from the bone marrow can differentiate into adipocytes and into other mesodermally derived lineages such as osteocytes and myocytes [15, 56]. Peroxisome proliferator-activated receptor γ (PPARγ) is a transcription factor and regarded as a master control gene that determines adipose differentiation, equivalent to Myf5 and MyoD in myogenesis, at least in this *in vitro* system [56]. However, early embryonic expression of *PPARγ* is not found in the paraxial mesoderm, but in the central nervous system, possibly mediating different functions [23]. These data suggest that evidence supporting the mesodermal origin of adipose tissue is relatively vague compared with that for muscle tissue.

Despite the limited knowledge of embryonic adipogenesis, adipose tissue is now a major focus of modern stem cell biology research. An increasing number of reports have suggested that MSC-like multipotent stem cells can be harvested from adipose tissue, so-called adipose-derived stromal cells (ASCs)(Fig. 5a) [152]. Since the first report by Zuk in 2001 [154], ASCs were shown to differentiate into multiple mesodermal lineages such as bone, muscle cartilage, and adipose tissue both *in vitro* and *in vivo* [34, 155]. ASCs were also shown to contribute to angiogenesis. An animal study showed that transplantation of ASC promoted angiogenesis in an ischemic limb [112]. Furthermore, ASCs seem to have a differentiation potential toward the neural lineage, such as neurons and glial cells [69, 70, 103]. As a source of stem cells for therapeutic use, adipose tissue has several advantages compared with other sources such as muscle or bone

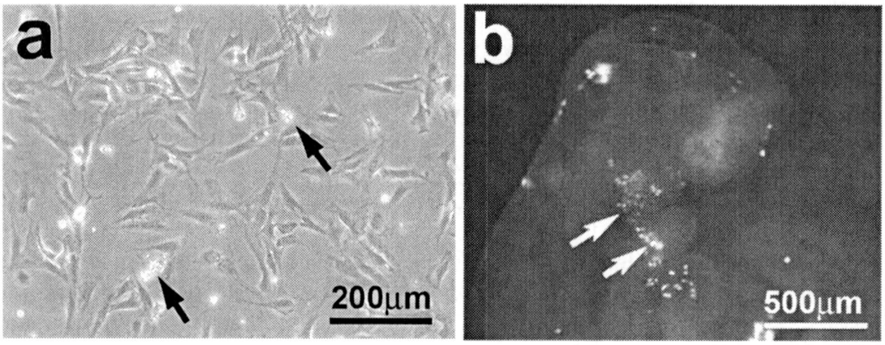

Fig. 5. Adipose-derived stromal cells (ASCs). (a) Fibroblast-like appearance of cultured ASCs, with lipid droplets within the cells (arrows). Bar=200 μm. (b) Human ASC-derived neurosphere cells can migrate within a mouse embryo after grafting in a whole embryo culture system, implying that ASCs have a neural (or neural crest-like) characteristic, in addition to differentiation potential toward mesodermal lineages. Bar=500 μm. Reproduced from ref. [97] with permission from the publisher.

marrow. It is easy to harvest large amounts of adipose tissue (together with the stem cells) with minimally invasive techniques such as liposuction. Harvesting itself is of value for the patients from the esthetic point of view. Therefore, numerous studies have been started for tissue engineering of various types of tissue, including bone, nerve and cardiovascular tissue, using ASCs.

ASCs may constitute a stem cell population that mediates homeostasis and turnover of adipose tissue, similar to the role of satellite cells in muscle. However, ASCs are obtained from the so-called stromal vascular fraction (SVF), a centrifuged pellet of enzymatically digested adipose tissue; thus, the localization of ASCs *in vivo* is still unclear. Furthermore, ASCs may constitute a mixture of various types of cells such as preadipocytes (adipose progenitor cells), small fragment of vessels and peripheral nerves, for example. Therefore, the question remains: where are ASCs (or the true population of adipose progenitors) localized and how do they function? Yamamoto *et al.* [150] reported that mouse SVF-derived cells positive for p75 neurotrophin receptor (p75-NTR: often used as a marker for MSCs) showed multilineage differentiation potential *in vitro*. In their report, the p75-NTR-positive cells were located in the deeper layer of adipose tissue *in vivo*, adjacent to the microvessels. More recently, Tang *et al.* [144] reported the existence of *PPARγ*-positive preadipocytes on the surface of small vascular fragments within the SVF. These *PPARγ*-positive cells were also positive for markers of vascular mural cells (vascular smooth muscle cell or pericytes). Considering the previously reported multipotency of mural cells [37], it is tempting to speculate that the true adipose progenitor and the multipotent population within ASCs may be the vascular mural cells, and this population may also be p75-NTR-positive. Nevertheless, these speculations need further investigation.

Furthermore, it is of note that p75-NTR is also used as a marker for the neural crest [90]. Because both the mural cells and the adipose tissue are derivatives of the neural crest in the cranial region [45, 79], we have hypothesized that, at least in part, some of the ASCs are of neural crest origin, rather than of mesodermal origin [15]. Indeed, we could obtain neurospheres, which are floating cultured aggregates of neural stem cells, from human ASCs [97]. When grafted into the mouse embryo, the ASC-derived neurosphere could migrate within the embryo just like neural crest cells (Fig. 5b). Therefore, it is a very fascinating question whether ASCs are included in the entity of adult neural crest stem cells, similar to stem cells within the dermis and cardiac muscles [49, 148].

Several studies have evaluated adipose tissue engineering using ASCs. Autogenous fat transplantation surgery has been performed mainly for esthetic purposes, such as breast augmentation. However, the conventional method, in which suctioned adipose tissue was washed and injected, was frequently associated with absorption and long-term failure. Matsumoto *et al.* [83] reported that human fat transplantation in combination with a suspension of concentrated ASCs in severe combined immunodeficiency mice resulted in massive adipose tissue formation compared with transplantation without ASCs. ASCs seemed to contribute to angiogenesis in their report, and promoted the survival of transplanted adipose cells. Other groups have suggested the use of ASCs for adipose tissue engineering in combination with various biomaterials or cytokines [58]. In summary, although ASCs still need to be fully characterized, tissue engineering using ASCs offers a very promising option for future DTI treatment.

3.3 Regeneration of Vessels: Endothelial Progenitor Cells?

Angiogenesis is a very important mechanism in the conventional framework of wound healing. The formation of good granulation tissue largely depends on angiogenesis. Because angiogenesis is indispensable for the endogenous repair mechanisms against DTI, we will briefly describe the mechanisms of angiogenesis and vessel regeneration.

The mechanisms involved in vascular formation were classically categorized into two steps [118]: 1) Vasculogenesis is the *de novo* formation of vessels from progenitor cells such as mesodermal hemangioblasts; and 2) Angiogenesis is a step that involves remodeling and maturation of pre-existing vessels by sprouting or septum formation. Various molecules are involved in vasculo-angiogenesis, including vascular endothelial growth factor (VEGF)/Flk-1 signaling, angiopoietin (Ang)-1/Tie-2 signaling, and EphrinB2/EphB4 signaling are the most important.[67, 118, 151] VEGF mediates both vasculogenesis and angiogenesis, and its main receptor Flk-1, which is also known as VEGFR2, is expressed in the vascular endothelium [50, 88]. Ang-1 is predominantly mediates angiogenesis process such as vascular sprouting and pericyte recruitment [36, 140]. Its receptor Tie-2 is also expressed in the vascular endothelium. EphrinB2 is an arterial marker and EphB4 is a venous marker; both regulate the arterio-venous distinction [149]. Accumulating data also suggest that Hedgehog (Hh) signaling plays a pivotal role in embryonic and adult angiogenesis, which will be discussed in more detail later in this chapter [98, 100]. Several other molecular markers are frequently assessed, including platelet endothelial cell adhesion molecule (PECAM-1; CD31) as a marker for the endothelium and α-smooth muscle actin (α-SMA) as a marker for pericytes/smooth muscle cells.

It was once proposed that vasculogenesis is only found in embryogenesis, and vascular formation in adult organisms, as seen in wound healing, is solely dependent on angiogenesis from pre-existing vessels,. However, the identification of the putative endothelial progenitor cells (EPCs) by Asahara *et al.*[6], about one decade ago, drastically changed this paradigm. They described that EPCs are derivatives from the bone marrow and circulate within the vessels in adults. They are recruited to the region of vessel formation, and neo-vessels generated in this manner can be categorized as vasculogenesis [10], again recapitulating the embryonic process in adults. Accordingly, EPCs can be regarded as stem/progenitor cells in vessels. The recruitment of EPCs towards the wounded site may be triggered mainly by local hypoxia and hypoxia-inducible factor-1 (HIF-1) is upregulated and activated in the hypoxic tissue. HIF-1 acts as a transcription factor for downstream genes including *vegf*, to stimulate vasculo-angiogenesis [72, 128]. We will discuss the role of HIF-1 in more detail in the next section. The contribution of bone-marrow derived cells to adult vasculo-angiogenesis was clearly established in several *in vivo* studies of wound healing that used animals with genetically labeled bone-marrow. In such experiments, the animals were often irradiated by X-rays to destroy their bone-marrow tissue, which was replaced by subsequent transplantation of genetically labeled bone marrow from other transgenic animals [7, 51]. These findings support the idea that endogenous recruitment of putative EPCs, or at least any kinds of bone-marrow derived cells, is an integral part of the optimum wound healing process. It is reasonable to consider that similar mechanisms also function in DTI. Indeed, numerous pre-clinical and clinical trials of therapeutic angiogenesis have been performed by injection of purified putative EPCs to promote angiogenesis within ischemic limbs or after myocardial infarction, for example [44]. This type of approach may also applicable to the treatment of DTI.

However, a recent review by Sievekin and Ng [132] claimed that the identity and function of EPCs, MSCs or other bone-marrow derived cells are now very controversial. For example, it appears that EPCs are not a single population of cells, but instead contain at least two subtypes of cells, namely early EPCs and late outgrowth endothelial cells (OECs) [74, 132]. It is still unclear whether EPCs are actually differentiated into the endothelium, or instead into another lineage (such as mural cells) acting as a source of local cytokine release. Recent studies have suggested that the early EPCs cannot be incorporated in the endothelium, but in the mural cells those secrete several angiogenic factors, and the late OECs constitute a real population for endothelial differentiation [132]. Furthermore, even the embryonic origin of MSCs, which were conventionally considered to be of mesodermal origin, is now a focus of much debate. Recent studies using transgenic mice with genetic markers of neural crest origin (*Wnt-1* or *P0-Cre*/Floxed-*GFP* or *-lacZ* mice) showed that at least some subpopulations of the bone marrow cells or MSCs may have a neural crest origin [101, 143]. Considering these findings together with our previous finding of the neural crest-like characteristics of ASCs [97] and the putative mural cell origin of ASCs [144], as described above, it is very tempting to speculate that some of the putative EPCs are, in fact, derived from the neural crest-derived bone marrow cells, which are integrated into mural cells in angiogenesis; thus, sharing common features with ASCs [15].

Finally, we must briefly mention the possible clinical application of cytokines and growth factors that promote angiogenesis in ulcers, including DTI. PDGF-BB is currently used in clinical settings in the USA [133]. In Japan, another type of recombinant growth factor, basic FGF (FGF-2), is approved and commercially available. Basic FGF is now widely used by wound care specialist nationwide, even in home care. It strongly induces angiogenesis to enhance good granulation tissues, and we believe that this growth factor is relevant for treatment of DTI. However, it should be considered, when using basic FGF for wound care, that FGF and other growth factors are physiologically multifunctional. The same growth factors are repeatedly used in temporospatially specific functions from the embryonic stages to the adult. Such multimodality is, in part, regarded as a recapitulation of the embryonic process. Accordingly, we can refer to these groups of growth factors as "stem molecules", in a similar sense to somatic stem cells [98]. Therefore, unpredicted side effects could be associated with the use of these growth factors. Indeed, we reported over-ossification in deep PUs with exposed bone in which basic FGF was used for several months [95]. The formation of bone is included in the physiological functions of basic FGF; thus, we should be vigilant for such side effects.

4 Biochemical and Molecular Markers for Detecting DTI

In the section above, we described how the deep tissue is damaged and which endogenous repair/regenerating systems are stimulated. Because DTI is difficult to assess by direct inspection, evidence at the cellular and molecular level is required for future clinical evaluation and decision-making for early intervention in DTI.

In this section, we will introduce candidate biochemical and molecular markers to detect and evaluate DTI. Furthermore, our groups have already established an *in vivo* animal model of DTI using rats. We will also describe our experimental results in terms of measuring these DTI markers in our model.

4.1 Candidate Molecular Markers for DTI

What kinds of molecules can we exploit as markers for DTI? We have now gained some detailed information about the mechanisms involved in tissue damage and the endogenous repair/regeneration processes that respond to tissue damage in DTI. These mechanisms are very complicated and all of the molecules described above are indispensable. There may also be extensive cross-talk in the molecular signaling pathways, within these mechanisms. Thus, simply focusing on each deep tissue or each signaling pathway may not be sufficient or practical from the clinical point of view.

Instead, we suggest a more global approach, by dividing the process of DTI into three sequential steps: 1) tissue damage; 2) triggering adaptation with endogenous healing/regenerating process; and 3) progressing the healing/regenerating process. Accordingly, biochemical and molecular markers could be selected to represent these three steps. In the following sections, we discuss three examples of biochemical and molecular markers that represent each step, including creatine phosphokinase (CPK) for step 1, HIF-1 for step 2 and Hh signaling for step 3. It should be remembered that our understanding of DTI mechanisms is still incomplete; thus, our proposal described here is still preliminary.

4.1.1 CPK: A Marker for Muscle Tissue Damage

CPK is an enzyme that mediates the transfer of energy-rich phosphate from creatine phosphate to adenosine diphosphate, to form adenosine triphosphate (ATP) [124]. Because the ATP formed in this pathway is the main source of energy for skeletal and cardiac muscles, CPK is an essential enzyme for muscle contraction. CPK is a dimer that comprises two distinct subunits, termed M and B. Depending on the subunits, CPK can form three distinct isoform: CK-MM (skeletal muscle), CK-MB (cardiac muscle) and CK-BB (brain). In human adult skeletal muscle, 99% of CPK activity is based on the CK-MM isoform. In cardiac muscle, 70–85% of CPK activity is dependent on CK-MM, and the remaining activity is dependent on CK-MB. CK-BB is predominantly found in brain tissue [124]. The molecular size of CPK is so large that CPK cannot be released from the muscle tissue unless the cytoplasmic membrane is injured. Based on these characteristics, serum CPK is one of the most widely used markers for muscle damage in clinical settings, including muscle injury, various types of myositis, and myocardial infarction.

Although serum CPK represents leakage of CPK from damaged muscle into the circulation, without any functional role in the serum, the notion of CPK as a specific marker for muscle injury is clinically established and should be appropriate for DTI. The serum CPK level can be routinely examined using blood samples taken in hospitals worldwide. Thus, the serum CPK level is the first promising marker for DTI, in terms of non-invasive sampling and widely used methodology. Indeed, Hagisawa *et al.* [61] reported elevated serum CPK levels in a pig PU model. However, we should remember that the serum CPK level may be affected by generalized or localized muscle damage, depending on the total mass of the damaged muscle. For example, the serum CPK level is even affected by the physical exercise [24], in which the damage itself is very slight but the total body muscle damage is reflected by elevated serum CPK levels. By contrast, muscle damage may be localized in most DTI cases, and we are concerned that the localized muscle damage in DTI is not sensitively reflected by the serum CPK level. Therefore, the wound exudate fluid may provide a good biofluid for the measurement of CPK to estimate localized muscle damage. We will describe our results of wound fluid CPK levels obtained using our DTI model later in this chapter.

4.1.2 HIF-1: A Marker for Tissue Adaptation to Hypoxia

As stated in section 2, ischemia is an important pathway involved in tissue breakdown in DTI. Ischemia results in local hypoxia, which may cause tissue damage and cell death. At the same time, local hypoxia, if at an intermittent level, also triggers tissue repair/regeneration, particularly angiogenesis. This process is mainly dependent on HIF-1 [72, 128, 129] (Fig. 6).

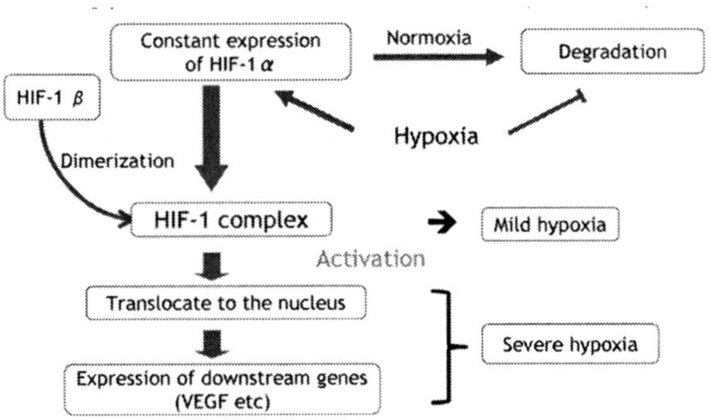

Fig. 6. Schema for Hypoxia-Inducible Factor-1 (HIF-1) activation during hypoxia

The HIF-1 pathway is extremely sensitive to hypoxia, reacting within a few minutes. The HIF-1 protein is a heterodimer composed of α and β subunits. HIF-1β, also known as aryl carbon nuclear translocator (ARNT), is continually expressed and is stably maintained regardless of oxygen availability [68]. Hif-1α is also continuously expressed but, under normoxic conditions, the HIF-1α protein is immediately degraded by prolyl hydroxylase and von Hipple-Lindau protein [65, 66]. Under hypoxic condition, the HIF-1α protein becomes stabilized by the reduced degradation, and then translocates from the cytoplasm to the nucleus, where it is dimerized with HIF-1β to form an activated HIF complex [72]. The activated HIF complex then acts as a transcription factor, associating with hypoxia response elements (HREs) in the regulatory regions of target genes to induce their expression. HIF-1 targets include genes that mediate processes such as angiogenesis (e.g., VEGF), erythropoiesis (e.g., erythropoietin and transferrin), cell survival/proliferation (e.g., IGF-2), vascular tone (e.g., nitric oxide syntase-2 and endothelin-1) and other metabolic pathways [72, 128, 129]. Thus, HIF-1 is considered to be the central effector of the cellular response to hypoxic conditions, mainly triggering the repair/regenerative process. Of particular interest, the activity of HIF-1 is mainly regulated by degradation at the post-translational protein level and the messenger RNA (mRNA) level of the *Hif-1a* and *Hif-1b* genes is considered to be mostly constitutive [72]. However, some studies showed upregulation of *Hif-1a* mRNA in hypoxic tissues [13]. A recent paper by Deudero *et al.* [38] showed that upregulation of *Hif-1a* mRNA through VEGF induced oxidative stress. Because VEGF is the main downstream molecule of HIF-1, this report suggests positive feedback resulting in *Hif-1a* mRNA upregulation in severe, prolonged hypoxia.

HIF-1 is a very promising marker for hypoxia in DTI. For example, Zhang *et al.* [153] used HIF-1 as a marker for muscle ischemia in their flap model. Expression of the HIF-1α protein was markedly increased in ischemic wound tissue. Another study used HIF-1 to determine the presence of ischemia after unilateral femoral artery ligation, and showed a profound increase in HIF-1α in the ischemic groups compared with the control group [87]. We will show later the data for changes in protein localization and gene upregulation of HIF-1 in our DTI model.

4.1.3 Hh Signaling: A Novel Marker for the Synergistic Regenerating Process of Angiogenesis and Myogenesis?

As stated in sections 3.1–3.3, a number of molecules as well as various stem/progenitor cells are involved in the regenerating process in DTI and interact in a very complicated manner. Notably, however, the entire tissue repair system in living organisms progresses in a miraculously coordinated fashion. Therefore, there must be a molecular conductor that synergistically orchestrates multiple molecular pathways in multiple types of tissues. We used the term "stem molecules" [98] in subsection 3.3, when we introduced FGF. Growth factors such as FGF and signaling pathways share common features with stem cells in terms of: 1) They play pivotal roles in embryogenesis; 2) Their functions are multimodal according to time and space; 3) They have several functions in adult tissues to recapitulate the embryonic process; and 4) They are sometimes carcinogenic in adults. Examples of these stem molecules include the FGF family, the TGF-β superfamily (including the bone morphogenetic protein [BMP] family), the Wnt family, and the Hh family. Of these, we have focused on Hh signaling as a putative coordinator of various healing process in DTI.

The mammalian Hh family contains three members of ligands, Sonic hedgehog (Shh), Indian hedgehog (Ihh) and Desert hedgehog (Dhh). Shh is one of the most important, well-known "morphogens" in the field of developmental biology. Shh mediates the morphogenesis of the central nervous system, limb bud, face, intestine and hair follicles for example, and shows temporospatially specific functions [32]. Ihh play a pivotal role in endochondral ossification of bone [31], and Dhh is involved in the formation of Schwann cells [110].

The molecular signaling cascades of Hh family members are very complicated, and are described in more detail elsewhere [98-100], but we have provided a brief outline of Hh signaling (Fig. 7). Hh proteins are secreted and diffused, acting as ligands for the cell surface receptor Patched-1 (Ptch-1). Ptch-1 inhibits the activity of another cell surface protein, Smoothened (Smo), in the absence of the ligands Shh, Ihh or Dhh. Once the ligands bind to Ptch-1, the suppression of Smo is recovered and activated. After the activation of Smo, via a signal transduction pathway comprising several molecules, the transcription factor Gli (Gli1, 2, 3) is translocated to the nucleus and activates the expression of target genes. These targets include *Ptch-1* and *Gli* themselves; thus, the expression level of the *Ptch-1* gene can be regarded as an indicator of the functional level of Hh signaling [82]. Alkaloid cyclopamine is an inhibitor of Hh signaling that targets Smo [28]. We previously showed facial anomalies in mice embryos exposed to cyclopamine [92], as a mild phenocopy of *Shh*-knockout mice [29]. Despite its importance in embryogenesis, the roles of Hh signaling in adult tissue have remained unclear, until recently. Carcinogenic effects of Shh signaling were reported in a subpopulation of people with skin cancer (basal cell carcinoma) or brain tumors (medulloblastoma) [19] but the physiological functions in adult were largely unknown.

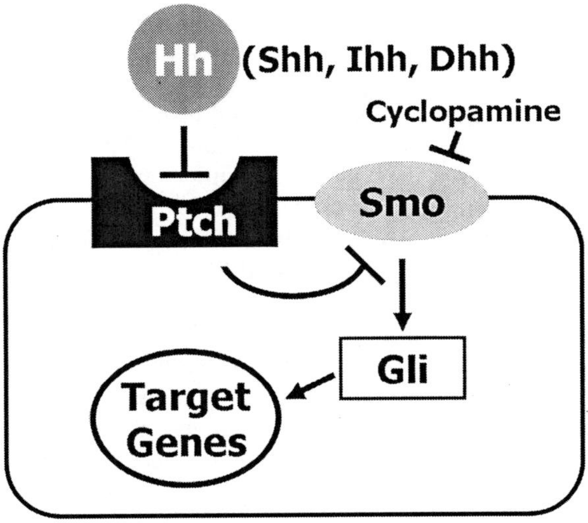

Fig. 7. Schema for Hedgehog (Hh) signaling

The involvement of Hh signaling in angiogenesis has been the focus of several recent studies [98, 100]. Previous studies revealed that Shh signaling induces embryonic angiogenesis, either directly or indirectly via the upregulation of other angiogenic factors such as VEGF and Ang-1 in adjacent tissues or cells [77, 93, 94]. Furthermore, previous studies, including ours, revealed that Ihh signaling mediates vasculogenesis and angiogenesis within the mouse yolk sacs [26, 41, 91]. Meanwhile, Colnot *et al.* [33] reported that angiogenesis penetrating into the ossification site in bone is impaired in *Ihh*-deficient mouse embryos. This vascular phenotype seems to be autonomous to endothelial cells and is not mediated by environmental angiogenic factors. This is because a co-grafting experiment with Ihh-null vessels and surrounding wild-type tissues showed similar vascular defects [33].

Is angiogenesis also mediated by Hh signaling in adult tissues? Pola *et al.* [114] first reported that the administration of Shh protein induces angiogenesis in ischemic limbs and the cornea in adult mice. Kusano *et al.* [76] also reported the induction of nerve vessels and restoration of nerve functions in rat diabetic neuropathy by systemic injection of Shh protein, suggesting a therapeutic angiogenic potential of exogenous Shh. In these studies, VEGF expression was upregulated by Shh signaling in the corneal stromal cells and nerve fibroblasts, respectively. Surace *et al.* [139] showed that retinal angiogenesis due to ischemia was associated with upregulation of Shh signaling and VEGF, which was antagonized by the Hh inhibitor cyclopamine, suggesting an involvement of endogenous Shh signaling in adult angiogenesis induced by hypoxia.

These data strongly support the possibility that Hh signaling is involved in the process of wound healing via angiogenesis. Indeed, Asai *et al.* [7] reported that endogenous Shh signaling is activated in the mouse skin wound model, and that transfection of naked *Shh* DNA promoted wound healing via VEGF-dependent vasculogenesis, i.e. by recruiting the putative EPCs. A more recent paper by Le *et al.* [78] demonstrated an indispensable role of endogenous Hh signaling in wound healing. They reported that skin wound healing and vascularization in the granulation tissue were significantly reduced in rats treated with the Hh inhibitor cyclopamine compared with the control rats without cyclopamine treatment.

Therefore, what about the repair process in DTI? Recent studies have suggested that Shh signaling is also involved in muscle regeneration after ischemia or injury. Pola *et al.* [115] reported the upregulation of Shh expression in regenerating muscles in ischemic limbs in mice. Notably, the administration of Shh-neutralizing antibody inhibited reactive angiogenesis and VEGF upregulation in this model, indicating pivotal roles of endogenous Shh signaling in this process. This paper was based on the concept that Shh signaling promotes myogenesis in an indirect manner via Shh-induced angiogenesis. However, more interestingly, Straface *et al.* [137] recently reported that Shh directly enhances muscle regeneration after injury, in addition to its angiogenic effects. They showed that the number of activated satellite cells and the protein levels of Myf5 and MyoD were significantly reduced in the injured rat muscles treated with cyclopamine compared with muscles in control animals without cyclopamine treatment. Furthermore, *Ptch-1* gene expression was observed in the Myf5- or MyoD-positive satellite cells in injured muscle tissue. These data demonstrate that satellite cells are a direct target of endogenous Shh signaling and that Shh signaling can promote myogenesis directly, independent of its angiogenic potency.

Overall, we have developed a hypothesis that Hh signaling exerts a dual role in the repair potential in DTI, namely in angiogenesis and myogenesis, acting in a coordinated manner. This hypothesis prompted us to conduct our own *in vivo* experiments, as described below.

4.2 Upregulation of Markers in Our Own Rat DTI Model

Now we have some understanding of the molecular pathways involved in DTI, although it is undoubtedly incomplete. To further elucidate the mechanisms involved, further studies are needed using *in vitro* and *in vivo* approaches. Next, we will describe our results obtained using our own rat *in vivo* model of DTI.

Several *in vivo* animal models of DTI have been developed. Linder-Ganz *et al.* [80] examined ischemia in a DTI model using thermography. In their model, the upper layer of the skin was resected to leave only muscle tissue, onto which pressure was applied. Thus, this model does not represent clinical DTI, which has intact surface skin. Another animal DTI model was reported by Stekelenburg *et al.* [134, 135] who used magnetic resonance imaging (MRI) to identify the deep tissue damage in combination with histological analysis to estimate the presence of ischemia. Although their data suggested that ischemia was involved in DTI in their model, their findings are at the morphological level, and do not reveal the biochemical and molecular changes. To address this point, we conducted the following studies using our own DTI model, and we measured changes in the above-mentioned biochemical and molecular markers.

4.2.1 Our DTI Model: Modification of Sugama's Model

One of the authors of this chapter (Dr Sugama) has developed an original device to induce full-thickness PUs in rat skin [138]. Briefly, after shaving the skin in the flank region, two 2-cm incisions (5 cm apart) that extended to the subperitoneum were created using a scalpel and a metal plate was inserted subperitoneally under general anesthesia (intraabdominal pentobarbital injection). Pressure was applied by lowering a cylinder with a contact area of 3 cm^2 with varying intensity and duration. After relieving the pressure, the metal plate was removed and the incisions were sutured. Thus, pressure was applied both on the skin and on the deep tissues in this model. Under the condition described in the original paper, the surface skin damage was not severe, and additional bacterial inoculation was needed to create full-thickness PUs. We focused on the deep tissue damage, and inoculation was not performed. Thus, the application of pressure using Sugama's device without inoculation can be regarded as an appropriate animal model for DTI (Fig. 8a). In addition, we artificially induced superficial skin loss using sandpaper to collect exudate fluids, before pressure application. The pressured lesions were then covered with transparent films to collect the exudate fluids.

4.2.2 Elevation of Serum and Exudate CPK

We have already conducted and published a pilot study for serum and exudate CPK measurement in our rat DTI model [122]. In this study, we divided the animals (12–14-week-old Wistar rats) into a control group or into two groups that were exposed to 10- or 20-kg pressure loading for 6 hours. A sham operation (incision and plate insertion without pressure loading) was performed in the control group. Blood samples (0.2 ml) were obtained from the tail vein before wounding and for 3 days after wounding. Exudate samples (30–80 μl) were aspirated from the wound sites underneath the transparent film on days 2 and 3 after loading. Histological analysis using hematoxylin/eosin (HE) staining showed more severe muscle damage in the 20-kg loading group than in the 10-kg group. The level of CPK in the serum and exudate was determined by a commercial laboratory (SRL Inc., Tokyo Japan) and was elevated in both the 10-kg and 20-kg groups. Thus, this was the first study to use the exudate CPK level as a marker for DTI *in vivo*. However, the differences between the 10-kg and 20-kg groups are unclear, possibly because both of the pressure conditions are too great for the rat muscle tissue.

Therefore, in the next study, we changed the conditions and exposed the rats to 1- or 10-kg pressure loading for 4 hours [123]. Under this condition, HE staining revealed more severe muscle damage with greater infiltration of inflammatory cells in the 10-kg group than in the 1-kg group (Figs. 8c–h). Of note, muscle degeneration was evident on day 1 in the 1-kg group, which had recovered to a near normal appearance on day 3 (Figs. 8d,g). Because there were no marked differences in the superficial skin appearance in this condition (data not shown), these finding support the fundamental hypothesis of DTI that muscle is more susceptible to pressure than the skin.

The exudate CPK level is shown in Fig. 8b [123]. It was significantly higher in the 10-kg group compared with the control and 1-kg groups; however, there was

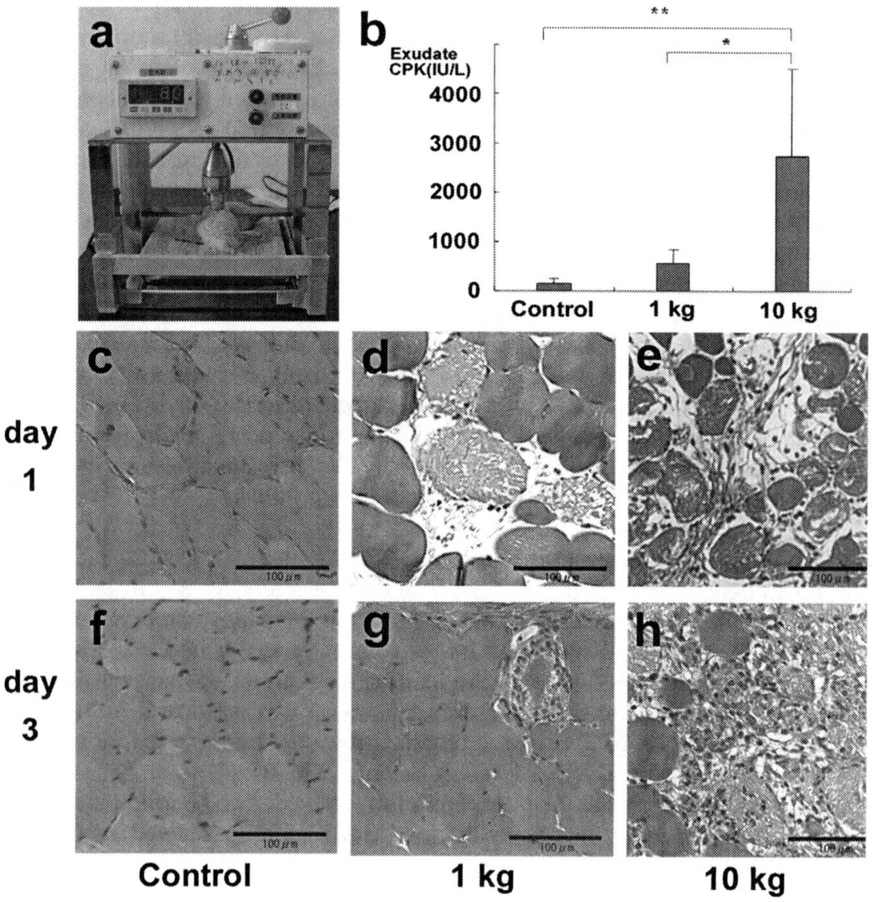

Fig. 8. (a) Sugama's device used in our DTI model. (b) Comparison of exudate CPK levels among the three groups. One-way ANOVA was performed followed by Fisher's least significant difference (LSD) *post hoc* test. *P<0.01 and **P<0.001; n = 11 per group. (c–h) Transverse slices of HE-stained muscle cross-sections. (c–e) day 1. (f–h) day 3. Bar=100 μm.

no significant difference between the control and 1-kg groups. Furthermore, there was no difference in the serum CPK level among the three groups (data not shown) in this condition. As stated in the former section, the serum CPK level is frequently used by clinicians as a marker for muscle damage. However, the present finding suggests that the exudate CPK level is a more sensitive marker than serum CPK, and better reflects the local muscle damage in DTI.

4.2.3 Upregulation of HIF-1 Protein Translocation and Gene Expression

HIF-1 is a hypoxia marker and acts as a trigger for the tissue repair/regeneration process. We consider that the milder pressure condition (1 or 10 kg for 4 hours) is more appropriate for HIF-1 analysis, because the endogenous regenerative potential may be anticipated with mild tissue damage.

HIF-1 activity can be assessed in terms of HIF-1 protein localization. Therefore, we performed immunohistochemical analysis using the anti-HIF-1α antibody. On day 1, the HIF-1α protein expression in the muscle cells at a basal level was very weak in the control group (Fig. 9a). The number of HIF-1α positive cells in the muscle and within the interstitial spaces between the muscle fibers increased in both the 1- and 10-kg groups (Figs. 9b,c). Magnified views showed the HIF-1α was localized to the cytoplasm in most of the HIF-1α-positive cells in all three groups (insets in Figs 9a–c). On day 3, the number of HIF-1α-positive cells remained increased in the 1- and 10-kg groups compared with the control groups (Figs. 9d–f). Of note, higher magnification of the tissue sections showed that nuclear localization was remarkably increased in the 10-kg group (inset in Fig. 9f), indicating that severe hypoxia induced HIF-1α translocation into the nucleus.

The changes in gene expression of *Hif-1a* were also examined by reverse transcription polymerase chain reaction (RT-PCR; Fig. 10). Total RNA was extracted from the control and the damaged muscle tissues, and complementary DNA (cDNA) was obtained for each sample using reverse transcriptase. The amount of PCR product obtained from cDNA using the target gene primers reflects the amount of mRNA and, thus, the expression level of the target gene. RT-PCR for 18S ribosomal RNA (rRNA) was simultaneously performed as an internal control among samples. The amount of each PCR product was represented as an intensity of the band on gel electrophoresis, and semi-quantification of the data was performed using Image-J software. The obtained values were adjusted to that of 18S rRNA from the same sample.

Hif-1a gene expression was weak in the control group because HIF-1 acts even in normoxic conditions to maintain tissue oxygen homeostasis. The gene expression level was unchanged in the 1-kg group on days 1 and 3, suggesting constitutive expression of *Hif-1a* in this condition [72]. However, *Hif-1a* expression was significantly upregulated in the 10-kg group on both days 1 and 3 (Fig. 10).

There was a distinct discrepancy between the *Hif-1a* gene expression level and the HIF-1α protein level. Although the number of HIF-1α-positive cells was increased in both groups on day 1, *Hif-1a* gene upregulation was not detected in our RT-PCR experiment. This finding is quite reasonable, because the increased protein level of HIF-1α on day 1 can be attributed to the inhibition of HIF-1α protein degradation under mild hypoxia, without changes in gene expression. The upregulation of *Hif-1a* gene expression in both groups on day 3 was surprisingly high, considering that HIF-1 activation is mainly dependent on its degradation and its mRNA level is known to be constitutive [72]. This finding can, at least in part, be interpreted that severe hypoxia caused positive feedback of *Hif-1a* gene upregulation via the downstream molecule VEGF, as suggested by Deudero *et al.* [38]. This hypothesis may be supported by the nuclear translocation of HIF-1α protein observed in the 10-kg group on day 3, where transcription of the downstream genes, including *vegf*, may be promoted.

Of interest, the hypoxic condition remained or was even enhanced on day 3. This may be in contrast to the conventional story that ischemia and hypoxia only

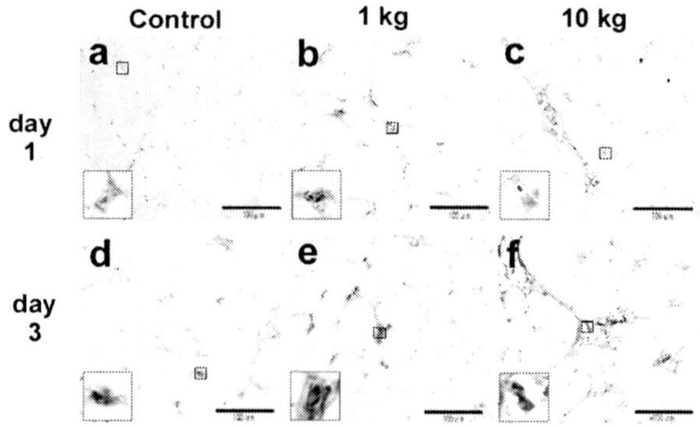

Fig. 9. HIF-1α immunohistochemistry on days 1 and 3 in the control and 1- and 10-kg pressure loading groups. The small box in the image for the 10-kg group on day 3 shows nuclear localization, while the small boxes in the other images indicate cytoplasmic localization. The nuclei appear to be stained, but this is because of Hematoxylin counterstaining. Bar=100 μm. The boxed areas are shown at higher magnification in the inset.

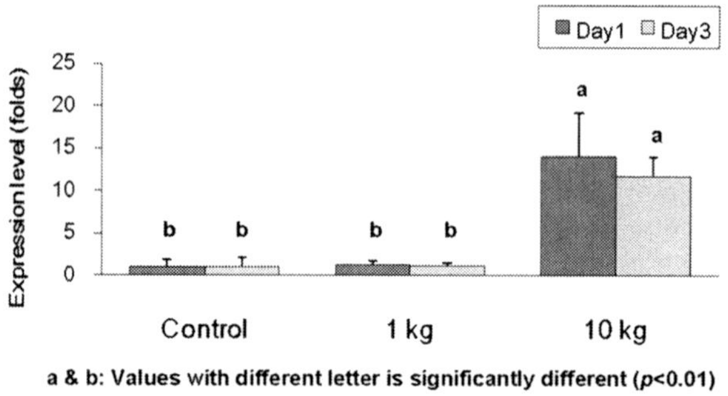

Fig. 10. Assessment of *Hif-1a* mRNA expression by RT-PCR. One-way ANOVA was performed followed by Student-Newman-Keuls (SNK) *post hoc* test. n=3 per group.

occur during the pressure loading phase, and are followed by circulatory reperfusion after the release of the load. Indeed, previous experiments involving vessel ligation showed that the hypoxia returned immediately to normoxia after the ligation was released [20, 130]. Further studies are needed to clarify why hypoxia was prolonged or enhanced, even after releasing the load in our DTI model.

4.2.4 Upregulation of Hh Signaling Gene Expression

The gene expression of *Shh* and *Ptch-1* was also examined by RT-PCR. As stated above, the expression level of *Ptch-1* can be regarded as a marker for Hh signaling activity, because the downstream transcription factor Gli regulates *Ptch*-1 expression itself [82]. In this study, the rat embryos (embryonic day 12) were used as a positive control for confirming appropriate condition of RT-PCR.

The RT-PCR results showed that *Ptch-1* gene expression was significantly upregulated in the 10-kg group on days 1 and 3, indicating that Hh signaling was activated in the 10-kg group (Fig. 11). The expression level of *Ptch-1* was similar in the 1-kg group and the control group. Although previous reports showed upregulation of Shh in ischemic or damaged skeletal muscle [115, 137], we could not detect *Shh* expression in any conditions in our RT-PCR experiments (data not shown).

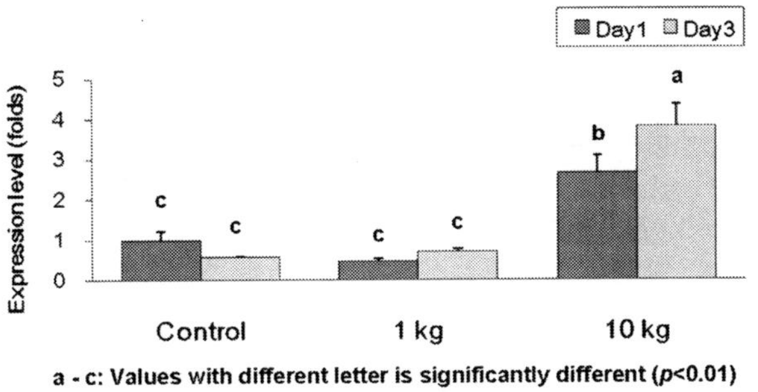

Fig. 11. Assessment of *Ptch-1* mRNA expression by RT-PCR. One-way ANOVA was performed followed by Student-Newman-Keuls (SNK) *post hoc* test. n=3 per group.

Our finding of *Ptch-1* upregulation without *Shh* expression can be interpreted in several ways. Because the Shh protein can diffuse over a wide area [99], there is a possibility that *Shh* upregulation occurred in adjacent tissue, such as adipose tissue, and the Shh protein was transferred to the muscle tissue by diffusion. Another hypothesis is that *Ptch-1* upregulation was induced by Hh ligands other than Shh. Ihh mediates embryonic angiogenesis in the yolk sac and in the developing limb bone [33, 91]. Because the molecular mechanisms of adult bone fracture healing are similar to those of embryonic skeletal development [48], there is a possibility that Ihh can also mediate angiogenesis in adult tissues.

Altogether, we successfully revealed that Hh signaling is activated in damaged muscle tissue in our DTI model. This constitutes the first step to elucidating the molecular events during DTI formation and repair. Nevertheless, questions remain. What is the ligand for Ptch-1 in our DTI model? How does Hh signaling contribute to the activation of muscle satellite cells or angiogenesis? Our results have opened an avenue towards a fascinating new field of PU research.

5 Conclusion

In this article, we have described the current understanding of DTI from the standpoint of both basic and clinical sciences. We first suggested that the concept of DTI is clinically a confused mixture of the general pathophysiological understanding of deep PUs and several specific conditions such as those seen in immobilized patients, Asian patients with bone protrusion, or ischial bursitis in patients with SCI. We consider that this confusion is one of the reasons why DTI is a focus of controversy in the current PU classification schemes. To overcome this problem, we need a sound understanding of the pathophysiology of DTI at a molecular and cellular level. The next part of this article was dedicated to a review of the recent progress made in understanding the molecular and cellular mechanisms of damage, repair and regeneration of deep tissues. There may be four pathways involved in deep tissue breakdown: ischemia, lymphatic impairment, reperfusion and deformation. Some of the somatic stem/progenitor cells might mediate the regeneration of deep tissues to recapitulate embryonic morphogenesis, Possible examples include satellite cells in muscle regeneration, ASCs in adipose tissue regeneration and EPCs in vascular regeneration. Because these processes involve changes in various signaling molecules, we focused on the CRK, HIF-1 and Hh signaling pathways as candidate markers for DTI in the last part of this article. Furthermore, using our own DTI rat model, we showed upregulation of these molecules at the mRNA and/or the protein level.

DTI is a real revolution as a concept regarding PUs, but the lack of fundamental knowledge of its mechanisms has led to serious confusion and controversy among clinicians. Learning from Shea's landmark studies and his attitude as a clinical scientist [131], we must further deepen our pathophysiological understanding of this intriguing phenomenon at the molecular and cellular level. We also understand the possible concerns of some readers that these approaches may cause a discrepancy between basic and clinical sciences. Basic science at the molecular level may seem too precise and mechanistic from the clinicians' eyes, as a kind of reductionism that all the vital phenomena within living humans can be deconstructed into mere chain reactions of devitalized substances. However, this is not true. If our molecular understanding of clinical problems can be enhanced, we can, paradoxically, confront the miracles of nature. For example: How does tissue repair or regeneration progress in such an integrated manner? How do stem molecules, such as Hh signaling, coordinate morphogenesis and regeneration? This level of insight may have a profound influence on clinicians in their daily practice within a spiritual sense.

At the same time, we should also highlight the importance of producing clinically accessible tools that are based on the knowledge gained from basic science. As seen in our evaluation of the exudate CPK level, exudate fluid can offer a valuable tool for monitoring DTI. Traditionally, the wound exudate was only used for visual assessment based on clinical experience, and was then discarded. Unfortunately, the activity of HIF-1 or Shh signaling pathways cannot currently be assessed using wound fluid, thus requiring tissue biopsy and modern biological techniques. However, there is an exciting future when our molecular understanding and technologies can be integrated into tools with more easily accessible biomarkers. Our understanding of DTI is developing into a cycle that progresses from the bedside to the bench and again to the bedside. Each turn of the cycle will be at a deeper level to enhance our understanding of DTI.

References

1. Akisue, T., Yamamoto, T., Marui, T., et al.: Ischiogluteal bursitis: multimodality imaging findings. Clin. Orthop. Relat. Res. 406, 214–217 (2003)
2. Ankrom, M.A., Bennett, R.G., Sprigle, S., et al.: Pressure-related deep tissue injury under intact skin and the current pressure ulcer staging systems. Adv. Skin Wound Care 18, 35–42 (2005)
3. Aoi, N., Yoshimura, K., Kadono, T., et al.: Ultrasound assessment of deep tissue injury in pressure ulcers: possible prediction of pressure ulcer progression. Plast. Recontr. Surg. 124, 540–550 (2009)
4. Aronovitch, S.A.: Intraoperatively acquired pressure ulcer prevalence: a national study. J. Wound Ostomy Continence Nurs. 26, 130–136 (1999)
5. Aronovitch, S.A.: Intraoperatively acquired pressure ulcers: are there common risk factors? Ostomy Wound Manage 53, 57–69 (2007)
6. Asahara, T., Murohara, T., Sullivan, A., et al.: Isolation of putative progenitor endothelial cells for angiogenesis. Science 275, 964–967 (1997)
7. Asai, J., Takenaka, H., Kusano, K.F., et al.: Topical sonic hedgehog gene therapy accelerates wound healing in diabetes by enhancing endothelial progenitor cell-mediated microvascular remodeling. Circulation 113, 2413–2424 (2006)
8. Atit, R., Sgaier, S.K., Mohamed, O.A., et al.: Beta-catenin activation is necessary and sufficient to specify the dorsal dermal fate in the mouse. Dev. Biol. 296, 164–176 (2006)
9. Bader, D., Bouten, C., Colin, D., et al.: Pressure Ulcer Research: Current and Future Perspectives. Springer, Berlin (2005)
10. Barber, C.L., Iruela-Arispe, M.L.: The ever-elusive endothelial progenitor cell: identities, functions and clinical implications. Pediatr. Res. 59, R26–R32 (2006)
11. Bates-Jensen, B.M., Vredevoe, D.L., Brecht, M.L.: Validity and reliability of the Pressure Sore Status Tool. Decubitus 5, 20–28 (1992)
12. Bates-Jensen, B.M., Guihan, M., Garber, S.L., et al.: Characteristics of recurrent pressure ulcers in veterans with spinal cord injury. J. Spinal Cord Med. 32, 34–42 (2009)
13. Bergeron, M., Yu, A.Y., Solway, K.E., et al.: Induction of hypoxia-inducible factor-1 (HIF-1) and its target genes following focal ischaemia in rat brain. Eur. J. Neurosci. 11, 4159–4170 (1999)
14. Berlowitz, D.R., Brienza, D.M.: Are all pressure ulcers the result of deep tissue injury? A review of the literature. Ostomy Wound Manage 53, 34–38 (2007)
15. Billon, N., Monteiro, M.C., Dani, C.: Developmental origin of adipocytes: new insights into a pending question. Biol. Cell 100, 563–575 (2008)
16. Biundo, J.: Bursitis. In The Merck Manuals Online Medical Library (2008), `http://www.merck.com/mmpe/sec04/ch040/ch040b.html#sec04-ch040-ch040a-790` (accessed April 27, 2009)
17. Black, J., Black, S.: Deep tissue injury. Wounds 15, 380 (2003)
18. Black, J., Baharestani, M.M., Cuddigan, J., et al.: National Pressure Ulcer Advisory Panel's updated pressure ulcer staging system. Adv. Skin Wound Care 20, 269–274 (2007)
19. Borzillo, G.V., Lippa, B.: The Hedgehog signaling pathway as a target for anticancer drug discovery. Curr. Top Med. Chem. 5, 147–157 (2005)

20. Bosco, G., Yang, Z.J., Nandi, J., et al.: Effects of hyperbaric oxygen on glucose, lactate, glycerol and anti-oxidant enzymes in the skeletal muscle of rats during ischaemia and reperfusion. Clin. Exp. Pharmacol. Physiol. 34, 70–76 (2007)
21. Bouten, C.V., Oomens, C.W., Baaijens, F.P., et al.: The etiology of pressure ulcers: skin deep or muscle bound? Arch. Phys. Med. Rehabil. 84, 616–619 (2003)
22. Braden, B.J., Maklebust, J.: Preventing pressure ulcers with the Braden Scale: an update on this easy-to-use tool that assesses a patient's risk. Am. J. Nurs. 105, 70–72 (2005)
23. Braissant, O., Wahli, W.: Differential expression of peroxisome proliferator-activated receptor-alpha, -beta, and -gamma during rat embryonic development. Endocrinology 139, 2748–2754 (1998)
24. Brancaccio, P., Maffulli, N., Limongelli, F.M.: Creatine kinase monitoring in sport medicine. Br. Med. Bull. 81-82, 209–230 (2007)
25. Brooks, S.V.: Current topics for teaching skeletal muscle physiology. Adv. Physiol. Educ. 27, 171–182 (2003)
26. Byrd, N., Grabel, L.: Hedgehog signaling in murine vasculogenesis and angiogenesis. Trends Cardiovasc Med. 14, 308–313 (2004)
27. Campbell, R.M.: Treatment of the pressure sore. In: Converse, J.M. (ed.) Reconstructive Plastic Surgery. The trunk, genitourinary system and experimental approaches, vol. 5. W.B. Saunders, Philadelphia (1964)
28. Chen, J.K., Taipale, J., Cooper, M.K., et al.: Inhibition of Hedgehog signaling by direct binding of cyclopamine to Smoothened. Genes Dev. 16, 2743–2748 (2002)
29. Chiang, C., Litingtung, Y., Lee, E., et al.: Cyclopia and defective axial patterning in mice lacking Sonic hedgehog gene function. Nature 383, 407–413 (1996)
30. Cho, K.H., Lee, S.M., Lee, Y.H., et al.: Non-infectious ischiogluteal bursitis: MRI findings. Korean J. Radiol. 5, 280–286 (2004)
31. Chung, U.I., Schipani, E., McMahon, A.P., et al.: Indian hedgehog couples chondrogenesis to osteogenesis in endochondral bone development. J. Clin. Invest. 107, 295–304 (2001)
32. Cohen Jr., M.M.: The hedgehog signaling network. Am. J. Med. Genet. A 123, 5–28 (2003)
33. Colnot, C., de la Fuente, L., Huang, S., et al.: Indian hedgehog synchronizes skeletal angiogenesis and perichondrial maturation with cartilage development. Development 132, 1057–1067 (2005)
34. Cowan, C.M., Shi, Y.Y., Aalami, O.O., et al.: Adipose-derived adult stromal cells heal critical-size mouse calvarial defects. Nat. Biotechnol. 22, 560–567 (2004)
35. Daniel, R.K., Priest, D.L., Wheatley, D.C.: Etiologic factors in pressure sores: an experimental model. Arch. Phys. Med. Rehabil. 62, 492–498 (1981)
36. Davis, S., Aldrich, T.H., Jones, P.F., et al.: Isolation of angiopoietin-1, a ligand for the TIE2 receptor, by secretion-trap expression cloning. Cell 87, 1161–1169 (1996)
37. Dellavalle, A., Sampaolesi, M., Tonlorenzi, R., et al.: Pericytes of human skeletal muscle are myogenic precursors distinct from satellite cells. Nat. Cell Biol. 9, 255–267 (2007)
38. Deudero, J.J., Caramelo, C., Castellanos, M.C., et al.: Induction of hypoxia-inducible factor 1alpha gene expression by vascular endothelial growth factor. J. Biol. Chem. 283, 11435–11444 (2008)
39. Donnelly, J.: Should we include deep tissue injury in pressure ulcer staging systems? The NPUAP debate. J. Wound Care 14, 207–210 (2005)

40. Doughty, D., Ramundo, J., Bonham, P., et al.: Issues and challenges in staging of pressure ulcers. J. Wound Ostomy Continence Nurs. 33, 125–130 (2006); quiz 131-122
41. Dyer, M.A., Farrington, S.M., Mohn, D., et al.: Indian hedgehog activates hematopoiesis and vasculogenesis and can respecify prospective neurectodermal cell fate in the mouse embryo. Development 128, 1717–1730 (2001)
42. Edsberg, L.E.: Pressure ulcer tissue histology: an appraisal of current knowledge. Ostomy Wound Manage 53, 40–49 (2007)
43. Eming, S.A., Brachvogel, B., Odorisio, T., et al.: Regulation of angiogenesis: wound healing as a model. Prog. Histochem. Cytochem. 42, 115–170 (2007)
44. Emmerich, J.: Current state and perspective on medical treatment of critical leg ischemia: gene and cell therapy. Int. J. Low. Extrem. Wounds 4, 234–241 (2005)
45. Etchevers, H.C., Vincent, C., Le Douarin, N.M., et al.: The cephalic neural crest provides pericytes and smooth muscle cells to all blood vessels of the face and forebrain. Development 128, 1059–1068 (2001)
46. European Pressure Ulcer Advisory Panel, Pressure ulcer treatment guideline (1998), `http://www.epuap.org/gltreatment.html` (accessed April 26, 2009)
47. European Pressure Ulcer Advisory Panel and National Pressure Ulcer Advisory Panel. Pressure ulcer prevention and treatment: Clinical practice guideline (Inaugural draft). In: Presented at 11th National NPUAP Biennal Conference. Arlington, VA, USA (2009)
48. Ferguson, C., Alpern, E., Miclau, T., et al.: Does adult fracture repair recapitulate embryonic skeletal formation? Mech. Dev. 87, 57–66 (1999)
49. Fernandes, K.J., McKenzie, I.A., Mill, P., et al.: A dermal niche for multipotent adult skin-derived precursor cells. Nat. Cell Biol. 6, 1082–1093 (2004)
50. Ferrara, N., Carver-Moore, K., Chen, H., et al.: Heterozygous embryonic lethality induced by targeted inactivation of the VEGF gene. Nature 380, 439–442 (1996)
51. Galiano, R.D., Tepper, O.M., Pelo, C.R., et al.: Topical vascular endothelial growth factor accelerates diabetic wound healing through increased angiogenesis and by mobilizing and recruiting bone marrow-derived cells. Am. J. Pathol. 164, 1935–1947 (2004)
52. Gawlitta, D., Li, W., Oomens, C.W., et al.: The relative contributions of compression and hypoxia to development of muscle tissue damage: an in vitro study. Ann. Biomed. Eng. 35, 273–284 (2007)
53. Gawlitta, D., Oomens, C.W., Bader, D.L., et al.: Temporal differences in the influence of ischemic factors and deformation on the metabolism of engineered skeletal muscle. J. Appl. Physiol. 103, 464–473 (2007)
54. Gefen, A.: How much time does it take to get a pressure ulcer? Integrated evidence from human, animal, and in vitro studies. Ostomy Wound Manage 54, 26–28, 30–25 (2008)
55. Gefen, A.: Deep tissue injury from a bioengineering point of view. Ostomy Wound Manage 55, 26–36 (2009)
56. Gesta, S., Tseng, Y.H., Kahn, C.R.: Developmental origin of fat: tracking obesity to its source. Cell 131, 242–256 (2007)
57. Gilbert, S.: Developmental Biology, 8th edn. Sinauer, Sunderland (2006)
58. Gomillion, C.T., Burg, K.J.: Stem cells and adipose tissue engineering. Biomaterials 27, 6052–6063 (2006)

59. Gonda, K., Nakaoka, T., Yoshimura, K., et al.: Heterotopic ossification of degenerating rat skeletal muscle induced by adenovirus-mediated transfer of bone morphogenetic protein-2 gene. J. Bone. Miner. Res. 15, 1056–1065 (2000)
60. Guihan, M., Garber, S.L., Bombardier, C.H., et al.: Predictors of pressure ulcer recurrence in veterans with spinal cord injury. J. Spinal Cord Med. 31, 551–559 (2008)
61. Hagisawa, S., Ferguson-Pell, M.W., Palmieri, V.R., et al.: Pressure sores: a biochemical test for early detection of tissue damage. Arch. Phys. Med. Rehabil. 69, 668–671 (1988)
62. Hausman, D.B., Lu, J., Ryan, D.H., et al.: Compensatory growth of adipose tissue after partial lipectomy: involvement of serum factors. Exp. Biol. Med. (May-wood) 229, 512–520 (2004)
63. Huard, J., Li, Y., Fu, F.H.: Muscle injuries and repair: current trends in research. J. Bone Joint Surg. Am. 84-A, 822–832 (2002)
64. Imai, K., Kadowaki, T., Aizawa, Y.: Standardized indices of mortality among persons with spinal cord injury: accelerated aging process. Ind. Health 42, 213–218 (2004)
65. Ivan, M., Kondo, K., Yang, H., et al.: HIFalpha targeted for VHL-mediated destruction by proline hydroxylation: implications for O2 sensing. Science 292, 464–468 (2001)
66. Jaakkola, P., Mole, D.R., Tian, Y.M., et al.: Targeting of HIF-alpha to the von Hippel-Lindau ubiquitylation complex by O2-regulated prolyl hydroxylation. Science 292, 468–472 (2001)
67. Jain, R.K.: Molecular regulation of vessel maturation. Nat. Med. 9, 685–693 (2003)
68. Kallio, P.J., Pongratz, I., Gradin, K., et al.: Activation of hypoxia-inducible factor 1alpha: posttranscriptional regulation and conformational change by recruitment of the Arnt transcription factor. Proc. Natl. Acad. Sci. USA 94, 5667–5672 (1997)
69. Kang, S.K., Putnam, L.A., Ylostalo, J., et al.: Neurogenesis of Rhesus adipose stromal cells. J. Cell. Sci. 117, 4289–4299 (2004)
70. Kang, S.K., Shin, M.J., Jung, J.S., et al.: Autologous adipose tissue-derived stromal cells for treatment of spinal cord injury. Stem. Cells Dev. 15, 583–594 (2006)
71. Kanno, N., Nakamura, T., Yamanaka, M., et al.: Low-echoic lesions underneath the skin in subjects with spinal-cord injury. Spinal Cord 47, 225–229 (2009)
72. Ke, Q., Costa, M.: Hypoxia-inducible factor-1 (HIF-1). Mol. Pharmacol. 70, 1469–1480 (2006)
73. Kim, S.M., Shin, M.J., Kim, K.S., et al.: Imaging features of ischial bursitis with an emphasis on ultrasonography. Skeletal Radiol. 31, 631–636 (2002)
74. Kovacic, J.C., Moore, J., Herbert, A., et al.: Endothelial progenitor cells, angioblasts, and angiogenesis–old terms reconsidered from a current perspective. Trends Cardiovasc. Med. 18, 45–51 (2008)
75. Kuang, S., Rudnicki, M.A.: The emerging biology of satellite cells and their therapeutic potential. Trends Mol. Med. 14, 82–91 (2008)
76. Kusano, K.F., Allendoerfer, K.L., Munger, W., et al.: Sonic hedgehog induces arteriogenesis in diabetic vasa nervorum and restores function in diabetic neuropathy. Arterioscler. Thromb. Vasc. Biol. 24, 2102–2107 (2004)
77. Lawson, N.D., Vogel, A.M., Weinstein, B.M.: Sonic hedgehog and vascular endothelial growth factor act upstream of the Notch pathway during arterial endothelial differentiation. Dev. Cell 3, 127–136 (2002)

78. Le, H., Kleinerman, R., Lerman, O.Z., et al.: Hedgehog signaling is essential for normal wound healing. Wound Repair Regen. 16, 768–773 (2008)
79. Le Lievre, C.S., Le Douarin, N.M.: Mesenchymal derivatives of the neural crest: analysis of chimaeric quail and chick embryos. J. Embryol. Exp. Morphol. 34, 125–154 (1975)
80. Linder-Ganz, E., Gefen, A.: The effects of pressure and shear on capillary closure in the microstructure of skeletal muscles. Ann. Biomed. Eng. 35, 2095–2107 (2007)
81. Linder-Ganz, E., Yarnitzky, G., Yizhar, Z., et al.: Real-time finite element monitoring of sub-dermal tissue stresses in individuals with spinal cord injury: toward prevention of pressure ulcers. Ann. Biomed. Eng. 37, 387–400 (2009)
82. Marigo, V., Tabin, C.J.: Regulation of patched by sonic hedgehog in the developing neural tube. Proc. Natl. Acad. Sci. U. S. A. 93, 9346–9351 (1996)
83. Matsumoto, D., Sato, K., Gonda, K., et al.: Cell-assisted lipotransfer: supportive use of human adipose-derived cells for soft tissue augmentation with lipoinjection. Tissue Eng. 12, 3375–3382 (2006)
84. Matsuzawa, Y.: Therapy Insight: adipocytokines in metabolic syndrome and related cardiovascular disease. Nat. Clin. Pract. Cardiovasc. Med. 3, 35–42 (2006)
85. Menetrey, J., Kasemkijwattana, C., Fu, F.H., et al.: Suturing versus immobilization of a muscle laceration. A morphological and functional study in a mouse model. Am. J. Sports Med. 27, 222–229 (1999)
86. Michel, C.: Trasport of fluid and solutes in tissues. In: Bader, D., Bouten, C., Colin, D., et al. (eds.) Pressure Ulcer Research. Springer, Berlin (2005)
87. Milkiewicz, M., Pugh, C.W., Egginton, S.: Inhibition of endogenous HIF inactivation induces angiogenesis in ischaemic skeletal muscles of mice. J. Physiol. 560, 21–26 (2004)
88. Millauer, B., Wizigmann-Voos, S., Schnurch, H., et al.: High affinity VEGF binding and developmental expression suggest Flk-1 as a major regulator of vasculogenesis and angiogenesis. Cell 72, 835–846 (1993)
89. Miller, G.E., Seale, J.: Lymphatic clearance during compressive loading. Lymphology 14, 161–166 (1981)
90. Morrison, S.J., White, P.M., Zock, C., et al.: Prospective identification, isolation by flow cytometry, and in vivo self-renewal of multipotent mammalian neu-ral crest stem cells. Cell 96, 737–749 (1999)
91. Nagase, M., Nagase, T., Koshima, I., et al.: Critical time window of hedge-hog-dependent angiogenesis in murine yolk sac. Microvasc. Res. 71, 85–90 (2006)
92. Nagase, T., Nagase, M., Osumi, N., et al.: Craniofacial anomalies of the cultured mouse embryo induced by inhibition of sonic hedgehog signaling: an animal model of holoprosencephaly. J. Craniofac. Surg. 16, 80–88 (2005)
93. Nagase, T., Nagase, M., Yoshimura, K., et al.: Angiogenesis within the developing mouse neural tube is dependent on sonic hedgehog signaling: possible roles of motor neurons. Genes Cells 10, 595–604 (2005)
94. Nagase, T., Nagase, M., Yoshimura, K., et al.: Defects in aortic fusion and craniofacial vasculature in the holoprosencephalic mouse embryo under inhibition of Sonic hedgehog signaling. J. Craniofac. Surg. 17, 736–744 (2006)
95. Nagase, T., Hisatomi, T., Koshima, I., et al.: Heterotopic ossification in the sacral pressure ulcer treated with basic fibroblast growth factor: coincidence or side effect? J. Plast. Reconstr. Aesthet. Surg. 60, 327–329 (2007)

96. Nagase, T., Koshima, I., Maekawa, T., et al.: Ultrasonographic evaluation of an unusual peri-anal induration: a possible case of deep tissue injury. J. Wound Care 16, 365–367 (2007)
97. Nagase, T., Matsumoto, D., Nagase, M., et al.: Neurospheres from human adipose tissue transplanted into cultured mouse embryos can contribute to cranio-facial morphogenesis: a preliminary report. J. Craniofac. Surg. 18, 49–53 (2007); discussion 60–61
98. Nagase, T., Nagase, M.: Time windows of hedgehog signaling in the craniofacial and vascular development: Analyses using mouse whole embryo culture system. In: Grachevsky, N. (ed.) Signal Transduction Research Trends. NOVA Science Publishers, Hauppauge (2007)
99. Nagase, T., Nagase, M., Machida, M., et al.: Hedgehog signaling: A biophysical or biomechanical modulator in embryonic development? Ann. N. Y. Acad. Sci. 1101, 412–438 (2007)
100. Nagase, T., Nagase, M., Machida, M., et al.: Hedgehog signalling in vascular development. Angiogenesis 11, 71–77 (2008)
101. Nagoshi, N., Shibata, S., Kubota, Y., et al.: Ontogeny and multipotency of neural crest-derived stem cells in mouse bone marrow, dorsal root ganglia, and whisker pad. Cell. Stem Cell 2, 392–403 (2008)
102. National Pressure Ulcer Advisory Panel, Updated stageing system: Pressure ulcer stages revised by NPUAP (2007), `http://www.npuap.org/pr2.htm` (accessed April 26, 2009)
103. Ning, H., Lin, G., Lue, T.F., et al.: Neuron-like differentiation of adipose tissue-derived stromal cells and vascular smooth muscle cells. Differentiation 74, 510–518 (2006)
104. Nola, G.T., Vistnes, L.M.: Differential response of skin and muscle in the experimental production of pressure sores. Plast. Reconstr. Surg. 66, 728–733 (1980)
105. Ohura, T.: Risk factors for pressure ulcers of elderly people. Jpn. J. Pressure Ulcers 4, 397–405 (2002) (in Japanese)
106. Ohura, T., Ohura, N.J.: Pathogenetic mechanisms and classification of undermining in pressure ulcers- Elucidation of relationship with deep tissue injury. Wounds 18, 329–339 (2006)
107. Oshima, H., Rochat, A., Kedzia, C., et al.: Morphogenesis and renewal of hair follicles from adult multipotent stem cells. Cell 104, 233–245 (2001)
108. Paget, J.: In the wards. Clinical Lecture on bed-sores. The Student's Journal and Hospital Gazette 10, 144–146 (1873)
109. Pajvani, U.B., Trujillo, M.E., Combs, T.P., et al.: Fat apoptosis through targeted activation of caspase 8: a new mouse model of inducible and reversible lipoatrophy. Nat. Med. 11, 797–803 (2005)
110. Parmantier, E., Lynn, B., Lawson, D., et al.: Schwann cell-derived Desert hedgehog controls the development of peripheral nerve sheaths. Neuron 23, 713–724 (1999)
111. Perry, R.L., Rudnick, M.A.: Molecular mechanisms regulating myogenic determination and differentiation. Front Biosci. 5, D750–D767 (2000)
112. Planat-Benard, V., Silvestre, J.S., Cousin, B., et al.: Plasticity of human adipose lineage cells toward endothelial cells: physiological and therapeutic perspectives. Circulation 109, 656–663 (2004)
113. Poissonnet, C.M., Burdi, A.R., Garn, S.M.: The chronology of adipose tissue appearance and distribution in the human fetus. Early Hum. Dev. 10, 1–11 (1984)

114. Pola, R., Ling, L.E., Silver, M., et al.: The morphogen Sonic hedgehog is an indirect angiogenic agent upregulating two families of angiogenic growth factors. Nat. Med. 7, 706–711 (2001)
115. Pola, R., Ling, L.E., Aprahamian, T.R., et al.: Postnatal recapitulation of embryonic hedgehog pathway in response to skeletal muscle ischemia. Circulation 108, 479–485 (2003)
116. Prins, J.B., O'Rahilly, S.: Regulation of adipose cell number in man. Clin. Sci. (Lond) 92, 3–11 (1997)
117. Reyne, Y., Nougues, J., Vezinhet, A.: Adipose tissue regeneration in 6-month-old and adult rabbits following lipectomy. Proc. Soc. Exp. Biol. Med. 174, 258–264 (1983)
118. Risau, W.: Mechanisms of angiogenesis. Nature 386, 671–674 (1997)
119. Rohrich, R.J., Broughton, G., Horton, B., et al.: The key to long-term success in liposuction: a guide for plastic surgeons and patients. Plast. Reconstr. Surg. 114, 1945–1952 (2004); discussion 1953
120. Rubayi, S., Montgomerie, J.Z.: Septic ischial bursitis in patients with spinal cord injury. Paraplegia 30, 200–203 (1992)
121. Sanada, H., Moriguchi, T., Miyachi, Y., et al.: Reliability and validity of DESIGN, a tool that classifies pressure ulcer severity and monitors healing. J. Wound Care 13, 13–18 (2004)
122. Sari, Y., Nakagami, G., Kinoshita, A., et al.: Changes in serum and exudate creatine phosphokinase concentrations as an indicator of deep tissue injury: a pilot study. Int. Wound J. 5, 674–680 (2008)
123. Sari, Y.: Hypoxia Is Involved in Deep Tissue Injury Formatio. In A Rat Model. Master thesis. University of Tokyo Graduate School of Medicine (2009)
124. Schneider, C.M., Dennehy, C.A., Rodearmel, S.J., et al.: Effects of physical activity on creatine phosphokinase and the isoenzyme creatine kinase-MB. Ann. Emerg Med. 25, 520–524 (1995)
125. Schultz, G.S., Sibbald, R.G., Falanga, V., et al.: Wound bed preparation: a systematic approach to wound management. Wound Repair Regen. 11(suppl. 1), S1–S28 (2003)
126. Schultz, G.S., Barillo, D.J., Mozingo, D.W., et al.: Wound bed preparation and a brief history of TIME. Int. Wound J. 1, 19–32 (2004)
127. Seale, P., Sabourin, L.A., Girgis-Gabardo, A., et al.: Pax7 is required for the specification of myogenic satellite cells. Cell 102, 777–786 (2000)
128. Semenza, G.L.: Regulation of tissue perfusion in mammals by hypoxia-inducible factor 1. Exp. Physiol. 92, 988–991 (2007)
129. Sen, C.K.: Wound healing essentials: let there be oxygen. Wound Repair Regen. 17, 1–18 (2009)
130. Setala, L.P., Korvenoja, E.M., Harma, M.A., et al.: Glucose, lactate, and pyruvate response in an experimental model of microvascular flap ischemia and reperfusion: a microdialysis study. Microsurgery 24, 223–231 (2004)
131. Shea, J.D.: Pressure sores: classification and management. Clin. Orthop. Relat. Res. 112, 89–100 (1975)
132. Sieveking, D.P., Ng, M.K.: Cell therapies for therapeutic angiogenesis: back to the bench. Vasc. Med. 14, 153–166 (2009)
133. Steed, D.L.: Clinical evaluation of recombinant human platelet-derived growth factor for the treatment of lower extremity ulcers. Plast. Reconstr. Surg. 117, S143–S149 (2006); discussion S150–151S

134. Stekelenburg, A., Oomens, C.W., Strijkers, G.J., et al.: Compression-induced deep tissue injury examined with magnetic resonance imaging and histology. J. Appl. Physiol. 100, 1946–1954 (2006)
135. Stekelenburg, A., Strijkers, G.J., Parusel, H., et al.: Role of ischemia and deformation in the onset of compression-induced deep tissue injury: MRI-based studies in a rat model. J. Appl. Physiol. 102, 2002–2011 (2007)
136. Stekelenburg, A., Gawlitta, D., Bader, D.L., et al.: Deep tissue injury: how deep is our understanding? Arch. Phys. Med. Rehabil. 89, 1410–1413 (2008)
137. Straface, G., Aprahamian, T., Flex, A., et al.: Sonic hedgehog regulates angiogenesis and myogenesis during post-natal skeletal muscle regeneration. J. Cell. Mol. Med (2008), doi:10.1111/j.1582-4934.2008.00440.x
138. Sugama, J., Sanada, H., Nakatani, T., et al.: Pressure-induced ischemic wound healing with bacterial inoculation in the rat. Wounds 17, 157–168 (2005)
139. Surace, E.M., Balaggan, K.S., Tessitore, A., et al.: Inhibition of ocular neovascularization by hedgehog blockade. Mol. Ther. 13, 573–579 (2006)
140. Suri, C., Jones, P.F., Patan, S., et al.: Requisite role of angiopoietin-1, a ligand for the TIE2 receptor, during embryonic angiogenesis. Cell 87, 1171–1180 (1996)
141. Tachibana, T., Matsui, Y., Sugama, J., et al.: Report from Academic and Educational Group in JSPU: Revision of DESIGN. Jpn. J. Pressure Ulcers 10, 586–596 (2008) (in Japanese)
142. Tajima, F., Mitsushima, T., Ito, N., et al.: Points for prevention and diagnosis of pressure ulcers in spinal cord injury patients. Orthop. Surg. Traumatol. 46, 861–868 (2003) (in Japanese)
143. Takashima, Y., Era, T., Nakao, K., et al.: Neuroepithelial cells supply an initial transient wave of MSC differentiation. Cell 129, 1377–1388 (2007)
144. Tang, W., Zeve, D., Suh, J.M., et al.: White fat progenitor cells reside in the adipose vasculature. Science 322, 583–586 (2008)
145. Taylor, R., James, T.: The role of oxidative stress in the development and persistence of pressure ulcer. In: Bader, D., Bouten, C., Colin, D., et al. (eds.) Pressure Ulcer Research. Springer, Berlin (2005)
146. Thomas, D.R., Rodeheaver, G.T., Bartolucci, A.A., et al.: Pressure ulcer scale for healing: derivation and validation of the PUSH tool. The PUSH Task Force. Adv. Wound Care 10, 96–101 (1997)
147. Toma, J.G., Akhavan, M., Fernandes, K.J., et al.: Isolation of multipotent adult stem cells from the dermis of mammalian skin. Nat. Cell Biol. 3, 778–784 (2001)
148. Tomita, Y., Matsumura, K., Wakamatsu, Y., et al.: Cardiac neural crest cells contribute to the dormant multipotent stem cell in the mammalian heart. J. Cell Biol. 170, 1135–1146 (2005)
149. Wang, H.U., Chen, Z.F., Anderson, D.J.: Molecular distinction and angiogenic interaction between embryonic arteries and veins revealed by ephrin-B2 and its receptor Eph-B4. Cell 93, 741–753 (1998)
150. Yamamoto, N., Akamatsu, H., Hasegawa, S., et al.: Isolation of multipotent stem cells from mouse adipose tissue. J. Dermatol. Sci. 48, 43–52 (2007)
151. Yancopoulos, G.D., Davis, S., Gale, N.W., et al.: Vascular-specific growth factors and blood vessel formation. Nature 407, 242–248 (2000)
152. Yoshimura, K., Shigeura, T., Matsumoto, D., et al.: Characterization of freshly isolated and cultured cells derived from the fatty and fluid portions of liposuction aspirates. J. Cell. Physiol. 208, 64–76 (2006)

153. Zhang, Q., Chang, Q., Cox, R.A., et al.: Hyperbaric oxygen attenuates apoptosis and decreases inflammation in an ischemic wound model. J. Invest. Dermatol. 128, 2102–2112 (2008)
154. Zuk, P.A., Zhu, M., Mizuno, H., et al.: Multilineage cells from human adipose tissue: implications for cell-based therapies. Tissue Eng. 7, 211–228 (2001)
155. Zuk, P.A., Zhu, M., Ashjian, P., et al.: Human adipose tissue is a source of multipotent stem cells. Mol. Biol. Cell 13, 4279–4295 (2002)
156. Zulkowski, K., Langemo, D., Posthauer, M.E.: Coming to consensus on deep tissue injury. Adv. Skin Wound Care 18, 28–29 (2005)

Proteomic Approaches for Studying the Phases of Wound Healing

Laura E. Edsberg

Center for Wound Healing Research, Daemen College, 4380 Main Street, Amherst, NY 14266

Abstract. Proteome level information is necessary to understand the function of specific cell types and their roles in health and disease. Proteomics is a rapidly developing field with a wide range of applications in wound healing. The ability to use proteomics to assess the wound healing process would have many benefits, including earlier evidence of healing and better understanding of how different treatments affect the wound at the protein level. The basis of what is known about the chronic wound proteome is based on results from a broad collection of studies utilizing a number of different proteomic techniques on fluids and tissues from wounds with different etiologies. The identification of biomarkers associated with healing or delayed healing in chronic wounds could have great significance in the use of current treatments, as well as in the development of new therapeutic interventions.

1 Introduction

Proteomics is the study of protein structures and functions (1). Proteome level information is necessary to understand the function of specific cell types and their roles in health and disease. The biochemical function of a protein is based on both the tertiary molecular structure and the linear amino acid sequence initially coded by its gene. The human proteome is comprised of 200,000 to 500,000 different proteins, compared to the 2200 genes which code proteins in humans (2). This large number of proteins is the result of differences in protein structures due to interactions with protein enzymes and special molecules (3).

Proteomics is comprised of a number of areas of study, including protein separation, protein identification, protein quantification, protein sequence analysis, structural proteomics, interactional proteomics, post-translational modifications, and cellular proteomics (3). Proteins can be expressed at levels or in forms that mRNA analysis cannot predict. Also, proteomics provides a system to evaluate the complex interaction between the cell's functional pathway and the environment around the cell, which is independent of changes at the mRNA level (4). Proteomics is a rapidly developing field with a wide range of applications in wound healing. The identification of biomarkers or molecular markers for predicting pressure ulcer development or predictors of healing or lack of healing is of great interest.

A. Gefen (Ed.): Bioengineering Research of Chronic Wounds, SMTEB 1, pp. 343–362.
springerlink.com

The identification of biomarkers associated with healing or delayed healing in chronic wounds could have great significance in the use of current treatments, as well as in the development of new therapeutic interventions. Currently, common predictors of wound healing are based solely on clinical appearance of the wound. The depth and size of the wound over time is most often used to assess wound progress with an increase in size indicating a lack of healing. The presence and assessment of differences in various tissue types, including eschar, granulation, and slough, and the presence and amount of exudate have all been used as a predictor of healing in wounds.

The ability to use proteomics to assess the wound healing process would have many benefits, including earlier evidence of healing and better understanding of how different treatments affect the wound at the protein level. As proteomics becomes more common in wound analysis and is eventually used as a bedside tool, it will change how treatments are evaluated and assessed. The most efficacious treatment for each individual wound may be based on proteomic data in the future.

In order to use proteomics to assess wound healing, the first question that must be answered is which fluid and techniques to analyze. Other medical conditions have utilized bile, saliva, cerebrospinal fluid, and dialysate. The most common or obvious choices for samples for chronic wound proteomics are tissue, blood, serum, plasma, drainage fluid, and wound fluids. The fluid or tissue from the surface of the wound is used with the assumption that it is representational of the extracellular environment of the wound. Many researchers in the field of chronic wound research have examined wound fluid and tissue from a mixture of wounds with a variety of techniques.

Techniques for the collection of wound fluid vary between researchers and types of wounds utilized for the study, with methods including the use of occlusive dressings, porous dextranomer beads, surgical drainage tube or bottle collection, wound swabbing, and laser capture microdissection (5). During the process of collection, proteins may bind to the collection devices; some techniques may allow proteins to gather away from the wound bed, losing the opportunity to have a spatial/temporal relationship between the sample and the wound. Additionally, samples may have high abundance proteins present at such great levels that less abundant proteins are masked.

The populations examined and time course of study varies widely between studies. The basis of what is known about the chronic wound proteome is based on results from a broad collection of studies utilizing a number of different proteomic techniques on fluids and tissues from wounds with different etiologies.

2 Sample Handling and Storage

Following the choosing of a fluid or tissue for analysis, preservation of the sample must be considered. Wysocki (6) showed that proteins begin to degrade in 10 seconds or less in wound fluid samples. Protease inhibitors may be necessary to preserve samples, and to limit physical degradation and post-collection changes. Additionally, samples must be either analyzed immediately or frozen for preservation, and thaw cycles must be limited to further protect proteins present for analysis.

3 Protein Concentration Measurement

After collecting the wound fluid sample, but prior to processing the protein samples for isolation, separation, or quantification, it is necessary to calculate the protein concentration measurement. It is important that each technique begin with the same amount of protein so that differences in amounts of protein present or trends in proteins quantities are not dependent on the varying amounts of protein initially loaded. Additionally, depending on the type of wound, very little protein may be present and analysis may be limited by the quantity collected, thus usage of the minimal sample for each technique will allow more experiments to be conducted. There are multiple techniques or assays, including the Bradford Assay, to measure protein concentration. The choice of protein assay is typically based on availability of protein to assay, need for accuracy, convenience, and potential interfering agents.

Moseley et al. (7) reported significantly greater protein concentrations in acute wound fluid versus chronic wound fluid from chronic venous ulcer wounds ($p=0.0005$). In our laboratory, we have found no significant correlation between protein concentration and healing or lack of healing in pressure ulcers over time (manuscript in preparation). We have noted differences in protein concentration depending on the type of wounds from which the sample was collected (manuscript in preparation). James et al. (8) used the Biuret method to measure total protein in the exudate from the leg ulcers of four subjects. The single subject that healed had an increase in the total protein of the exudate and two of the non-healing subjects had total protein levels that dropped over time, while the third subject's exudate protein level stayed low throughout. Trengove et al. (9) also noted this increase in total protein in the exudates of healing leg ulcers. This difference in the total protein values in the sample relative to wound healing may be a function of the different etiologies of leg ulcers, acute wounds, and pressure ulcers.

After the protein sample is collected and the amount of protein has been quantified, other techniques are employed to separate, quantify, and identify the proteins present in the samples.

4 Separation of Proteins

4.1 SDS-PAGE

One basic technique used to separate proteins is sodium dodecyl sulfate polyacrylamide gel electrophoresis (SDS-PAGE). The sample collected is a mix of proteins and biomolecules. The SDS is a strong detergent with a negative charge and when the sample is incubated in it, cell membranes dissolve and cell proteins are denatured, soluabilized, and given negative charges (10). The sample is then placed on a gel in a DC electric field and the proteins migrate toward the anode (+) at different rates depending on their molecular weights and negative charge distribution. The end result is a series of bands, each containing proteins of the same molecular weight.

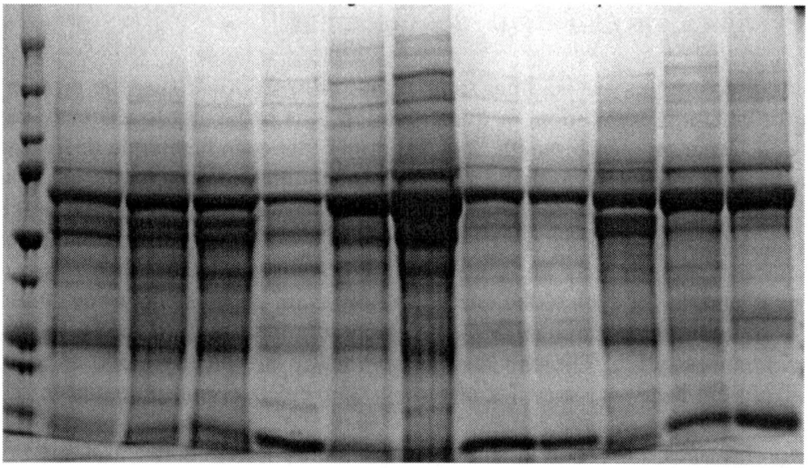

Fig. 1. 1D gel of wound fluid samples

SDS-PAGE separates proteins with different molecular weights, but two different proteins with the same molecular weight can migrate through the gel at the same rate. Since the density of the bands present on the gel will vary with the amount of each protein present in the sample, the same amount of protein must be loaded if gels are to be compared.

Jaynes et al. (11) utilized SDS-PAGE to analyze wound fluid proteins from the pressure ulcers of two subjects. Four bands, two of low molecular weight (10kD and 15kD) and two of high molecular weight (55kD and 80kD) had obvious density changes. The high molecular weight bands decreased as a Stage III sacral pressure ulcer healed, but this decrease in density was not observed in the non-healing Stage IV heel ulcers present in the other subject. The lower molecular weight bands fluctuated in intensity during the course of the study and were thought to contain growth factors.

4.2 Western Blot

In order to confirm the identity of the proteins in the bands, a standard of that protein and a confirming Western Blot test are typically conducted. A Western Blot involves isolating the test antigen in question on the PAGE and then transferring it to a nitrocellulose surface. In order to prevent interactions between the antibody used for detection and the membrane, non-specific proteins must be blocked. Blocking is typically done by placing a dilute solution protein like bovine serum albumin or non-fat dry milk with a small amount of detergent, often Tween. The protein in the solution attaches to the membrane everywhere except where the target proteins have attached. Thus, the antibody must attach at the binding sites of the target protein. A modified antibody with a linked reporter enzyme is exposed to the substrate. Antibodies specific to the fixed antigen will complex with the antigen and a reaction of the bound enzyme-labeled antibody results in a colored band in the location where the sample was placed, if present. This test is qualitative, not quantitative (10).

Wysocki (6) evaluated wound fluid from venous leg ulcers, surgical mastectomy wounds, blisters, and blood from healthy volunteers to compare fibronectin degradation and matrix metalloproteinases in local wound environments and blood. SDS-PAGE and Western Blotting were used to separate the proteins and identify intact and degraded fibronectin and examine fibronectin proteolysis. Fibronectin mediates cell-cell and cell-matrix interactions, functions as a structural protein, and it also interacts with cells found in wound healing, including platelets, fibroblasts, neutrophils, and epidermal cells. Fibronectin is present and active during each stage of wound healing (6). The degradation products of fibronectin were detected in chronic wound fluid at 93kD and 125kD and when intact fibronectin was added to chronic wound fluid, degradation occurred within 10 seconds or less in samples treated with intact fibronectin; within 15 minutes, all high-molecular weight fibronectin polypeptides had been degraded to lower-molecular weight forms (6). The fibronectin in blister fluid, mastectomy wound fluid had little evidence of degradation, and the blood-derived and plasma-derived serum had no evidence of fibronectin degradation (6). Grinnell et al. (12) compared fibronectin and vitronectin in chronic leg ulcers, surgical mastectomy wounds, and blister fluid. The surgical wound and blister fluid contained intact fibronectin and vitronectin, whereas three of the leg ulcer samples had complete degradation of the vitronectin and locally derived fibronectin. Vitronectin and fibronectin are important adhesion proteins in wound healing and the breakdown present in chronic wound fluid may be part of the delayed healing in chronic wounds.

Moses et al. (13) utilized SDS-PAGE and Western Blot to evaluate proteinase activity in wound fluids collected from pigs with partial thickness wounds over seven days. Bands with molecular weights similar to active and latent forms of MMP-2 (72kD) were detected in wound fluid collected on days 1-7 (13). The greatest number of active MMP-2 species were detected on days 4-6. The latent form of MMP-9 was the band with the greatest density noted throughout, but the active form was never detected (13).

Schmidtchen et al. (14) evaluated fluid collected from the venous ulcers of five subjects and utilized SDS-PAGE and Western Blotting to evaluate plasma and wound fluid. The wound fluid was collected over four hours and when bands were compared to those in plasma, all wound fluid bands appeared less dense. Anti-chymotrypsin (an antiproteinase) was identified using Western Blotting (14).

Moseley et al. (7) evaluated the oxidative stress biomarkers, total protein carbonyl content, malondialdehyde content, and total antioxidant capacities in wound fluid from 10 patients with acute wounds and 12 patients with chronic venous ulcers using SDS-PAGE and Western Blot analysis. Reactive oxygen species (ROS) are generated by inflammatory cells during healing, and the balance of oxidant and antioxidants is important for normal cellular functioning. No significant differences were present in the total protein carbonyl content (from protein oxidation) of acute versus chronic wound fluid, but following reducing ($p<0.02$) and non-reducing ($p<0.001$) conditions, acute wound fluid had significantly denser bands of protein carbonyl content. No significant differences in malondialdehyde levels or total antioxidant capacities were noted between acute and chronic wound fluids (7), but when the values were corrected for total wound fluid protein, chronic wounds had significantly higher antioxidant capacities. Decreased skin nonenzymatic antioxidants as a result of aging may make individuals more susceptible to the elevated oxidative stress that is part of normal wound healing, but the increased antioxidant capacities found in these samples indicate a potentially adaptive environment within the wound bed (7,15).

The combination of SDS-Page and Western Blots allows the initial analysis of the proteome, but since multiple proteins may be contained in the bands, the density and variation between samples may increase as a function of multiple proteins versus any single protein. Also, Western Blot allows confirmation of the presence of a specific protein, but not quantification, which clearly is needed if biomarkers or molecules significant to the healing of chronic wounds are to be identified.

4.3 2D-PAGE

Another technique for protein separation which overcomes some of the limitations of SDS-PAGE is two-dimensional polyacrylamide gel electrophoresis (2D-PAGE). 2D-PAGE has been one of the most widely used techniques for protein separation and quantification (16). The process adds a second dimension of movement on the gel to further separate the proteins. Proteins are first separated by charge by loading the sample onto a gel with a pH gradient. A voltage is applied to the gel and the proteins migrate across the gel until they reach the point on the gel where the protein charge is the same as the surrounding pH. This is the isoelectric point (pI) of the protein. The gel is then softened in both a denaturing solution, which causes the proteins to unfold, and a detergent, which has a strong negative charge and binds to all the proteins. An electric field is again applied to the gel, but at a 90 degree angle from the first separation. The proteins migrate through the gel and the lower molecular weight proteins move the furthest. Gels are stained to visualize the spots and then the gels are analyzed using 2D gel analysis software. Spots are chosen and identified using mass spectrometry or 2D Western Blot (16).

In our laboratory, we have analyzed samples from 50 pressure ulcers over 42 days utilizing 2D-PAGE. Analysis of gels is based on the presence of the spot, as well as the density of the spot. Gels can be difficult to compare when run at different times, even when run under the same conditions (pH, time, etc), due to slight fluctuations in runs of samples. Multi-gel set-ups allow up to 12 gels to be run simultaneously. The complexity of samples collected from chronic wounds makes the analysis of2D gels daunting. We have noted up to 850 unique spots on the gels, but the number varies depending on the sample and some gels have only 350 spots present (manuscript in preparation). The large number of spots in the wound fluid makes identification of molecules of interest challenging and is complicated by the fact that a single spot on a 2D gel can contain multiple proteins. Thus, the spot may change in density between samples or time points, but that change can be the result of a number of proteins, depending on the gel. To confirm the proteins present in a spot, the spot must be cut from the gel and then the identity of the proteins contained in the spot can be determined using mass spectrometry.

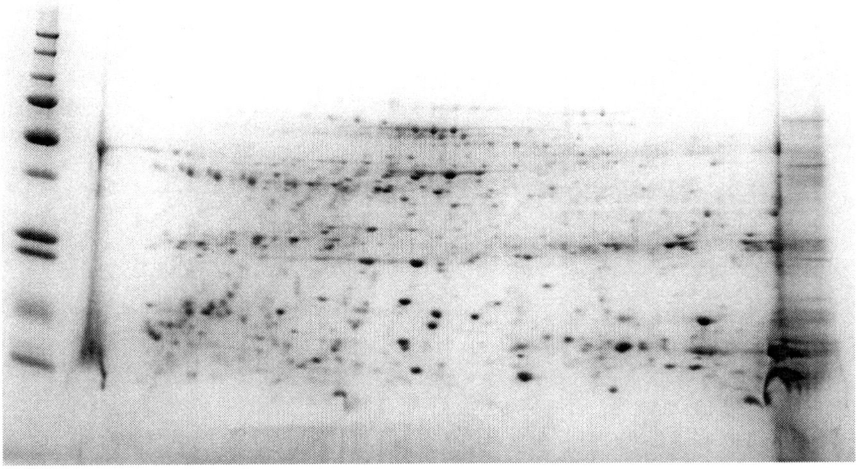

Fig. 2a. 2D gel of wound fluid from a pressure ulcer with 490 spots present

Fig. 2b. The same 2D gel as Figure 2a, but with 485 spots cut out for analysis

Fernandez et al. (17) utilized 2-D gel electrophoresis and mass spectrometry to evaluate wound fluid from chronic venous ulcers. Fibrinolytic and coagulation proteins were identified. Samples were pooled for analysis and high-abundant proteins (albumin, etc) were removed. In fluid collected from pressure ulcers in our laboratory, we did not note significant amounts of high-abundant proteins (manuscript in preparation), but the difference between the etiology of venous leg ulcers and pressure ulcers must be noted. The location from which the sample was

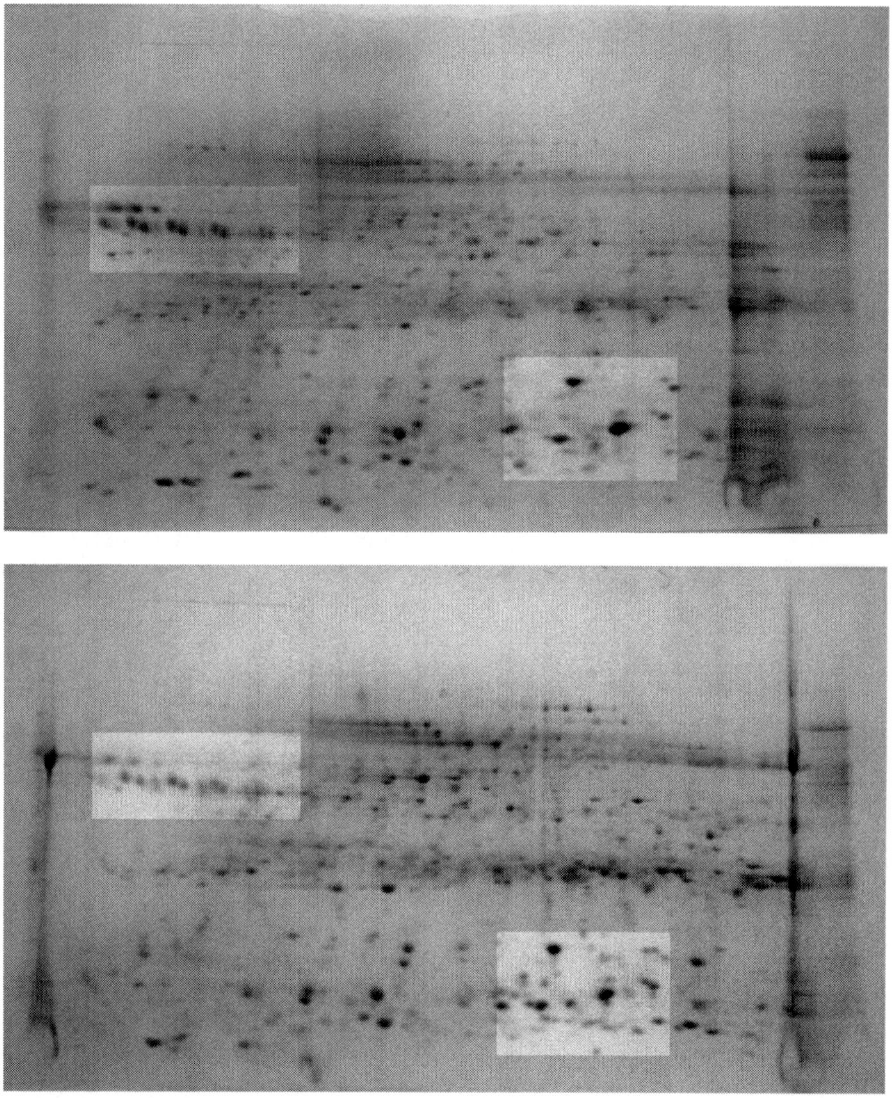

Fig. 3. Pressure ulcer wound fluid 2D gels with differing spot densities highlighted

collected and content of the particular samples will dictate the need to remove abundant proteins prior to analysis.

4.4 Mass Spectrometry

Matrix-assisted laser desorption/ionization (MALDI) is an ionization technique used in mass spectrometry. Time of flight (TOF) mass spectrometry is most often used with this technique.

Toman et al. (18) used 2D-PAGE and matrix assisted laser desorption/ionization time-of-flight (MALDI-TOF) mass spectrometry to study wounds in a rat model at different time points. Tissue from the wounds was collected, frozen and ground into a powder prior to analysis. 26 distinct proteins were identified, including the major spots in the samples. The level of hemoglobin was the most significant change over time and was elevated in the wound samples at five minutes to three hours after the wound was created, but was near control levels by 12 hours (18). Oh et al. (19) used 2D gels and MALDI-TOF to evaluate the collagen and collagen-related proteins present in a line of human skin fibroblasts. 80 proteins were identified on the 2D gels, which were speculated to be isoforms or posttranslational modifications (19). The proteins identified included collagen alpha 1 (I) chain precursor, collagen alpha 1 (III) chain precursor and other collagen-related structures, as well as metabolic enzymes.

The blister fluid of 11 patients was examined by Volden et al. (20). Two subjects were subjected to ultraviolet light to create erythema, and then suction blisters were created on those sites. Blister fluid from individuals without erythema was also obtained. Fluid was collected using a syringe and compared with serum samples collected. Several hundred spots were present in the 2D gels. Protein patterns that were obtained using 2D-PAGE were similar between blister fluid and serum from the same individual. Blister fluid had hemoglobin present, and increased alpha-antitrypsin dimmer and actin, and two unknown spots compared to the serum (20). The hemoglobin was thought to be a product of the syringe collection technique. 100 low-molecular weight compounds were also identified in this study using gas chromatography and mass spectrometry (20). Macdonald et al. (21) evaluated blister fluid from normal skin and non-lesional skin from an individual with plaque psoriasis before and after IL-1beta intradermal application using 2D-PAGE and MALDI-TOF. 670 protein spots were present in the samples and compared, and nine proteins were identified. Vitamin D binding protein was present in the greatest amounts in the psoriasis subject both with and without IL-1beta treatment (21). Haptoglobin was absent in the IL-1beta-treated psoriasis subject sample, and present in low values in the fluid from the psoriasis subject without treatment. It was present with the greatest density in both normal samples.

Chojnacki et al. (22) used multidimensional protein identification technology (MudPIT), which combines multidimensional liquid chromatography with ESI-MS/MS, to identify a variety of growth factors and inflammatory proteins in the fluid from diabetic foot ulcers. After removing the 20 most abundant proteins, EGF, PDGF-BB, KGF-2, and TGF-beta were identified, as well as calgranulin A, B, and C (22).

4.5 2D-DIGE

2-D fluorescence difference electrophoresis (2-D DIGE) is a technique used to compare different samples on the same gel simultaneously by labeling the samples with a fluorescent dye prior to 2D-PAGE. Up to three different samples can be separated with the same gel, thus two samples and a standard can be on the same gel. The three samples on the gel are clearly present in different dyes, allowing differences to be more easily evaluated. Pollins et al. (23) evaluated protein extracts from normal and burned skin, and identified temporal patterns of expression of cytoskeletal and heatshock proteins using 2D-DIGE.

5 Protein Quantification

The quantification of different proteins identified following separation is necessary to understand the role of these proteins in the wound and to correlate changes in concentrations with wound outcomes. Both the presence and the levels present are critical to understanding the biochemical changes associated with healing or lack of healing in wounds.

5.1 iTRAQ

In our laboratory we use multiplexed isobaric tagging technology (iTRAQ^{TM}), which involves isobaric tagging of peptides. This enables the simultaneous identification and quantification of peptides using tandem mass spectrometry (24). Samples can be run in a 4-plex, allowing four different samples to be run simultaneously with sample 1 compared to 2, sample 1 compared to 3, and sample 1 compared to 4. This technique easily allows comparisons over time or between subjects. Additionally, 8-plexes are now becoming more common to allow even larger groups of samples to be run in an experiment.

iTRAQ identifies the most abundant proteins, and typically the least abundant includes most cytokines, but as the proteome of chronic wounds is being constructed all of the proteins present are important in providing a complete view of the wound environment. In conjunction with other techniques to quantify smaller proteins, a more complete view of the wound protein profile is becoming possible (manuscript in preparation).

5.2 ELISA

Enzyme-linked immunoabsorbent assay (ELISA) is used to quantify a specific antibody or antigen. This technique goes beyond the Western Blot and allows for greater specificity, sensitivity, and quantification of proteins.

Trengove et al. (25) analyzed wound fluid from chronic leg ulcers using ELISA kits and found that bFGF, EGF, PDGF, and TGF-beta had differing concentrations in both healing and non-healing chronic leg ulcer fluid, but no clear trend. Il-1, IL-6, and TNF-alpha were present in high levels in both groups, but IL-1 and TNF-alpha decreased significantly in wounds that were healing (25). In a later study, Trengove et al. (26) collected fluid from 26 patients with 27 venous ulcers on admission and again after two weeks. ELISA assays were used to measure IL-1alpha, IL-6beta, IL-6, TNF-alpha, TGF-beta, PDGF, EGF, and FGF. IL-1alpha, IL-1beta and TNF-alpha had much greater concentrations in non-healing versus healing wound fluid. Again, these levels decreased significantly ($p<0.05$) during healing. IL-6 levels were not significantly different between healing and non-healing fluid samples, but bioactivity was greater in non-healing wounds. The levels of PDGF, EGF, BFGF, and TGF-beta did not differ over the 2 week time period (26). Interestingly, the size of the ulcer and the level of individual cytokines were not directly related (26).

Barone et al. (27) measured IL-1alpha, TIMP-1, MMP-1, and TIMP-1/MMP-1 complex using sandwich ELISA assays on fluid collected from pressure ulcers and surgical dehiscenses, and compared the results to acute wound fluids collected post-mammoplasty, abdominoplasty, or flap, and before and after treatment with hydroactive dressings. A significant decrease in IL-1alpha after one week of treatment with hydroactive dressings was reported and the wounds decreased in area and volume as well during this time period (27). A significantly greater concentration of IL-1alpha was present in chronic versus acute wounds, and collagenase activity was greater in chronic wounds prior to treatment and not detectable in acute wounds (27). Harris et al. (28) collected fluid from healing and non-healing leg ulcers, and used ELISA to measure PDGF-AB, GM-CSF, IL-1alpha, IL-1beta, IL-6, and Bfgf. IL-1ALPHA, IL1-BETA, and IL-6 were also measured using bioassays. Cytokine levels were generally higher in fluid from nonhealing ulcers, but no significant difference in cytokines or collagenase was found between healing and non-healing leg ulcers as others have reported (25,26,27). However, fluid was only collected at a single time point and not over a period of time with multiple collections as in most other studies.

Aiba-Kojima et al. (29) collected and compared drainage fluid from post-surgical wounds, fluid from seroma, and serum platelet-poor and platelet-rich plasma. EGF, PDGF, TGF-beta, VEGF, and HGF concentrations in the serum platelet-rich plasma were significantly greater than in the platelet-poor plasma. Drainage fluid collected on days 0 or 1 had greater concentrations of b-FGF, EGF, PDGF, and TGF-beta as compared to both drainage fluid collected on days 5-7 and serum platelet-poor plasma. VEGF, HGF, and KGF in both drainage fluid time points and seroma fluid were much greater than in serum platelet-poor plasma. Seroma fluid had high levels of TGF-beta, VEGF, HGF, and KGF; however, bFGF, EGF, and PDGF were not detectable (29). In drainage fluid from days 0-1, IL-6 was greatest, then decreased with time, but IL-8 had the opposite trend. These findings in surgical wound fluid differ from the findings of Trengove et al. (25), where the IL-6 levels did not change as chronic leg ulcers healed, whereas IL-1 did decrease as those wounds healed.

Aiba-Kojima et al. (29) compared drainage fluid from post-surgical wounds, fluid from seroma, and serum platelet-poor and platelet-rich plasma and found the concentration of MMP-8 was greater than either MMP-1 or MMP-13 in drainage fluid, peaked on days 2-3 and then decreased over time. Nwomeh et al. (30) used ELISA to measure MMP-1, MMP-8, and TIMP-1 in healing acute experimental wounds in humans, as well as non-healing pressure ulcers and venous ulcers, and reported that MMP-8 was significantly greater in healing wounds than MMP-1. In chronic wounds MMP-1 and MMP-8 were present in greater amounts than in the healing wounds. MMP-8 was always present in the higher levels than MMP-1 and both MMP-1 and MMP-8 were present in greater amounts than TIMP-1 in the samples (30).

Rayment et al. (31) collected fluid from nine subjects with chronic venous ulcers being treated with compression therapy. MMP-9 was the protease present in the greatest quantities in all samples and was elevated significantly in chronic wound fluid versus acute wound fluid. The Pressure Ulcer Scale of Healing (PUSH) tool, in which scores range from 0 to 17 with 0=healed, was used to group wounds. The greatest PUSH scores (>=12) had a significantly greater ($P<0.01$) MMP-9 level versus the acute wound and serum samples (31). Yager et al. (32) evaluated fluid from surgical wounds and pressure ulcers, and found that the pressure ulcer fluid samples had elevated levels of MMP-2 and MMP-9. Pressure ulcer fluid samples also had elevated levels of collagenoytic activity, as well as elevated levels of collagenase inhibitor complexes. Ladwig et al. (33) measured levels of the pro and activated forms of MMP-2 and MMP-9 in wound fluid and tissue samples of subjects in a pressure ulcer clinical trial, and also reported elevated MMP-9 levels in pressure ulcers. Greater concentrations of MMP-2 and MMP-9 were found in the majority of wound fluid samples. Four times as much MMP-9 was present vs. MMP-2 in these samples. MMP-9 and TIMP-1 levels were significantly higher ($p<0.05$) on day 0 in patients with poorly healing wounds (33). The MMP-9/TIMP-1 ratio was significantly lower in samples from patients with healing pressure ulcers and the MMP-9/TIMP-1 ratio was four-fold greater in patients with poorly healing pressure ulcers versus those that healed well ($p=0.05$) (33).

Trengove et al. (34) compared acute wound fluid collected after radical mastectomy and fluid collected from various types of chronic wounds and venous ulcers. The levels of TIMP-1 were greater in acute wounds versus chronic wounds and had a negative correlation with protease levels ($p=0.02$). Bullen et al. (35) also found lower levels of TIMP-1 in chronic versus healing wounds when comparing mastectomy wound fluid to venous ulcer wound fluid, in contrast to the findings of Ladwig et al (33).

In a study of acute wounds created in healthy individuals, MMP-1, MMP-8, and TIMP-1 levels were tracked over time. Nwomeh et al. (36) created four full-thickness skin biopsies in the sacral and trochanteric region of healthy individuals and collected wound fluid over eight days and biopsied the wounds on days 2, 4, 14, and 28. MMP-1, TIMP-1, and MMP-1/TIMP-1 complexes were measured using ELISA assays. Levels of TIMP-1 were elevated on day 1 post-wounding, and declined over time. MMP-1/TIMP-1 complex increased over time and plateaued on day 7. MMP-1 levels from wound fluid were very low initially, but peaked on day 7. Tissue biopsy levels peaked on day two. MMP-8 was present in levels 100- to 200-fold greater than MMP-1 in exudates and tissue. MMP-8 levels fluctuated, but peaked on day 4 in the wound fluid and day 2 in the tissue samples (36).

In these acute wounds in healthy subjects, TIMP-1 decreased over time, whereas others have reported lower TIMP-1 levels in chronic wounds (34,35). These results are also interesting in contrast to the findings in cultures. Cook et al. (37) cultured chronic wound fibroblasts from venous ulcers and uninvolved fibroblasts cultures from the same individuals, and using ELISA assays showed increased TIMP-1 and TIMP-2 levels, and decreased MMP-1 levels in the chronic wound fluid.

In patients with acute traumatic wounds or chronic wounds from people with diabetes, Lobman et al. (38) examined the concentrations of MMP-1, MMP-8, and TIMP-2 using ELISA assays. MMP-1 was 65-fold greater in the biopsies of the chronic wounds versus the traumatic wounds ($p<0.001$). Average MMP-8 values were two-fold ($p<0.002$) and MMP-9 average concentrations were 14-fold greater ($p=0.027$). TIMP-2 was lower in diabetic wounds ($p<0.007$) (38). These results in the wounds of people with diabetes of MMP-1 concentrations greater than MMP-8 and MMP-9 levels differ from the findings in venous leg ulcers and pressure ulcers (30,31).

Mouës et al. (39) utilized ELISA assays to measure albumin, pro and activated MMP-9, TIMP-1 and the total MMP-9/TIMP-1 ratio in wounds treated with topical negative pressure versus gauze over 10 days. Both acute and chronic wounds of a wide variety of etiologies were sampled. Albumin levels were greater in the acute versus chronic wounds, but not different between therapies. Pro-MMP-9 and the MMP-9/TIMP-1 ratio were lower in the wounds treated with topical negative pressure on day 10, but, interestingly, did not differ between acute and chronic wounds (39). Kilpadi et al. (40) also looked at the fluid collected from wounds treated with negative pressure wound therapy. Stage III and IV pressure ulcers were evaluated over 7 days for TNF-a, IL-1b, MMP-3, MMP-9, and TIMP-1. The concentrations of MMP-3, MMP-9, and the MMP-3/TIMP-1 ratio were significantly lower in samples taken on days 1, 3, and 7 versus day 0 ($P<0.05$) (40). No information regarding clinical outcomes or changes indicative of healing or non-healing was given.

Baker et al. (41) collected wound fluid from 58 patients over 5-8 days following colo-rectal surgery. MMP-1, MMP-3, MMP-8, TIMP-1, and TIMP-2 were measured using ELISA assays. Differing levels of MMPs and TIMPs were correlated with post-operative complications and the severity of the complications. MMP-2 and MMP-9 were positively correlated with complications, while TIMP-2 and TIMP-1 showed a negative correlation. Differing levels of MMPs and TIMPs were observed with tumor pathology, but no consistent trends were identified (41). The total MMP-9 and MMP-2 levels had a significant negative correlation on days 1-5, but MMP-8 and MMP-9 had a positive correlation on days 1-5. Many variables existed within this patient population, including surgery performed, duration of surgery, pre-operative radiotherapy, and tumor pathology. In another study, Baker et al. (42) used ELISA assays to evaluate MMP-1, MMP-3, TIMP-1, TIMP-2, IL-6, IL-1beta, TNF-alpha, EGF, PDGF, VEGF, TGF-beta, and bFGF levels in fluid collected from the wounds of 50 breast and colorectal surgery patients. MMP-1, MMP-3, TIMP-1, IL-6, IL-1beta, and TNF-alpha were significantly greater in the samples collected from wounds post-colorectal surgery versus breast wound fluid. These findings are interesting in light of the greater concentration of IL-1beta and TNF-alpha reported in chronic leg ulcer and pressure ulcer fluid versus acute wounds. EGF, PDGF, TGF-beta1, and bFGF were greater in the breast wound fluid and VEGF did not significantly differ between the two groups.

The results of these studies by Baker et al. (41,42) elucidate differences between the biochemical profile of the wound fluid and type of surgery. In order to compare chronic versus acute wound healing, both types of fluid are necessary, but it may be much more important than previously thought to choose a single type of acute wound for comparison. Additionally, the biochemical profile of that type of wound during healing must be more completely understood before it is used for comparison. Multiple studies have used wound fluid collected from a variety of surgical wounds and grouped them together as representative of the acute wound biochemical profile during healing, but with a variety of diagnoses requiring these surgeries initially, it is difficult to compare across groups. The differences in the acute wound "control" group may be as vast when analyzed using proteomic techniques as the differences seen in chronic wounds of varying etiologies, which makes a complex problem even more daunting.

5.3 Immunohistochemistry

Immunohistochemistry is the process of labelling proteins in the cells of a tissue section using antibodies to bind specific antigens. This technique allows researchers to identify the location of proteins within the tissue and might be of value when looking at various locations within the wound and wound margin, as well as healthy tissue.

Galkowski et al. (43) examined tissue from the margin of diabetic foot ulcers and normal nondiabetic foot skin. The expression of MCP-1, IL-8, IL-10, CCR2A, CXCR1, IL10R, and IL-15 in the keratinocytes of a diabetic foot ulcer was similar to normal skin. TGF-beta1 and TGF-betaR1 had increased expression in the suprabasal layer of diabetic epidermis ($p<0.05$). Also, diabetic foot ulcer keratinocytes had elevated expression of GM-CSF ($p<0.05$) and increased expression of EGF ($p<0.05$). The vascular endothelial of the diabetic ulcer had significantly lower IL-10, IL-15, and TGF-beta1 expression (43). Proinflammatory factors were expressed at higher levels in diabetic ulcer keratinocytes and ECM versus normal foot skin, but receptors were low as well (43). Stojadinovic et al. (44) compared normal skin from reduction mammoplasty, debrided tissue, and punch biopsies. C-myc and beta-catenin were noted in the nuclei of cells from the non-healing edge of chronic wounds versus acute wound edges, and researchers hypothesized that they may be indicative of delayed healing.

5.4 Bioassay and Enzyme Assays

Bioassays, which include zymograms are used to determine the biological activity of a sample as opposed to the presence or concentration of the protein. Zymography measures proteolytic activity via an electrophoretic technique based on SDS-PAGE using a sodium dodecyl sulfate gel with a protein substrate that is degraded by the proteases during incubation (45). White bands appear on a dark gel and the band intensity is related to the amount of protease loaded (45). Enzyme assays are techniques to measure enzyme activity.

Trengove et al. (26) collected fluid from 26 patients with 27 venous ulcers on admission and again after two weeks and found significantly greater levels of biologically active cytokines IL-1 and IL-6 in non-healing versus healing wounds ($p<0.05$). In another study, Trengove et al. (25) noted reduced activity in IL-1 and IL-6 activity as measured by bioassays in healing chronic leg ulcers.

In fluid from nine subjects with chronic venous ulcers being treated with compression therapy, collagen type I and type IV zymography showed a high level of protease activity in chronic wound fluid versus serum and acute wound fluid (31). The main band in both was at 102kD, which may be indicative of the same protease. In type IV zymography, fewer low molecular weight bands were noted and less activity was noted overall compared to type I (31). MMP-9 was the protease present in the greatest quantities in all samples and was elevated significantly in chronic wound fluid versus acute wound fluid (31). Bullen et al. (35) also found elevated MMP-9 levels, as well as MMP-2 in chronic venous ulcer wounds and a number of lower molecular weight proteolytic species versus mastectomy wound fluid. Results were similar in a study by Wysocki et al. (46) comparing fluid from chronic leg ulcers and surgical mastectomy sites using gelatin-zymography were similar. Acute wound fluid had levels of MMP-2 and MMP-9 five- to 10-fold greater than serum, and chronic wound fluid had levels of MMP-2 and MMP-9 five- to 10-fold greater than mastectomy fluid (46).

In another study, Wysocki et al. (6) extended this work by utilizing zymography to evaluate MMP-2 and MMP-9 in blood-derived serum, platelet-derived serum, blister fluid, chronic wound fluid, and mastectomy fluid. Platelet-derived blood serum contained MMP-2 versus blood-derived serum, which contained a majority of MMP-9. The expression of MMP-2 was twice as high and MMP-9 was five times as great in fluid from mastectomies versus blood-derived and platelet derived serum. Chronic wound fluid had three to five times more MMP-2 and 5 to 20 times as much MMP-9 compared to mastectomy fluid (6).

The overexpression of MMPs in chronic wounds is a hallmark of delayed healing. The elevated activity and expression of MMP-2 and MMP-9 were consistent in each of these studies comparing venous leg ulcers to surgical wounds (6,35,46).

Wysocki et al. (47) examined blood from normal volunteers and wound fluid collected from venous leg ulcers and mastectomies for plasminogen activator and plasminogen activator inhibitor-1, as well as the activity of both. Plasminogen is important in normal wound healing. The mean levels of urokinase plasminogen activator, plasminogen activator inhibitor-1, and MMP-9 were twice as high in chronic wound fluid versus acute wound fluid. The levels of urokinase plasminogen activator and plasminogen activator inhibitor-1 were four to five times greater than in serum and the MMP-9 was 32 times greater than in blood-derived serum. MMP-9 and urokinase plasminogen activator decreased as the chronic wound healed. When enzymatically active early in the chronic wound healing process, urokinase plasminogen activator was not inhibited by plasminogen activator inhibitor-1, and MMP-9 was very elevated. When plasminogen activator is in a complex with plasminogen activator inhibitor-1, MMP-9 levels decrease as healing continues. This complex is apparently necessary to restore the enzyme/inhibitor balance required for a wound to heal.

Trengove et al. (34) compared acute wound fluid collected after radical mastectomy with fluid collected from various types of chronic wounds and venous ulcers. Protease activity was measured utilizing an Azocoll assay to detect metalloproteinases and serine proteases. The mean level of MMP activity was significantly greater (30-fold $p<0.001$) in the mixed-etiology chronic wounds, but no differences could be determined between types of chronic wounds due to the small numbers in each group. Azocasenin assay results showed a decrease in protease levels in healing chronic venous leg ulcers and neutrophil elastase activity was decreased in acute vs. mixed chronic wounds. No significant changes over time were noted. The degradation of EGF was significantly greater in acute versus chronic wounds (34).

Moses et al. (13) used a radiometric enzyme assay to evaluate the time course of collagenase or MMP inhibition over 10 days in wound fluid collected from pigs after partial thickness wounds were created. The greatest amount of collagenase inhibition was noted on day 0 and decreased to 0 on day 8, then rose slightly again over the final two days (13). MMP-9 was the most prominent band identified using 1D gels, but the active form was never detected in the samples analyzed.

Lobman et al. (38) used gelatin zymography to compare samples from diabetic patients with chronic wounds with samples from non-diabetic subjects with acute traumatic wounds. Pro-MMP-2 was three fold ($p=0.041$) and MMP-2 active was six fold greater ($p=0.033$) in chronic versus traumatic wound fluid samples. The elevated MMP-2 activity was also seen in chronic diabetic foot ulcer versus traumatic foot injury wound fluid. Cook et al. (37) cultured chronic wound fibroblasts from venous ulcers and uninvolved fibroblasts cultures from the same individuals and found that MMP-2 activity was lower in the chronic wound cultures.

Baker et al. (42) examined the fluid collected from the wounds of 50 breast and colorectal patients. Gelatin zymography was used to identify active MMP-9 and MMP-2 bands in a greater portion of colorectal wound samples than breast wound samples. The difference in activity levels relative to the type of surgical wound is interesting to note, considering the wide variety of surgical wounds which have been used as the acute or control wound in published studies. Often surgical wounds of mixed etiologies have been pooled to serve as the control or acute group, and these results indicate that all surgical wounds do not all heal with the same levels of MMP-2 and MMP-9 activity.

5.5 Antibody Arrays/Protein Expression Microarrays

Protein microarrays have molecules of proteins, antibody proteins, oligopeptides, aptamers, or carbohydrates arranged in a grid pattern on glass, ceramic, or other inert material surfaces (10). Antibody microarrays bind monoclonal antibodies or chemical probes to the surface (10). Microarrays use small quantities of sample, but can analyze many samples simultaneously – thus using far less sample than separate ELISA assays for each protein. The microarrays can be used to assist in identifying potential targets by looking for the presence of different molecules or by quantifying the amount of specific proteins present.

In our laboratory we have used antibody arrays to allow an overview of the protein expressions in a sample. Using these expression levels, we search for targets of interest for future experiments (manuscript in preparation). We have also used human inflammation antibody arrays and human matrix metalloproteinase antibody arrays to measure multiple cytokines or antigens simultaneously. Custom arrays are readily available as well and allow the minimum use of samples for the molecules of the greatest interest.

Brem et al. (48) have used microarrays to explore analyze the expression patterns of the genome. The gene expression pattern of a biopsy from a venous stasis ulcer at the nonhealing edge and the adjacent nonulcerated skin were compared. mRNA levels of four genes specific for two cell types (fibroblast and keratinocyte) suggest that there is a recognizable pattern of gene expression for these two locations in the wound. The researchers further suggest use of a similar technique to guide debridement (48).

6 Conclusion

The healing wound is a dynamic and complex environment. Proteomic techniques provide an opportunity to better understand the protein profile of acute, chronic, healing, and nonhealing wounds of all etiologies. Researchers have utilized a number of current techniques to evaluate the protein profiles of chronic wounds over time. As the field of proteomics is very rapidly growing and changing, so too are the techniques available. The challenge in the field is experimental design, appropriate proteomic technique, and better understanding of the acute wounds used for comparison.

Much of the research to date has focused on identification of differing levels of MMP-1, MMP-2, MMP-8, MMP-9, TIMP-1, TIMP-2, TNF-alpha, interleukins, growth factors, and a few other proteins of interest. As a result of these studies it is understood that MMP-2 and MMP-9 are elevated in chronic wounds. The current wound proteome knowledge base has not been enough to develop strategies and therapies for treating chronic and nonhealing wounds to end the suffering and complications that occur as a result of these wounds. Future work must look towards identifying and tracking additional proteins present and their levels in various wounds over time. Proteomic techniques that can be used to search for proteins of potential interest, which may serve as biomarkers of healing in chronic wounds, must be employed. Obviously more must be learned about the chronic wound and acute wound proteome. Studies comparing surgical wounds elucidate just how complex a healing wound is, and the diagnoses and co-morbidities in all subjects impact the unique protein profile of each wound, making trends and protein identification even more challenging.

Additionally, the range of sample models, and the inability to compare between studies limits our ability to synthesize the findings across studies. The techniques described in this chapter are a glimpse into the rapidly changing proteomic landscape. In order to identify changes present in healing versus nonhealing and chronic wounds or to evaluate the impact of a treatment on the clinical outcome of a wound, these techniques and new developing and emerging techniques will be necessary.

References

1. Tyers, M., Mann, M.: From genomics to proteomics. Nature 422, 193–197 (2003)
2. ENCODE. Identification and analysis of functional elements in 1% of the human genome by the ENCODE pilot project. Nature 447(7146), 799–816 (2007)

3. Northrup, R.B., Connor, A.N.: Introduction. In: Neuman, M.R. (ed.) Introduction to Molecular Biology, Genomics and Proteomics for Biomedical Engineers, pp. 1–11. CRC Press Taylor and Francis Group, Boca Raton (2009)
4. Wagner, P.D., Srivastava, S.: The Promise of Proteomics: Biology, Applications, and Challenges. In: Srivastava, S. (ed.) Informatics in Proteomics, pp. 1–15. CRC Press Taylor and Francis Group, Boca Raton (2005)
5. Nanney, L.B., Caldwell, R.L., et al.: Novel approaches for understanding the mechanisms of wound repair. Journal of Investigative Dermatology Symposium Proceedings 11, 132–139 (2006)
6. Wysocki, A.B.: Wound fluids and the pathogenesis of chronic wounds. J. WOCN 23, 283–290 (1996)
7. Moseley, R., Hilton, J.R., et al.: Comparison of oxidative stress biomarker profiles between acute and chronic wound environments. Wound Repair Regen. 12, 419–429 (2004)
8. James, T.J., Hughes, M.A., et al.: Simple biochemical markers to assess chronic wounds. Wound Repair Regen. 8, 264–269 (2008)
9. Trengove, N.J., Langton, S.R., et al.: Biochemical analysis of wound fluid from nonhealing healing chronic leg ulcers. Wound Repair Regen. 4(2), 234–239 (1996)
10. Northrup, R.B., Connor, A.N.: Some Instrumental Methods Used in Genomics, Proteomics, and Forensic Science. In: Neuman, M.R. (ed.) Introduction to Molecular Biology, Genomics and Proteomics for Biomedical Engineers, pp. 283–327. CRC Press Taylor and Francis Group, Boca Raton (2009)
11. Jaynes, C.D., Fries, K., Brogan, M., Karch, J.E., Baird, K., Edsberg, L.E.: Development of a protocol for biochemical analysis of wound fluids. Acute Care Perspectives 12(1), 11–15 (2003)
12. Grinnell, F., Ho, C.-H., Wysocki, A.: Degradation of fibronectin and vitronectin in chronic wound fluid: Analysis by cell blotting, immunoblotting, and cell adhesion assays. J. Invest. Dermatol. 98, 410–416 (1992)
13. Moses, M.A., Marikovsky, J.W., et al.: Temporal study of the activity of matrix metalloproteinases and their endogenous inhibitors during healing. Journal of Cellular Biology 60, 379–386 (1996)
14. Schmidtchen, A.: Chronic ulcers: A method for sampling and analysis of wound fluid. Acta. Derm. Venereol. 79, 291–295 (1999)
15. Rasik, A.M., Shukla, A.: Antioxidant status in delayed healing type of wounds. Int. J. Exp. Path. 81, 257–263 (2000)
16. Chen, G., Beer, D.G.: Protein Expression Analysis. In: Srivastava, S. (ed.) Informatics in Proteomics, pp. 227–254. CRC Press Taylor and Francis Group, Boca Raton (2005)
17. Fernandez, M.L., Broadbent, G.K., et al.: Development of an enhanced proteomic method to detect prognostic and diagnostic markers of healing in chronic wound fluid. British Journal of Dermatology 158, 281–290 (2008)
18. Tarran, S.L.S., Craft, G.E., et al.: The use of proteomics to study wound healing: a preliminary study for forensic estimation of wound age. Medicine, Science, and the Law 47(2), 134–140 (2007)
19. Oh, J.E., Krapfenbauer, K., Lubec, G.: Proteomic identification of collagens and related proteins in human fibroblasts. Amino Acids 27, 305–311 (2004)
20. Volden, G., Thorsrud, A.K., et al.: Biochemical composition of suction blister fluid determined by high resolution multicomponent analysis (capillary gas chromatography-mass spectrometry and two-dimensional electrophoresis). The Journal of Investigative Dermatology 75, 421–424 (1980)

21. Macdonald, N., Cumberbatch, M., et al.: Proteomic analysis of suction blister fluid isolated from human skin. Clinical and Experimental Dermatology 31, 445–448 (2006)
22. Chojnacki, C., Steinsstraber, L., et al.: Proteome analysis of chronic wound fluids via MudPIT. In: 11th Annual Meeting on Surgical Research, Saar-brucken, Germany, Langenbeck's Archives of Surgery, November 2007, vol. 392, pp. 769–838 (2007)
23. Pollins, A.C., Friedman, D.B., et al.: Proteomic investigation of human burn wounds by 2D-difference gel electrophoresis and mass spectrometry. Journal of Surgical Research 142, 143–152 (2007)
24. Aggarwal, K., Choe, L.H., et al.: Shotgun proteomics using the iTRAQ isobaric tags. Briefings in Functional Genomics and Proteomics 5(2), 112–120 (2006)
25. Trengove, N., Bielefeldt-Ohmann, H., et al.: Cytokine profiles of wound fluid from chronic leg ulcers. Wound Repair and Regeneration 2(3), 228 (1994)
26. Trengove, N.J., Bielefeldt-Ohmann, H., Stacey, M.C.: Mitogenic activity and cytokine levels in non-healing and healing chronic leg ulcers. Wound Rep. Reg. 8, 13–25 (2000)
27. Barone, E.J., Yager, D.R., et al.: Interleukin-1alpha and collagenase activity are elevated in chronic wounds. Plastic and Reconstructive Surgery 102, 1023–1027 (1998)
28. Harris, I.R., Yee, K.C., et al.: Cytokine and protease levels in healing and non-healing venous leg ulcers. Experimental Dermatology 4, 342–349 (1995)
29. Aiba-Kojima, E., Tsuno, N.H., et al.: Characterization of wound drainage fluids as a source of soluble factors associated with wound healing: comparison with platelet-rich plasma and potential use in cell culture. Wound Rep. Regen. 15, 511–520 (2007)
30. Nwomeh, B.C., Liang, H.-X., et al.: MMP-8 is the predominant collagenase in healing wounds and nonhealing ulcers. Journal of Surgical Research 81, 189–195 (1999)
31. Rayment, E.A., Upton, Z., Shooter, G.K.: Increased matrix metalloproteinase-9 (MMP-9) activity observed in chronic wound fluid is related to the clinical severity of the ulcer. British Journal of Dermatology 158, 951–961 (2008)
32. Yager, D.R., Zhang, L.-Y., et al.: Wound fluids from human pressure ulcers contain elevated matrix metalloproteinase levels and activity compared to surgical wound fluids. J. Invest. Dermatol. 107, 743–748 (1996)
33. Ladwig, G.P., Robson, M.C., et al.: Ratios of activated matrix metalloproteinase-9 to tissue inhibitor of matrix metalloproteinase-1 in wound fluids are inversely correlated with healing pressure ulcers. Wound Rep. Reg. 10, 26–37 (2002)
34. Trengove, N.J., Stacey, M.C., et al.: Analysis of acute and chronic wound environments: the role of proteases and their inhibitors. Wound Rep. Reg. 7, 442–452 (1999)
35. Bullen, E.C., Long, M.T.: Tissue inhibitor of metalloproteinases-1 is decreased and activated gelatinases are increased in chronic wounds. J. Invest. Dermatol. 104, 236–240 (1995)
36. Nwomeh, B.C., Liang, H.-X., et al.: Dynamics of the matrix metalloproteinases MMP-1 and MMP-8 in acute open human dermal wounds. Wound Rep. Reg. 6, 127–134 (1998)
37. Cook, H., Stephns, P., et al.: Defective extracellular matrix reorganization by chronic wound fibroblasts is associated with alterations in TIMP-1, TIMP-2, and MMP-2 activity. J. Invest. Dermatol. 115, 225–233 (2000)
38. Lobmann, R., Ambrosch, A., et al.: Expression of matrix-metalloproteinases and their inhibitors in the wounds of diabetic and non-diabetic patients. Diabetologia 45, 1011–1016 (2002)

39. Mouës, C.M., van Toorenenbergen, A.W., et al.: The role of topical negative pressure in wound repair: Expression of biochemical markers in wound fluid during wound healing. Wound Rep. Reg. 16, 488–494 (2008)
40. Kilpadi, D.V., Stechmiller, J.K., et al.: Composition of wound fluid from pressure ulcers treated with negative pressure wound therapy using V.A.C. therapy in home health or extended vare patients: A pilot study. Wounds 18(5), 119–126 (2006)
41. Baker, E.A., Leaper, D.J.: Profiles of matrix metalloproteinases and their tissue inhibitors in intraperitoneal drainage fluid: Relationship to wound healing. Wound Rep. Reg. 11, 268–274 (2003)
42. Baker, E.A., Leaper, D.J.: Proteinases, their inhibitors, and cytokine profiles in acute wound fluid. Wound Rep. Reg. 8, 392–398 (2000)
43. Galkowski, H., Wojewodzka, U., et al.: Chemokines, cytokines, and growth factors in keratinocytes and dermal endothelial cells in the margin of chronic diabetic foot ulcers. Wound Repair Regeneration 14, 558–565 (2006)
44. Stojadinovic, O., Brem, H., et al.: Molecular pathogenesis of chronic wounds. American Journal of Pathology 167, 59–69 (2005)
45. Leber, T.M., Balkwill, F.R.: Zymography: A single-step staining method for quantitation of proteolytic activity on substrate gels. Anal. Biochem. 249(1), 24–28 (1997)
46. Wysocki, A.B., Staiano-Coico, L., et al.: Wound fluid from chronic leg ulcers contains elevated levels of metalloproteinases MMP2- and MMP-9. J. Invest. Dermatol. 101, 64–68 (1993)
47. Wysocki, A.B., Kusakabe, A.O., et al.: Temporal expression of urokinase plasminogen activator, plasminogen activator inhibitor and gelatinase-B in chronic wound fluid switches from a chronic to acute wound profile with progression to healing. Wound Rep. Reg. 7, 154–165 (1999)
48. Brem, H., Stojadinovic, O., et al.: Molecular markers in patients with chronic wounds guide surgical debridement. Mol. Med. 13(1-2), 30–39 (2007)

Bioengineering Techniques in Wound Assessment

Marco Romanelli* and Valentina Dini

Department of Dermatology
Wound Healing Research Unit
University of Pisa
Pisa, Italy

* Corresponding Author
Department of Dermatology
University of Pisa
Via Roma, 67
56126 Pisa, Italy
tel +39-050-992436
fax +39-050-551124
m.romanelli@med.unipi.it

1 Introduction

Wound measurement is essential in assessing the progress of wound healing in clinical practice. Such measurement provides objective data that assists in establishing wound healing progress, alerts the clinician to evidence of deterioration, enhances communication between healthcare providers as well as healthcare providers and patients, aids in the selection of advanced treatment modalities and provides a concise account of the patient's progress.

Imaging cutaneous ulcers to detect the progression of a disease is a routine part of medical practice. Although imaging technology has continuously evolved over the years in all fields of medicine, its direct application to cutaneous disorders has increased only in recent years. In fact, only over the past decade significant research has been undertaken to further develop techniques for specifically examining the skin. Advances in both the technology of imaging and computer systems have greatly supported this process and brought it closer to the clinical area (1). Assessment of any wound should begin with the determination of the extent of the wound. Because the extent of a wound is a dynamic process, it requires repeated systematic assessment. The total wound extent is based on the wound dimensions and the tissue level involved. The clinical evaluation of the extent of the tissue involvement due to a skin lesion and, moreover, the way a lesion evolves over time are often assessed according to the common sense and memory of the clinician. Evaluations are in general performed on the basis of clinical experience and using very basic, low-tech equipments to make objective measurements. The determination of the extent of a wound may also be accomplished by non-invasive and invasive technologies. Non-invasive wound assessment includes the measurement of perimeter, maximum dimensions of length and width, surface area, volume, amount of undermining, and determination of tissue

A. Gefen (Ed.): Bioengineering Research of Chronic Wounds, SMTEB 1, pp. 363–380.
springerlink.com

viability (2). Invasive methods may be necessary to quantify the extent of a wound. The tissue involvement in a wound must be defined from its surface to its depth and may vary depending on the organs involved. The total wound extent should be determined by means of the integration of the maximum possible amount of available data.

A wound can be further described through the use of various parameters, which include the following: duration, blood flow, oxygen, infection, oedema, inflammation, repetitive trauma and/or insult, innervation, wound metabolism, nutrition, previous wound manipulation, and coexisting systemic factors. These parameters are clues to the definition of the cause, pathophysiology, and status of the wound but we consider also fundamental a complete and careful history and physical examination.

The use of skin imaging techniques to improve the management of wounds remains a novel area for most practitioners, since the traditional approach continues to be used for clinical inspection. The techniques used to obtain an effective wound assessment are currently based on the use of transparent acetate sheets, which are applied to the ulcer so as to measure its perimeter manually. The depth of the lesion is measured by positioning a q-tip inside it or by filling the lesion cavity with hypoallergenic material to produce a cast, which is then measured to obtain the volume of the lesion. The main goal of current research is to create a system that monitors the qualitative and quantitative evolution of wounds with an easy-to-use technological system, which is able to produce an objective evaluation of the wound status and which allows the evolution of the wound to be monitored by means of measurable attributes (3). Dedicated wound photography, high frequency ultrasound assessment, laser Doppler perfusion imaging, confocal microscopy, transcutaneous oxymetry, pH measurement and magnetic resonance imaging are some of the techniques that are currently available and being used to specifically examine different types of wounds.

This chapter reviews current non-invasive technology and the relevance of such methods to clinical practice in wound healing.

2 Wound Video Image Analysis

The visual nature of wounds lends itself well to surface imaging. Illustrative documentation of wounds has been an established practice in dermatology over the years. Innovations in photographic techniques have greatly improved the capacity to create and record images of wounds, which can be used for a wide variety of goals – in some cases increasing the amount of information that is derived from naked-eye inspection.

Traditional film photography can give an accurate and objective representation of the condition of a wound and is a relatively inexpensive and an easy method for visually documenting a wound's condition. Serial photography can be particularly useful in monitoring patients affected by acute and chronic wounds with images taken at different times allowing comparison and helping detect early changes, thereby reducing unnecessary therapies (Fig. 1). These images provide a visual record of changes in the appearance of the surface of the wound during the healing

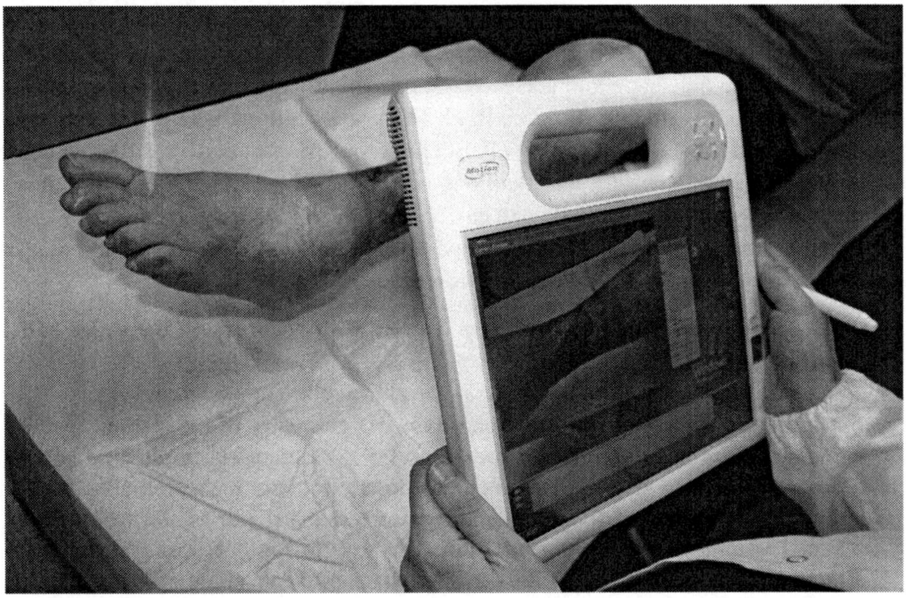

Fig. 1. A portable digitizing tablet for wound assessment and picture analysis

process, but gives no indication as to the dimensional changes occurring deep within the wound (4).

Digital photography uses a film-free digital camera that captures an image and then stores it in an electronic memory (5-6). The ability to store images in this format allows them to be easily displayed on a computer monitor or transferred to other storage media. The versatility of digital imaging has made it ideal for recording and storing images of the skin (7-8).

Clinical applications are numerous and changes in skin lesions can readily be documented and monitored through serial imaging. Digital photography is also useful in the relatively new area of telemedicine (9). High-quality images of the skin can be sent via internet to a distant centre of excellence for advice on management, without the need for the centre experts to physically examine the patients.

The evolution of lesions in a more objective manner include systems based on the acquisition of 2D images or video streams. In these studies, a characterization of the tissue status is reconstructed from the segmentation of the RGB image, and some 2D measurements are inferred from this segmentation (e.g. the wound area computed in the 2D image projection space). Calibration specimens are in general placed in proximity of the lesion, to allow colour calibration and reconstruction of linear measurements.

Laser scanning systems were introduced some years ago and have been adopted to produce very accurate 3D digital models in many different applications: film and advertising industries, industrial quality control, rapid prototyping, cultural heritage, etc. The accuracy of the scanning systems has improved in the last few years and prices have also decreased, making these devices affordable for a wider community of potential users. The integration into a single system of capabilities that can capture the shape and surface reflection characteristics (i.e. colour) makes 3D scanning an invaluable resource in all those applications where it is necessary to sample both surface attributes (10). The acquisition speed of some of these scanning devices (generally less than one second to take a range maps) is suitable for scanning human beings, who can easily stay still for such a short time.

From a purely technical point of view, an integrated system for wound assessment has to support the following functionalities:

- 3D geometrical acquisition of skin lesions with accuracy in the order of 0.2 millimetres;
- acquisition of the colour attribute of the skin section considered;
- mapping of the colour information on the 3D geometry of the lesion;
- characterization of significant measures for the objective monitoring of various types of skin lesions. This includes both automatic measurements, such as the surface extent of the lesion, its perimeter, the depth of the lesion in selected locations, the volume delimited by the lesion surface, and the hypothetical skin surface, as well as any point-to-point distance requested by the medical user;
- segmentation of the colour attribute and integration of the segmented interpretation with the numeric measures taken by the 3D lesion geometry. Colour characterization is important since areas with different colours correspond to different states of the tissue;
- storage of all the acquired/computed data in a database, organized on a *per patient* and *per examination* base.

Obviously all of the above functionalities have to be provided to the prospective clinical user with an easy-to-use guide, which should allow an intuitive use of the system as much as possible. The final system chosen must be extremely reliable, since diagnosis and medical treatment will depend on the system characterization of the lesion status.

3 Shape Characterization

The main clinical parameters involve the measurement of lengths, surface extension and volume. Point-to-point distances can be obtained very easily. Points pairs are selected on the 3D model through a point-and-click technique: the segment connecting the points is displayed immediately, together with the associated length. The perimeter of the wound is firstly specified by the user with an interactive approach and then refined by the system. Selection of the border of the lesion is performed in a semi-automatic manner: first, the user draws a closed poly-line covering the border of the lesion; then the system improves the fit of this border with the 3D data by considering the shape and the colour gradient of the selected areas. The initial poly-line is refined, with the creation of new points whose coordinates are located in 3D space according to surface curvature and colour gradient estimation. The method used is based on the snake approach. The user has complete control over the shape defined by this refinement phase and can modify the location of any point which delimits the lesion boundary. Once the perimeter of the lesion has been selected, the assessment of its length is straightforward.

Once the wound perimeter is known, the corresponding wound bed is isolated and measured to obtain the wound surface. Perimeter, surface and volume measurements are obviously performed on the 3D mesh, and therefore take into account any possible roughness.

An innovative device combines a digital camera and structured lighting in the form of 2 laser beams to automatically correct for image scale and skin curvature, allowing rapid and accurate measurements of the wound surface area and depth. Non-invasive wound assessment includes the measurement of perimeter, maximum dimensions of length and width, surface area, volume and determination of tissue viability. Recently we have published a paper (11) in order to validate a non-contact portable computer system used to collect and document image-based clinical wound assessment data: The Aranz Medical Silhouette® (New Zealand) (Fig.2).

The ARANZ Medical Silhouette® Mobile is a secure electronic, hand held, noncontact wound measurement and documentation device that is used to accurately measure both wound surface and depth. The device consists of a scanner head that attaches to a standard Personal Digital Assistant (PDA). The scanner comprises a digital camera used to capture an image of the wound and structured lighting that includes 2 fan laser beams. The device's software runs on the PDA allowing the user to operate the scanner, and performs all the necessary computation at patient's side. The process begins with the user identification on the PDA software via login procedure, selecting the appropriate patient, and recording the location of the wound on the body. A photographic image on the wound is taken to capture the wound surface area. The operator traces the wound margin on the displayed on the PDA's touch screen using the PDA stylus. The structured lighting enables a 3-dimensional model of the patient to be created, thereby compensating for the skin curvature; the distance and orientation of the camera from the wound and scale and perspective distortions in the image. An unwrapped or morphed photographic image is then created, akin to unwrapping the anatomy onto a flat surface, thus enabling an accurate and repeatable measurement of surface area to be made. Wound depth is captured by a second image where the laser beam is repositioned along the wound base with the profile of depth through the wound being recorded by the system. The system then computes wound volume from the measurements of wound surface area and depth. An historic data from earlier assessments are stored on the PDA, the results from the current assessment can be compared to those of the previous assessments in the form of graphs showing progress over time. Once data are collected on the PDA, it can be transferred to a database on the clinic's computer network.

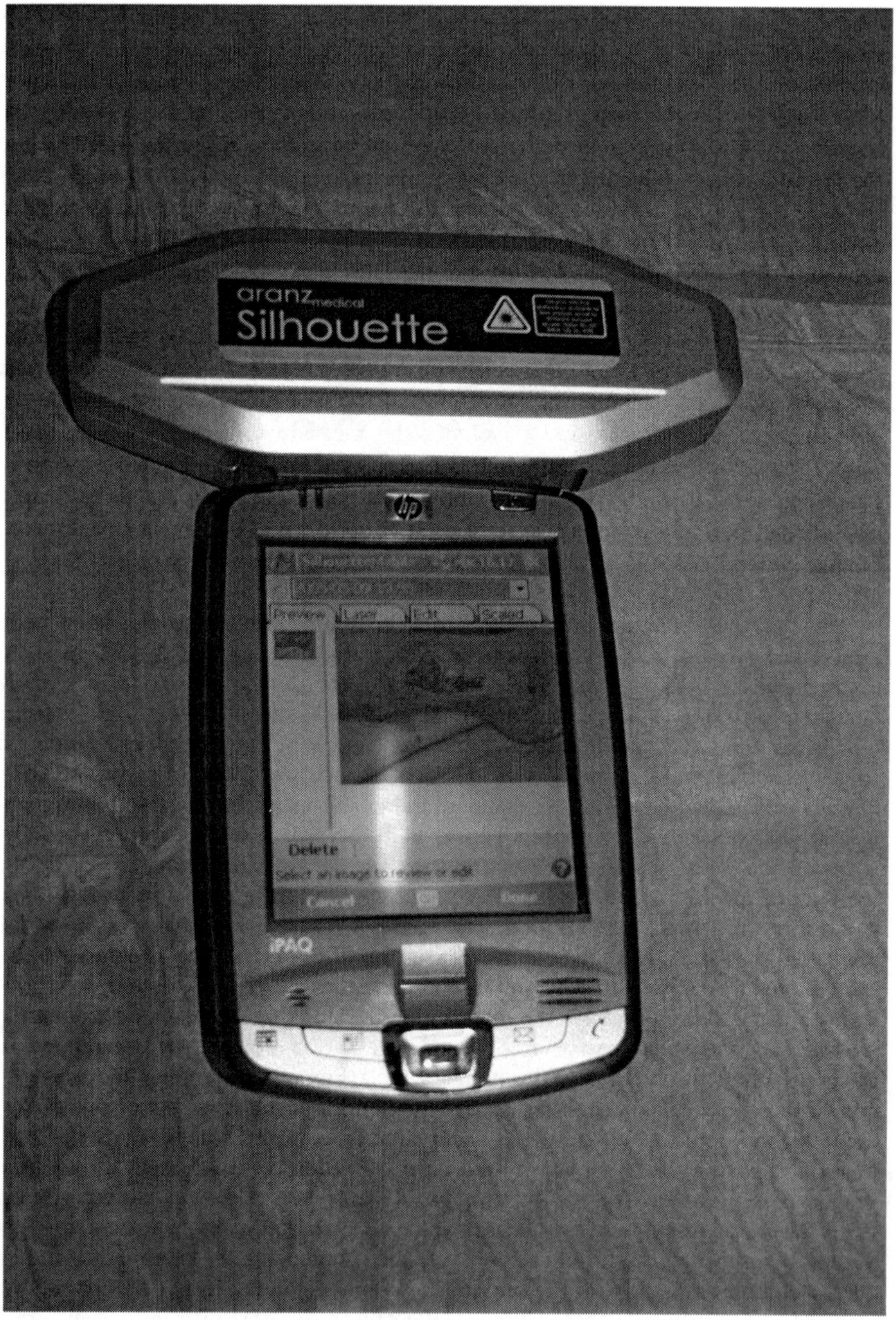

Fig. 2. Laser scanner with a personal digital assistant

4 Chromatic Assessment

Characterization of the chromatic data is another analysis of particular importance in wound assessment. The colour of different tissues can be linked to different conditions of the lesion. A segmentation step is therefore needed, in order to reduce the different colour shades in the lesion region to a few, user-defined colour clusters, which are directly linked to different clinical conditions of the tissue. Standard image segmentation techniques are therefore applied to the lesion image.

In general, three main classes are used by clinicians: black for a necrotic eschar, yellow for slough and red for granulation tissue. Once we have segmented the lesion region into a few classes, the system produces numerical data computed on the corresponding 3D mesh, such as: the perimeter, the surface area and the percentage of each class with respect to the wound size. As usual, the segmented chromatic characterization and the computed data can be saved in the database and can be used in the comparison of the status of the lesion at different intervals.

Recently we monitored the efficacy of debridement through the use of tristimulus colorimetric assessment in chronic wounds (12). In this study we were comparing a hydrogel and an enzymatic preparation, which were applied once a day for two weeks. The colorimetric evaluation was more accurate than clinical scoring in assessing granulation tissue and it has been shown to possess high reproducibility, with the advantage of avoiding the bias involved in clinical scoring (Fig.3).

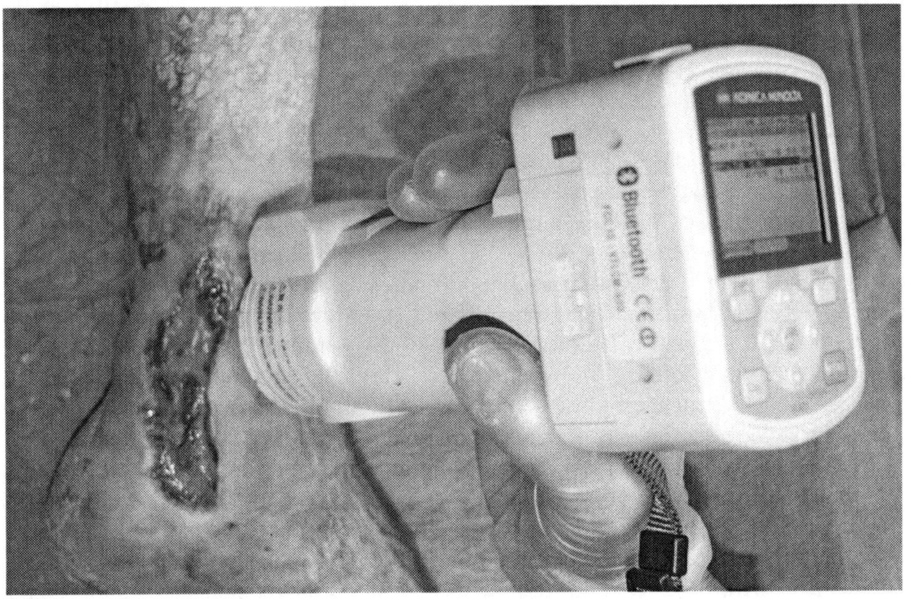

Fig. 3. Skin colour analysis with spectrophotometer

5 Ultrasonography

Although introduced in the 1950s in many medical fields, diagnostic ultrasonography (US) has only recently been used for dermatological studies. The US equipment available up to 15 years ago did not have sufficient resolution to achieve satisfactory results in skin studies because of low-frequency probes, but present-day units, with their improved technical characteristics, permit a specific application in dermatology (13). Electronic linear probes of 7.5-13 MHz, mostly equipped with color Doppler, sometimes need the interposition of a synthetic separator of 1.5 cm maximum thickness, to include the region of interest within the best focal range (0.5-3 cm). These probes may prove effectiveness in the study of flat and regular surfaces and provide a wider field of view than sectorial probes (14). Water bath sectorial mechanical probes with 10-20 MHz frequency have very superficial focalization and are excellent in the study of irregular surfaces, particularly the wound volume, thanks to their small size and small support. The main advantages of this method of imaging include its non-invasiveness, safety, high patient tolerability and relatively low cost (Fig. 4).

The technique of ultrasonography involves the detection of reflected sound waves through tissues that have inherently different acoustic properties. Images of the skin can be displayed in one-dimensional A mode (adequate for skin thickness measurements), two-dimensional B mode (which produces a vertical cross-section of the tissue being scanned), or C mode (which produce images horizontal to the skin surface). Computer processing can create three-dimensional images (15). Generally a higher ultrasound frequency provides a better resolution of the image. Resolution is defined as either axial resolution, which refers to the smallest thickness of a structure that can be measured and is directly dependent on the ultrasound frequency or lateral resolution, which refers to the width of the smallest structures that can be resolved and depends mainly on the geometric shape of the ultrasound beam.

Most common ultrasound units used in imaging abdominal organs use a frequency of 7.5 Mhz, which provides an axial resolution of only 3-5 mm. Units used in dermatology need to achieve much greater resolution to be able to visualize the skin and usually operate at a 20-25 MHz range, providing an axial resolution of 50-80 μm and a lateral resolution of 200-300 μm. Higher-frequency ultrasound imaging operating at frequencies between 40 and 100 MHz can provide an axial resolution of 17-30 μm and a lateral resolution of 33-94 μm (16,17,18). Although this improvement in resolution allows more detailed visualization of the upper skin layers, the higher-frequency ultrasound waves penetrate tissues poorly and therefore do not provide clear images of deeper skin layers and subcutaneous regions.

When ultrasound waves penetrate through the skin they are partially reflected at the boundaries between adjacent structures and `echoes' of various amplitudes are generated. The resulting ultrasound image that is produced consists of regions of varying echogenicity, which correlate to different histological regions of the skin. Generally, with 20 MHz ultrasound, the dermis and hypodermis can be distinguished easily, but the epidermis over most of the body, except on the palm and sole, cannot be visualized as a separate structure because it is too thin to be

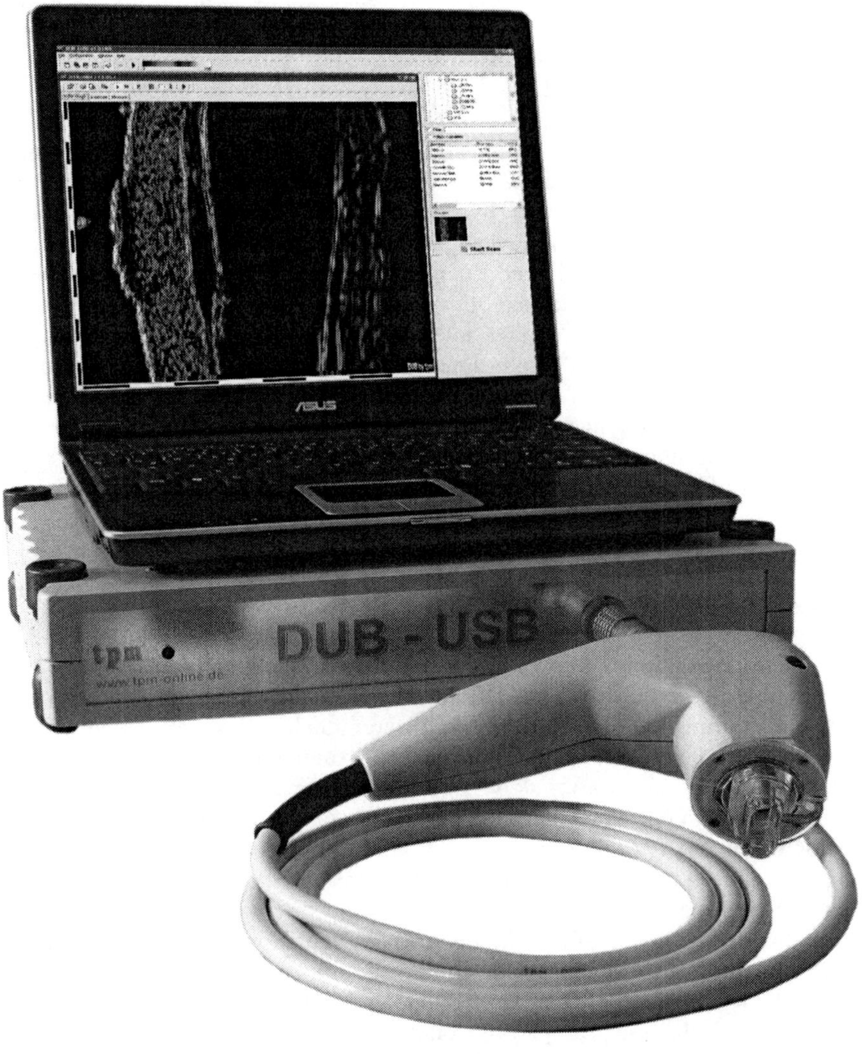

Fig. 4. High frequency ultrasound device for skin and wound assessment

resolved. Purpose-built 100 MHz high-frequency ultrasound units allow more detailed visualization of the upper skin layers, and in palmar skin can detect distinct structures such as the stratum corneum, hair follicles and eccrine sweat-gland ducts. Ultrasonic frequencies in excess of 500 MHz have been used on skin biopsy specimens to produce images of great detail; however, such high frequencies require high intensities that can damage living tissue and therefore are not useful in vivo. The main clinical value of ultrasonography has been in the objective monitoring of disease progression, with serial measurements allowing the effectiveness of treatment in patients to be assessed more accurately.

One of the main uses of cutaneous ultrasonography is in the measurement of skin thickness, which has been shown to be reliable and accurate. The actual method of evaluating skin thickness is very easy, painless and safe, and can be applied to any part of the body.

This function has been used to assess skin-thickness changes in a variety of skin diseases such as scleroderma and morphoea, photodamaged and chronologically aged skin, and in psoriasis. High frequency ultrasound is used to analyze the ultrastructure in chronic wounds, hypertrophic scars, keloids and normal surrounding skin (19). The parameters investigated are the depth between skin surface and the inner limit of the dermis and the tissue density. The depth measurement, expressed in mm, gives an estimate of wound volume and scar thickness. Compared to photography, US can be applied to easily monitor the progression of wound repair giving indications on dimensional changes related to wound volume (20). Ultrasonography is characterized by the high echogenicity of the dermis, which is sharp compared to a hypoechogenicity of subcutaneous fat. This technique allows an accurate determination of granulation, sloughy/necrotic tissue and the physical dimensions of ulcers, while also providing an index to the structural components of the ulcer (21).

6 Laser Doppler Systems

Skin microcirculation is known to consist of two functionally different networks: (a) the superficial, nutritive; and (b) the deeper, mainly thermoregulatory vascular bed. For examination of each network, different techniques have been established. The nutritional network of the skin can be examined by capillary microscopy, which has a short penetration depth of a few microns; the number of visible capillaries in the skin correlates with the degree of compensation of peripheral arterial occlusive disease (PAOD). In chronic critical limb ischemia the number of capillaries is reduced and in severe cases even avascular fields are found. A laser Doppler can be used for investigation of the mainly thermoregulatory bed, which is located in deeper layers of the skin. Nutritive circulation is also partly determined by this method. If this technique is combined with capillary microscopy, the thermoregulatory portion of the laser Doppler signal can be estimated. Signals measured with the laser Doppler are calculated values of the product of blood cell average speed and concentration. Directional effects are usually not considered, except in flux measurements of single vessels. In 1993 a further development of this technique was introduced: the laser Doppler imager (LDI). This device makes it possible to map the local distribution of the laser Doppler flux without direct contact, thus producing a two-dimensional, color-coded image (22).

6.1 Laser Doppler Flowmetry

Laser Doppler flowmetry is commonly used because it is a non invasive, simple, objective measurement, which evaluates cutaneous blood flow 1-2 mm under the skin surface and gives a continuous or near-continuous record. Monochromatic, coherent laser light is conducted by glass fibers to a probe, attached to the skin by means of adhesive discs. The movement of blood cells leads to a scattering of the laser light, generated by a low-powered helium-neon source, inducing a Doppler shift. The back-scattered signal containing data on flux, cell concentration and cell velocity is displayed on screens and the data may be recorded by a computer (23,24). This technique is useful in the evaluation of wound healing and it is used in stage 2 and stage 3 pressure ulcers for the constant evaluation of local skin microcirculation.

6.2 Laser Doppler Perfusion Imaging

Laser Doppler perfusion imaging (LDPI) is a technique that has been found increasing utility in skin research. It measures cutaneous perfusion by scanning a low-power laser beam over a region of skin (25,26). When monochromatic light interacts with a moving object such as a blood cell, a slight change in the frequency of the scattered light is induced according to the Doppler effect, whereas the light backscattered from non-moving tissue structures remains at the same frequency. The frequency shift is dependent on the average speed of the blood cells. At each measurement site, the backscattered laser light is detected by a photo-detector and the resulting signal is used to calculate the degree of tissue perfusion, which is expressed in arbitrary units. The scanning laser beam can penetrate to an average depth of 0.2 mm below the surface of the skin. An important feature of this technique is the fact that the instrument is positioned at 50 cm distance from the area under investigation, allowing an easy and reproducible assessment inside the wound bed (27).

The images obtained by LDPI are displayed on a computer monitor using various colours to depict the variations in perfusion that occur in different regions of the skin (28). The measurements obtained are objective and reproducible. The instrument has been used to evaluate the effects of postural changes on blood flow (29), as well as postural vasoregulation and mediators of reperfusion injury in venous ulceration (30). It has also been used to monitor changes in experimental skin wounds and island flaps (31) and to assess burn wound depth (32).

7 Transcutaneous Oxymetry

Transcutaneous oxymetry is a valid technique widely used in the evaluation of local skin microcirculation, nutrition and tissue ischemia (33). Many methods for non-invasive measurements of tissue O_2 are available. One of these is the $tcPO_2$ technique, based on the electrochemical reduction of oxygen, which is measured on the skin surface with a calibrated Clark electrode (34). The $tcPO_2$ measurement provides information about the tissue oxygenation in superficial skin layers. The post-heating reactive hyperemia responses of $tcPO_2$ can be used as a relative index of the vasodilatatory capacity of skin micro-vessels. The $tcPO_2$ values depend on skin microcirculation, arterial PO_2, oxygen consumption in skin tissue and oxygen diffusion through the skin itself. The values evaluated represent the partial pressure of oxygen diffusing from the capillaries and give data on the oxygenation of superficial skin layers (35). The $tcPO_2$ and $tcPCO_2$ values have been used for monitoring the evolution of leg ulcers and improving their management, but we have to consider that these values can be determined by many local factors such as blood flow, thickness of epidermis, conductivity of the gases, and the production and consumption of such gas in situ. Other perfusion analyses such as arteriography, capillaroscopy, plethysmography and videomicroscopy are considered difficult to perform and too invasive (36).

8 pH Measurement

pH measurement is defined as the negative logarithm of the activity of hydrogen ions in aqueous solution, used to express acidity and alkalinity on a scale of 0 to 14. The pH value of normal skin ranges from about 4.8 to 6.0 in relation to the presence of the so-called "acid mantle", while the interstitial fluid shows neutral values (37). Several studies have shown that the acid mantle plays an important role as a regulating factor in stratum corneum homeostasis in maintaining the integrity of barrier function. The role of pH in wound healing has proven to be of fundamental importance, and prolonged chemical acidification of the wound bed has been shown to increase the healing rate in chronic venous leg ulcers. The mechanism of interaction between acidic pH and the wound healing process is related to the potential to increase tissue oxygen availability through oxygen dissociation and to reduce the histotoxicity of bacterial end products, thus stimulating the wound's healing process. pH values in chronic venous leg ulcers and in pressure ulcers were found to be alkaline or neutral if compared with the normal surrounding skin (38). The change of value is in accordance with the stage of the ulcer, moving to an acidic state during the healing process. Two significant methods are widely used for measuring cutaneous pH: the colorimetric technique and the glass electrode potentiometric measurement. The most common pH instrument is a flat glass electrode connected to a meter and applied to the skin, with one or two drops of bi-distillated water interposed between the electrode and the skin. The use of a flat electrode is important in order to provide appropriate contact with the skin surface. The electrode is applied onto the skin at intervals of 10 seconds until stabilization of the reading. Measurements are performed at a room temperature that is below 23°C and a relative humidity of less than 65%, because sweat can influence the results. Readings should be taken 12 hours after the application of detergents or creams to the skin.

Measurement of pH is a non-invasive technique, simple, easy to use and provides important information about changes in wounds.

9 Confocal Microscopy

Confocal microscopy (CM) has recently become quite commonly used in the dermatological field. The basic principle uses a light source and a lens to focus on a specific plane within the sample of tissues. The returning light from this focal point is detected by the instrument and used to create an image that is a composite of a large number of imaged points. The final image acquired is clear, because the system is trained to detect mainly the light that is directly backscattered from the focal point and to exclude any scattered and reflected light from out-of-focus planes, thus minimizing image blur. The light source used can be either intense visible light or near-infrared light, as used with the video-rate laser-scanning confocal microscope.

The main advantage of CM is that it can allow the skin to be evaluated in its native state either in vivo, or when freshly biopsied (ex vivo) without the fixing, sectioning and staining that is necessary for routine histology.

Confocal microscope imaging of normal skin in vivo gives clear images of the cellular layers of the epidermis and upper dermal region. Beyond the dermo-epidermal junction, at a depth of 100-150 μm, blood flow in the capillary loops within each dermal papilla can be evaluated at the highest resolution; erythrocytes, leucocytes and platelets can be distinguished in relation to their relative sizes and shapes. Further imaging at depths of 100-350 μm below the stratum corneum can show a network of fibres and bundles in the papillary dermis and superficial reticular network that represent the collagen network. Skin appendages such as sebaceous glands, hair shafts and sweat-gland ducts can also be seen (39).

The quality of images obtained by CM can depend on the type of confocal microscope used and the ability of the operator. Traditional confocal microscopes require the use of fluorescent dyes in order to achieve adequate tissue contrast in creating a clear image. For in vivo skin examination, some commercially available confocal microscopes can achieve image contrast entirely through the detection of reflected light from within the skin, whereas others require an intradermal injection of a fluorescent contrast agent before different skin-cell layers can be clearly visualized .

The completely non-invasive nature and high-resolution capability of CM has made it a useful instrument in skin research. Some skin conditions that have been studied in vivo with CM and have been reported in the literature include solar keratoses, psoriasis, amelanotic melanoma and allergic contact dermatitis, where dendritic cells resembling activated Langerhans cells have been directly visualized (40,41). Confocal microscopy has been shown to characterize the pattern of neovascularization and reinnervation in a model of human skin equivalent grafted in a pig (42), confirming that angiogenesis occurs first and act as an influencing and guiding factor on innervation in experimental wound healing (43).

Currently, confocal microscopes are costly instruments and, although available commercially, their use is mainly confined to research. However, as the technology of CM improves, and smaller, more affordable instruments become available, the technique could have immense potential as a diagnostic tool in wound healing, enabling clinicians to characterize wound parameters, image skin lesions and diagnose them without the need for biopsy, and to define the margins of skin lesions prior to any intended excision (44).

10 Magnetic Resonance Imaging

Magnetic resonance imaging is a valuable diagnostic technique that can be used to image a variety of body tissues (45). The basis of the MRI technique is the varying response of different molecules when they are perturbed within a magnetic field. Atoms with an odd number of protons will respond strongly to the presence of a magnetic field. A radio-frequency pulse is used to knock spinning protons out of alignment with the magnetic field. When the pulse is turned off, the proton will return to its previous state at a rate dependent on its interaction with its environment. Magnetic resonance imaging utilizes the abundance of hydrogen in the human body as its signal source (46). A series of encoded radio-frequency pulses are transmitted, and the returning signals from the excited atoms are collected. A magnitude image of these signal intensities is then displayed by the computer after data processing. Depending on the timing of the sequence used, the strength of the signal received from a given tissue will vary because of the different molecular relaxation times.

Initial research in the dermatological field of MRI involved the study of cutaneous melanocytic lesions. Subsequent trials demonstrated that clinically useful MRI scans of any skin lesion, whether or not it contains melanin, can be obtained with excellent resolution (47,48). Tumor depths measured by MRI correlate well with postoperative histologic measurements (49). Thus, MRI should be of help to dermatologic, plastic, hand, and ear, nose and throat surgeons in the preoperative determination of tumor size, underlying tissue involvement, and exact location of the tumor. In addition, postoperatively, MRI can define and possibly detect skin cancer recurrences under skin flaps and grafts. MRI procedures can plan surgical intervention or to perform closed drainage of poorly localized abscess cavities. MRI procedures may accelerate the decision for surgery or closed drainage in patients with signs of severe infected lesions and significantly improve survival. In infected wounds, MRI provided excellent anato-moradiologic correlations by precisely defining the extent of infection (50). The results of this study confirmed the correlation of MRI parameters with histologic and physiologic data on the wound-healing process, demonstrating once more the usefulness of MRI in monitoring and assessing the time-dependent changes of angiogenesis during the wound healing process. One important aspect of the wound healing phases is represented by angiogenesis, which is a process characterized by new vessel growth toward and into the tissue. Without new vessels, which ensure an adequate supply of blood, oxygen and nutrients to the wound area, wound healing cannot proceed. Although MRI imaging of the skin is mainly a research tool, we consider MRI to be a non-invasive, safe procedure which can be used in vivo to assess wound healing process. MRI provides a non-invasive quantitative assay for the wound-healing process, but its use in clinical settings requires further development.

11 TEWL

The management of moisture is an essential aspect of wound bed preparation, which focuses on optimizing conditions to promote endogenous healing. Proper moisture balance reduces the risk-of maceration and of drying out of the wound. Maceration is a largely under estimated problem and one of the causes of delayed wound healing in chronic wounds. According to the instrumental TIME principles of wound assessment the objective evaluation of the surrounding skin water content and waterloss represents also an essential aspect in order to promote and maintain the concept of wound bed preparation.

Transepidermal water loss (TEWL) measurements correlate significantly with absolute rates of water loss assessed gravimetrically indicating the intension of quantifying the amount of evaporating water at the skin surface as a maker for barrier function was reached. Different approaches exist to assess TEWL. The method most currently used is based on the estimation of the water gradient through an open chamber, providing continous measurements in ambient air, with little alterations of the microclimate overlying the skin surface. As an example of an open chamber system, the Tewameter® (Courage+Khazaka Electronic, Koln, Germany) is well-known. It is based on Fick's Law of diffusion and represents a standard instrument for the evaluation of TEWL. Criticisms on this traditional open system are related to effects of ambient and body-induced airflows near the probe, probe size, the limitation in measurement sites and application/probe angles[i]. Other important factors to consider during TEWL measurement with an open chamber method are air convection, room temperature and ambient humidity. In order to obtain full benefit from the biophysical technique applied, the importance of validation and use of controlled conditions needs to be emphasized. For obvious reasons, objective measurements of skin properties by instrumentation are preferable above subjective assessments. However, such devices allow quick, accurate and reproducible measurements. The Vapometer®, a new device used in our study , is an example of a closed unventilated chamber system. A recent paper (51) has compared two different measuring tools to assess TEWL *in vivo* which can be summarized that factors affecting TEWL values measured with the Tewameter® are also relevant for the VapoMeter®. Especially in the very low range values, the VapoMeter gave accurate results, this observation was attributed to the fact that an open chamber system records stabilized TEWL values- after at least 30s- whereas the closed chamber system computes values based on the initial progressively increasing rate within the chamber. Advantages of VapoMeter could be the rapid assessments, being of interest when repeated measurements or large numbers of test subjects are involved and it can be used in different positions. However a limitation is the principle of single point measurement, making continuous readings impossible. Also, the short measurement time makes the device more sensitive to variations due to skin movement, tremor, surface moisture, etc. Our result demonstrate the correlation between TEWL values and intensity of maceration, we can consider the trans epidermal water loss as an objective parameter monitoring perilesional skin maceration. In this study we found TEWL measurement a simple, rapid, valid and reproducible method which can be used daily in clinical practice and research.

12 Conclusions

Wound healing assessment is becoming a more and more sophisticated section in wound management due to the introduction of different types of equipment, which are able to monitor non-invasively the various phases of tissue repair. With the availability of biomedical engineering technologies, wound measurement instruments are rapidly evolving, as evidenced by the considerable amount of data produced in recent literature. The wide range of clinical and biochemical parameters to be assessed represents the main challenge for the years to come, while caregivers will hopefully be provided with user-friendly tools to be used routinely and safely. A better reproducibility on the part of the various devices and a reduction in costs it is to be expected in the near future, allowing widespread diffusion of the techniques in question among the end users. In clinical use, applications of these technologies will be differentiated from basic research and particular emphasis will be placed on therapeutic control and the prevention of recurrences.

References

1. Corcuff, P., Pierard, G.E.: Skin Imaging: State of art at the dawn of year 2000. Curr. Prob. Dermatol. 26, 1–11 (1998)
2. van Rijswijk, L.: Wound assessment and documentation. Wounds 8(2), 57–69 (1996)
3. Romanelli, M., Gaggio, G., Coluccia, M., Rizzello, F., Piaggesi, A.: Technological advances in wound bed measurements. Wounds 14(2), 58–66 (2002)
4. Perednia, D.A.: What dermatologists should know about digital imaging. J. Am. Acad. Dermatol. 25, 89–108 (1991)
5. Ratner, D., Thomas, C.O., Bickers, D.: The uses of digital photography in dermatology. J. Am. Acad. Dermatol. 41, 749–756 (1999)
6. Bittorf, A., Fartasch, M., Schuler, G., Diepgen, T.L.: Resolution requirements for digital images in dermatology. J. Am. Acad. Dermatol. 37, 195–198 (1999)
7. Kvedar, J.C., Edwards, R.A., Menn, E.R., Mofid, M., Gonzalez, E., Dover, J., Parrish, J.A.: The substitution of digital images for dermatological physical examination. Arch. Dermatol. 133, 161–170 (1997)
8. Price, M.A., Goldstein, G.D.: The use of a digital imaging system in a dermatologic surgery practice. Dermatol. Surg. 23, 31–32 (1997)
9. Krupinski, E.A., LeSueur, B., Ellsworth, L., Levine, N., Hansen, R., Silvis, N., et al.: Diagnostic accuracy and image quality using a digital camera for teledermatology. Telemed. J. 5, 257–263 (1999)
10. Romanelli, M., Dini, V., Bianchi, T., Romanelli, P.: Wound assessment by 3 dimensional laser scanner. Arch. Dermatol. 143(10), 1333–1334 (2007)
11. Romanelli, M., Dini, V., Rogers, L.C.: Clinical evaluation of a wound measurement and documentation system. Wounds 20(9), 258–264 (2008)
12. Romanelli, M.: Objective measurement of venous ulcer debridement and granulation with a skin color reflectance analyzer. Wounds 9, 122–126 (1997)
13. Gropper, C.A., Stiller, M.J., Shupack, J.L., Driller, J., Rorke, M., Lizzi, F.: Diagnostic high-resolution ultrasound in dermatology. Int. J. Dermatol. 32, 243–262 (1993)
14. Tumball, D.H., Starkoski, B.G., Harasiewicz, K.A., Semple, J.L., From, L., Gupta, A.K., Sander, N., Goster, F.S.: A 40-100 MHz B-scan ultrasound backscatter microscope for skin imaging. Ultrasound Med. Biol. 21, 79–88 (1995)
15. Stiller, M.J., Drilller, M.D., Shupack, J.L., Gropper, C.G., Rorke, M.C., Lizzi, F.L.: Three dimensional imaging for diagnostic ultrasound in dermatology. J. Am. Acad. Dermatol. 29, 171–175 (1993)
16. Fornage, B.D., McGavran, M.R., Duvic, M., Waldron, C.A.: Imaging of the skin with 20 MHz US. Radiology 189, 69–76 (1993)

17. Harland, C.C., Bamber, J.C., Gusterson, B.A., Mortimer, P.S.: High frequency, high resolution B-scan ultrasound in the assessment of skin tumours. Br. J. Dermatol. 128, 525–532 (1993)
18. Gniadecka, M., Quistorff, B.: Assessment of dermal water by highfrequency ultrasound: Comparative studies with nuclear magnetic resonance. Br. J. Dermatol. 135, 218–224 (1996)
19. Whiston, R.J., Melhuish, J., Harding, K.G.: High resolution ultrasound imaging in wound healing. Wounds 5(3), 116–121 (1993)
20. Dyson, M., Moodley, S., Verjee, L., Verling, W., Weinman, J., Wilson, P.: Wound healing assessment using 20MHz ultrasound and photography. Skin Res.Technol. 9, 116–121 (2003)
21. Rippon, M.G., et al.: Ultrasound assessment of skin and wound tissue; comparison with histology. Skin Res. Technol. 4, 147–154 (1998)
22. Wardell, K., Nilsson, G.: Laser Doppler imaging of skin. In: Serup, J., Jemec, B.E. (eds.) Handbook of Non-Invasive Methods and the Skin, pp. 421–428. CRC Press, London (1995)
23. Bircher, A., De Boer, E.M., Agner, T., Wahlberg, J.E., Serup, J.: Guidelines for measurement of cutaneous blood flow by laser Doppler flowmetry. Contact Dermatitis. 30, 65–72 (1994)
24. Wardell, K., Andersson, T., Anderson, C.: Analysis of laser Doppler perfusion images of experimental irritant skin reactions. Skin Res. Technol. 2, 149–157 (1996)
25. Moller, H., Bjorkner, B., Bruze, M., Lundqvist, K., Wollmer, P.: Laser Doppler perfusion imaging for the documentation of flare-up in contact allergy to gold. Contact Dermatitis. 41, 131–135 (1999)
26. Sommer, A., Veraart, J., Neumann, M., Kessels, A.: Evaluation of the vasoconstrictive effects of topical steroids by laser Doppler perfusion imaging. Acta. Derm. Venereol. 78, 15–18 (1998)
27. Vongsavan, N., Mattews, B.: Some aspects of the use of laser Doppler flow meters for recording tissue blood flow. Exp. Physiol. 78, 11–18 (1993)
28. Bray, R., Forrester, K., Catherine, L., Mc Arthur, R., Tulip, J., Lindsay, R.: Laser Doppler imaging of burn scars: a comparison of wavelength and scanning methods. Burns (29), 199–206 (2003)
29. Svedman, C., Cherry, G.W., Ryan, T.J.: Postural changes in the circulation of venous leg ulcer patients studied with the laser Doppler imager. J. Invest. Dermato. 98, 640a (1992)
30. He, C.F., Cherry, G.W., Arnold, F.: Postural vasoregulation and mediators of reperfusion injury in venous ulceration. J. Vasc. Surg. 25, 647–653 (1997)
31. Arnold, F., He, C.F., Jia, C.Y., Cherry, G.W.: Perfusion imaging of skin island flap blood flow by a scanning laser-Doppler technique. Br. J. Plast. Surg. 48, 280–287 (1995)
32. Niazi, Z.B.M., Essex, T.J.H., Papini, R., et al.: New laser Doppler scanner: A valuable adjunct in burn depth assessment. Burns 19, 485–489 (1993)
33. Sheffield, P.J.: Measuring tissue oxygen tension: a review. Undersea Hyper Med. 25, 179–184 (1998)
34. Nemeth, A.J., Eaglstein, W.H., Falanga, V.: Clinical parameters and transcutaneous oxygen measurements for the prognosis of venous ulcer. J. Am. Acad. Dermatol. 20, 186–194 (1989)
35. Romanelli, M., Katz, M.H., Alvarez, A.F., Eaglstein, W.H., Falanga, V.: The effect of topical nitroglycerin on transcutaneous oxygen. Br. J. Dermatol. 124, 354–357 (1991)

36. Takiwaki, H., Nakanishi, H., Shono, Y., Arase, S.: The influence of cutaneous factors on transcutaneous pO2 and pCO2 at various body sites. Br. J. Dermatol. 125, 243–247 (1991)
37. Dikstein, S., Zlogorski, A.: Skin surface hydrogen ion concentration (pH). In: Leveque, J.L. (ed.) Cutaneous investigation in health and disease: Non invasive methods and instrumentation, pp. 59–78. Marcel Dekker, New York (1988)
38. Glibbery, A.B., Mani, R.: pH in leg ulcers. Int. J. Microcirc. Clin. Exp. 2, 109 (1992)
39. Rajadhyaksha, M., Gonzalez, S., Zavislan, J.M., Anderson, R.R., Webb, R.: In vivo confocal scanning laser microscopy of human skin. Advances in instrumentation and comparison with histology. J. Invest. Dermatol. 113, 293–301 (1999)
40. Aghassi, D., Anderson, R.R., Gonzalez, S.: Confocal laser microscopic imaging of actinic keratoses in vivo: A preliminary report. J. Am. Acad. Dermatol. 43, 42–48 (2000)
41. Gonzalez, S., Rajadhyaksha, M., Rubinstein, G., Anderson, R.R.: Characterization of psoriasis in vivo by reflectance confocal microscopy. J. Med. 30, 337–356 (1999)
42. Ferretti, A., Boschi, E., stefani, A., Saturnino, S., Romanelli, M., lemmi, M., Giovannetti, A., Longoni, B., Mosca, F.: Angiogenesis and nerve regeneration in a model of human skin equivalent transplant. Life Sciences 73, 1985–1994 (2003)
43. Gu, X.H., terenghi, G., Kangesu, T., Navsaria, H.A., Springaal, D.R., Leigh, I.M., Green, C.J., polka, J.M.: Regeneration pattern of blood vessels and nerves in cultured keratinocyte grafts assessed by confocal laser scanning microscopy. Br. J. Dermatol. 132, 376–383 (1995)
44. Vardaxis, N.J., Brans, T.A., Boon, M.E., Kreis, R.W., Marres, L.M.: Confocal laser scanning microscopy of porcine skin: implications for human wound healing. J. of Anatomy 190, 601–611 (1997)
45. Conolly, S., Macovski, A., Pauly, J., Schenck, J., Kwong, K., Chesler, D., Hu, X., Chen, W., Patel, M., Ugurbil, K.: Magnetic resonance imaging. In: Bronzino, J.D. (ed.) The Biomedical Engineering Handbook, pp. 1006–1014. CRC Press, Florida (1995)
46. Bittoun, J., Saint-Jalmes, H., Querleux, B.G., Darrasse, L., Jolivet, O., Idy-Peretti, I., Wartski, M., Richard, S.B., Leveue, J.: In vivo high resolution MR imaging of the skin in a whole-body system at 1.5 T. Radiology 176, 457–460 (1990)
47. Richards, S., Querleux, B., Bittoun, J., Jolivet, O., Idy-Peretti, I., de Lacharriere, O., Leveque, J.L.: Characterisation of the skin in-vivo by high resolution magnetic resonance imaging: Water behaviour and age-related effects. J. Invest. Dermatol. 100, 705–709 (1993)
48. Idy-Peretti, I., Bittoun, J., Alliot, F., Richard, S.B., Querleux, B.G., Cluzan, R.V.: Lymphedematous skin and subcutis: In vivo high resolution magnetic resonance imaging evaluation. J. Invest. Dermatol. 110, 782–787 (1998)
49. El-Gammal, S., Hartwig, R., Aygen, S., Bauermann, T., El-Gammal, C., Altmeyer, P.: Improved resolution of magnetic resonance microscopy in examination of skin tumors. J. Invest. Dermatol. 106, 1287–1292 (1996)
50. Helbich, T.H., Roberts, T.P.L., Rollins, M.D., Shames, D.M., Turetschek, K., Hopf, H., Muhler, M., Hunt, T.K., Brasch, R.C.: Noninvasive assessment of wound healing angiogenesis with contrast-enhanced MRI. Acad. Radiol. 9(Suppl. 1), S145–S147 (2002)
51. De Paepe, K., Houben, E., Adam, R., et al.: Validation of the VapoMeter, a closed unventilated chamber system to assess transepidermal water loss vs. the open chamber Tewameter®. Skin Res. and Technol. 11, 61–69 (2005)

Optical Non-invasive Characterization of Chronic Wounds

Michael Neidrauer and Elisabeth S. Papazoglou*

School of Biomedical Engineering, Science & Health Systems
Drexel University
3141 Chestnut Street, Philadelphia, PA 19104, USA
esp25@drexel.edu

* Corresponding Author.

Abstract. The health burden of chronic wounds is increasing at an alarming rate corresponding to the increase in the elderly and diabetic population. It is estimated that approximately 1% of the total health care costs in the western world are likely to be used for the management of chronic leg ulcers. This chapter reviews the optical methodologies that have been used mostly in a pre-clinical, research setting for the characterization of chronic wounds. Chronic wounds include diabetic foot ulcers, venous leg ulcers and pressure ulcers that do not heal within 4 weeks. The technologies covered include:

a) Optical coherence tomography (OCT) for structural imaging of wound tissue.
b) Laser Doppler perfusion imaging for monitoring cutaneous blood flow.
c) Spectroscopic methods for monitoring metabolic status at superficial depths (less than 1 millimeter), such as Diffuse Reflectance Spectroscopy.
d) Spectroscopic methods for monitoring metabolic status deep into the tissue (up to several centimeters of penetration depths) such as Diffuse Photon Density Wave methodology.
e) Hyperspectral imaging for creating two-dimensional maps of spectroscopic data over the surface of the skin.
f) Orthogonal polarization spectral imaging for visualizing cutaneous microcirculation *in vivo.*
g) Thermal imaging which for measuring the temperature of tissue and correlating it to inflammatory or microcirculation state.

The principles of each methodology are summarized with brief schematics and the key studies that demonstrate the use of each technology in assessing the healing status of chronic wounds in human studies are summarized and discussed.

List of Abbreviations

CCD - Charge-coupled device; CVI - Chronic venous insufficiency; DPDW - Diffuse photon density wave; DRS - Diffuse reflectance (or remittance) spectroscopy; HSI - Hyperspectral (or multi-spectral) imaging; LDF - Laser Doppler flux; LDPI - Laser Doppler perfusion imaging; LDPM - Laser Doppler perfusion monitoring; OCT - Optical coherence tomography; OPS - Orthogonal polarization spectral imaging; PS-OCT - Polarization-sensitive optical coherence tomography; SaO_2 - Oxygen saturation; SPC – Sphingosylphosphorylcholine; TAPS – Tetraacetylphytosphingosine

A. Gefen (Ed.): Bioengineering Research of Chronic Wounds, SMTEB 1, pp. 381–404.
springerlink.com

1 Introduction

Chronic wounds such as pressure ulcers, ischemic ulcers, venous leg ulcers, and diabetic foot ulcers constitute a major challenge in clinical practice. There are over 5-7 million Americans afflicted with these types of wounds, with an annual treatment cost of more than 20 billion dollars, with a large percentage occurring in the growing elderly population (Samson et al. 2004). Diabetes affects 13 million people every year (5.2% of the U.S. population) with approximately 650,000 new cases diagnosed annually (Skyler and Oddo 2002). It is estimated that 15% of diabetic patients will develop an ulcer of the feet or ankles at some time during their disease course. In addition, the US cost of managing pressure ulcers is estimated to be about $1.4 billion annually. This cost is likely to rise due to an aging population, particularly the increase in the number of octogenarians (Singer and Clark 1999). It is estimated that approximately 1% of the total health care costs in the western world are likely to be used for management of chronic leg ulcers (Nelzen 2000).

A wide variety of advanced chronic wound treatments such as topical growth factors, bioengineered skin equivalents, negative pressure wound therapy, and hyperbaric oxygen therapy are commercially available and clinical studies of these products have shown some evidence of improved healing compared to standard of care (Sibbald et al. 2007). However, the effectiveness of each treatment is not the same in all patients. For example, there is evidence that growth factors and cell therapies improve healing in diabetic neuropathic foot ulcers compared to standard care; however, only roughly 30-50% of patients who received these treatments healed by 12-20 weeks of care (Margolis et al. 2002). Clinicians often need to try several different treatment options before finding one that is effective for a particular wound. During the prolonged healing process of a chronic wound, rapid and accurate evaluation of healing progress is critical so that unsuccessful treatments can be discontinued and alternate treatments initiated as soon as possible (Goldman and Salcido 2002; Jessup 2006). Hence, there is a need to develop non-invasive technologies that would enable improved diagnosis of chronic wounds and be able to reduce the current high cost of their treatment.

The rapid development of opto-electronic and semiconductor components has led to an increasing interest in optical technologies for non-invasive medical diagnostics. A significant disadvantage of optical methods compared to other non-invasive medical diagnostic modalities like ultrasound, x-ray, and magnetic resonance imaging is that optical methods cannot penetrate tissue beyond several centimeters. However, this issue is less of a concern when working with cutaneous chronic wounds, and in recent years several optical technologies have been applied to the characterization of chronic wounds in research and clinical environments.

In this chapter, we summarize progress made toward improved characterization of chronic wounds (diabetic foot ulcers, venous leg ulcers, pressure ulcers) using non-invasive optical methods. Technologies covered include: optical coherence tomography for structural imaging of wound tissue, laser Doppler perfusion imaging for monitoring cutaneous blood flow, spectroscopic methods for monitoring metabolic status at both superficial (less than 1 millimeter) and deep (up to several centimeters) penetration depths, hyperspectral imaging for creating spatial maps of spectroscopic data, orthogonal polarization spectral imaging for visualizing cutaneous microcirculation, and thermal imaging for visualizing cutaneous temperature distributions.

2 Optical Coherence Tomography

Optical Coherence Tomography (OCT) is a non-invasive imaging modality that uses low coherence interferometry to create high resolution cross-sectional images of structural features in skin at depths of up to 1-2 mm. In most OCT systems, light of low coherence length is delivered to the skin through one arm of a fiber optic Michelson interferometer that terminates with a lens assembly that focuses light onto the skin (see Fig. 1). As a result, OCT imaging is completely non-invasive and requires no contact to be made between the skin and the optical system, although some OCT systems do require light contact to be made between the skin and a window at the end of the probe. Light that is reflected and backscattered to the surface of the skin is combined with light from the reference arm of the interferometer and registered by a photodetector. The reference arm of the interferometer contains a mirror that is moved mechanically to control the axial scan depth. Multiple axial scans are combined to form two or three dimensional images of scattering objects within the skin. The wavelength of the light source is typically in the near infrared region (800-1800nm), with wavelengths between 1300nm and 1800nm providing the best penetration depth (Schmitt 1999). The resolution of the images is typically on the order of 10-20 microns, but recent advances have led to resolutions approaching 1 micron (Schmitt 1999; Gambichler et al. 2005).

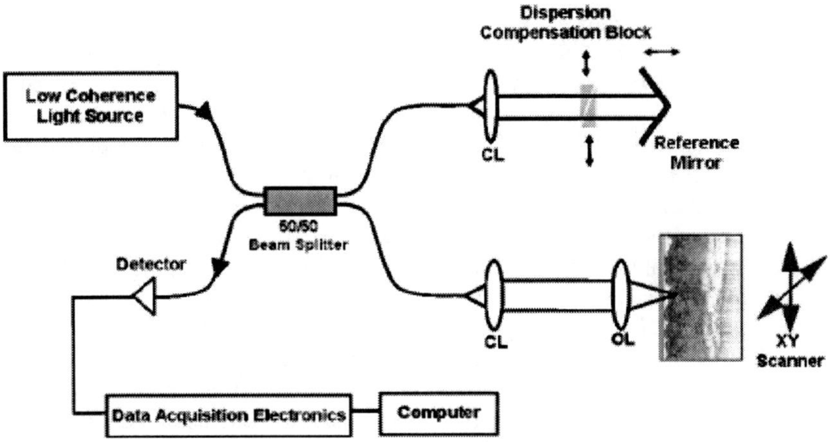

Fig. 1. Schematic of the OCT system described in Cobb et al. 2006. CL = collimating lens; OL = objective lens. Reprinted from (Cobb et al. 2006).

Imaging of human chronic wounds with OCT has not yet been reported, but several studies of experimentally-created animal wounds have been conducted (Cobb et al. 2006; Singer et al. 2007; Wang et al. 2008). Structures visible in OCT images of acute murine wounds such as the epithelial layer, granulation tissue, blisters, collagen matrix, and regions of high inflammatory response have been qualitatively validated against histological micrographs from the same wound beds. In particular, the superficial layer of epithelial cells in a healing wound exhibits high scattering and appears very bright in OCT images, allowing researchers to monitor the migration and development of epithelium over time (Cobb et al. 2006; Wang et al. 2008). Serial imaging of full-thickness wounds allowed changes in wound morphology to be followed throughout the healing process, and in one study the integration of collagen implants into healing wound tissue could be visualized using OCT (Wang et al. 2008). Algorithms for automated analysis of OCT images have been developed and they can calculate the size of a wound and identify the location of the dermal-epidermal junction, the formation of which may be used as an indicator of epithelialization in a healing wound (Cobb et al. 2006). In a study of partial-thickness wounds in porcine skin, which is more similar to human skin than rodent skin, OCT was used to visualize and measure the thickness of re-epithelialized skin (Singer et al. 2007).

Polarization-sensitive optical coherence tomography (PS-OCT) is a variation of OCT that provides not only structural images, but also information regarding polarization changes due to tissue birefringence (Schmitt 1999; Park et al. 2001; Gambichler et al. 2005; Oh et al. 2006). Normal skin exhibits birefringence due to the interlaced structure of collagen fibers and causes an alteration in the polarization state of light that interacts with it. This alteration in polarization, or phase retardation angle, can be detected and quantified using PS-OCT systems and is indicative of the quantity and structure of collagen within the imaged tissue.

Oh et al. 2006 used a PS-OCT system with a 1310 nm light source to quantify temporal changes in collagen birefringence in rabbit ear skin wounds. There were three groups of wounds: (1) wounds treated with sphingosylphosphorylcholine (SPC) to accelerate healing, (2) wounds treated with tetraacetylphytosphingosine (TAPS) to impair healing, and (3) untreated wounds that underwent normal healing. Birefringence was quantified using the change in phase retardation angle over depths ranging from 0.05 to 0.20 mm. Wounds in the SPC accelerated healing group showed 30% more birefringence than untreated wounds, while wounds in the TAPS impaired healing group showed 50% less birefringence than untreated wounds. The authors compared their OCT images with histological micrographs, and concluded that the differences in birefringence were due to differences in both the quantity and orientation of collagen fibers within each group of wounds (Oh et al. 2006). PS-OCT has also been used to quantify collagen damage *ex vivo* in burned human skin by measuring the loss of birefringence (Pierce et al. 2004). Similar measurements were conducted *in vivo* using burned animal skin, and subsequent histological analysis of the same skin revealed that the magnitude of birefringence loss measured using PC-OCT corresponded to the depth of collagen damage (Park et al. 2001; Srinivas et al. 2004).

The translation of OCT and PS-OCT to clinical assessment of human chronic wounds is still uncertain due to the lack of published human studies in this area and the size and complexity of human chronic wounds compared to experimental acute wounds in animals. However, OCT technology is rapidly improving and the arrival of real-time, non-contact imaging may make studies of human chronic wounds more feasible in the near future.

3 Laser Doppler Perfusion Monitoring and Imaging

Laser Doppler perfusion monitoring (LDPM) relies on frequency shifts of an incident light beam to determine a quantitative index that is related to microvascular blood flow or perfusion (Humeau et al. 2007). Coherent laser light at near infrared wavelengths is delivered to the skin by an optical fiber (see Fig. 2a). Photons that are scattered by moving particles undergo shifts in frequency (Doppler shifts) that are proportional to the velocity of the scattering particles. Light that is backscattered to the surface is transported by a second optical fiber to a photodetector, where a spectrum of frequency shifts is detected. It is assumed that the primary source of moving particles within tissue is red blood cell circulation. Algorithms have been developed to determine Laser Doppler Flux (LDF) indices that are proportional to the average velocity and number of red blood cells within the probed tissue volume (Rajan et al. 2009). The calculated values of LDF reflect the movement of cells in any direction, so that the signals originating from cells moving in opposing directions do not cancel each other out (Khan and Newton 2003). The volume of tissue that is measured with LDPM depends on the wavelength of light used, the optical properties of tissue, and the separation distance between the source fiber and detector fiber. Typically, LDPM probes have a source-detector separation of 0.25 mm, wavelengths ranging from 630-830nm, and measure tissue depths of roughly 0.5-1.0 mm and volumes of about 1 mm^3 (Rajan et al. 2009).

Laser Doppler perfusion imaging (LDPI) systems operate on the same principle as LDPM, except that light is delivered to and from the tissue using lenses and mirrors rather than optical fibers (see Fig. 2b). As a result, LDPI systems are non-contact and two-dimensional maps of LDF can be generated by scanning the laser over a region of tissue. Commercially available LDPI systems can create a 256x256-pixel blood flow image in 4-5 minutes with a spatial resolution of approximately 1 mm (Rajan et al. 2009).

Jünger et al. 1996 used a laser Doppler perfusion monitoring (LDPM) device with a 633nm light source to measure changes in laser Doppler flow (LDF) at the center of 12 venous leg ulcers while the legs of the patients underwent elevation and arterial occlusion. Similar tests were performed on the intact skin of the ankles of 12 healthy control subjects. Changes in LDF during leg elevation and arterial occlusion were significantly reduced in patients with leg ulcers compared to controls, indicating that LDPM could be used to distinguish healthy microcirculatory responses from diseased responses (Jünger et al. 1996).

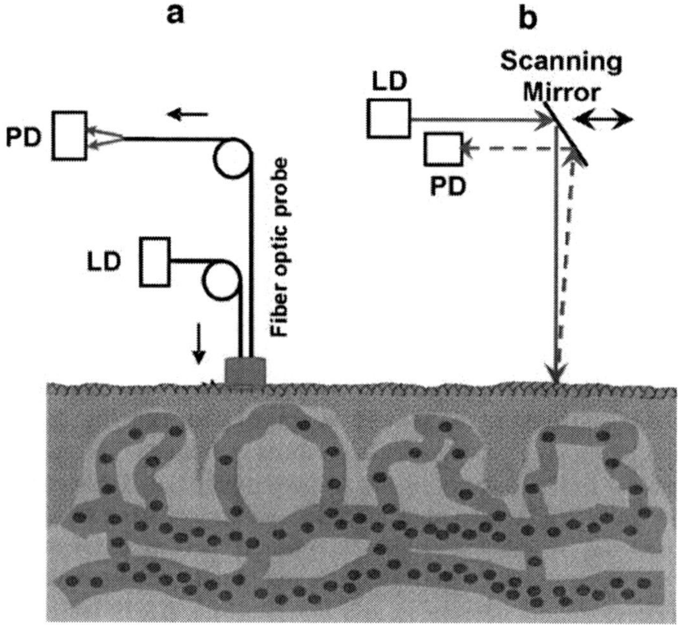

Fig. 2. Schematic of the optical systems used for (a) laser Doppler perfusion monitoring and (b) laser Doppler perfusion imaging. LD = laser diode, PD = photodetector. Reprinted from (Rajan et al. 2009).

A series of studies were published over the past decade by Gschwandtner and Ambrozy (Gschwandtner et al. 1999; Gschwandtner et al. 1999; Gschwandtner et al. 2001; Ambrozy et al. 2009) in which LDPI with a 632 nm laser was used to quantify microcirculatory disturbances in venous and arterial ulcers at depths of 1-2 mm. Groups of patients were recruited having venous leg ulcers, ischemic leg ulcers, and ulcers caused by a combination of peripheral arterial disease and chronic venous insufficiency. An LDF image was created for each ulcer and image analysis was used to calculate the mean LDF of four locations: granulation tissue within the ulcer, non-granulation tissue within the ulcer, intact skin adjacent to the ulcer, and intact skin distant from the ulcer. As expected, ischemic leg ulcers overall showed less LDF than venous leg ulcers. In venous leg ulcers (n=15), mean LDF in granulation tissue was significantly greater than mean LDF in the non-granulation tissue and the skin adjacent to the ulcer. However, in ischemic leg ulcers (n=15), LDF was greater in intact skin compared to either granulation or non-granulation tissue within the ulcers (Gschwandtner et al. 1999; Gschwandtner et al. 1999; Gschwandtner et al. 2001). In combination venous/arterial ulcers (n=17), significantly greater values of LDF were found in both the granulation tissue and in the intact skin adjacent to the ulcer than in the non-granulation tissue within the ulcer. An example of a laser Doppler perfusion image of a mixed arterial/venous ulcer is shown in Fig. 3 (Ambrozy et al. 2009). These studies indicate that insufficient blood flow to localized areas of the wound may be

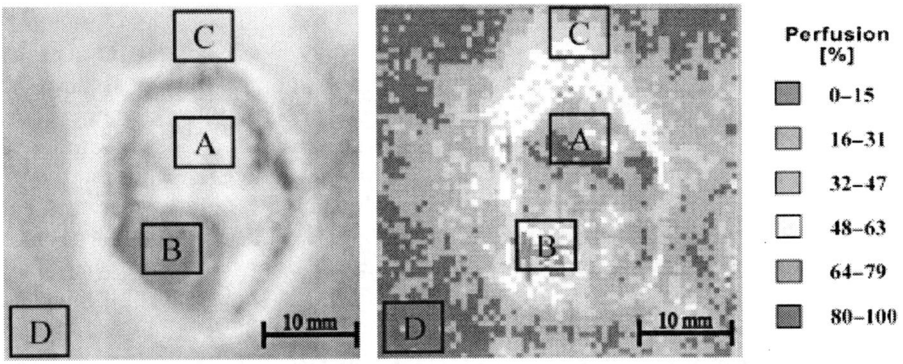

Fig. 3. Macroscopic photograph and corresponding laser Doppler perfusion image of a mixed arterial/venous ulcer. Four areas of interest are marked with capital letters: A, non-granulation tissue area; B, granulation tissue area; C, adjacent skin area; D, a more distant skin area. Reprinted from (Ambrozy et al. 2009).

preventing the formation of granulation tissue. Based on these results, one could envision the possibility of using LDPI measurements to quantify the formation of granulation tissue in venous and arterial leg ulcers, and the authors (Ambrozy et al. 2009) point out that a longitudinal study is needed to monitor changes in perfusion during the healing process.

A longitudinal study was published in 2005 by Mlacak et al. that compared changes in LDF measured before and after healing in venous leg ulcers. Nine subjects with venous leg ulcers and nine healthy control subjects participated in this study. Laser Doppler perfusion monitoring probes were fastened to the intact skin adjacent to venous leg ulcers, and values of LDF were recorded at rest, after arterial occlusion, during heating to 44°C, and during venous occlusion. The same tests were repeated after each ulcer has healed, and control subjects were also measured twice at comparable time intervals. The values of LDF measured at rest in subjects with venous leg ulcers were approximate six times greater than those measured in healthy subjects. Changes in LDF during leg elevation and arterial occlusion were significantly reduced ($p<0.008$) in patients with leg ulcers compared to controls, which is in agreement with the findings of Junger et al. 1996. However, within the group of patients with venous leg ulcers, no significant differences were observed between the tests conducted before and after healing. This suggests that blood perfusion in the skin adjacent to venous leg ulcers does not change in a manner that can be detected by LDPM, and indicates that while LDPM may be an appropriate tool for quantifying microcirculatory pathology near venous leg ulcers, it might not be a good tool for monitoring changes that occur during the healing process of the ulcers themselves.

Laser Doppler perfusion measurements have been conducted in chronic wound types other than venous and arterial leg ulcers. Timar-Banu et al. 2001 used LDPM with a 785nm diode laser to assess perfusion in chronic wounds of mixed etiology (4 diabetic foot ulcers, 10 venous leg ulcers, and 1 pressure ulcer). The authors found increased values of LDF in granulation tissue compared to intact skin, similar to the results reported by Gschwandtner and Ambrozy (Gschwandtner et al. 1999; Gschwandtner et al. 1999; Gschwandtner et al. 2001; Ambrozy et al. 2009) in venous and arterial leg ulcers (Timar-Banu et al. 2001). In a study published by Newton et al. 2001, 13 diabetic foot ulcers were scanned with an LDPI system having a 633nm light source. Intact skin on the contralateral limbs of the same subjects was used as a control. Average LDF values at the periphery of the diabetic foot ulcers were approximately 50% greater than those of control skin; however, the difference was not significant because of large inter-subject variance within each group of subjects. The high variability may have been related to heterogeneity of diabetic foot ulcers, as some were full-thickness wounds while others were partial thickness wounds, 5 of the 13 limbs were neuropathic, and there was considerable variation in the size and duration of the ulcers studied (Newton et al. 2001). It would be interesting to see if limb elevation, heating, or arterial occlusion tests similar to those used in venous leg ulcers by Jünger (Jünger et al. 1996) and Mlacak (Mlacak et al. 2005) would lead to significant differences in LDF between diabetic foot ulcers and control skin.

There are some potential limitations of using Laser Doppler to assess wound healing in a clinical environment. Manufacturers of LDPM and LDPI systems use their own proprietary algorithms for calculating relative perfusion indices. The values of LDF are not calculated in absolute units but in arbitrary units, so it is difficult to compare results obtained with different systems. In LDPI systems, specular reflection of the light beam may occur due to curvature of the feet and presence of moisture on the surface of the wound, resulting in a loss of signal (Khan and Newton 2003). Furthermore, the patient is required to remain motionless for the duration of each LDPI measurement in order to avoid motion artifacts. The image acquisition time of 4-5 minutes may make this difficult for some patients and result in non-reliable data.

4 Diffuse Reflectance Spectroscopy

Diffuse Reflectance (or Remittance) Spectroscopy (DRS) is an optical method that uses light at visible and near infrared wavelengths (400 to 1500nm) to measure hemoglobin concentration and oxygenation of blood in superficial capillaries (less than 1 mm depth). This method is sometimes referred to as tissue light-guide spectrophotometry. As shown in Fig. 4, DRS devices typically use optical fiber bundles to deliver light from a white light source to the skin. When light enters the skin, it can be reflected, scattered, or absorbed. Light that is reflected from the skin surface (specular reflection) and light that interacts with scattering particles within the skin and then returns to the surface (diffuse reflection) is transported to a spectrometer by optical fibers located near the light source fibers. The geometry of the source and detector fibers within the probe varies among different DRS systems. Some probes contain a random mix of source and detector fibers while others have a fixed geometry, often with the source fiber(s) in the center and detector fibers immediately surrounding the source. The spectrometer disperses light into a spectrum, and a photodetector array or CCD chip is used to convert the spectrum to an electronic signal (Mourant and Bigio 2003).

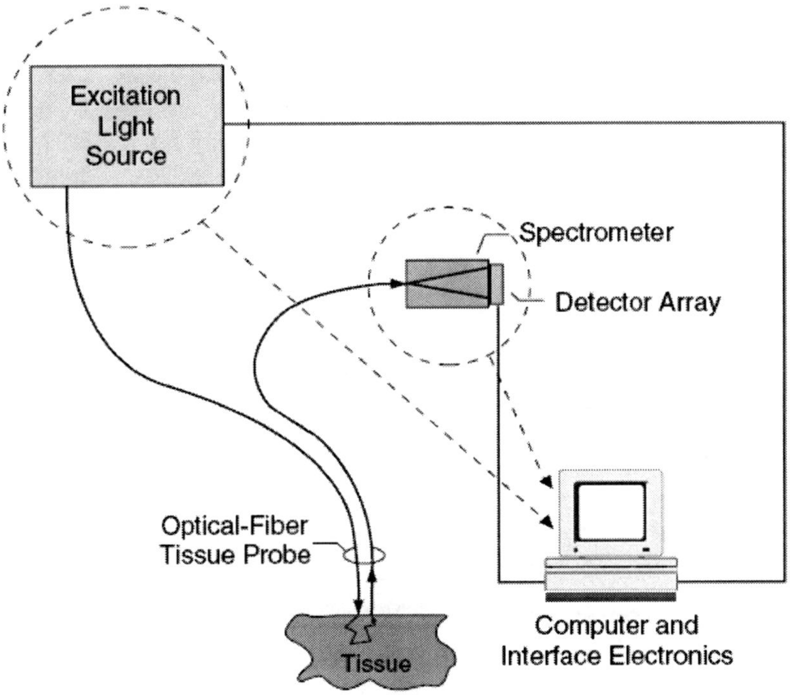

Fig. 4. Schematic of a typical diffuse reflectance spectroscopy system. The excitation light source is typically a bulb or arc-lamp emitting broadband visible and near infrared radiation. Reprinted from (Mourant and Bigio 2003).

The intensity of the reflectance spectrum is altered by absorbing molecules (chromophores) within the skin. DRS intensity spectra are often converted to absorbance spectra by measuring a white reflectance standard that is defined as having 100% reflectance at all wavelengths. The ratio of the measured intensity signal, $I(\lambda)$, to the reference intensity signal, $I_{ref}(\lambda)$, can be used to calculate an absorbance spectrum using the following equation (Stamatas et al. 2004):

$$A(\lambda) = -\log_{10}[\, I(\lambda) \,/\, I_{ref}(\lambda) \,] \qquad (1)$$

The main chromophores of interest in wound healing studies are oxyhemoglobin, deoxyhemoglobin, and water; their absorption spectra are shown in Fig. 5. The relative contribution of each chromophore to the diffuse reflectance spectrum gathered from a wound or intact skin can be found by fitting the measured DRS spectrum to a reference spectrum from each chromophore (Stamatas et al. 2004). This provides the relative concentration of each molecule within the volume of tissue probed. Alternatively, specific wavelength bands associated with a molecule of interest can be analyzed directly from DRS spectra. For example, changes in reflectance or absorbance at 550-575nm may be used to identify changes in hemoglobin concentration because these wavelengths are associated with high hemoglobin absorption (see Fig. 5) (Wollina et al. 2002).

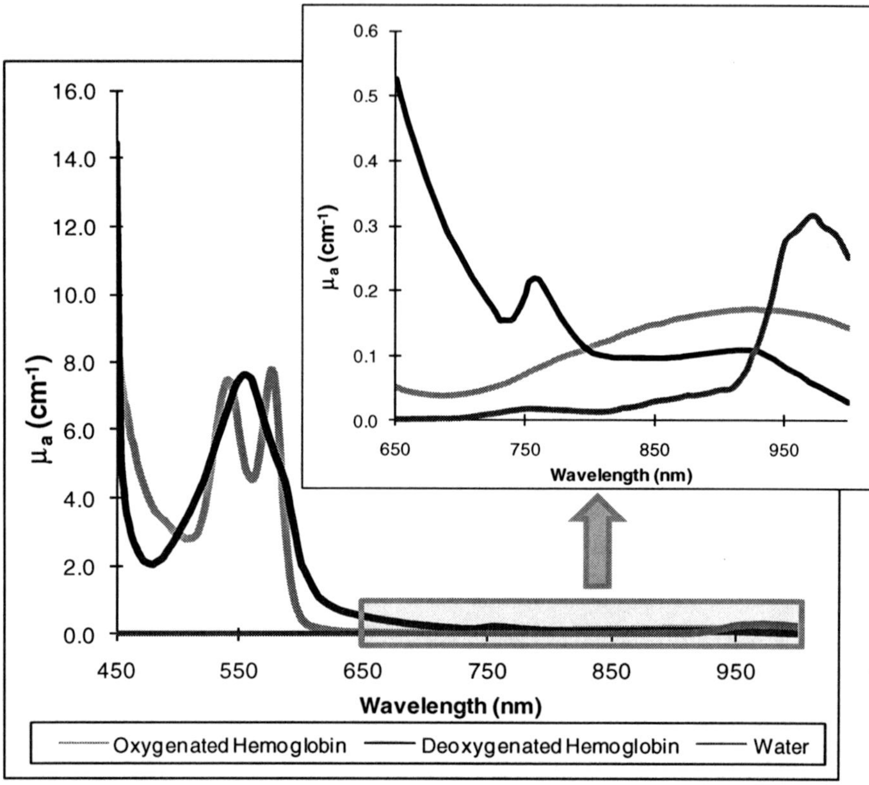

Fig. 5. Absorption of hemoglobin and water in tissue at visible and near infrared wavelengths, assuming 14mM tissue hemoglobin concentration and 70% water. Hemoglobin absorption coefficients from (Prahl 1999) and water absorption coefficients from (Querry et al. 1978).

In a study by Schmidt et al. 2001, DRS spectra from 29 chronic venous and arterial leg ulcers were obtained at wavelengths in the visible and near infrared range (400-1600nm) using a non-contact fiber optic probe. Non-contact measurements were made by holding the end of the probe at a distance of 1 mm from either the wound surface or the skin immediately adjacent to the wound, eliminating the possibility of contamination or changes in cutaneous microcirculation due to the pressure of the probe. Clinical scores based on the color, consistency, and amount of granulation tissue apparent in each wound were recorded. The authors reported characteristic differences in the spectra of high-scored wounds and low-scored wounds, with high-scored wounds generally having low remittance intensity at wavelengths corresponding to high hemoglobin absorption (450-600 nm). This indicates higher concentrations of hemoglobin in the high-scored wounds compared to the low-scored wounds, perhaps because the high wound scores have better granulation tissue which is highly perfused with blood. However, simple remittance intensity values at wavelengths 450-600 nm did not correlate well with wound score because of high variability among measurements of the same wound (relative standard deviation 5-30%). A more complex multivariate analysis was performed in an attempt to classify the DRS spectra based on clinical wound scores. Cross-validation analysis showed that the DRS spectra could predict the clinical wound scores with an accuracy of 69% (Schmidt et al. 2001).

The same research group used DRS to monitor the effect of biosurgical debridement (maggot therapy) on chronic leg or foot ulcers of mixed etiology in 30 patients. These wounds were scanned with a DRS device before and after bio-debridement, which typically lasted 3-4 days. Characteristic differences between the pre- and post-bio-debridement spectra were observed, however these differences were not quantified. Low remittance intensity was observed at wavelengths 548-575nm in post-treatment wounds compared to pre-treatment wounds. This wavelength range corresponds to maximum hemoglobin absorption, and it was concluded that increased hemoglobin concentration in treated wounds was probably a result of increased granulation tissue formation after bio-debridement (Wollina et al. 2002). Although the spectral differences from before and after bio-debridement were not quantified in this paper, one could envision the development of quantifiable index from these data that would be used to measure changes in granulation tissue caused by biosurgical debridement.

Rajbhandari, et. al. 1999 published a study in which serial DRS measurements of 21 neuropathic diabetic foot ulcers were taken over a period of nine months. The device used in this study operated in the wavelength range 502-628nm, and had a fiber-optic probe that came in contact with the wound through a sterile piece of transparent film. Whereas the previously described DRS studies used measurements of relative hemoglobin concentration to compare chronic wounds, Rajbhandari et al. report measurements of microvascular oxygen saturation. Oxygen saturation (SaO_2) is typically defined as the ratio of oxyhemoglobin to total hemoglobin, where total hemoglobin is defined as the sum of oxygenated and deoxygenated hemoglobin levels. A curve fitting procedure was used to fit each wound spectrum to reference spectra of oxygenated hemoglobin and deoxygenated hemoglobin, and the mean values of microvascular SaO_2 were calculated for each wound at intervals of 2-6 weeks. Wounds that healed within the 9-month period of the study were compared to wounds that did not heal. In healed wounds (n=13), values of SaO_2 were initially elevated and then decreased significantly (p=0.018) over the course of healing. In non-healing wounds (n=8), values of SaO_2 did not change significantly over the course of the study. Six of the eight non-healing wounds were classified as ischemic by standard clinical diagnostic methodology (ankle-brachial pressure index), and in these wounds the values of SaO_2 measured by DRS were significantly lower than in the healed wounds (44 vs. 56%) as would be expected due to poor blood perfusion. The other two non-healing wounds became infected during the study, and the values of SaO_2 became elevated in these wounds prior to clinical recognition of infection. The likely cause of increased SaO_2 in infected wounds was increased blood flow to the ulcer due to inflammatory response. Since increased levels of SaO_2 were measured using DRS prior to the infections becoming clinically apparent, this technology may have potential clinical utility by providing early warning of infection (Rajbhandari

1999). Furthermore, the decreasing levels of SaO_2 observed over time in healing wounds suggest that it may be possible to use DRS to monitor healing progress in diabetic foot ulcers, which could lead to improved evaluation of the effectiveness of treatment efficacy and the prediction of wound closure.

In addition to chronic wounds, DRS operating at wavelengths 500-1100nm has been used to monitor healing of surgical skin flaps in animal models. Blood flow to a group of flaps was temporarily occluded to create a state of ischemia, while a control group flaps remained non-ischemic. Both groups were measured with DRS and laser Doppler perfusion monitoring (LDPM) throughout the procedure. A significant difference in DRS measurements of SaO_2 was reported between ischemic and non-ischemic skin flaps, while very little difference was observed in LDPM measurements, indicating that DRS may be more reliable than LDPM in predicting arterial insufficiency (Payette et al. 2005). In another study, DRS spectra obtained from 41 human burns were used to predict whether or not healing would occur within 14 days. Rather than calculating relative hemoglobin concentrations and SaO_2, an artificial neural network system was developed to predict the time of healing from DRS reflectance spectra. The authors reported sensitivity of 75% and specificity of 97% in predicting whether or not healing would occur within 14 days (Yeong et al. 2005). Although not conducted in chronic wounds, these studies demonstrate that DRS may have clinical value in assessing wound ischemia and in predicting wound closure.

As pointed out by Schmidt et al. 2001, one limitation of using DRS to assess chronic wounds is that there is a high degree of variability among DRS spectra obtained from a single wound. Changes of the surface appearance due to bleeding, dried blood, or wetness from wound dressings will significantly affect the capability of DRS to provide information about the wound status and oxygenation. Another limitation for clinical environments is that ambient light can adversely affect results, so calibrations must be performed in each examination room prior to DRS measurement.

5 Deep Tissue Spectroscopy

In spectroscopic devices, the depth of tissue probed depends on the distance between the light source fiber(s) and the detector fiber(s) within the probe, as well as on the wavelength of incident light. As the source-detector separation distance increases, the average depth of penetration increases. The studies described in the previous section were conducted using very small source-detector separations, resulting in penetration depths on the order of several hundred microns. However, some researchers have used larger probes to penetrate tissue up to several millimeters (Beckert et al. 2004; Cross et al. 2007). The wavelength of light used for illumination can also affect significantly the penetration depth of spectroscopic measurements. As shown in Fig. 5, the absorption of light by hemoglobin at near infrared wavelengths (700-1000nm) is roughly an order of magnitude lower than its absorption at visible wavelengths (400-700nm). As a result, near infrared light is able to penetrate up to several centimeters beneath a tissue surface. As with DRS, a measured spectrum can be fitted to reference spectra for oxyhemoglobin, deoxyhemoglobin, and water in order to calculate oxygen saturation (SaO_2) and the relative concentration of each chromophore.

Beckert et al. 2004 used a commercially available spectrophotometer (Oxygen to See, LEA Medizintechnik, Germany) to measure SaO_2 in diabetic foot ulcers of patients who were clinically ischemic. The authors state that their system used wavelengths ranging from 500-800nm and was capable of measuring SaO_2 at depths of 2mm and 6mm. Technical specifications from the device manufacturer indicate that these penetration depths were achieved by using a larger probe size compared to those used by conventional DRS devices. The results from both measurement depths showed significantly increased values of SaO_2 and hemoglobin concentration in healed wounds compared to non-healed wounds, indicating that this technology may be helpful in predicting wound healing in diabetic patients with ischemic foot ulcers (Beckert et al. 2004).

Cross et al. 2007 have developed a spectroscopic device for assessing burn depth. The device detects light in the 500nm to 1100nm wavelength range and has a probe with four detection fibers at distances of 1.5, 3, 4.5 and 6 mm from the source fiber, which provide measurements of SaO_2 and relative hemoglobin concentration from four different tissue depths. Significant differences in SaO_2 between superficial and full-thickness burns were observed at the 3 deepest locations, and in relative hemoglobin concentration at all 4 detector locations (Cross et al. 2007). The ability to provide spectroscopic data at several depths may prove to be valuable in assessing chronic wound healing as well as burns.

6 Near Infrared Diffuse Photon Density Wave Methodology

Papazoglou et al. (Papazoglou et al. 2008; Papazoglou et al. 2009) used Diffuse Photon Density Wave (DPDW) methodology at Near Infrared wavelengths to monitor changes in hemoglobin concentration in diabetic foot ulcers. DPDW methodology allows the determination of tissue oxyhemoglobin and deoxyhemoglobin concentrations at depths of up to several centimeters. In contrast to the near infrared devices described in the previous section in which the intensity of light illumination was constant, a light source that was intensity-modulated at a frequency of 70 MHz was delivered to the wounds by an optical fiber, as shown in Fig. 6. Diode lasers emitting light at wavelengths of 685, 780, and 830 nm were used as the light source. The probe consisted of one source and four detector optical fibers spaced at 4, 8, 12, and 16 mm distance from the source fiber that deliver remitted light to the photodetectors. Because the light is intensity modulated, it is possible to register both the amplitude and phase shift of the detected signals. A diffusion-based model of light propagation in tissue was used to calculate the absorption and reduced scattering coefficients (μ_a and μ_s') from the detected values of amplitude and phase shift (Papazoglou et al. 2006). The absorption coefficients calculated at each wavelength were fitted to the extinction coefficients of oxy- and deoxy-hemoglobin at each wavelength to determine absolute values of the concentration of each chromophore within the volume of tissue investigated. The spacing of the detector fibers enabled a penetration depth of approximately 4-6 mm.

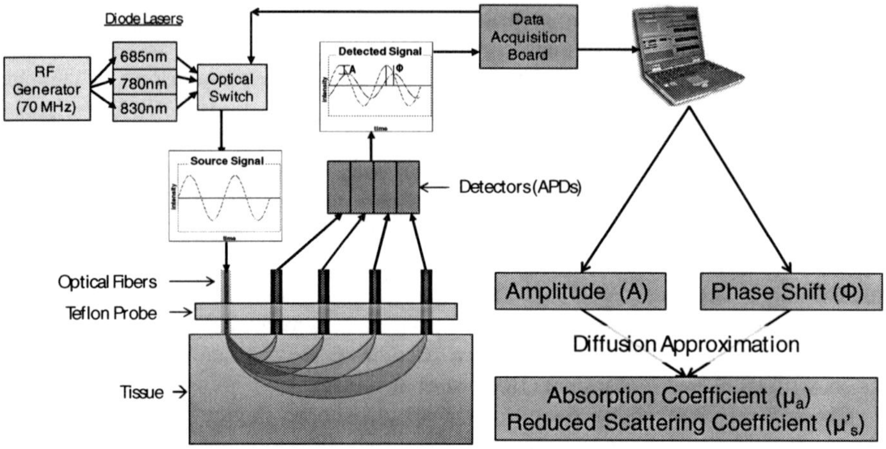

Fig. 6. Schematic of the frequency-domain near infrared device used in (Papazoglou et al. 2008; Papazoglou et al. 2009).

The main advantage of this method over the spectroscopic methods discussed in the previous sections is that absolute instead of relative values of hemoglobin concentration can be calculated at depths of several millimeters. The main disadvantage is the low spatial resolution of these measurements, since the signals from all four detectors must be combined to obtain an average value of hemoglobin concentration throughout the tissue volumes probed by each detector fiber.

In a recently completed pilot study, 12 diabetic foot ulcers were monitored using DPDW methodology on a weekly or biweekly basis for 20 weeks or until the wounds were healed (Papazoglou et al. 2008; Papazoglou et al. 2009). Five of the twelve wounds healed before the end of the study, and in all five wounds gradual decreases in oxygenated and total hemoglobin concentrations were observed between the first and last measurements. In all seven non-healing wounds, the rates of change were nearly zero or slightly positive. A statistically significant difference ($p<0.05$) between healing and non-healing wounds was obtained for the rates of change in both oxygenated and total hemoglobin concentration. These results indicate the ability of DPDW to differentiate healing from non-healing diabetic foot ulcers, and may have the potential to be used as a predictor of wound healing potential.

7 Hyperspectral Imaging

Hyperspectral (or multi-spectral) imaging (HSI) is a spectroscopic method used to obtain two-dimensional maps of cutaneous oxygenation. The principle behind HSI is similar to Diffuse Reflectance Spectroscopy, except that instead of obtaining a spectrum from a point source, many spectra are captured over a wide spatial area (Gillies et al. 2003; Vo-Dinh 2004; Khaodhiar et al. 2007; Stamatas and Kollias 2007). HSI systems typically consist of a broadband light source with visible and near infrared wavelengths that is used to illuminate the skin and a camera with a tunable optical filter to capture images from the illuminated skin. Tunable optical filters allow the rapid selection of many narrow bands of wavelengths by an electronically controlled tuner. In HSI systems the tunable filter scans a series of wavelengths while a stack of images is gathered by the camera. Software is used to ensure that the images are spatially aligned, and then a reflectance spectrum consisting of the intensity values registered at each pixel for each wavelength can be analyzed (see Fig. 7). Relative absorbance spectra can be calculated by comparing the reflectance spectra to white references. Each absorbance spectrum can be fit to reference curves for oxygenated and deoxygenated hemoglobin, and two dimensional maps of hemoglobin oxygen saturation (SaO_2) can be generated (Martinez 2002; Gillies et al. 2003). Because HSI is a reflectance-based modality, its penetration depth is limited to several hundred microns beneath the surface of the skin. The accuracy of HSI when measuring wounds directly may be compromised if the surface is covered with a reflective fluid, causing a large percentage of the incident light to undergo specular reflection. Light that is specularly reflected from the surface has not propagated through the wound tissue, and therefore does not contain any spectral information about the wound composition. For this reason, some researchers have analyzed the intact skin immediately surrounding the wound rather than the center of the wound (Khaodhiar et al. 2007).

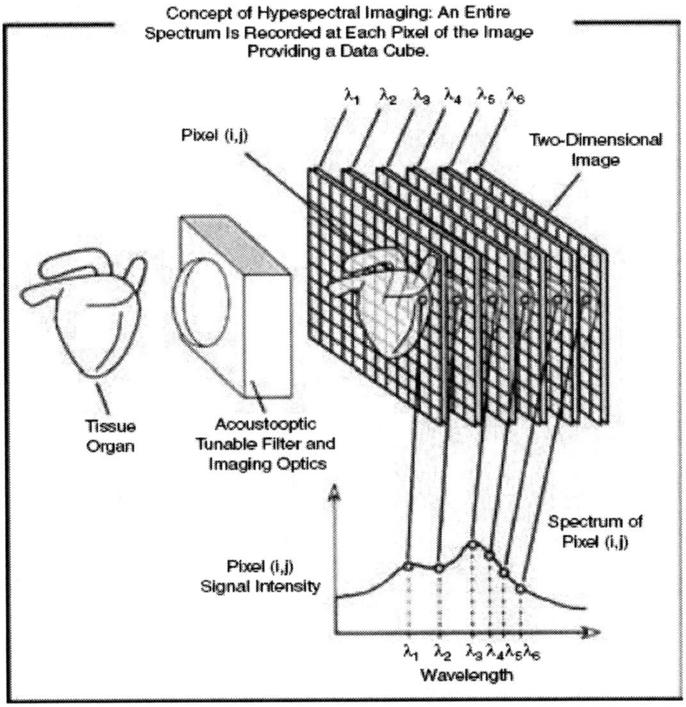

Fig. 7. Hyperspectral data cube. The HSI approach provides a "data cube" of spectral information, which consists of a series of optical images recorded at various wavelengths of interest. Reprinted from (Vo-Dinh 2004).

Hyperspectral imaging for the analysis of chronic wounds was first reported in 2002 (Martinez 2002). In this proof-of-concept study, oxygen saturation images were generated for one diabetic foot ulcer at several time points during the healing process. Changes in the spatial variability of oxygen saturation were observed over time. In a later study (Greenman et al. 2005), hyperspectral imaging was used to compare oxy-hemoglobin, deoxy-hemoglobin, and oxygen saturation in diabetic patients with peripheral neuropathy (n=51), diabetic patients without peripheral neuropathy (n=36), and non-diabetic control patients (n=21). Reduced levels of both oxy-hemoglobin and SaO_2 were observed in the skin of neuropathic diabetic patients compared to non-neuropathic diabetic patients and non-diabetic patients ($p<0.03$). Deoxy-hemoglobin levels were not significantly different across the three groups. This study demonstrated that HSI has the potential to be used for the early identification of diabetic feet at high risk of ulceration (Greenman et al. 2005).

In a study published in 2007 (Khaodhiar et al. 2007), tissue oxy-hemoglobin and deoxy-hemoglobin were quantified in the skin immediately surrounding 21 foot ulcers in Type I diabetic patients using a commercially available hyperspectral imaging system. Among the subset of ulcers that had healed after 6 months (n=14), it was observed that both oxyhemoglobin and deoxyhemoglobin levels were significantly greater than those observed in non-healing ulcers (n=7). An example of a hyperspectral image of a foot ulcer is shown in Fig. 8. A "healing index" based on oxyhemoglobin and deoxyhemoglobin levels was developed to simplify the clinical interpretation of the hyperspectral data. Values of the healing index measured during each subject's first visit were used to predict whether or a not wound would heal within 6 months with a sensitivity of 93% and a specificity of 86% (Khaodhiar et al. 2007), demonstrating the potential of HSI to predict healing of diabetic foot ulcers.

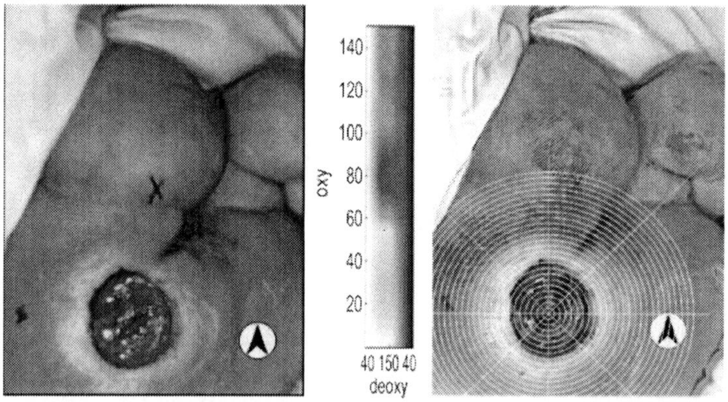

Fig. 8. An example of a diabetic foot ulcer imaged using color photography (left) and hyperspectral technology (left). The color bar indicates oxy-hemoglobin values along the vertical color scale and deoxy-hemoglobin values along the shorter horizontal brightness scale. The radial grid pattern was added to the hyperspectral image to help with quantification of oxy- and deoxy- hemoglobin levels at various locations around the wound. Reprinted from (Khaodhiar et al. 2007).

Hyperspectral imaging has been used to analyze the depth of cutaneous burns in both animal (Sowa et al. 2001) and human studies (Levasseur et al. 2005). Increased levels of oxygenation and hemoglobin were observed in superficial wounds compared to normal skin, while decreased levels were observed in full-thickness wounds compared to normal skin, and it was shown that first, second, and third degree burns could be distinguished using HSI data (Levasseur et al. 2005). HSI was also used in a proof-of-concept study to analyze oxygenation changes in surgical wounds resulting from amputation (Zuzak et al. 2007).

8 Orthogonal Polarization Spectral Imaging

Orthogonal polarization spectral (OPS) imaging is a method used to directly visualize microcirculation *in vivo* without the use of contrast agents or dyes. Monochromatic light (typically with wavelength = 548 nm) is linearly polarized and used to illuminate the tissue through an objective lens, as shown in Fig. 9. Light that is reflected and scattered from the tissue is collected by the same objective lens, and the image passes through a polarizer that is oriented orthogonally to the illumination polarization. Light that is specularly reflected from the tissue surface will have the same polarization as the incident light and will be rejected by the second polarizer. Only light that has undergone multiple scattering events within the tissue and has lost is original polarization state passes through the polarizer to a CCD video camera that captures video images in real-time. The illumination spot size is typically around 1 mm, and the penetration depth of *in vivo* measurements is approximately 200 microns (Groner et al. 1999). The wavelength of incident light is usually 548 nm, which is the wavelength at which both oxyhemoglobin and deoxyhemoglobin absorb equally. Therefore, red blood cells appear dark (due to high absorption) on a bright background when they are within the field of view, regardless of the oxygenation state of hemoglobin. The magnification used in OPS systems provides images with a resolution of approximately 1 micron per pixel, which allows individual erythrocytes to be visualized and tracked as they move through the microvasculature (Groner et al. 1999; Milner et al. 2005).

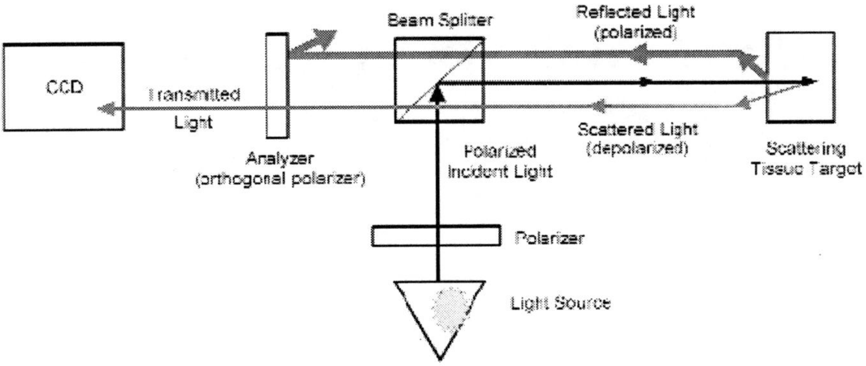

Fig. 9. Schematic of an OPS system. The light source is typically monochromatic with wavelength of 548nm. Reprinted from (Groner et al. 1999).

A study published in 2002 by Langer et al. showed that OPS imaging could be used to quantify the microvascular characteristics of experimental acute wounds in hairless mice. Wounds with a diameter of 2.5mm were created on the ears of 8 mice, and 20 seconds of OPS video were recorded at three locations on each wound. Measurements were obtained on days 1, 4, 7, 10, and 15 after the wound surgery. Video processing software was used to calculate the diameter of arterioles and venules, red blood cell velocity in arterioles and venules, and the functional capillary density within each field of view. Changes in these parameters were observed over the course of healing. The same OPS system was used to obtain serial measurements from the edge of a single human chronic wound over eight weeks (see Fig. 10), showing that the use of OPS in clinical environment may be feasible (Langer et al. 2002).

OPS was used by Virgini-Magalhães et al. 2006 to image the intact skin of 87 lower limbs of patients having chronic venous insufficiency (CVI), a condition that often leads to the formation of venous ulcers. Morphological and functional microcirculatory parameters were quantified from OPS data and compared to clinical evaluation of CVI severity. In patients with more severe CVI, decreased values of functional capillary density were observed, and increases in the capillary bulk, the dermal papilla diameter, the percentage of irregularly shaped capillaries were observed, indicating a decline in capillary function and an increase in tissue edema. A microcirculatory index was developed by the authors that took into account all of the OPS derived parameters. This microcirculatory score correlated well with the clinical system currently used to classify CVI patients (CEAP system – based on clinical observations), indicating that it may be possible to use OPS imaging to objectively quantify the progression of CVI and identify those patients most at risk of developing venous ulcers (Virgini-Magalhães et al. 2006).

Milner et al. 2005 used OPS to image burn damage *in vivo* in 12 human patients, and a quantifiable difference was seen between the dermal microvasculature of superficial vs. deep burns. Large coagulated vessels were visible in deep burns, in contrast to the functioning capillaries that were visible in superficial burns. These results further demonstrate that OPS imaging may be both feasible and valuable in a clinical environment (Milner et al. 2005).

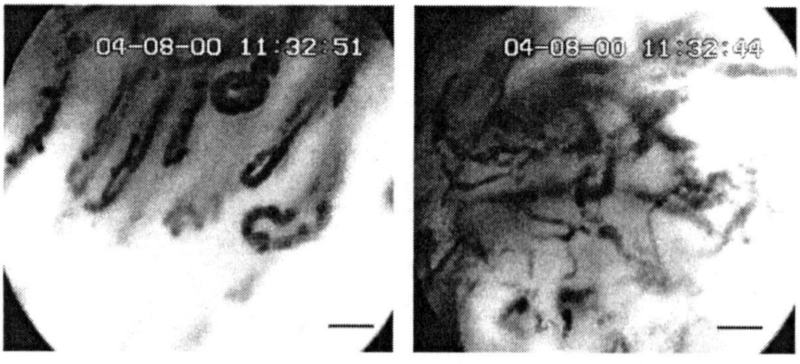

Fig. 10. OPS images from a human chronic wound using a probe equipped with a 5X lens. Scale bar = 50 μm. Reprinted from (Langer et al. 2002).

There are a few potential limitations of using OPS imaging as a diagnostic tool for chronic wounds in clinical environment. OPS image acquisition is fast (real time), but the analysis of video is time consuming because significant manual input is required to calculate circulatory parameters (Virgini-Magalhães et al. 2006). However, the automation of analysis could be accelerated with improved image analysis software (Christ et al. 2002). Another potential issue is that OPS may not be able to image patients with high melanin content. Virgini-Magalhaes et al. 2006 indicated that patients with dark skin (Fitzpatrick skin phototypes 5 and 6) were excluded from their study owing to limitation of the technique.

9 Thermal Imaging

Thermal imaging is a method of measuring the thermal energy emitted by any object with a temperature above absolute zero. In contrast to the other optical methods described in this chapter, thermal imaging requires no external source of illumination because the measured radiation is emitted naturally from the skin or wound under examination. The emissive behavior of human skin is similar to a physical black body at wavelengths ranging from 2-14 μm, and the emissive power (E_b) of a black body is strongly dependent upon temperature (T) according to the Stefan-Boltzmann law: $E_b = \sigma T^4$, where σ is the Stefan-Boltzmann constant (Jones 1998). The wavelength at which skin (at a temperature of 300 K) exhibits maximum emissive power is approximately 10 μm, and modern thermal imaging systems employ detectors that are sensitive in the 8-10μm wavelength range (Fauci et al. 2001). The design of thermal imaging systems is similar to that of CCD cameras in that an array of detectors is placed in the focal plane of a lens. However, glass absorbs strongly at wavelengths near 10μm, so the lenses used in thermal imaging systems must be made from materials such as germanium (Jones 1998). The response of thermal detector elements must be calibrated using objects of known temperature in order for an imaging system to provide absolute measurements of temperature. After calibration, thermal imaging systems can provide maps of temperature values with a spatial resolution of approximately 2-4 mm (Jones and Plassmann 2002) and temperature resolution near 0.1°C (Jones 1998). Radiation emitted from deep tissue does not reach the skin surface because of the high absorption coefficient of tissue at infrared wavelengths. Therefore, the temperature maps provided by thermal imaging systems are superficial, reflecting the temperatures at the surface of the epidermis (Jones 1998).

The use of thermal imaging in the assessment of chronic wounds has been reported for over 35 years. Verhonick et al. 1972 described the use of infrared thermal imaging to visualize temperature increases in the skin of healthy subjects after applying pressure to bony prominences where pressure ulcers are most likely to occur. Although the temperature changes were not quantified, this study showed that thermal imaging had the potential to identify skin regions most susceptible to developing pressure ulcers (Verhonick et al. 1972). Newman and Davis 1981 captured infrared thermal images of the intact sacral skin of 91 patients admitted to a geriatric assessment ward. Eleven of these images showed diffuse regions of increased temperature at the sacrum, and 5 of the 11 patients with increased skin temperatures developed a sacral pressure ulcer within 10 days. None of the 63 patients who had normal thermograms developed pressure ulcers within 10 days, showing that thermography may be capable of providing early indication of sacral pressure sore risk (Newman and Davis 1981).

Armstrong et al. 1997 conducted a retrospective study to determine whether skin temperature measured with infrared thermometry could be correlated with diabetic foot ulceration. The feet of 44 subjects with diabetic neuropathic foot ulcers and 78 subjects with diabetic neuropathy but no foot ulceration were measured using an infrared skin temperature probe. The infrared device functioned in a similar manner to infrared imaging systems; however it measured the mean temperature across a 1 cm^2 area of skin rather than creating a temperature map of a wider skin area. No temperature differences were observed between the left and right feet of non-ulcerated neuropathic subjects. On the other hand, significantly increased temperatures (5.6ºF) were observed in the skin of ulcerated feet compared to the non-ulcerated contralateral limbs of the same subjects ($p<0.0001$). Furthermore, after their ulcers had completely healed and temperature differences between feet had resolved, 5 subjects experienced reulceration. Importantly, on the clinical visit immediately prior to reulceration, skin temperature on the pre-ulcerative limbs increased by an average of 7.1ºC. This study provides evidence that skin temperature increases measured with infrared thermometry may be useful in predicting the formation of diabetic foot ulcers (Armstrong et al. 1997).

Harding et al. 1998 conducted a study to determine whether thermal imaging could be used to identify patients with soft tissue infection and osteomyelitis (bone infection) adjacent to diabetic foot ulcers. Thermal images of the feet of 49 patients with diabetic foot ulcers were captured using a commercially available thermal imaging camera (Aga Thermovision 782 I-R). The presence of osteomyelitis was determined in each patient using standard clinical radiological methods that included x-ray radiography, magnetic resonance imaging, and isotope bone scans. To analyze the thermal images, the authors classified an image as a "positive thermogram" if the temperature of a diffuse area of skin on the sole of the ulcerated foot was at least 0.5ºC greater than the temperature in the contralateral foot. 21 of the 26 subjects (81%) who has positive thermograms tested positive for osteomyelitis, and all of the 23 subjects (100%) with negative thermograms tested negative for osteomyletis (Harding et al. 1998).

Elevated local skin temperature has been reported in several recently published case studies of venous leg ulcers (Sayre et al. 2007; Bagavathiappan et al. 2008). Sayre et al. 2007 reported a case in which a subject with chronic venous disease was monitoring skin temperature with an infrared thermometer on a daily basis for 46 days. During this period, the subject developed two leg ulcers on sites that were being monitored. In the days immediately preceding ulceration, a sudden drop and then increase in temperature were observed, suggesting the possibility of an ischemic event followed by reperfusion effect that may have caused inflammation and skin damage (i.e. ischemia-reperfusion injury). Bagavathiappan et al. 2008 acquired thermal images of the limbs of two non-diabetic patients with a history of vascular disorders and ulceration using an infrared camera. In both patients, regions having temperatures elevated by 0.7 – 1.0 ºC were identified as regions of abnormal blood flow. Although the number of patients in these studies precludes any statistically conclusions, they show that infrared imaging has the potential to help in the diagnosis and prediction of the location of venous leg ulcers.

10 Conclusions

The above review of the relevant literature of using non-invasive optical methods to assess wound healing reveals that although several technologies exist that could assist in evaluating the status of a wound, the heterogeneity of the population with chronic wounds and the length of time required to properly complete human studies have limited the studies available or make definite conclusions from such studies difficult to reach. There is no doubt that metabolic information will be critical along with structural details to help clinicians evaluate new and / or existing wound treatments and customize such treatments. Portable technologies could be beneficial, allowing the patients to receive continuous monitoring at home and not being evaluated only during visits to the clinic. One can envision some of the simpler, less expensive technologies providing a first order assessment at home, and some of the more expensive and sophisticated ones being reserved for use on-site at wound centers. There is a clear need for well designed human studies that provide comparative evaluation of some of these technologies. This would encourage their adoption by wound centers which would help to accelerate the rate of data acquisition and further refine the best technologies. The cost of treatment, in the form of active dressings or hyperbaric oxygen therapy or vacuum assisted devices, is significant and application of suboptimal treatment may lead to amputations. Given this cost scenario, commercialization of diagnostic optical devices and their incorporation into the standard of care could help to improve the quality of life for patients and contain costs by giving timely critical information to the clinical decision makers.

References

Ambrozy, E., Waczulikova, I., et al.: Microcirculation in mixed arterial/venous ulcers and the surrounding skin: clinical study using a laser Doppler perfusion imager and capillary microscopy. Wound repair and regeneration 17(1), 19 (2009)

Armstrong, D.G., Lavery, L.A., et al.: Infrared dermal thermometry for the high-risk diabetic foot. Physical Therapy 77(2), 169 (1997)

Bagavathiappan, S., Saravanan, T., et al.: Investigation of peripheral vascular disorders using thermal imaging. The British Journal of Diabetes & Vascular Disease 8(2), 102 (2008)

Beckert, S., Witte, M.B., et al.: The Impact of the Micro-Lightguide O2C for the Quantification of Tissue Ischemia in Diabetic Foot Ulcers. Diabetes Care 27(12), 2863–2867 (2004)

Christ, F., Bauer, A., et al.: Different optical methods for clinical monitoring of the microcirculation. Eur. Surg. Res. 34, 145–151 (2002)

Cobb, M.J., Chen, Y., et al.: Noninvasive assessment of cutaneous wound healing using ultrahigh-resolution optical coherence tomography. Journal of Biomedical Optics 11, 064002 (2006)

Cross, K.M., Leonardi, L., et al.: Clinical utilization of near-infrared spectroscopy devices for burn depth assessment. Wound Repair & Regeneration 15(3), 332 (2007)

Fauci, M.A., Breiter, R., et al.: Medical infrared imaging–differentiating facts from fiction, and the impact of high precision quantum well infrared photodetector camera systems, and other factors, in its reemergence. Infrared Physics and Technology 42(3-5), 337–344 (2001)

Gambichler, T., Moussa, G., et al.: Applications of optical coherence tomography in dermatology. Journal of Dermatological Science 40(2), 85–94 (2005)

Gillies, R., Freeman, J.E., et al.: Systemic effects of shock and resuscitation monitored by visible hyperspectral imaging. Diabetes Technology & Therapeutics 5(5), 847–855 (2003)

Goldman, R.J., Salcido, R.: More than One Way to Measure a Wound: An Overview of Tools and Techniques. Advances in Skin & Wound Care 15(5), 236–243 (2002)

Greenman, R.L., Panasyuk, S., et al.: Early changes in the skin microcirculation and muscle metabolism of the diabetic foot. The Lancet 366(9498), 1711–1717 (2005)

Groner, W., Winkelman, J.W., et al.: Orthogonal polarization spectral imaging: a new method for study of the microcirculation. Nature Medicine 5, 1209–1213 (1999)

Gschwandtner, M.E., Ambrozy, E., et al.: Microcirculation in venous ulcers and the surrounding skin: findings with capillary microscopy and a laser Doppler imager. European Journal of Clinical Investigation 29(8), 708 (1999)

Gschwandtner, M.E., Ambróźy, E., et al.: Microcirculation is similar in ischemic and venous ulcers. Microvascular Research 62(3), 226–235 (2001)

Gschwandtner, M.E., Ambróźy, E., et al.: Laser Doppler imaging and capillary microscopy in ischemic ulcers. Atherosclerosis 142(1), 225–232 (1999)

Harding, J.R., Wertheim, D.F., et al.: Infrared imaging in diabetic foot ulceration. In: Proceedings of the 20th Annual International Conference of the IEEE Engineering in Medicine and Biology Society, Hong Kong, China (1998)

Humeau, A., Steenbergen, W., et al.: Laser Doppler perfusion monitoring and imaging: novel approaches. Medical and Biological Engineering and Computing 45(5), 421–435 (2007)

Jessup, R.L.: What is the Best Method for Assessing the Rate of Wound Healing?: A Comparison of 3 Mathematical Formulas. Advances in Skin Wound Care 19(3), 138–146 (2006)

Jones, B.F.: A reappraisal of the use of infrared thermal image analysis in medicine. IEEE Transactions on Medical Imaging 17(6), 1019–1027 (1998)

Jones, B.F., Plassmann, P.: Digital infrared thermal imaging of human skin. IEEE Engineering in Medicine and Biology Magazine 21(6), 41–48 (2002)

Jünger, M., Klyscz, T., et al.: Disturbed blood flow regulation in venous leg ulcers. Int. J. Microcirc. 16, 259–265 (1996)

Khan, F., Newton, D.J.: Laser Doppler Imaging in the Investigation of Lower Limb Wounds. The International Journal of Lower Extremity Wounds 2(2), 74 (2003)

Khaodhiar, L., Dinh, T., et al.: The use of medical hyperspectral technology to evaluate microcirculatory changes in diabetic foot ulcers and to predict clinical outcomes. Diabetes Care 30(4), 903–910 (2007)

Langer, S., Born, F., et al.: Orthogonal Polarization Spectral Imaging Versus Intravital Fluorescent Microscopy for Microvascular Studies in Wounds. Annals of Plastic Surgery 48(6), 646 (2002)

Levasseur, M., Leonardi, L., et al.: Near infrared hyperspectral imaging: the road traveled to a clinical burn application. In: Proc. SPIE, vol. 5969, 59691O1-9 (2005)

Margolis, D.J., Allen-Taylor, L., et al.: Diabetic Neuropathic Foot Ulcers: The association of wound size, wound duration, and wound grade on healing. Diabetes Care 25(10), 1835 (2002)

Martinez, L.: A Non-Invasive Spectral Reflectance Method for Mapping Blood Oxygen Saturation in Wounds. In: Proceedings of the 31st Applied Imagery Pattern Recognition Workshop (2002)

Milner, S.M., Bhat, S., et al.: Observations on the microcirculation of the human burn wound using orthogonal polarization spectral imaging. Burns 31(3), 316–319 (2005)

Mlacak, B., Blinc, A., et al.: Microcirculation disturbances in patients with venous ulcer before and after healing as assessed by laser Doppler flux-metry. Archives of Medical Research 36(5), 480–484 (2005)

Mourant, J.R., Bigio, I.J.: Elastic-scattering spectroscopy and diffuse reflectance. In: Vo-Dinh, T. (ed.) Biomedical Photonics Handbook. CRC Press, Boca Raton (2003)

Nelzen, O.: Leg ulcers: Economic aspects. Phlebology 15(3-4), 110–114 (2000)

Newman, P., Davis, N.H.: Thermography as a predictor of sacral pressure sores. Age and Ageing 10(1), 14–18 (1981)

Newton, D., Leese, G., et al.: Microvascular abnormalities in diabetic foot ulcers. The Diabetic Foot 4(3), 141–146 (2001)

Oh, J.T., Lee, S.W., et al.: Quantification of the wound healing using polarization-sensitive optical coherence tomography. Journal of Biomedical Optics 11, 041124 (2006)

Papazoglou, E.S., Neidrauer, M., et al.: Non-invasive assessment of diabetic foot ulcers with diffuse photon density wave methodology: A pilot human study. Journal of Biomedical Optics (2009) (accepted Pending Revision)

Papazoglou, E.S., Weingarten, M.S., et al.: Assessment of diabetic foot ulcers with diffuse near infrared methodology. In: 8th IEEE International Conference on BioInformatics and BioEngineering (BIBE), Athens, Greece (2008)

Papazoglou, E.S., Weingarten, M.S., et al.: Optical properties of wounds: diabetic versus healthy tissue. IEEE Transactions on Biomedical Engineering 53(6), 1047–1055 (2006)

Park, B.H., Saxer, C., et al.: In vivo burn depth determination by high-speed fiber-based polarization sensitive optical coherence tomography. Journal of Biomedical Optics 6, 474 (2001)

Payette, J.R., Kohlenberg, E., et al.: Assessment of Skin Flaps Using Optically Based Methods for Measuring Blood Flow and Oxygenation. Plastic and Reconstructive Surgery 115(2), 539 (2005)

Pierce, M.C., Sheridan, R.L., et al.: Collagen denaturation can be quantified in burned human skin using polarization-sensitive optical coherence tomography. Burns 30(6), 511–517 (2004)

Prahl, S.: Optical absorption of hemoglobin. Oregon Medical Laser Center, Portland, Oreg (1999), `http://omlc.ogi.edu/spectra/hemoglobin/index.html` (retrieved May 15, 2009)

Querry, M.R., Cary, P.G., et al.: Split-pulse laser method for measuring attenuation coefficients of transparent liquids: application to deionized filtered water in the visible region. Applied Optics 17(22), 3587–3592 (1978)

Rajan, V., Varghese, B., et al.: Review of methodological developments in laser Doppler flowmetry. Lasers in Medical Science 24(2), 269–283 (2009)

Rajbhandari, S.M.: Early identification of diabetic foot ulcers that may require intervention using the micro lightguide spectrophotometer. Diabetes Care 22(8), 1292–1295 (1999)

Samson, D.J., Lefevre, F., et al.: Wound-healing Technologies: Low-level Laser and Vacuum-assisted Closure. Evidence Report/Technology Assessment No. 111 (Prepared by the Blue Cross and Blue Shield Association Technology Evaluation Center Evidence-based Practice Center, under Contract No. 290-02-0026) Agency for Healthcare Research and Quality, Rockville, MD (2004)

Sayre, E.K., Kelechi, T.J., et al.: Sudden increase in skin temperature predicts venous ulcers: A case study. Journal of Vascular Nursing 25(3), 46–50 (2007)

Schmidt, W.D., Liebold, K., et al.: Contact-Free Spectroscopy of Leg Ulcers: Principle, Technique, and Calculation of Spectroscopic Wound Scores. Journal of Investigative Dermatology 116, 531–535 (2001)

Schmitt, J.M.: Optical coherence tomography (OCT): a review. IEEE Journal of selected topics in quantum electronics 5(4), 1205–1215 (1999)

Sibbald, R.G., Orsted, H.L., et al.: Best Practice Recommendations for Preparing the Wound Bed: Update 2006. Advances in Skin & Wound Care 20(7), 390–405 (2007)

Singer, A.J., Clark, R.A.F.: Cutaneous Wound Healing. New England Journal of Medicine 341(10), 738–746 (1999)

Singer, A.J., Wang, Z., et al.: Optical coherence tomography: a noninvasive method to assess wound reepithelialization. Academic Emergency Medicine 14(5), 387–391 (2007)

Skyler, J.S., Oddo, C.: Diabetes trends in the USA. Diabetes Metab. Res. Rev. 18(3), S21–S26 (2002)

Sowa, M.G., Leonardi, L., et al.: Near infrared spectroscopic assessment of hemodynamic changes in the early post-burn period. Burns 27(3), 241–249 (2001)

Srinivas, S.M., de Boer, J.F., et al.: Determination of burn depth by polarization-sensitive optical coherence tomography. Journal of Biomedical Optics 9, 207 (2004)

Stamatas, G.N., Kollias, N.: In vivo documentation of cutaneous inflammation using spectral imaging. Journal of Biomedical Optics 12, 051603 (2007)

Stamatas, G.N., Zmudzka, B.Z., et al.: Non-Invasive Measurements of Skin Pigmentation In Situ. Pigment Cell Research 17(6), 618 (2004)

Timar-Banu, O., Beauregard, H., et al.: Development of Noninvasive and Quantitative Methodologies for The Assessment of Chronic Ulcers and Scars in Humans. Wound Repair and Regeneration 9(2), 123–132 (2001)

Verhonick, P.J., Lewis, D.W., et al.: Thermography in the study of decubitus ulcers: preliminary report. Nursing Research 21(3), 233 (1972)

Virgini-Magalhães, C.E., Porto, C.L., et al.: Use of microcirculatory parameters to evaluate chronic venous insufficiency. Journal of Vascular Surgery 43(5), 1037–1044 (2006)

Vo-Dinh, T.: A hyperspectral imaging system for in vivo optical diagnostics. IEEE Engineering in Medicine and Biology Magazine 23(5), 40–49 (2004)

Wang, Z., Pan, H., et al.: Assessment of dermal wound repair after collagen implantation with optical coherence tomography. Tissue Engineering Part C: Methods 14(1), 35–45 (2008)

Wollina, U., Liebold, K., et al.: Biosurgery supports granulation and debridement in chronic wounds-clinical data and remittance spectroscopy measurement. International journal of dermatology 41(10), 635 (2002)

Yeong, E.K., Hsiao, T.C., et al.: Prediction of burn healing time using artificial neural networks and reflectance spectrometer. Burns 31(4), 415–420 (2005)

Zuzak, K.J., Perumanoor, T.J., et al.: A Multimodal Reflectance Hyperspectral Imaging System for Monitoring Wound Healing in Below Knee Amputations. In: IEEE Engineering in Medicine and Biology Workshop, Dallas, TX (2007)

Regenerative Wound Healing via Biomaterials

Anshu B. Mathur

Tissue Regeneration and Molecular Cell Engineering Labs (TRAMCEL),
Department of Plastic Surgery, The University of Texas M.D. Anderson Cancer Center,
Houston, TX
amathur@mdanderson.org

Abstract. Regenerative tissue engineering has the potential to revolutionize reconstructive approaches by providing either prefabricated tissue or responsive biomaterials with patient-specific geometry. The question to ask is how regeneration and repair are controlled *in vivo* and if a responsive biomaterial system can drive these processes? Does the cellular control lie at the cell-biomaterial nano-interface and do we have the tools to study this? The chemical and structural parameters and molecular linkages of the extracellular matrix that contribute to the internal mechanics of the cell and regulate a remodeling of an implanted biomaterial at the nano-interface have to be identified. This chapter is focused on introducing this concept of nano-scale regulated tissue regeneration by identifying numerous parameters of the biomaterial scaffold and cells so that the cell remodels the biomaterial without the addition of any growth factors, or other regulatory molecules by being influenced by composition, intermolecular linkages, nanostructure, nanomechanics of the biomaterial, and the biological/ chemical/ mechanical balance at the cell-biomaterial interface.

1 Introduction

Most currently available biomaterials used in areas of reconstruction, repair, and artificial devices are not designed to regenerate tissue but to cover or fill defects and/ or perform a mechanical function. Restoration of large volumes of tissue requires methods that establish an extensive and patent vascular network within the engineered tissue at the same time minimizing recurrence of disease and chronic wound healing during regeneration. Materials or Biologics that promote tissue formation, preserve volume and shape, and shorten the time of wound healing would be considered progress toward an effective clinical approach.

The design and development of biodegradable materials that will reconstruct and regenerate the native tissue without scar formation requires engineering of the composition, architecture, and mechanical properties, which control the initial inflammatory response, cell conductivity or infiltration, generation of the neo-extracellular matrix (ECM), degradation, mechanical properties of the remodeled or regenerated tissue, vascularization, and differentiation of different layers of the remodeled tissue e.g. musculofascia, bone.

A regenerative or responsive biomaterial is defined as one that recruits wound healing and precursor cells to form new viable tissue *in vivo*. The ideal way to

A. Gefen (Ed.): Bioengineering Research of Chronic Wounds, SMTEB 1, pp. 405–424.
springerlink.com

generate new tissue is to stimulate and guide the body's *in vivo* capacity to regenerate within a biologically recognizable environment.

2 Evolution of Biomaterials

Paradigms in the area of biomaterials have developed beginning from the age where polyester, polytetrafluoroethylene, and polyurethanes were kings of biomaterials and for artificial devices they still are. These are synthetic biomaterials whose *in vivo* biocompatibility and biostability response is signified by hydrolytic cleavage by inflammatory cells and fibrous capsule or scar formation that is body's response to blocking a foreign intruder[2, 26]. It should be noted that under current FDA standards, these are considered biocompatible biomaterials. In reality, the definition of biocompatibility has evolved to highlight materials that not only fill space but assist the body in regenerating tissue to develop permanent solutions to reconstruction reducing abnormal chronic wound healing.

Comparatively, the newer biodegradable biomaterials such as polylactic acid (PLA) and polyglycolic acid (PGA), also have a polyester chemistry, hydrolytically degrade, and create large amounts of byproducts that lower the local pH of the microenvironment, thereby influencing the wound healing response of the body. Biodegradable polymers such as PGA have been used for tissue engineering of artificial arteries since they provide a biocompatible surface for the cells to proliferate although do not provide a mechanically robust matrix which would enable the cells to withstand the *in vivo* shear and compressive loads and also result in dedifferentiation of cells [16].

Polymers such as polyethylene glycol (PEG) are good hydrogels but lack the chemistry for cellular interface unless ligand presenting entities are introduced [15]. Classical tissue engineers prepare cell seeded biomimetic hydrogel such as ligand modified PEG and study the cell response by measuring proliferation and differentiation[15]. In this situation, at most the cells can adhere to the adhesion ligands and perhaps secrete a protease whose cleavage sequence is incorporated within the gel. This does not make the hydrogel completely responsive. In this particular case, the gel has all the characteristics of a synthetic non-degradable biomaterial and only two characteristics out of many of being a biologically responsive biomaterial. This work has added value by showing that cells need more than a few signals in a tissue engineered scaffold to behave the way they do *in vivo*.

None of the established biomaterials are what would be labeled as "RESPONSIVE" to the cell. In essence, from the initial inflammatory reaction of the body to scar formation, the process of wound healing is controlled by cellular reaction to the biomaterial. A RESPONSIVE biomaterial would have a topography, architectural assembly, biomolecular composition, and chemistry that would give a cell, control of the biomaterial at the cell-material interface. This interface is the physical link that the cell has with the biomaterial as the cell adheres and migrates through the biomaterial. The cytokine production, growth factor release and other chemical mediators that are released by the cell are controlled in a feedback loop at the cell-substrate interface [31].

3 Development of Biologics for Regenerative Wound Healing

Matrices obtained from biological sources are available for clinical use, such as human acellular dermal matrix, HADM (Alloderm®, LifeCell Corp., Branchburg, NJ), small intestine submucosa or SIS (Surgisis® Cook Surgical, Bloomington, IN)[6, 5, 3, 4, 28] and porcine acellular dermal matrix, PADM (Strattice®, Life-Cell Corp., Branchburg, NJ). These matrices are derived from the extracellular matrices (ECM) of animal or human sources and fall in the category of Biologics. The antigenicity of these matrices is removed by various treatments that alter the structure and mechanical properties of the treated-Biologic. A limitation of off-the-shelf Biologics is that the composition, biological, and mechanical property could inconsistently vary with different donor sources. More importantly, the structure and biological activity of ECM derived matrices is dependent on the native ECM architecture that it is derived from and would be difficult to engineer as per clinical application.

By engineering the composition and structure of the matrix, properties such as molecular composition, flexibility, tensile strength, thickness, degradation rate, and elasticity, the remodeling behavior can be controlled according to the clinical need of the repair site per patient[13]. Two biologically derived biomaterials have shown promise for their ability to form engineered structures and potential regenerative capacity. Silk fibroin and chitosan are both highly biocompatible, cell adhesion promoting, and responsive/ dynamic materials [13, 36] whose local mechanical properties can be controlled by varying the blend ratios[1, 19, 44, 37, 33, 30, 10, 32]. The blend of silk fibroin (SF), a fibrous protein, and chitosan (CS), a glycosaminoglycans-mimic component, imparts distinct properties to the SFCS blends.

Silk fibroin is obtained from the silk fiber, which is spun by silkworms and spiders as a fibroin core surrounded by a sericin hydrophobic coating. Fibroin, from the silkworm *Bombyx Mori*, is comprised of a heavy chain that is 325 kDa and light chain of 25 kDa. The structural sequence consists of a crystalline domain or heavy chain mostly comprising of glycine, alanine, serine, and tyrosine amino acids repeats (GAGAGSGAAG$[SG(AG)_2]_8$Y and short amorphous or light chain domain although less well defined consists of bulkier side chain amino acids as aspartic acid. The amino acid sequence enables the silks to form an anti-parallel β-pleated sheet secondary structure. This structural configuration gives silk its unique mechanical properties that include high strength with flexibility. They also have high compressibility and resistance to axial compressive deformations, which most high strength synthetic fibers lack [22].

The regeneration of silk fibroin into foams, films, membranes, fibers and gels makes it a versatile material [26]. Fibroin films support adhesion, growth and morphology of anchorage dependent cells such as the fibroblast similar to collagen [30]. While human osteoblast-like cells produced bone on fibroin films and enhanced by RGD modification, human anterior cruciate ligament fibroblast proliferated on fibroin fibers [1, 42]. Its permeablity to oxygen and water, relatively low thrombogenecity, and low inflammatory response is advantageous for tissue growth [1, 37, 33, 23]. The ability of silk fibroin to form liquid crystalline fibrils and its proteolytic susceptibility provides it unique characteristics as a responsive/dynamic yet supportive matrix [1, 23, 29, 12].

Although, liquid crystalline and fibril forming properties of silk fibroin are desirable in construction of support matrices, *in vivo* the ECM also has a glycosaminoglycan component that stores water and gives tissues their compressive properties. Chitosan is a glucosamine that is obtained from deacetylated chitin that is readily available from sea shells and serves as a good glycosaminoglycans mimic. Chitosan is non toxic and has low thrombogenicity, although it does cause RBC agglutination [35]. The rate of degradation of chitosan can be controlled by the degree of deacetylation as the rate of degradation decreases as the degree of deacetylation increases [45].

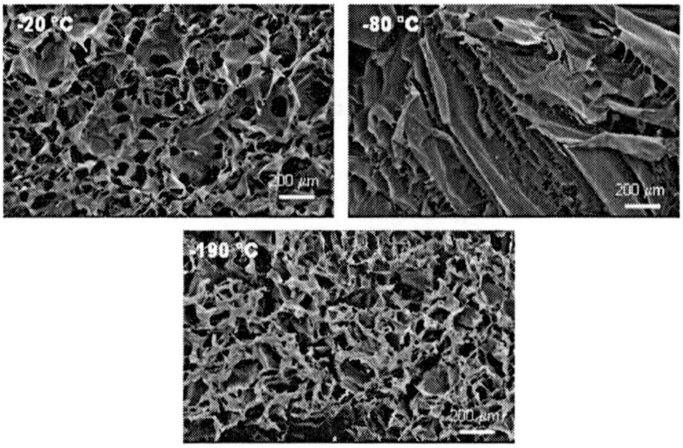

Fig. 1. SFCS blend scaffolds processed at freezing temperatures of -20°C, -80°C, and -190°C exhibit variable microstructure from large pores to parallel sheets to vertical porous sheets.

The structural configuration of SFCS blend scaffolds was characterized using SEM imaging. The processing conditions of the SFCS scaffolds regulate the type of structure obtained (Fig. 1). The structure varied from porous to assembled sheets with micro-fibril extensions depending on the processing freezing temperature. The 75:25 blend of SFCS show an *in vivo* basement membrane like assembly when processed at a temperature of -80°C. The major similarity was the large micro-sheets assembled in a parallel array to each other with intermingled fibrillar extensions[14]. At the edges of the sheets are fibrils with buds (on the order of a micron in diameter) (Fig. 2). As the SF content is increased in the blends, the percent fibrils, ultimate tensile strength, and elastic modulus of SFCS increases. The differences in the composition of the SFCS blends affect their initial and final mechanical properties pre-implantation [14] and will ultimately affect their *in vivo* remodeling behavior. While the matrices such as HADM are labeled as Biologics, biomaterials such as SFCS that are similar to HADM may be categorized and labeled as "Engineered Biologics".

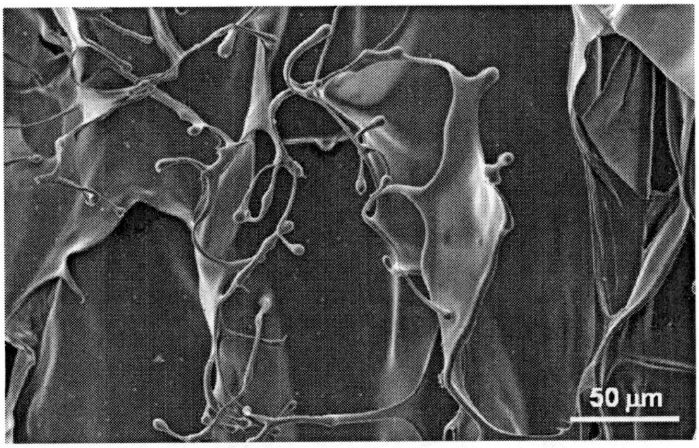

Fig. 2. The parallel sheets assembled in three-dimensions of the SFCS scaffolds processed at -80°C have micro-fibril extensions at its edges.

4 Regeneraive Capacity of Engineered Biologics

The regenerative capacity of SFCS blend scaffold was examined for regeneration of two types of tissues, musculofascia and bone. The SFCS blend scaffolds were investigated for abdominal wall musculofascia repair in an incisional ventral hernia repair model and compared with HADM [9]. HADM was present at

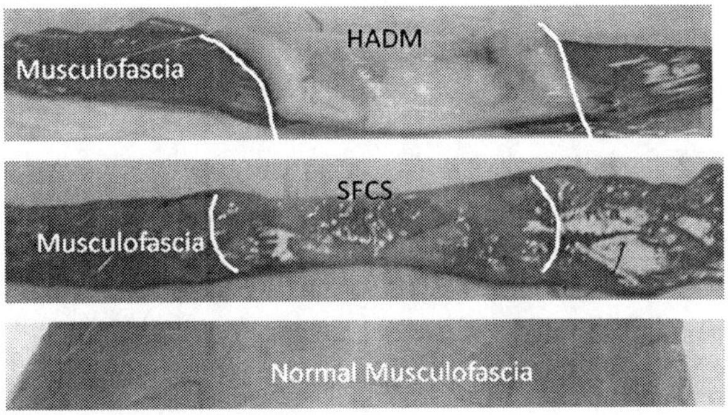

Fig. 3. Abdominal wall musculofascia repair with an off-the-shelf biologic, HADM and engineered biologic, SFCS shows gross differences in the repair site after 4 weeks as compared to the normal abdominal wall musculofascia. While HADM is still grossly visible as a thicker matrix with clear edges, there is no gross presence of SFCS and the interface integrated with the native musculofascia.

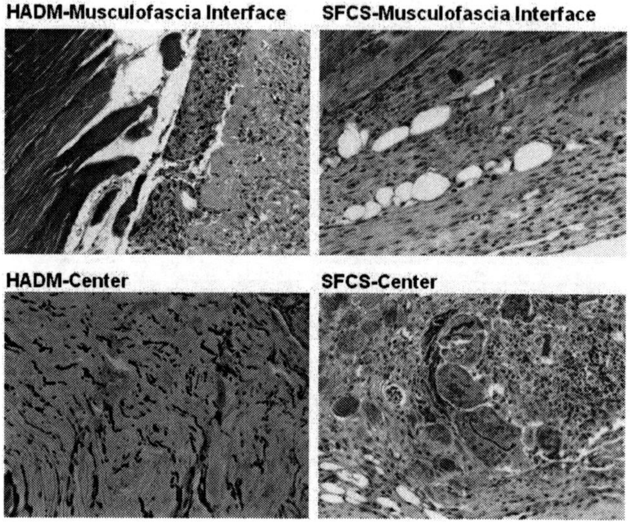

Fig. 4. Histological examination of 4 mm cross-section of the HADM and SFCS repair site shows distinctly separate zones of remodeled HADM adjacent to the native muscle at the interface. On the other hand, musculofascia repaired with SFCS was seamlessly integrated and remodeled the musculofascia interface during the remodeling process. The center of HADM interface remained mostly intact as evident by the presence of collagen and elastin of the original HADM, whereas only a few strands of SFCS scaffold remained by 4 weeks. The residual SFCS strands were surrounded by the cells that are part of the foreign body regenerative response.

the implant site after 4 weeks, while there was no gross evidence of SFCS. The edges of the HADM were curled and distinctly separated from the native musculofascia. On the other hand, gross examination of the abdominal wall repaired with SFCS showed a seamless musculofascia with well integrated healed interface with the regenerated tissue. In addition, the SFCS regenerated tissue had similar visible characteristics to the intact normal native abdominal wall musculofascia that suffered no incisions or repairs (Fig. 3).

Histological examination of the repair site indicated that connective tissue cells infiltrated the SFCS and new collagen matrix was deposited. The musculofascia interface was cellularized forming a dense connective and well vascularized tissue. While no differentiated muscle structures (skeletal muscle) resembling the abdominal wall anatomy were evident, there was no necrosis or obvious scar tissue either. Similar, cell infiltration and matrix deposition was evident in HADM at the musculofascia interface although the degradation area for HADM is markedly less than SFCS blend since the center of implant had much less cellular infiltration as compared to SFCS (Fig. 4)[13].

Fig. 5. Representative micro-CT images of the regenerated bone in bone graft and SFCS filled chambers implanted over the periosteum-latissumus dorsi in an *in vivo* sheep model. The bone volume resulting from the SFCS scaffolds was statistically similar to the clinically relevant bone graft control.

The bone regenerating capacity of SFCS biomaterial system in an *in vivo* sheep model was recently reported [36]. Periosteal pieces (1x3cm) were harvested from the ribs and grafted as autografts onto the latissimus dorsi muscle, with the cambium layer facing away from the muscle. Chambers (inner volume- 3.75 cm^3) containing SFCS scaffolds were implanted on top of the grafted periosteum over the latissimus dorsi muscle and harvested at days 7, 21, 41, 63, and 84. An empty chamber and chambers filled with bone graft (clinically relevant control) were negative and positive controls for all time points, respectively. Micro-CT scans were conducted of the chambers at the harvest time points at 46 μm and 92 μm sections/slices. Bone volume significantly increased from day 7 to day 41 in SFCS scaffolds. Beyond day 41, there was no significant difference in bone volume up to day 84 although it appears from micro-CT scan images that there was a slight decrease at day 63 due to compaction/ contraction of the bone in the chamber. Representative micro-CT images of bone regeneration in bone graft, SFCS, and empty chambers at 84 days are shown in Fig. 5. While new bone formed in SFCS scaffolds, bone graft filled chambers showed degradation of the bone graft to form new remodeled and compact bone. The internal architecture of the regenerated bone in SFCS looked similar to the internal architecture of the normal rib bone, both showing tortuous bone organization. SFCS scaffold supported regeneration of vascularized and differentiated bone without the addition of any growth factors, cells or chemical agents. The maximum volume of regenerated bone was 890 mm^3 as early as day 41[36].

5 Importance of Nanostructures in Nature

5.1 Structure of the ECM

ECMs provide the support and mechanical strength to cells and thus play a major role in cell morphogenesis during regenerative wound healing. The global role of ECM components, such as collagen and glycosaminoglycans is well established to provide structural support and mechanical integrity to the tissues. Collagen molecules in particular are also sensitive to proteases that perturb the mechanical synergy of the tissue. The cell interfaces the ECM via the adhesion proteins and mechanically couples itself to the ECM molecules such as collagen, laminin, nidogen, fibronectin, and heparan sulfate proteoglycan, which form an interactive yet complex architecture.

The composition and architecture of ECMs in tissues varies with tissue type and function. While the cornea, skin, and bone are both constructed from type I collagen, the properties and function of the three tissues vary. The orthogonal lattice of narrow collagen-I fibrils in the cornea result in a transparent matrix and at the same time sheets of collagen-I fibrils gives the skin its opaque properties. The differences in the assembly of collagen I fibrils leads to variable tissue properties and function. In another scenario, cartilage and bone have similar collagen-II arrangement of an interlocking weave, except in bone they mineralize and in cartilage they don't. Similar fibril arrangements are found in cornea and the blood vessel wall, although the cornea is composed of collagen I and the vessel wall is composed of collagen-III [21].

In general, collagen fibrils with a diameter of 30-100 nm bundle together to make a collagen fiber with a diameter of 1-20 μm [46]. The length of the fibrils is around a 1 μm. The transparent ECM of the vitreous body consists of very long collagen fibrils that have an average diameter of 10-20 nm, where the longer collagen fibril diameter is 16 nm, and shorter network forming fibrils are around 7 nm in diameter[8]. The stiff architecture of this ECM is intriguing since the collagen fibrils are spaced apart yet linked to each other via proteoglycans that also support collagen fibril assembly or fibrillogenesis. Moreover, proteoglycans are key to regulated three dimensional assembly of protein nano-fibrils in to a matrix.

During the process of wound healing and regeneration, the deposition of new matrix and the diameter of the *in vivo* fibrillar collagen is dependent upon the interactions with proteoglycans that space the fibrils apart. According to Scott[39], protein of the proteoglycan binds to the collagen fibrils connecting the two fibrils via the anionic glycosaminoglycan (AGAG) bridge. Ionic interactions between the proteoglycan chains and collagen fibrils have been proposed as the linking mechanism[38]. Anionic glycosaminoglycans, chondroitin sulfate, dermatan sulfate, and keratin sulfate, linked to proteoglycans such as decorin, bridge collagen fibrils at 65 nm intervals non-covalently. An increase in proteoglycan decorin concentration decreases the collagen fibril diameter. The lack of this proteoglycan results in thicker and unorganized fibril assembly by controlling the rate and degree of collagen fibrillogenesis[11]. This repeat module bridge consists of collagen fibrils and maintains the overall tissue shape[40].

This intricate arrangement of collagen with the proteoglycans provides the ECM with its unique tissue specific mechanical properties. The AGAG to fiber ratio in specific tissues varies with their mechanical function. For example, a weak gel of the vitreous humor has <1% fibers while, the strong tendons have 70% fibers. The elastic nature of collagen fibers allows for transmitting and resisting tension and the compressive forces are deflected by AGAGs. The compressive properties of the AGAGs are generated from their ability to swell in water and create interfibrillar spaces between collagen fibrils. The AGAG bridges are hypothesized to convert local compression into tensile strain via a sliding proteoglycan-collagen filament model.

The role of decorin in maintaining skin integrity by regulating collagen structures is evident in *in vivo* models [11]. Decorin knockout mice *($Dcn^{-/-}$)* were viable but were found to have fragile skin with a threefold reduction in tensile strength. The collagen morphology was also abnormal with coarser and irregular fiber outlines in the absence of decorin. Bulges formed along the collagen fibrils increasing the fibril diameter and possibly decreasing the possibility for fibril fusion. Decorin's role in collagen fibrillogenesis is evident during cardiac fibrosis following a myocardial infarction [51]. The left anterior descending artery was ligated in wildtype and decorin-null mice producing left anterior wall infarcts. A wider distribution of collagen fibril sizes with less organized scars were found in the decorin-null mice producing larger scars, more right ventricular remote hypertrophy and more left ventricular dilation than in the wildtype mice at 14 days. At 4 and 8 weeks post-ischemia, the decorin-null mice had depressed left ventricular systolic function. The absence of decorin lead to improper fibrotic evolution of infarctions forming abnormal scar tissue.

In order to target engineering of regenerative wound healing, the ECM component architecture such as collagen fibril size is an important parameter to regulate via perhaps cellular engineering techniques.

5.2 Cellular Interface with ECM

The fine features of the ECM are complimented by nano-scale entities of the cell at the cell-matrix interface. Imaging of the apical cell surface by application of a nanonewton contact force allows one to see the structure of the cortical actin network, which is an organized polygonal mesh that networks the cell membrane with actin cables (Fig. 6) [34].

The cell-matrix distance (31±19 nm) is bridged by anchoring transmembrane proteins and components of the glycocalyx [27] [52]. The core proteins of the glycocalyx are predicted to be the mechano-transducers that transmit the extracellular stresses to the cytoskeletal structures of the cell as the core proteins link the hexagonal lattice of the surface layer to cortical actin intracellularly[50] (Fig. 7). At the cytoskeletal foci, there are 27 core proteins that have bush like fibers emanating from them and that take a maximum compressive force of 0.3 pN or 300 femtonewton. The flexural rigidity of the core proteins in the glycocalyx is 700 pN.nm^2 as compared to 15x10^3 pN.nm^2 to 73x10^3 pN.nm^2 for filamentous actin that is anchoring the core protein to the cell. Thus, the cortical actin is stiffer than the core protein and thus provides support as the core protein senses and deforms under stress. For a 100 nm compressive indentation, the resisting stress of the fibers is 1.6 kPa (Note: This is the apparent elastic modulus reported for endothelial cells adherent to glass surface during nanometer indentations using the AFM) [24].

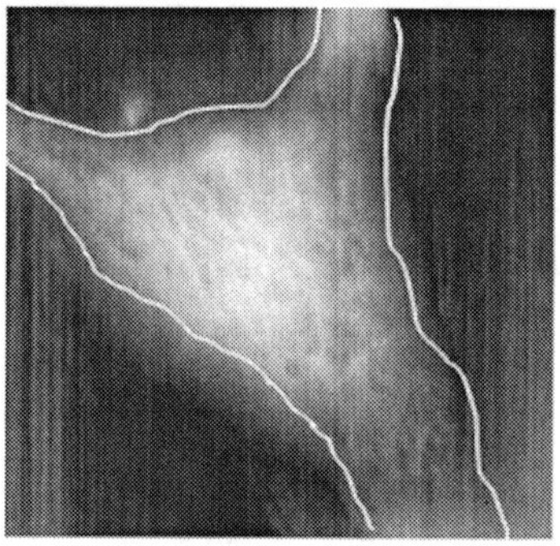

Fig. 6. The apical surface of an endothelial cell imaged with the atomic force microscope (AFM) shows the polygonal architecture of the cortical actin network.

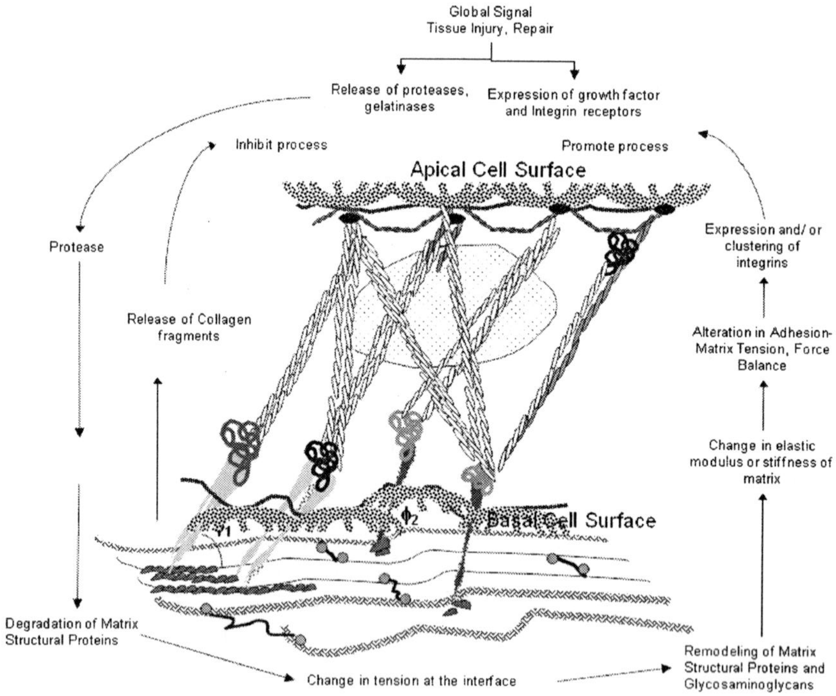

Fig. 7. A schematic of the structural components of the cell that maintain mechanical force balance of the cell via feedback signaling.

While the mechanical contribution of the glycocalyx is evident, the detailed structural parameters reflect the length scale at which the glycocalyx molecules transmit the mechanical forces from the cell exterior surface to the cell interior and beyond. Endothelial cell glycocalyx consists of proteoglycans and glycoproteins residing in the extracellular environment. It has a thickness of 0.4 to 0.5μm or about 15 to 20% of the radius of the smallest capillary[47]. Squire et al.[43] and Weinbaum et al.[50] proposed a quasi-periodic structure for the endothelial surface layer or glycocalyx characterized by a 20 nm periodicity of the glycoproteins and 80-120 nm spaced hexagonal lattice (Figs. 6 and 7). The thickness of the glycocalyx was shown to be 100-200 nm by Squire et al.[43], a value that differs from observations by Vink and Duling (400-500 nm)[47]. The difference may be due to the methodology. Squire et al. presents data and a detailed schematic of the nanostructure of the glycocalyx as a three dimensional layered structure with 20 nm spacing in all directions. The hexagonal lattice spacing of 80-120 nm was shown to be anchored into the membrane cortical cytoskeletal structures via the glycoproteins[50]. Nano-filamentous (7-15 nm diameter) ordered network structure of the glycocalyx has been most recently reported for epithelial cells of the small intestine[17].

The structural intricacy of the glycocalyx and its mechanical nature shows that the entire cell machinery is linked by nano-structured components, which make the cell mechanically responsive, adaptable, and sensitive to external parameters at femtoscale sensitivity.

The success of the cell-biomaterial/ matrix interface from a cellular perspective reflects on how molecules bound to the cell membrane and physically linked to the matrix respond to mechano-chemical signals initiated by the matrix. A model for cell-biomaterial interface implicates the tension generated by cells at the cell-matrix/material interface due to the affect on integrin-fibronectin assembly[48]. This model implies that as the cell forms links to the biomaterial, it maintains a certain tension thereby sending a signal via the cytoskeleton to the internal cell machinery to inhibit or promote function such as the formation of blood vessels by endothelial cells. It is well established that the integrin-fibronectin assembly is based on a chemical or tri-peptide linkage of RGD (arginine-glycine-aspartate) on the fibronectin molecule. While the specificity of binding is based on the chemistry of RGD amino acids, the signaling generated from this chemical link is transferable into a mechanical signal and thus the **composition or chemistry** of the biomaterial is critical in maintaining this mechanical balance.

Balaban et al [7] proposed two mechanisms that may control the mechano-sensing ability of the cell at the cell-substrate interface. According to this model, the organization and assembly of the focal adhesion complex molecules is disrupted by the force experienced at this interface. In addition, the applied force unfolds binding domains of integrins and affects their binding capacity. The force per focal adhesion having a size in the range of 2.5-5 μm^2 was found to be on the order of 10 nN and a stress of 5.5±2 nN/μm^2 and force per molecule of integrin around 1 pN. Although these mechanisms are suggested as either-or scenarios, the reality could be closer to a combination of the two mechanisms and perhaps more complex than that. Since the measurements to reach this model were conducted on cells adherent to 1 μm deep silicone elastomer, the chemical specificity and nano-fibrillar topography were two factors ignored. Forces at the cell-focal adhesion interface on a silicone elastomer are on the order of nanonewtons, whereas the sensitivity on native ECMs may be far superior than that due its nano-scale structure. The distribution of the forces at the cell-substrate interface is driven by the nano-scale fibrillar features of the ECM via the integrin interface, which was not addressed by this model.

The mechanical force balance that exists at the cell focal adhesion-substrate interface can be explained by another mechanical model based on the hypothesis that deformation of cells by force application results in change of conformation and/ or organization of focal adhesion proteins lying in a narrow region. Based on continuum elasticity theory, the focal adhesions are treated as infinite, thin plate since their thickness is on the order of 100 nm and the width on the order of microns. The adhesions (thin films) are modeled as particles or protein complexes that attach via integrin springs to the extracellular matrix and to each other via another set of springs. The front edge of the adhesions is sensitive to compression thus acts as a mechanosensor. It was predicted that the stressed focal adhesions grow in the direction of the applied force and that this growth is a function of extracellular **matrix rigidity and the strength of the adhesion**. Cells adherent to a flexible matrix were grouped as those in which the integrin molecules that are linked to the ECM cannot resist the pulling force and thus form fibrillar adhesions. In this force regime, the adhesion contact at the cell-material inteface moves rather than deforms. Indeed protein aggregates involved in fibrillar adhesions are labeled as tension-insensitive, although an appropriate label may be "tension transmitters", since their fibrillar architecture permits them to transfer loads to the adjacent extracellular matrix fibrils of collagen. The suggestion that large cytoskeletal forces did not result in growth of the adhesions as in the case of focal adhesions but a stick-slip motion of focal adhesions is interesting since this phenomena is associated with long fibrillar adhesions. The question that comes to mind is whether the transmission of force (amount and type) from fibrillar adhesions to the fibrillar ECM components is coupled based on the morphology or structure of the adhesions?

The impact of the cell-ECM mechanical coupling on the measured stiffness of the matrix or a biomaterial surface was evidenced by a study conducted by stretching the fibroblast populated matrix or FPM. Stress-strain curves were obtained for fibroblast populated matrices with or without dissipated fibroblast cytoskeleton. Since the stress-strain curves reported had large amounts of hysteresis, the absolute numbers obtained for stiffness may be treated as apparent ones. Despite the hysteresis, relative numbers are comparable within the same FPM system that was tested. The stiffness of the FPM was lower when the fibroblast cytoskeleton was dissipated versus a matrix where the cytoskeleton was actively intact and linked to the ECM. This implies that the deformation of a matrix with cells whose cytoskeleton is dissipated yields higher stiffness than a matrix that does not contain cells. Why? The cell-ECM coupling is still intact and allows the cells to remodel the matrix and re-establish the cell-matrix mechanical balance, thus contributing to the overall mechanical properties of the FPM. These experiments revealed that as the concentration of the cells is increased in the matrix, the extent of stiffening by collagen remodeling is increased. The increase in stiffness due to the high cell number could be associated with having a separate phase within the FPM formed by cell-cell contacts coupled in with cell-matrix contacts, yielding a composite value for the stiffness.

When studying cell behavior using *in vitro* cell culture methods, it is becoming increasingly evident that the cells morphology and other characteristics may vary in a three-dimensional gel type environment versus a hard surface of a flask. Since the cells survive and can mimic *in vivo* characteristics under certain conditions, it reflects on the adaptive nature of cells. A cell adherent on a stiff and relatively flat substrate will form adhesions that have different characteristics as compared to cells adherent to a malleable substrate[49]. As a matter of fact, the entire stress bearing assembly is affected by the substrate to which the cells are adherent to. If we assume that the focal adhesions or the cell-substrate adhesions are the load sensing, bearing, and transferring entities in a cell, the mechanism of load transfer can be reflected upon. The evidence of transfer of applied load to focal adhesions is seen in AFM experiments, where a localized load (300-500 pN) is applied to cell using the AFM probe and a remodeling of the focal adhesions is seen in response to the applied load (Fig. 8)[27].

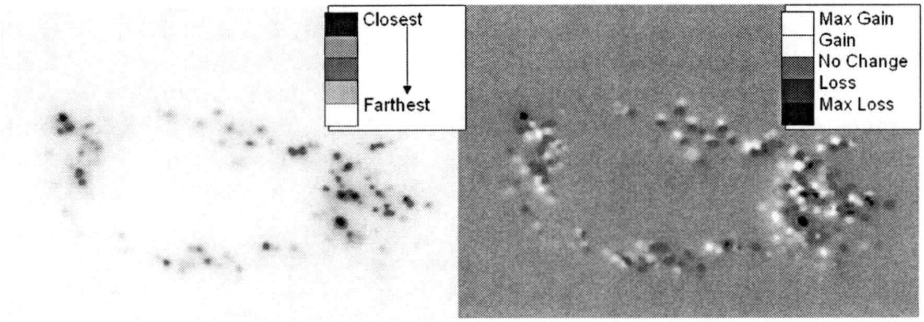

Fig. 8. The apical surface of endothelial cells adherent to fibronectin coated glass surface was perturbed with an nanoprobe with forces in the 300-500 pN regime. A representative image of the focal adhesion arrangement as imaged with total internal reflection fluorescence microscopy before the force was applied shows that the adhesive regions are spread around the nuclear region of the cell in a peripheral arrangement. The focal adhesions closest to the substrate appear in blue-black and the region farthest from the substrate appears in yellow. After application of the force for 1 minute with the atomic force microscopy nanoprobe and imaging after 5 minutes of the removal of the force shows areas of loss and gain of focal adhesions. There was a global response of the adhesions and a contractual movement of the head and tail region towards the nucleus.

6 Balancing Forces: Cell-Biomaterial "Responsive" System

We assume that the success or failure of a biomaterial is controlled by the cells at the cell-substrate nano-interface since cell proliferation, differentiation into mature tissue, and guidance of blood vessel development is dependent on the formation, stability, and the force balance at this interface that is affected by the chemical composition, microstructure, viscous, tensile, and compressive properties, of the biomaterial. A single cell must be given local control of its microenvironment at the cell adhesion nano-interface (~30-100 nm) in order to regenerate and repair tissue. This control is achieved at the cell-biomaterial interface where the cell forms its adhesions and perceives the biomaterial. The only way this control can be achieved is by using a responsive biomaterial system that is defined as one whose chemical composition provides cell-binding specificity which allows for cellular response to the biochemical factors (e.g. proteases) secreted by cells during regeneration, whose chemical composition permits the storage of water and growth factors, whose fibril nano-structure and mechanical properties allow for the attachment of cells at cell-material interface and create a mechanical force balance between adhesion proteins and the matrix nano-fibrils, and whose structured 3-D assembly from nano to micro to macro can bear the loads endured by the tissues *in vivo*.

Whether a cell is able to modify a component of the matrix/ biomaterial is dependent on its adhesion strength to the matrix and whether it has the internal cytoskeletal strength to remodel the matrix. Does the cell want to maintain internal stiffness/ force or maintain a force balance at the interface? We assume cells have a critical force balance that they need to maintain at the cell-substrate interface in order remain viable and functional (ΔE_i). The viability force could be slightly different from the functional force since all cells that are viable in a 2-D culture are not necessarily functional as they are in a 3-D tissue environment. As the degree of functionality increases from a 2-D stiff substrates to *in vivo* like behavior, the ability to achieve and ease of achieving the critical force balance also increases (Fig. 9).

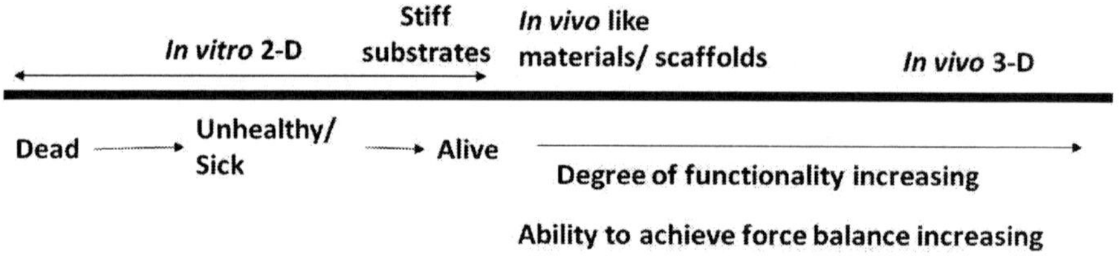

Fig. 9. A schematic showing the evolution of cell behavior depending on its culturing environment *in vivo* and *in vitro*.

What does the achievement of this force balance depend on? Properties of the substrate, properties of the cells (type, morphology, etc)?

Bischofs and Schwarz hypothesize that "cells prefer maximal effective stiffness in their environment" (Fig. 10). This is not entirely a true statement although it points in the right direction.

$\Delta E_{ideal} = \Delta E_i$ (In the case of ideal force balance at the cell-material interface to make cells functional)

E_1= Stiffness of the cell
E_2= Stiffness of the matrix
E_2-E_1= $\Delta E_{calculated}$ = balancing modulus or stiffness= ΔE_c
If $\Delta E_c > 0$, that is a stiff matrix as compared to the cell
E_1 would have to approach E_2 in order to reach ΔE_i

Bischofs and Schwarz also suggest that "cells prefer fiber under the largest tension". The question is why? That would mean that the cells need to input less energy to achieve the force balance, if ΔE_i is close to ΔE_c. This energy is the differential energy that would be contributed towards the cytoskeletal and adhesion remodeling in order to achieve ΔE_i.

In a reverse sense: On a soft substrate, the cell would have to produce stress fibers to achieve a minimal mechanical tension in order to remain adherent.

Question: In which case does the cell have more control of the substrate/ matrix soft or stiff?

On a stiff surface such as glass or tissue culture plastic, the stiffness is orders of magnitude higher than the cells, molecules of the glass have unknown chemistry and are tightly bound such that the cell has no control over the substrate topography, feature size, mechanical properties. It does have the ability to secret small amount of ECM components on top of the stiff surface, such as collagen, fibronectin to make the composition that it interfaces on this substrate adherent.

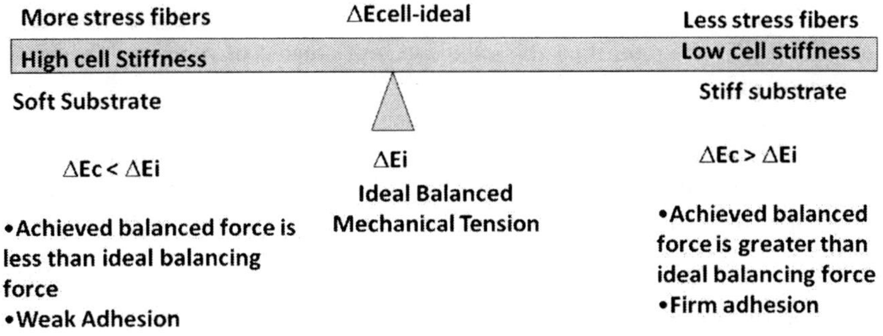

Fig. 10. A schematic of the force balance scenarios for cells adherent to soft versus stiff substrates relative to the ideal cell-material force balance.

The response of adhesion assembly on stiff glass surfaces has been studied. Response of focal adhesions on a stiff substrate can be explained by a couple of mechanisms. On a glass slide or a stiff substrate, where the adhesions have no direct link to the substrate except to the adsorbed proteins lying on the surface of the stiff substrate, the focal adhesions will move in response to the applied force globally resulting in the net movement of the cell in some cases. If the integrin bonds with adsorbed protein remain intact or strong, the integrin-protein complex would move within the membrane (within bound regions of cortical actin domain) which also appears as the net movement of adhesions over the substrate as the substrate remains stiff and immobile.

On soft substrate, the cell has the capacity to remodel the matrix to different extents depending on the characteristics of the soft matrix or biomaterial.

Soft Matrix/ Biomaterial characteristics that have to be considered are as follows.

A. Chemistry of the matrix is recognizable by the cell for binding.
B. For a known recognizable chemistry, the matrix is biodegradable, biologically derived, but different structure
   (i) Isotropic matrix- no fibril, no orientation
   (ii) Isotropic matrix with fibrils, no orientation- non-woven type
   (iii) Anisotropic matrix with fibrils and orientation
C. Degradable or Non-degradable- Chemistry dependant
D. Remodelable or not- Soft will generally be remodelable although to different extents depending on the fibril size and mechanical properties
   - 1 mm size fibrils, 1 um size fibrils, 1 nm size fibrils will affect binding affinity at the cell-material interface
   - As you go from mm to nm, the ease of remodeling increases, and propensity to remodel increases.
   - Structural or architectural dependence

A piece by piece analysis of these parameters will lead to a model for the "sensitivity of the cell for the cell-material interface". If the biomaterial is flexible and the structural elements of the biomaterial are bound or linked to the integrins with a bond that is stronger than the intermolecular forces that bind the adsorbed protein to the stiff substrate, then the adhesion will consist of integrin-fibronectin fibril-biomaterial fibril complex. As the force is applied over the cell the transfer of stress would now involve the lateral displacement of the biomaterial fibrils in response to the applied stress. If the stress is successfully transferred to the load bearing elements of the tissue, substrate or extracellular matrix components or biomaterial, then the adhesions remain stable and strong, stress fibers may increase or become prominent and the cell stiffness may increase. The response becomes a function of the molecular scale properties of the initial adhesion/ anchorage assembly[25].

7 Engineered Fibrous Scaffolds for Regenerative Repair

During *in vivo* embryonic development the no-scar regeneration of tissues is dependent on the fibrillar collagen assembly guided by the proteoglycans. Thus, our goal as tissue engineers should be to mimic that property of the ECM in our engineered scaffolds in order to promote tissue-specific regenerative repair.

While the nano/micro fibril assembly, deposition, and growth is a critical step for controlling matrix architecture, it is a limiting step for tissue engineering applications due to the challenges of assembling such a complex structure in three-dimensions. Fibroin has an integrated crystalline and amorphous milieu due to its amino acid sequence which can be engineered by processing it in the nematic liquid crystalline state which orients molecular structure at various levels. Silk fibroin is a protein that forms nanofibrils on the order of 50-100 nm due to its liquid crystalline nature [29, 18]. The blending of the SF and CS polymers creates a microenvironment similar to what is seen in embryonic development where collagen molecules in the midst of proteoglycans form nano-fibrils and guide tissue development. The processing of low viscosity liquid crystalline fibroin solution initiates structural ordering of molecules rendering them in an extended state and globally aligning them to promote chemical, physical, and mechanical order. This self-assembly of SF amino acids can be embedded within a three-dimensional scaffold of SFCS with a blended GAG mimic, such as CS. The ability to align molecules in a liquid state enables the ordering of molecules in a solid state, since less reorganizing is required upon solidification by, for example rapid solvent loss. It has been reported that the liquid crystalline behavior of silk solutions is dependent upon the initial concentration of molecules, which controls the number of interacting molecules by hydrogen bonding and the rate at which the water is lost. The two cooperative factors control the initial number of interacting fibroin and water molecules via hydrogen bonding in solution affecting the reformation of hydrogen bonding between fibroin polymer chains upon orientation, as a solid[22].

Thus, SF derived blend microenvironment provides the SF molecules platform to self-assemble in engineered structures within a three-dimensional scaffold[14]. This enables us to engineer structure and scale-up, to regulate the overall regenerative capacity per our clinical need[13].

Acknowledgments. This work was supported by the United States National Institutes of Health/ National Institute on Aging, Musculoskeletal Transplant Foundation, United States Department of Defense, Gillson-Logenbaugh Foundation, Anderson Foundation, Department of Plastic Surgery, The University of Texas M. D. Anderson Cancer Center.

References

1. Altman, G.H., Diaz, F., Jakuba, C., Calabro, T., Horan, R.L., Chen, J., Lu, H., Richmond, J., Kaplan, D.L.: Silk-based biomaterials. Biomaterials 24, 401–416 (2003)
2. Anderson, J.M., Hiltner, A., Wiggins, M., Schubert, M.A., Collier, T.O., Kao, W.J., Mathur, A.B.: Recent Advances in Biomedical Polyurethane Biostability and Biodegradation. Polymer International 46, 163–171 (1998)
3. Badylak, S.: The extracellular matrix as a scaffold for tissue reconstruction. Cell and Developmental Biology 13, 377–383 (2002)
4. Badylak, S.: Xenogeneic extracellular matrix as a scaffold for tissue reconstruction. Transplant Immunology 12, 367–377 (2004)
5. Badylak, S., Kokini, K., Tullius, B., Simmons-Byrd, A., Morff, R.: Morphologic study of small intestinal submucosa as a body wall repair device. Journal of Surgical Research 103, 190–202 (2002)

6. Badylak, S., Kokini, K., Tullius, B., Whitson, B.: Strength over time of a resorbable bioscaffold for body wall repair in a dog model. Journal of Surgical Research 99, 282–287 (2001)
7. Balaban, N.Q., Schwarz, U.S., Riveline, D., Goichberg, P., Tzur, G., Sabanay, I., Mahalus, D., Safran, S., Bershadsky, A., Addadi, L., Geiger, B.: Force and focal adhesion assembly: a close relationship studied using elastic micropatterned substrates. Nature Cell Biology 3, 466–472 (2001)
8. Bos, K.J., Holmes, D.F., Meadows, R.S., Kadler, K.E., McLeod, D., Bishop, P.N.: Collagen fibril organization in mammalian vitreous by freeze etch/ rotary shadowing electron microscopy. Micron 32, 301–306 (2001)
9. Butler, C.E., Navarro, F.A., Orgill, D.P.: Reduction of abdominal adhesions using composite collagen-GAG implants for ventral hernia repair. Journal of Biomedical Materials Research: Applied Biomaterials 58, 75–80 (2001)
10. Chiarini, A., Petrini, P., Bozzini, S., Pra, I.D., Armato, U.: Silk fibroin/ poly(carbonate)-urethane as a substrate for cell growth: *in vitro* interactions with human cells. Biomaterials 24, 789–799 (2003)
11. Danielson, K.G., Baribault, H., Holmes, D.F., Graham, H., Kadler, K., Iozzo, R.V.: Targeted disruption of decorin leads to abnormal collagen fibril morphology and skin fragility. The Journal of Cell Biology 136(3), 729–743 (1997)
12. Foschi, D., Corsi, F., Cellerino, P., Rizzi, A., Morandi, E., Trabucchi, E.: Angiogenic effects of suture biomaterials. An experimental study in rats. Eur. Surg. Research 33(1), 16–20 (2001)
13. Gobin, A.S., Butler, C.E., Mathur, A.B.: Repair and regneration of the abdominal wall musculofascial defect using silk fibroin-chitosan blend. Tissue Engineering 12(12), 3383–3394 (2006)
14. Gobin, A.S., Froude, V.E., Mathur, A.B.: Structural and mechanical characteristics of silk fibroin and chitosan blend scaffolds for tissue regeneration. J. Biomed. Mat. Res. 74A(3), 465–473 (2005)
15. Gobin, A.S., West, J.L.: Effects of epidermal growth factor on fibroblast migration through biomimetic hydrogels. Biotechnology Progress 19(6), 1781–1785 (2003)
16. Higgins, S.P., Solan, A.K., Niklason, L.E.: Effects of polyglycolic acid on porcine smooth muscle cell growth and differentiation. Journal of Biomedical Materials Research 67(1), 295–302 (2003)
17. Horiuchi, K., Naito, I., Nakano, K., Nakatani, S., Nishida, K., Taguchi, T., Ohtsuka, A.: Three-dimensional ultrastructure of the brush border glycocalyx in the mouse small intestine: a high resolution scanning electron microscopic study. Archives of Histology and Cytology 68(1), 51–56 (2005)
18. Inoue, S., Magoshi, J., Tanaka, T., Magoshi, Y., Becker, M.: Atomic force microscopy: Bombyx mori silk fibroin molecules and their higher order structure. Journal of polymer science B: Polymer Physics 38, 1436–1439 (2000)
19. Inouye, K., Kurokawa, M., Nishikawa, S., Tsukada, M.: Use of Bombyx mori silk fibroin as a substratum for cultivation of animal cells. J. Biochem. Biophys. Methods 37, 159–164 (1998)
20. Iwasaki, S., Hosaka, Y., Iwasaki, T., Yamamoto, K., Nagayasu, A., Ueda, H., Kokai, Y., Takehana, K.: The modulation of collagen fibril assembly and its structure by decorin: An electron microscopic study. Archives of Histology and Cytology 71(1), 37–44 (2008)

21. Kadler, K.: Matrix Loading: Assembly of extracellular matrix collagen fibrils during embryogenesis. Birth Defects Research (Part C) 72, 1–11 (2004)
22. Kaplan, D., Adams, W., Farmer, B., Viney, C. (eds.): Silk Polymers Materials Science and Biotechnology. American Chemical Society, Washington (1994)
23. Li, J., Ogiso, M., Minoura, N.: Enzymatic degradation behavior of porous silk fibroin sheets. Biomaterials 24, 357–365 (2003)
24. Mathur, A.B., Collinsworth, A.M., Reichert, W.M., Kraus, W.E., Truskey, G.A.: Endothelial, cardiac muscle and skeletal muscle exhibit different viscous and elastic properties as determined by atomic force microscopy. Journal of Biomechanics 34(12), 1545–1553 (2001)
25. Mathur, A.B., Reichert, W.M., Truskey, G.A.: Flow and high affinity binding affect the elastic modulus of the nucleus, cell body, and the stress fibers of endothelial cells. Annals of Biomedical Engineering 35(7), 1120–1130 (2007)
26. Mathur, A.B., Tonelli, A.E., Rathke, T., Hudson, S.: The dissolution and characterization of bombyx mori silk fibroin in calcium nitrate-methanol solution and the regeneration of films. Biopolymers 42, 61–74 (1997)
27. Mathur, A.B., Truskey, G.A., Reichert, W.M.: Atomic force and total internal reflection fluorescence microscopy for the study of force transmission in endothelial cells. Biophysical Journal 78(4), 1725–1735 (2000)
28. Matthews, B.D., Pratt, B.L., Pollinger, H.S., Backus, C.L., Kercher, K.W., Sing, R.F., Heniford, B.T.: Assessment of adhesion formation to intra-abdominal polypropylene mesh and polytetrafluoroethylene mesh. Journal of Surgical Research 114, 126–132 (2003)
29. Miller, L., Putthanarat, S., Eby, R., Adams, W.: Investigation of the nanofibrillar morphology of silk fibers by small angle x-ray scattering and atomic force microscopy. International Journal of Biological Macromolecules 24, 159–165 (1999)
30. Minoura, N., Aiba, S., Gotoh, Y., Tsukada, M., Imai, Y.: Attachment and growth of cultured fibroblast cells on silk protein matrices. Journal of biomedical materials research 29, 1215–1221 (1995)
31. Miranti, C.K., Brugge, J.S.: Sensing the environment: a historical perspective on integrin signal transduction. Nature Cell Biology 4(4), E83–E90 (2002)
32. Morita, Y., Tomita, N., Aoki, H., Wakitani, S., Tamada, Y., Suguro, T., Ikeuchi, K.: Visco-elastic properteis of cartilage tissue regenerated wtih fibroin sponge. Bio-Medical Materials and Engineering 12, 291–298 (2002)
33. Panilaitis, B., Altman, G., Chen, J., Jin, H.-J., Karageorgiou, V., Kaplan, D.: Macrophage response to silk. Biomaterials 24, 3079–3085 (2003)
34. Pesen, D., Hoh, J.H.: Micromechanical architecture of the endothelial cell cortex. Biophysical Journal 88, 670–679 (2005)
35. Rao, S., Sharma, C.: Use of chitosan as a biomaterial: studies on its safety and hemostatic potential. Journal of biomedical materials research 34(1), 21–28 (1997)
36. Rios, C.N., Skoracki, R.J., Miller, M.J., Satterfield, W.C., Mathur, A.B.: *In vivo* bone formation in silk fibroin and chitosan blend scaffolds via ectopically grafted periosteum as a cell source: a pilot study. Tissue Engineering (February 2009) (in Press)
37. Santin, M., Motta, A., Freddi, G., Cannas, M.: *In vitro* evaluation of the inflammatory potential of the silk fibroin. J. Biomed. Mater. Res. 46, 382–389 (1999)
38. Scott, J.E.: The first and second 'laws' of chemical morphology, exemplified in mammalian extracellular matrices. European Journal of Histochemistry 46, 111–124 (2002)
39. Scott, J.E.: Elasticity in extracellular matrix 'shape modules' of tendon, cartilage, etc. A sliding proteoglycan-filament model. Journal of Physiology 553(Pt 2), 335–343 (2003)

40. Scott, J.E., Dyne, K.M., Thomlinson, A.M., Ritchie, M., Bateman, J., Cetta, G., Valli, M.: Human cells unable to express decoron produced disorganized extracellular matrix lacking "shape modules" (interfibrillar proteoglycan bridges). Experimental Cell Research 243, 59–66 (1998)
41. Scott, J.E., Parry, D.A.D.: Control of collagen fibril diameters in tissues. International Journal of Biological Macromolecules 14, 1–2 (1992)
42. Sofia, S., McCarthy, M., Gronowicz, G., Kaplan, D.: Functionalized silk-based biomaterials for bone formation. Journal of Biomedical Materials Research 54, 139–148 (2001)
43. Squire, J.M., Chew, M., Nneji, G., Neal, C., Barry, J., Michel, C.: Quasi-periodic substructure in the microvessel endothelial glycocalyx: a possible explanation for molecular filtering. Journal of Structural Biology 136, 239–255 (2001)
44. Sugihara, A., Sugiura, K., Morita, H., Ninagawa, T., Tubouchi, K., Tobe, R., Izumiya, M., Horio, T., Abraham, N., Ikehara, S.: Promotive effects of a silk film on epidermal recovery from full-thickness skin wounds. P.S. E. B. M. 225, 58–64 (2000)
45. Tomihata, K., Ikada, Y.: *In vitro* and *in vivo* degradation of films of chitin and its deacetylated derivatives. Biomaterials 18(7), 567–575 (1997)
46. Ushiki, T.: Collagen fibers, reticular fibers, and elastic fibers. A comprehensive understanding from a morphological viewpoint. Archives of Histology and Cytology 65(2), 109–126 (2002)
47. Vink, H., Duling, B.R.: Identification of distinct luminal domains for macromolecules, erythrocytes, and leukocytes iwthin mammalian capillaries. Circulation Research 79, 581–589 (1996)
48. Vogel, V., Baneyx, G.: The tissue engineering puzzle: a molecular perspective. Annual Review in Biomedical Engineering 5, 441–463 (2003)
49. Walpita, D., Hay, E.: Studying actin-dependent processes in tissue culture. Nature Reviews 3, 137–141 (2002)
50. Weinbaum, S., Zhang, X., Han, Y., Vink, H., Cowin, S.C.: Mechanotransduction and flow across the endothelial glycocalyx. Proceedings of the National Academy of Sciences 100(13), 7988–7995 (2003)
51. Weis, S.M., Zimmerman, S.D., Shah, M., Covell, J.W., Omens, J.H., Ross, J.J., Dalton, N., Jones, Y., Reed, C.C., Iozzo, R.V., McCulloch, A.D.: A role for decorin in the remodeling of myocardial infarction. Matrix Biology 24(4), 313–324 (2005)
52. Woods, A., Smith, C.G., Rees, D.A., Wilson, G.: Stages in specialization of fibroblast adhesion and deposition of extracellular matrix. European Journal of Cell Biology 32(1), 108–116 (1983)

Abdominal Wall Hernias and Biomaterials

Levi Procter[1], Erin E. Falco[2], John P. Fisher[3], and John S. Roth[4]

[1] University of Kentucky College of Medicine,
Division of General Surgery, Lexington, Kentucky
ldproc0@uky.edu
[2] University of Maryland, Department of Chemical and Biomolecular
Engineering, College Park, Maryland
efalco@umd.edu
[3] University of Maryland, Fischell Department of Bioengineering,
College Park, Maryland
jpfisher@umd.edu
[4] University of Kentucky College of Medicine, Division of General Surgery,
Director of Minimally Invasive Surgery, Lexington, Kentucky
jsroth2@emial.uky.edu

1 Introduction

Hernia is derived from the Latin word meaning "rupture or protrusion". A hernia is the protrusion of tissue or an organ through a defect or weakness in the surrounding walls. Abdominal wall hernias occur at sites lacking a covering with overlapping aponeuroses and fascia. Hernias can be present at birth (congenital), develop spontaneously over time or as a result of surgery or trauma. Areas on the ventral abdominal wall prone to hernia formation are located at the arcuate line, epigastric, inguinal and umbilical hernias. Iatrogenic hernias on the abdominal wall occur at sites of surgical incisions (incisional hernias).

Hernias are a common problem in the United States. Nearly 600,000 hernia repairs are performed each year making it one of the most common general surgical procedures formed annually. Approximately 90,000 of these hernia repairs are ventral or incisional hernia repairs (Candage, Jones et al. 2008). Usher (Usher, Ochsner et al. 1958), in 1958, pioneered the evolution of prosthetic ventral hernia repair evolving into a tension-free prosthetic repair being advocated in all abdominal wall hernias (Luijendijk, Hop et al. 2000; Burger, Luijendijk et al. 2004). Over the past 50 years, many techniques for hernia repair and prosthetic materials have been utilized. Although many options for hernia repair are available to the surgeon, the ideal hernia repair technique and prosthetic material remains a topic of considerable debate. A thorough understanding of abdominal wall anatomy, individual hernia characteristics, and patient factors must be considered when choosing the most appropriate procedure and mesh prosthesis for hernia repair.

A. Gefen (Ed.): Bioengineering Research of Chronic Wounds, SMTEB 1, pp. 425–447.
springerlink.com

2 Anatomy of the Anterior Abdominal Wall

The anterior abdominal wall is a complex structure composed of the rectus abdominis muscles anteriorly and three muscle groups laterally; internal and external obliques and transversus abdominis. The paired rectus abdominis muscles run from the xiphoid process to the pubic symphysis. The external oblique has obliquely oriented "hands-in-pockets" muscle fibers originating on the external and inferior surfaces of the costal margin and attach to the outer lip of the iliac crest inferiorly and fuse medially at the linea alba. The internal oblique is deep to the external oblique with the muscle fibers oriented 90° to the external obliques fusing medially at the linea alba. The transversus abdominis muscle is deep to the internal oblique fusing medially at the linea alba.

Muscles are enveloped by aponeuroses which are sheet-like fibrous membranes, connecting muscles to bone or other muscle groups. Aponeuroses can fuse with others to create strong tendinous structures. The aponeurosis of the external oblique fuses with the anterior layer of the internal oblique aponeurosis to form the anterior rectus sheath. The posterior rectus sheath is composed of the posterior aponeurotic layer of the internal oblique and the transversus abdominis aponeurosis. The aponeuroses of all of these muscle groups fuse in the midline at the linea alba. The rectus abdominis loses the posterior rectus sheath at the arcuate line, which is approximately 3 to 6-cm inferior to the umbilicus. All of the aponeurotic layers below the arcuate line travel anterior to the rectus abdominis muscles.

3 Ventral Hernias

A ventral hernia (VH) is any protrusion of tissue and/or organ through a defect in the anterior abdominal wall fascia. Incisional hernias represent 20% of all abdominal wall hernias, are the most common form of ventral hernia while umbilical hernias represent nearly ten percent of all ventral hernias(Mark A. Malangoni 2007). Epigastric and Spigelian hernias occur with a far lower frequency than other ventral hernias. See Figure 1 for general anatomy and hernia locations.

3.1 Evaluation for Ventral Hernia

VHs are diagnosed routinely during a history and physical exam. Performing maneuvers that increase intra-abdominal pressure help define the hernia boundaries and size. In the event of an unreliable physical exam, the use of diagnostic imaging (ultrasound or computed topographic scan) may be required for definitive diagnosis. The decision to repair depends on the presence of symptoms and the type of hernia. Small hernia defects increase the risk of incarceration and strangulation therefore deserve expedient elective repair however inguinal hernias in men are safe to clinically follow if minimally symptomatic(Fitzgibbons, Giobbie-Hurder et al. 2006). Relative indications for VH repair include pain, small defect, obstruction, decreased ability to eat, loss of abdominal domain and decreased quality of life. A hernia will not resolve (except

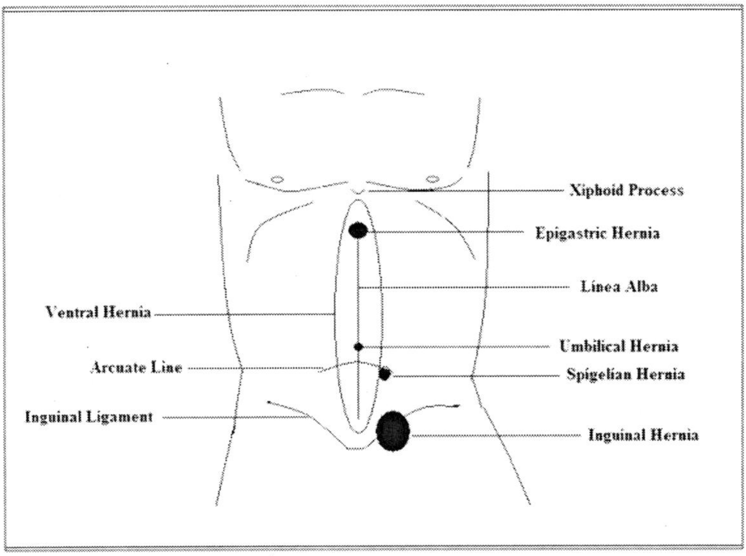

Fig. 1. This figure illustrates typical abdominal wall anatomy and location of common hernias. *Illustration by Levi Procter, MD.*

for an infantile umbilical hernia) without repair and will increase in size over time. The presence of a hernia is an indication for repair unless the operative risk is prohibitive. Absolute indications for repair are incarceration and strangulation.

3.2 Umbilical Hernia

There are three types of umbilical hernia (UH); adult, infantile and congenital (Muschaweck 2003). A congenital umbilical hernia (omphalocele) is present at birth and occurs by two unique mechanisms. Either the fetal gut fails to return intra-abdominally after week 8 or the ventral abdominal wall fails to fuse after week 8. An infantile umbilical hernia is small fascial defect, typically less than 1 cm with only 5% still apparent at 5 years of age therefore repair is often unnecessary(Muschaweck 2003). The management of congenital umbilical hernias includes primary surgical closure, staged approaches involving a series of operations and occasionally non-operative strategies.

Adult UHs occur in 2%(Jackson and Moglen 1970; Velasco M 1999) of the population and is an acquired hernia. Only 10% of UHs result from a persistent infantile umbilical hernia(Jackson and Moglen 1970). Time and increased intra-abdominal pressure canalize the umbilical channel. UHs occur more often in women and African Americans(Salameh 2008). Common causes for increased intra-abdominal pressure are obesity, chronic cough, intra-abdominal mass, multiple gestations and ascites(Muschaweck 2003). New UHs in adults should raise the suspicion for an intra-abdominal process (e.g. cancer) and appropriate workup is warranted. The UH defect is often small and is associated with a 17.7%

incarceration rate(Hjaltason 1981) therefore repair is indicated in both symptomatic and asymptomatic patients. Patients often complain of pain at the hernia site and poor cosmesis. Symptoms necessitate repair regardless of defect size. Expeditious repair should be considered in adults with an UH associated with thinning of the overlying skin, large hernias or unmanageable ascites. Ascites increase hernia recurrence(Kirkpatrick and Schubert 1988) however is not a contraindication to umbilical hernia repair (UHR). All options should be exhausted to control the ascites prior to UHR because the rupture of an umbilical hernia as a result of ascites is associated with significant morbidity and mortality. Elective repair should be considered after failure of maximal ascites control to prevent this complication (Kirkpatrick and Schubert 1988).

UHR can be performed open or laparoscopic. The classic vest-over pants repair described by Mayo in 1895, has fallen out of favor due to high recurrence rates (~20-30%) (Celdran, Bazire et al. 1995). Umbilical hernia repair with a synthetic prosthetic mesh has been advocated to dramatically reduce hernia recurrence. In a randomized clinical trial, Arroyo, demonstrated approximately 1% recurrence rate with a tension free synthetic prosthetic mesh repair (Arroyo, Garcia et al. 2001) and supported in other studies (Luijendijk, Hop et al. 2000; Arroyo Sebastian A 2002; Yavuz, Ipek et al. 2005). A tension free UHR with a prosthetic mesh has proven success, however, laparoscopic repair is gaining popularity among surgeons and patients. Laparoscopic hernia repair has risks not associated with the open repair. Laparoscopy requires general anesthesia, placement of intra-abdominal trocars increasing the risk of major visceral and vascular injuries and trocar site hernias. The location of the mesh prosthesis for optimal outcomes is debated. The mesh prosthesis placement during laparoscopic UHR can be intra-abdominal or pre-peritoneal. Intra-abdominal prosthesis placement increases the risk of visceral adhesion leading to bowel obstruction and/or enterocutaneous fistulae(Nguyen, Lee et al. 2000; Losanoff, Richman et al. 2002; Balique, Benchetrit et al. 2005). Laparoscopic pre-peritoneal prosthesis placement has demonstrated success in laparoscopic UHR with the potential to prevent the complications associated with intra-abdominal prosthesis placement however long-term data is lacking on the efficacy of this technique (Hilling, Koppert et al. 2008).

3.3 Epigastric Hernia

An epigastric hernia occurs anywhere in the midline spanning the xiphoid process superiorly and the umbilicus inferiorly and occurs in 3-5% of the population (Klinge, Prescher et al. 1997). Many theories exist to explain the development of epigastric hernias however no definitive etiology is known (Lange 1967; Askar 1978; Askar 1984; Larson and Vandertoll 1984; Axer, Keyserlingk et al. 2001; Korenkov, Beckers et al. 2001; Lang, Lau et al. 2002). The defect is commonly small and solitary although multiple defects occur in up to 20% of patients and are often just lateral to the midline(Muschaweck 2003). Physical exam can be difficult to differentiate from other masses such as lipoma or fibroma often requiring imaging modalities to confirm the diagnosis. Ultrasound is effective in diagnosis however a CT scan may be a useful adjunct in indeterminate studies. Repair is indicated after diagnosis due to the small defect size and high risk of incarceration (often preperitoneal fat). Open or laparoscopic approaches are appropriate. A tension free synthetic mesh prosthesis repair is recommended because primary suture repair has a 10%-20% recurrence rate (Muschaweck 2003). Laparoscopic repair often reveals adjacent fascial defects, which can be repaired simultaneously. Muschaweck reports a 0% recurrence rate utilizing a tension free prosthetic repair at 5 years (Muschaweck 2003).

3.4 Spigelian Hernia

A Spigelian hernia is named after Adriaan van den Spieghel, a Belgian anatomist, who described the anatomy in this region(Larson and Farley 2002). A Spigelian hernia occurs at the lateral edge of the anterior rectus fascia at, or caudal to, the arcuate line of Douglas. The arcuate line is a natural weakness in the anterior abdominal wall fascia (Spigelian fascia) and is prone to hernia formation due to the lack of the posterior rectus fascia. Spigelian hernias typically have a small defect therefore are high risk for incarceration or strangulation. This hernia is difficult to palpate because the defect is sheltered by external oblique aponeurosis thus minimizing or eliminating the "bulge". Diagnosis often requires imaging modalities such as CT, US or magnetic resonance imaging (MRI). Primary repair is efficacious. Long term data on laparoscopic repair, with or without a prosthetic mesh, is unavailable.

3.5 Inguinal Hernia

Inguinal hernias are the most common hernia among males and females but more common in males, 9:1. Inguinal hernias are bilateral in 20% of patients. Indirect and direct inguinal hernias are the two types and are defined by their relationship to the inguinal triangle (Hesselbach's triangle). Hesselbach's triangle is bound laterally by the inferior epigastric vessels, medially by the lateral edge of the rectus muscle and inferiorly by the inguinal ligament. A direct hernia occurs medial to the inferior epigastric vessels and indirect inguinal hernias lateral to the inferior epigastric vessels. Indirect hernias typically result from a patent process vaginalis and a weak transversalis fascia. An indirect hernia follows the spermatic cord structures exiting the internal ring. A direct hernia results from a weakness in the transversalis fascia.

3.5.1 Open Inguinal Hernia Repair

Numerous techniques for open inguinal hernia repair have been described. The *American College of Surgeons* recommends the open Lichtenstein inguinal hernia repair(Fitzgibbons 2000) which is the most commonly performed technique. Lichtenstein demonstrated impressive success using a tension-free mesh on-lay repair. Recurrence rates are < 1%(Nathan and Pappas 2003). No difference in hernia recurrence or complications has been demonstrated between open and laparoscopic inguinal hernia repair (Ridings and Evans 2000; McCormack, Scott et al. 2003). The optimum repair is dictated by the individual surgeon's experience.

3.5.2 Laparoscopic Inguinal Hernia Repair: Transabdominal Preperitoneal Approach (TAPP) and Totally Extra-Peritoneal Repair (TEP)

TAPP was first described by Arregui in 1991(Riazz AA 2004). TAPP utilizes laparoscopy to reflect the inguinal peritoneum to identify the hernia defect(s) and contents. Hernia contents are reduced and a prosthetic mesh is placed and affixed in this preperitoneal plane. The peritoneum is then re-approximated over the mesh to reduce mesh-viscera related complications.

TEP mimics the TAPP technique differing by the method required to access to the inguinal space. A pre-peritioneal space is created with pneumatic gas balloon insufflation. The hernia sac is identified, reduced, and mesh is placed over the inguinal defect. Pneumoperitoneum is released allowing the tissue planes to seal around the mesh. Advantages of this repair include avoidance of the need for peritoneal closure and intra-abdominal access.

Both TAPP and TEP require dissection in the pre-peritoneal space, reduction of hernia contents and mesh placement, however neither has demonstrated superiority (McCormack, Scott et al. 2003; McCormack K 2005). Generally accepted indications for laparoscopic inguinal hernia repair are recurrent inguinal hernias and bilateral inguinal hernias. Unilateral hernias are also repaired utilizing this technique.

3.6 Incisional Hernia

An incisional hernia is the most common complication following a laparotomy occurring in 2-11% of patients with recurrence rates ranging from 5-63%(Shell, de la Torre et al. 2008). IH repair is associated with significant morbidity and enormous health care costs. IHs enlarge over time increasing the loss of abdominal domain. Symptoms include changes in bowel habits, pain with eating, thinning of overlying skin, alterations in respiratory mechanics and decreased quality of life. No gold standard repair exists. Primary suture repair a recurrence rate in excess of 60%. The use of a synthetic mesh prosthesis in a tension-free manner has halved recurrence rates (Anthony, Bergen et al. 2000; Luijendijk, Hop et al. 2000). A Cochrane review of open incisional hernia repairs found that mesh prosthesis repair is superior to primary suture repair. Mesh prosthesis repair, however, is inferior in contaminated cases(den Hartog, Dur et al. 2008). The etiology of IH recurrence is multifactorial including patient comorbidities, surgical technique and post-operative complications (e.g. surgical site infection). Recent data is emerging that the individual's tissue response to the prosthesis and subsequent collagen synthesis may be a significant factor in IH recurrence (Jansen, Mertens Pr et al. 2004; Franz 2008). The most significant risk factor for IH recurrence is a surgical site infection.

Risk factors for IH include obesity, smoking, aneursymal disease (collagen metabolism defects), malnutrition, prostatism, chronic cough (chronic obstructive pulmonary disease), renal failure, malignancy and immunosuppression (Santora and Roslyn 1993; Luijendijk, Hop et al. 2000; Heniford, Park et al. 2003; Yahchouchy-Chouillard, Aura et al. 2003; Jansen, Mertens Pr et al. 2004). Surgical techniques play an important role in the development of IH. A significant reduction in IH has resulted from the use of a double stranded monofilament suture (e,g. PDS) ran in a continuous fashion.

Presence of an IH is indication for repair. Primary suture repair is indicated in a singular defect < 5cm. If the defect is > 5cm a tension-free mesh prosthetic repair results in nearly a 50% decrease in IH recurrence (Burger, Luijendijk et al. 2004) compared to a primary suture repair. Failure of a mesh prosthetic repair often results from separation of the lateral mesh prosthesis-fascia fixation. Identifying healthy fascia and fixation with at least 4 cm of mesh prosthesis-fascia overlap significantly reduces this complication.

If visceral exposure is present or possible then an appropriate mesh prosthesis must be selected. The method of IH repair depends upon the location, size, surrounding tissue, visceral exposure and the presence of contamination. IH repair can be performed open or laparoscopically. Open IH repair with a synthetic prosthetic mesh is generally performed using one of three general techniques; 1) primary suture repair of fascial components, 2) repair with mesh prosthesis placement, or 3) a separation of component (with or without a mesh prosthesis placement). The mesh prostheses repairs are defined based on the location of the mesh prosthesis in the abdominal wall referred to as on-lay, sub-lay or in-lay. An on-lay repair involves placement of a mesh prosthesis in the pre-fascial plane. This repair requires extensive lateral dissection, increasing the risk of seroma formation and wound infection. A sublay repair places a mesh prosthesis either sub-fascial or pre-peritoneal. The inlay uses the mesh prosthesis essentially as a bridge between fascial edges that cannot be closed through primary suture without tension. This repair has fallen out of favor due to high recurrence rates, which result from the tension related prosthesis-fascia separation. A separation of components repair isolates the individual layers of the lateral abdominal wall, incising them to allow tissue advancement to close the fascia without tension.

Superiority between laparoscopic or open incisional hernia repair is debated. Laparoscopic hernia repair (LHR) has less postoperative pain, earlier return to work or daily activities and shorter LOS. However, LHR requires greater surgeon skill, higher patient costs, operative duration, visceral and vascular injuries and requires general anesthesia (Wright, Kennedy et al. 1996; Champault, Rizk et al. 1997; Heikkinen, Haukipuro et al. 1997; Khoury 1998; Wellwood, Sculpher et al. 1998; Johansson, Hallerback et al. 1999; McCormack, Scott et al. 2003; Memon, Cooper et al. 2003). Exposure of the viscera to synthetic materials such as polypropylene (PP) or polyester is not without risk. These products can lead to viscera-to-mesh prosthesis adhesion with erosion, bowel obstructions and fistulization (Kaufman, Engelberg et al. 1981; Leber, Garb et al. 1998; Matthews, Pratt et al. 2003). These findings have led to intensive research to develop a composite or combination prosthetic product to prevent visceral tissue ingrowth while allowing non-visceral (parietal) tissue ingrowth, safety in contaminated fields and long term strength. Barrier coatings have been developed to lessen or prevent these visceral complications and show promise.

4 History of Prosthetic Hernia Repair

Prosthetic mesh is a device that acts as a bridge or covers an anatomic defect to prevent hernia recurrence. Egyptians used external wraps and bandages to control a hernia bulge or protrusion (Lyons AS 1987; JF 1995; FHG 1997; Lau 2002). Ancient Greeks used gold wire as suture followed by the use of silver strands woven into filigrees for the first hernia prosthesis(Goepel 1900). Tantalum prosthesis most closely resembled a true prosthesis as used today(Burke 1940). Stiffness, patient discomfort, soft tissue complications and gradual material resorption led to research into other materials, mainly plastics. Theodore Billroth is credited with envisioning the current practice of prosthesis repair. Billroth proposed, in 1878, "a device or artificial tissue that mimicked human fascia or tendon and strength would solidify hernia repair"(Read 2004). The ability to polymerize carbon compounds revolutionized hernia surgery through the development of plastics. Nylon was the first plastic material used as a suture and was later woven into a mesh prosthesis for hernia repair(Aquaviva 1944). Nylon was not suitable in hernia repair because it lost strength over time due to hydrolytic digestion and it required explantation if infected. Koontz et al, in 1959, proposed the search for a nonmetallic, synthetic, non-absorbable material that was resistant to infection(Koontz AR 1959). This led to the testing of many materials including PP, polytetrafluorethylene (PTFE), Dacron, Orlon, polyethylene, mylar and Marlex. The most commonly used prosthetic materials are discussed below.

5 Synthetic Prosthetic Meshes

The large number of available synthetic prosthetic mesh options illustrates the lack of a single best material. A synthetic mesh prosthesis should be biocompatible, strong, resistant to infection, non-immunogenic and minimal bio-reactivity. Two pathways exist for implanted prosthetic meshes, degradation or incorporation. Tissue incorporation is the goal and depends upon the material, density, filament type, three-dimensional construction, pore size, electric charge and compliance. All of these elements can be altered in prosthetic meshes to accommodate a wide range of mechanical and biological properties.

Mesh prostheses are classified by pore size. A pore size of at least 75 μm is macroporous, Type I. Macroporous prostheses permit significant tissue ingrowth including angiogensis, fibroplasia, collagen ingrowth and immune cell surveillance. Type II prostheses have a pore size < 10 μm and are termed micro-porous. Type III prostheses are macroporous and consist of microporous elements (filaments) to maximize the benefits of both pore sizes (parietal tissue in-growth and inhibition of visceral tissue in-growth). Type IV prosthetics are composed of a submicronic porous material however they are not used as hernia mesh prostheses (*Table 1*). Use of multifilament or monofilament prostheses affects pore size. Multifilament prostheses have smaller pore sizes, typically 10 μm or less thus inhibiting rich collagenous ingrowth and immune cell surveillance.

Table 1. This table lists the classes of prosthetic hernia mesh (Type I-III) including pore size and examples of mesh products

Prosthetic Mesh Classes		
Class	Pore Size (um)	Examples of Products (Manufacturers)
Type I	> 75, Macroporous and monofilament	Bard® (Davol, Inc), Prolene® (Ethicon), ProLite® (Atrium), Marlex® (C.R. Bard), Trelex®
Type II	< 10, Microporous	ePTFE, Dual-Mesh® and Surgical Membrane®(Gore-Tex)
Type III	Macroporous with multi-filamentous or microporous components	Braided Dacron (Mersilene®), braided PP(Surgipro®) and perforated PTFE (MycroMesh®)
Type IV	Submicronic pore size *(not a suitable prosthesis)*	Silastic and polypropylene sheeting

Porosity allows for fluid egress and confers significant host responses, infection tolerance and tissue in-growth. Up to 20% of implanted prosthetic meshes become infected (Leber, Garb et al. 1998) often requiring explantation of the mesh therefore resistance to infection is paramount. Mesh prosthetic pore size confers infection resistance. Bacteria are significantly smaller than immune cells. A pore size less than 50 μm blocks macrophage transit but allows bacteria passage therefore the body's immune cells are unable to clear a mesh prosthesis infection(Grevious, Cohen et al. 2006). Porosity may limit or prevent seroma formation however not all prostheses are porous (e.g. ePTFE). Pores < 75 μm prevent fibrous in-growth but allows histiocyte in-growth. Histocytic tissue is weak and non-ordered therefore the mesh prosthesis is not firmly incorporated into the surrounding tissue. A pore size > 75 μm allows for angiogenesis, fibroplasia, collagen synthesis (particularly collagen I) and immune cell surveillance permitting graft incorporation and infection resistance. *Table 2* outlines cell types and corresponding prosthetic mesh type.

Table 2. All available mesh prostheses are susceptible to bacteria cell passage however only Type I-II are passable by human immune cells. Type II-III mesh should be avoided in contaminated fields and if these prostheses become infected they must be explanated.

Immune and Bacterial Cell Diameter		
Cell	Pore (μM)	Susceptible Mesh Type
Macrophage	>50	Type I
Neutrophil	10-15	Type I-II
Bacteria	1	Type I-III

Mesh prosthesis compliance depends on the material, porosity, density and construction. Increasing pore size increases mesh compliance at the expense of strength. Less compliant mesh prostheses are typically denser, greater strength tolerance however a greater inflammatory reaction is common leading to significant scar formation, mesh prosthesis shrinkage, and mesh prosthesis erosion through skin or adjacent organs.

All synthetic implants are recognized as a foreign material initiating an inflammatory response in four phases. The initial phase is acute inflammation and the magnitude of the response depends on the type and amount of synthetic material present. Lightweight prosthetic meshes have less of an inflammatory response than their heavyweight counterparts (Klosterhalfen, Klinge et al. 1997; Klosterhalfen, Klinge et al. 1998; Klinge, Klosterhalfen et al. 1999). The inflammatory phase is followed by a chronic inflammation where the prosthetic implant is excluded from the surrounding tissue. Platelet derived growth factory (PDGF), transforming growth factor-B (TGF-B), insulin like growth factor (ILGF) and epidermal growth factor are generated at the mes prosthesis site promoting collagen scar formation. Ideally, the prosthetic mesh should allow three-dimensional in-growth of collagen.

All synthetic prosthetic meshes are sufficiently strong (Klinge, Klosterhalfen et al. 1998; Cobb, Burns et al. 2006; Binnebosel, Rosch et al. 2007). Assuming the abdomen is a sphere and applying Laplace's law the maximum force required to maintain abdominal integrity is 16 N/cm(Bachman and Ramshaw 2008). Most commonly used mesh prosthetics have a tensile strength of at least 32 N/cm (Bachman and Ramshaw 2008). Prosthetic mesh hernia repair failure often results from the separation of the mesh prosthesis-fascia interface; not the mesh prosthesis failing. Increased strength often involves greater material density and as discussed prior, has its own risks. Choosing a stronger mesh prosthetic product should proceed cautiously. Strength depends on filament type (multi- or mono-), woven or knitted, and the polymer type. Lightweight products have been developed to lessen the risks associated with high density (heavyweight) mesh prostheses and will be discussed below.

Mesh prostheses can be woven, knitted or cut from films by laser. Woven mesh prostheses (Fig. 3a) involve strands of the material oriented at ninety-degree angles to each other with individual strands coursing in an over-under-over manner. This configuration allows stretch only in the direction oblique to the ninety-degree intersection of the strands. Polypropylene (PP) was initially woven however when it was cut the mesh would unravel leading to the development of a machine-knitted PP mesh prosthesis. Knitting (Fig. 3b) interlaces thread in a series of connected loops, locking the threads in place permitting cutting without unraveling and stretch in any direction. Marlex®, Dacron® and Prolene® are examples of knitted mesh prostheses. Knitted mesh prostheses have greater flexibility and larger pores but aren't as strong compared to woven mesh prostheses (Cobb, Peindl et al. 2009). Cutting mesh prostheses from a sheet of material with a laser provides clean melted edges to prevent unraveling or fragments from breaking off mesh prostheses (Fig. 2).

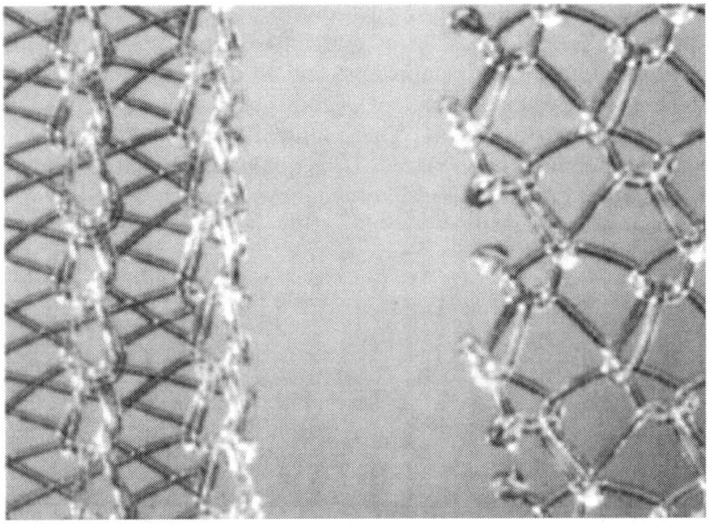

Fig. 2. The image on the left shows the edges of a laser cut edge compared to a non-laser cut edge on the right

5.1 Polypropylene (PP)

Polypropylene (PP) is a hydrophobic polymer of carbon atoms with alternating methyl moieties. PP was introduced in the early 1960s by Usher (Read 2004) as a mesh hernia prosthesis. This material is flexible, strong, easily cut (without fraying), readily incorporated by surrounding tissues and resists infection. PP remains the most popular material in prosthetic mesh hernia repair. The monofilament nature provides large pores facilitating fibrovascular in-growth, infection resistance and improved compliance.

5.2 Polytetrafluoroethylene (PTFE)

PTFE (Teflon®) is a chemically inert synthetic fluoropolymer developed by DuPont, Inc. PTFE has a high negative charge (due to fluorine) therefore water and oils do not adhere to it. This chemically inert plastic was made famous in the cooking world as a non-stick surface coating. These properties prompted researchers to evaluate it for surgical uses. Teflon chips that were placed into the peritoneal cavities of canines were found to be free-floating in the peritoneal cavity at successive intervals after implantation(Leveen and Barberio 1949; DeBord 1998). Minimal inflammatory changes were seen in the surrounding tissue and no proteins were attached to the Teflon surface. PTFE is not incorporated into human tissue and becomes encapsulated. Poor tissue incorporation increases hernia recurrence and an infected PTFE mesh prosthetic must be explanted. PTFE is microporous, which allows bacteria passage but prevents macrophage passage; therefore the body cannot clear the infection (Grevious, Cohen et al. 2006).

PTFE was improved upon through a process to expand PTFE (ePTFE). This method produced a uniform, fibrous and microporous structure with improved strength (*Figure 3c*). ePTFE remains inert and produces little inflammatory effects after implantation however it is not incorporated into tissue and has a high incidence of seroma formation. The minimal inflammatory response allows ePTFE to be placed directly on viscera. This quality permits its use in laparoscopic VHR or situations where visceral organ coverage is needed due to lack of abdominal wall in a clean case.

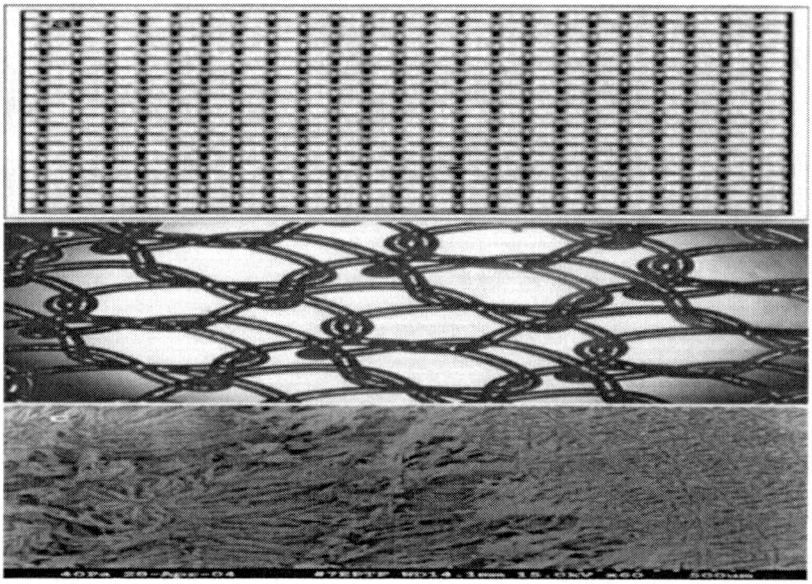

(a)

(b)

(c)

Fig. 3. Photomicrographs of woven (a), knitted (b), and expanded polymer (c) constructs (Permission Pending)

5.3 Polyester

Polyester is a carbon polymer of terepthalic acid and can be fashioned into strong fibers suitable to be woven into a prosthetic mesh. Dacron®, Parietex® and Mersiline® contain polyester. Polyester is hydrophilic and is degraded by hydrolysis. Polyester mesh prostheses were evaluated in randomized controlled trials (RCTs) versus PP mesh prostheses, implanted in a sub-lay technique, demonstrated no differences in hernia recurrences or complications. This illustrates the importance of surgical technique in preventing hernia recurrence and postoperative complications. Polyester mesh prosthetics have fallen out of favor due to extensive hydrolysis which weakens the prosthesis.

6 Combination Prostheses

Newer mesh prostheses combine materials to reach the best compromise for successful repair. These mesh prostheses are often composed of two different materials annealed together. Combination or composite mesh prostheses require a specific orientation relative to the viscera. The visceral side has a microporous surface to prevent visceral adhesions. The non-visceral side is often macroporous to allow parietal tissue in-growth. There are two categories, absorbable and permanent.

6.1 Absorbable Combination Prostheses

Barrier coated mesh prostheses are an attractive option for hernia repair. Barrier coatings allow for neoepithelialization of the mesh prosthesis before visceral adhesion. This lessens viscera-prosthesis related complications and can aid in tissue in-growth. Parietex® (Covidien, Inc) is a polyester prosthetic mesh developed as a three-dimensional weave to facilitate maximal tissue in-growth. This dual-facing prosthesis is coated on the visceral side with a collagen film to mitigate visceral. The collagen film is composed of glycerol, polyethylene glycol and porcine collagen. The hydrophilic compound coating promotes tissue in-growth by reducing scar encapsulation. These coatings are not amenable to modification (cannot be cut) and the requires hydration prior to placement.

Proceed® (Ethicon) is a synthetic, multi-layered mesh. Proceed® is composed of PP, oxidized regenerated cellulose (ORC) fabric (to decrease intestinal adhesions), and absorbable polydiaxanone between the two layers. Proceed® is lightweight, strong, low profile, compliant and color coded to distinguish correct mesh prosthesis-viscera placement. Like many PP mesh prostheses, resist bacterial growth and allow fluid egress, lessening seroma formation. Proceed® can be used in open or laproscopic ventral hernia repair. Contraindications are contaminated cases and inadequate hemostasis. Inadequate hemostasis permits tissue ingrowth to the ORC layer increasing visceral-mesh prosthesis adhesion complications. Proceed® does not grow, stretch or expand and therefore should be used with caution in infants, children, pregnant women or women who may become pregnant.

Sepramesh® (Bard, Inc) is a made of PP and bioresorbable fibers. The bioresorbable hydrogel coating (sodium hyaluronate and carboxymethylcellulose) promotes tissue in-growth while the hydrogel coating lessens visceral tissue adherence (Gonzalez, Rodeheaver et al. 2004). Sepramesh® lacks an ORC layer and polyester which makes it favorable in cases with non-ideal hemostasis. It is appropriate for use in open and laparoscopic VHR, is easily manipulated laparoscopically due to its intrinsic memory, and requires specific parietal to visceral placement. During laparoscopic VHR, this prosthesis requires hydration prior to insertion into the abdomen. The prosthesis should be rolled-up lengthwise with the hydrogel on the inside. Contraindications for using this prosthesis include contaminated cases and caution should be used in patients where tissue growth is likely because this prosthesis does not stretch or grow.

C-QUR® (Atrium Medical Corporation) is a thin, lightweight PP mesh coated with a bioresorbable omega-3 fatty acid. The coating mitigates visceral adhesion formation and significantly reduces tissue foreign body reaction allowing for maximal tissue in-growth and tissue strength. The non-polymeric bioresorbable coating reduces in-growth of dense acellular collagen. This mesh prosthesis is nearly translucent during laparoscopic placement providing the surgeon with the ability to fixate the prosthesis under direct vision to minimize the risk of damage to underlying structures. Contraindications to placement include contaminated cases and caution should be considered in patients where growth of tissue may be possible in the future.

6.2 Permanent Combination Prostheses

Combination prostheses are designed to take advantage of the properties of macro- and microporous mesh prostheses. A microporous prosthesis permits placement adjacent to viscera. The macroporous "rough side" promotes parietal tissue in-growth. These mesh prostheses are easily cut to fit specific applications. Efficacy has been demonstrated in animal models lessening visceral adhesions and complications (Gonzalez, Rodeheaver et al. 2004). These properties permit intra-peritoneal placement. DualMesh® (Gore-Tex) and Dulex® (CR Bard, Inc) are ePTFE mesh prostheses with distinct sides. Composix® (Davol, Inc) is composed of PP and ePTFE bonded together. This soft, low profile, macroporous prosthesis stimulates tissue ingrowth, and is 60% lighter than traditional PP. The thin microporous layer of ePTFE faces the viscera acting as a barrier to minimize visceral tissue attachment. Visceral adhesions have been recorded at the edge of the scaffold and in-between the layers (Gonzalez, Rodeheaver et al. 2004).

6.3 Absorbable Prostheses

Absorbable synthetic prosthetics are used in contaminated fields. Dexon® is made of polyglycolic acid which is soft, pliable, biodegradeable and resorbed over 90 days. Vicryl®, polyglactin 910, is similar to Dexon® but has less stretch. Both are cheap, woven mesh prostheses and neither is suitable for permanent hernia repair because they have > 50% recurrence rates. Absorbable mesh prostheses are used in contaminated cases where primary abdominal closure is not feasible. Patients with intra-abdominal catastrophes or abdominal compartment syndrome are susceptible to an extensive duration where their abdomen is open. Surgeons are tormented by the dilemma of when to close, what method and which mesh prosthetic material to facilitate closure. Ultimately, the decision is based on risk-benefit for each patient. The decision to close the abdomen with a prosthetic mandates the use of absorbable or biologic mesh prostheses. Absorbable mesh prostheses will allow abdominal wall closure at the expense of a persistent hernia or likely hernia recurrence. The high rate of recurrences implies a future single or multi-staged procedure to achieve an acceptable functional and cosmetic result. Biologic mesh prostheses will be discussed later.

6.4 Lightweight Or Heavyweight Prostheses

Abdominal wall mechanics, force tolerances and strength are important considerations in hernia repair. All non-biologic prostheses have sufficient strength. These prostheses can sustain at least 2-4 times the maximum tolerable forces to the abdominal wall (Klinge, Klosterhalfen et al. 1998; Cobb, Burns et al. 2006; Binnebosel, Rosch et al. 2007). In tissue models it was found that hernia recurrence, during forces greater than tolerable by the abdominal wall, occurred at the edge of the prosthesis where it was sutured to the surrounding tissue(Welty, Klinge et al. 2001). This data suggests that synthetic mesh prostheses are sufficiently strong to remain intact at supra-physiologic forces. When at least 4-cm of mesh prosthesis-tissue-overlap was tested there was a significant reduction in mesh prosthesis-tissue separation and no mesh prostheses failure(Binnebosel, Rosch et al. 2007). Technique is paramount in successful hernia repair.

Non-biologic mesh prostheses are grouped as heavyweight or lightweight. These groups are defined by the density of the material present in the prosthesis. Heavyweight meshes are at least 80-90 g/m^2 (Schumpelick, Klinge et al. 2004). Heavyweight meshes are thick and easy to handle. Surgeons feel reassured placing a strong material during the repair however superiority between heavyweight and lightweight prostheses is currently being investigated. Bellón et al demonstrated that there was no significant difference in tissue incorporation, inflammatory response or tensile strength between a heavyweight PP mesh and a lightweight PP mesh (Bellon, Rodriguez et al. 2009). The greater material density in a heavyweight prostheses can generate a greater foreign body reaction as well as decrease the compliance of the defect site(Bellon, Rodriguez et al. 2009). Klinge et al found that the lightweight PP mesh exhibited better tissue integration however these prostheses had different pore sizes leading to the conclusion that pore size may be a significant factor for tissue in-growth (Klinge, Klosterhalfen et al. 2002). Data is emerging that some previously thought inert materials, such as PP, are actually degraded (Costello, Bachman et al. 2007). The PP incites a chronic inflammatory response that degrades PP via oxidation, altering the mechanical properties of the mesh prosthesis. This data lends support for the use of lightweight prostheses in decreasing the inflammatory response.

Heavyweight mesh prostheses have many potential disadvantages. The greater foreign body load incites a greater inflammatory response leading to greater scar formation, decreased abdominal wall and graft compliance, decreased tissue incorporation, greater graft shrinkage (>50% with PP in some animal models)(Klinge, Klosterhalfen et al. 1998), increased pain and patient discomfort. Mesh prosthesis contraction occurs in all synthetic mesh prostheses to a certain degree (Gonzalez and Ramshaw 2003; Harrell, Novitsky et al. 2006; Bachman and Ramshaw 2008) however the greater the inflammatory response evoked by the prosthesis the greater the degree of contraction. Contraction can lead to hernia recurrence as the prosthesis pulls away from the surrounding repair.

Lightweight mesh prostheses are sufficiently strong to tolerate the maximal burst forces encountered by the human abdominal wall (Cobb, Burns et al. 2006). These prostheses contain less material and incite a lower inflammatory response(Klosterhalfen, Junge et al. 2005). A decreased inflammatory response results in better tissue incorporation, increased compliance of the prosthesis and abdominal wall, and decreased patient discomfort and pain. Numerous randomized prospective trials have evaluated lightweight versus heavyweight mesh in ventral hernia repair with equal outcomes in ventral hernia repair (VHR) recurrence therefore technique likely contributes to differences in recurrence (Junge, Klinge et al. 2002; Cobb, Kercher et al. 2005; Conze, Kingsnorth et al. 2005; Klosterhalfen, Junge et al. 2005; Schmidbauer, Ladurner et al. 2005).

Hernia repair success is largely dependent upon surgical technique. All prosthetic mesh materials available are sufficiently strong to be utilized for hernia repair. The choice between a lightweight and heavyweight prosthesis is multifactorial and superiority has yet to be proven.

7 Biologic Grafts

Biological grafts have been developed in hopes of achieving true graft recapitulation. They differ based on their source (human or animal), composition (dermal, pericardial or submucosa) and methods of processing (stripping, cross-linking). Biologic grafts are acellular collagen matrices implanted during hernia repair to facilitate native tissue incorporation. Graft recapitulation is not always complete, leading to chronic hernia recurrence. Certain patients demonstrate a decreased ratio of mature collagen type I-to-collagen type III in the hernia scar tissue, in the surrounding skin and connective tissues (Jansen, Mertens Pr et al. 2004; Franz 2008). This collagen ratio leads to a decrease in the tensile strength of the scar tissue formed within the defect. Additional reports have shown that patients with collagen metabolism disorders often have impaired collagen degradation and re-modeling processes which affect scar tissue strength and wound healing (Jansen, Mertens Pr et al. 2004; Franz 2008). The goal of biological grafts is to add a collagen matrix to aid in repair by creating functional tissue. An ideal biologic graft is biocompatible, strong, chemically inert, sterilizeable, non-carcinogenic and recapitulates as native tissue. The fundamental step for recapitulation is angiogenesis throughout the graft to overcome nutrient diffusion capabilities. Without a rich, extensive blood supply recapitulation is not possible. All biologic grafts attempt to utilize the body's own healing properties to recapitulate the graft.

Processing affects graft recapitulation. A processed biologic graft should allow cell growth factor stimulation while providing a structural framework to facilitate tissue growth. Stripping involves removal of all foreign materials and cells. Collagen cross-linking increases the strength of the biologic graft. As the density of collagen cross-linking increases there are decreases in cellular infiltration (decreased angiogenesis), increased fibroblast encapsulation and increased resistance to degradation by the body (Liang, Chang et al. 2004). Biologic grafts that undergo stripping or collagen cross-linking are less able to stimulate or retain cellular growth factors and signaling molecules leading to a decreased ability to promote and direct angiogenesis. Graft recapitulation cannot occur without ordered and extensive angiogenesis. This results in less graft incorporation and a residual foreign body. The density of collagen cross-linking that allows sufficient angiogenesis to sustain recapitulation is unknown, therefore, the use of collagen cross-linked grafts should be considered cautiously.

Not all biologic grafts are stripped, e.g. Surgisis® (Cook Medical). Surgisis®, a porcine submucosa product, is able to stimulate human fibroblasts (compared to stripped Surgisis®), PC-12 cell differentiation and FGF-2 growth factor release allowing for significant angiogenesis(Hiles 2009). Alloderm® (LifeCell) is an acellular human dermal matrix derived from cadaveric human skin and is non-crosslinked. This cadaveric skin undergoes a proprietary processing where the epidermis and accompanying cells are completely removed (stripped) leaving behind the 3-D dermal matrix composed of collagen (VI and fibrillar), hyaluronan, elastin, proteoglycans, fibronectin and vascular channels. Alloderm® is non-sterile, strong, and resists infection and adhesion formation making it an ideal product in contaminated fields. Alloderm® and other acellular collagen matrices are expensive and have little long term follow up for hernia recurrence. The short term hernia recurrence rates vary significantly from 16-80%(Blatnik, Jin et al. 2008) which is similar to polyglycolic with a recurrence rate of approximately 75% (Dayton M 1986) yet is 1/20th the cost of Alloderm® (Blatnik, Jin et al. 2008). Allomax® (Davol, Inc) is an acellular human dermal collagen that is stripped, non-crosslinked, sterile and requires hydration prior to use. FlexHD® (Ethicon), an acellular human dermal matrix derived from cadaveric human skin is similar to Alloderm® but doesn't require rehydration or refrigeration. This feature offers the potential for decreased operative duration (no need to wait as the product soaks) therefore decreasing costs.

Elective VHR has proven success with a synthetic mesh prosthesis placed in a tension-free technique. Patients with potential or gross contamination, enterocutaneous fistulae, simultaneous bowel resection or chronic non-healing wounds are absolute contraindications to synthetic mesh prosthesis placement. For these reasons a biologic graft is appealing. Primary closure at the time of initial operation or second look operation with a biologic prosthesis may prevent the need for highly morbid staged repairs. The proven efficacy of appropriate surgical technique combined with a synthetic prosthetic mesh obviates the need for biologic grafts in clean cases. Biologic grafts have reported success in VHR in contaminated fields of at least 75%; and up to 90% in clean cases however long term follow up is limited(Hiles, Ritchie et al. 2009).

9 Appropriate Prosthetic Choice

The ideal mesh prosthesis should be biocompatible, resistant to infection, inexpensive to manufacture and tolerate abdominal wall forces(Cobb, Peindl et al. 2009). Nearly all synthetic non-absorbable prostheses are over engineered for force tolerance in humans (Klinge, Klosterhalfen et al. 1998; Junge, Klinge et al. 2002; Cobb, Kercher et al. 2005; Cobb, Burns et al. 2006; Binnebosel, Rosch et al. 2007). The type of hernia and presence or absence of contamination determines prosthesis choice. Irrespective of the prosthesis chosen most hernia repairs fail due technical failures (Cobb, Burns et al. 2006; Binnebosel, Rosch et al. 2007). Common technical failures include insufficient mesh prosthesis-tissue overlap;

poor mesh prosthetic fixation and/or lack of mesh prosthetic graft-tissue incorporation.

Heavyweight prostheses have significant force tolerability (more than needed in humans) however the abundance of prosthetic material (greater density) incites a greater inflammatory response, more patient discomfort, decreased compliance and greater shrinkage. Decreased compliance increases the mesh erosion through the skin and/or adjacent structures. Lightweight versions of the common prosthetic materials offers the potential to decrease inflammation, infectious complications, increased compliance and improved patient comfort while supplying sufficient force tolerability.

PP and polyester are the most common prosthesis materials used. There is no evidence to suggest one is superior to the other in hernia repair or abdominal wall reconstruction. Both products are degraded over many years after implantation. PP is hydrophobic and degraded by oxidation. Polyester is hydrophilic and is degraded by hydrolysis.

The most appropriate prosthesis cannot be derived from the literature as RCT data does not exist. Hernia recurrence and clinical outcomes appear to be similar irrespective of mesh prostheses used therefore proper technique, surgeon preference and costs will largely determine prosthesis allocation. An algorithm of prosthesis choice is shown in *Figure 4*.

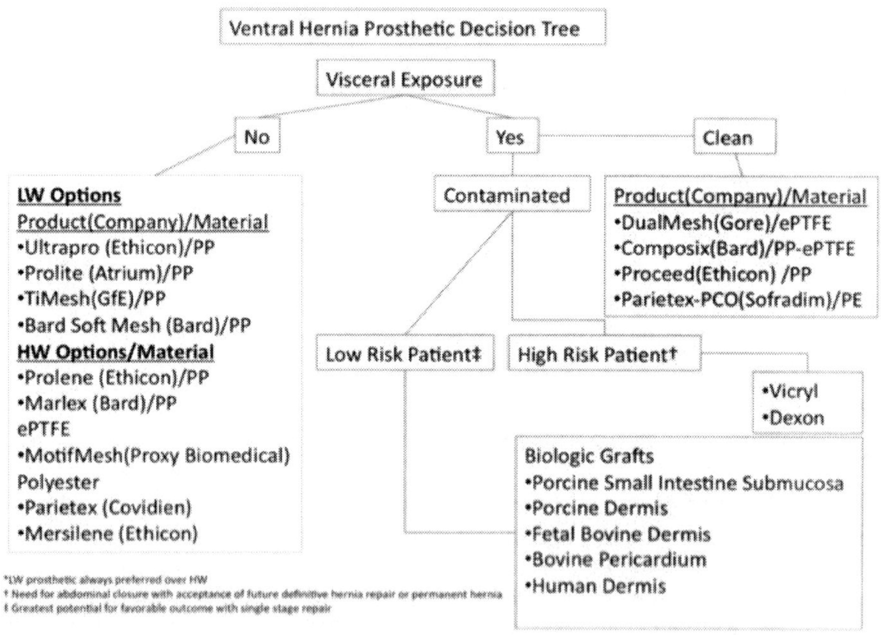

Fig. 4.

Table 3. ePTFE – expanded polytetrafluoroethylene, PP - polypropylene

Examples of Prosthetic Meshes Suitable for Intraperitoneal or Suspected Visceral Exposure	
Product/Manufacturer	Composition/Comments
Parietex Composite/Covidien®	Composed of polyester-collagen-polyethylene glycol-glycerol
C-Qur/Atrium®	PP with omega-3 fatty acid barrier coating/Lightweight and proprietary coating supports tissue ingrowth and decreased adhesive complications
Proceed/Ethicon®	PP-polydiaxone-oxidized regenerated cellulose/Lightweight
Sepramesh/Genzyme®	PP-carboxymethylcellulose-sodium hyaluronate-polyethylene glycol/Lightweight
Composix/Bard-Davol®	PP and ePTFE combination prosthesis/Heavyweight version is E/X and lightweight version is L/P
DualMesh, DualMeshPlus/W.L. Gore®	ePTFE
Dulex/Bard-Davol®	ePTFE
Parietex, Sofradim Corp®	polyester

This figure is only an example of the available prostheses and is no way an exhaustive list. *Table 3* is a non-exhaustive list of synthetic prosthetic meshes suitable in intraperitoneal placement or potential visceral exposure. Biologic prostheses lack long-term recurrence data, are expensive and shouldn't be utilized in elective hernia repairs or abdominal reconstructions. The efficacy of synthetic prostheses in elective hernia repair and abdominal wall reconstruction support avoidance of biologic prostheses until the biologic prostheses data is known. Consideration for a biologic prosthesis should be given in contaminated abdominal wall reconstructions or when the benefit of abdominal wall closure outweighs the risk of likely hernia recurrence in the future.

References

Anthony, T., Bergen, P.C., et al.: Factors affecting recurrence following incisional herniorrhaphy. World J. Surg. 24(1), 95–100 (2000); discussion 101

Aquaviva: Cure d'une volumineuse eventration par plaque de Crinofil. Extraits Bull. Soc. Chir. de Mareseille 17 (1944)

Arroyo, A., Garcia, P., et al.: Randomized clinical trial comparing suture and mesh repair of umbilical hernia in adults. Br. J. Surg. 88(10), 1321–1323 (2001)

Arroyo Sebastia´n A, P.r.F., Serrano, P., Costa, D., Oliver, I., Ferrer, R., Lacueva, J., Calpena, R.: Is prosthetic umbilical hernia repair bound to replace primary herniorrhaphy in the adult patient? (2002)

Askar, O.M.: A new concept of the aetiology and surgical repair of paraumbilical and epigastric hernias. Ann. R. Coll. Surg. Engl. 60(1), 42–48 (1978)

Askar, O.M.: Aponeurotic hernias. Recent observations upon paraumbilical and epigastric hernias. Surg. Clin. North. Am. 64(2), 315–333 (1984)

Axer, H., Keyserlingk, D.G., et al.: Collagen fibers in linea alba and rectus sheaths. I. General scheme and morphological aspects. J. Surg. Res. 96(1), 127–134 (2001)

Bachman, S., Ramshaw, B.: Prosthetic material in ventral hernia repair: how do I choose? Surg. Clin. North. Am. 88(1), ix, 101–112 (2008)
Balique, J.G., Benchetrit, S., et al.: Intraperitoneal treatment of incisional and umbilical hernias using an innovative composite mesh: four-year results of a prospective multicenter clinical trial. Hernia 9(1), 68–74 (2005)
Bellon, J.M., Rodriguez, M., et al.: Comparing the behavior of different polypropylene meshes (heavy and lightweight) in an experimental model of ventral hernia repair. J. Biomed. Mater. Res. B. Appl. Biomater. 89(2), 448–455 (2009)
Binnebosel, M., Rosch, R., et al.: Biomechanical analyses of overlap and mesh dislocation in an incisional hernia model in vitro. Surgery 142(3), 365–371 (2007)
Blatnik, J., Jin, J., et al.: Abdominal hernia repair with bridging acellular dermal matrix–an expensive hernia sac. Am. J. Surg. 196(1), 47–50 (2008)
Burger, J.W., Luijendijk, R.W., et al.: Long-term follow-up of a randomized controlled trial of suture versus mesh repair of incisional hernia. Ann. Surg. 240(4), 578–583 (2004); discussion 583-5
Burke: Corrosion of metals in tissues; and introduction to tantalum. Can. Med. Assoc. J. 43, 125–128 (1940)
Candage, R., Jones, K., et al.: Use of human acellular dermal matrix for hernia repair: friend or foe? Surgery 144(4), 703–709 (2008); discussion 709-11
Celdran, A., Bazire, P., et al.: H-hernioplasty: a tension-free repair for umbilical hernia. Br. J. Surg. 82(3), 371–372 (1995)
Champault, G.G., Rizk, N., et al.: Inguinal hernia repair: totally preperitoneal laparoscopic approach versus Stoppa operation: randomized trial of 100 cases. Surg. Laparosc. Endosc. 7(6), 445–450 (1997)
Cobb, W.S., Burns, J.M., et al.: Textile analysis of heavy weight, mid-weight, and light weight polypropylene mesh in a porcine ventral hernia model. J. Surg. Res. 136(1), 1–7 (2006)
Cobb, W.S., Kercher, K.W., et al.: The argument for lightweight polypropylene mesh in hernia repair. Surg. Innov. 12(1), 63–69 (2005)
Cobb, W.S., Peindl, R.M., et al.: Mesh terminology 101. Hernia 13(1), 1–6 (2009)
Conze, J., Kingsnorth, A.N., et al.: Randomized clinical trial comparing lightweight composite mesh with polyester or polypropylene mesh for incisional hernia repair. Br. J. Surg. 92(12), 1488–1493 (2005)
Costello, C.R., Bachman, S.L., et al.: Materials characterization of explanted polypropylene hernia meshes. J. Biomed. Mater. Res. B. Appl. Biomater. 83(1), 44–49 (2007)
Dayton, M.T., Buchele, B.A., Shirazi, S.S., Hunt, L.B.: Use of an absorbable mesh to repair contaminated abdominal wall defects. Arch. Surg. 121, 954–1040 (1986)
DeBord, J.R.: The historical development of prosthetics in hernia surgery. Surg. Clin. North. Am. 78(6), vi, 973–1006 (1998)
den Hartog, D., Dur, A.H., et al.: Open surgical procedures for incisional hernias. Cochrane Database Syst. Rev. (3), CD006438 (2008)
Bridgewater, F.H.G.: An historical perspective: reviewing the evolution of the surgical management of groin hernias. Churchill Livingstone, New York (1997)
Fitzgibbons Jr., R.J.: Management of an inguinal hernia: conventional? tension-free? laparoscopic? or maybe no treatment at all. In: Present to the General Sessions of the American College of Surgeons, 86th Annual Clinical Congress (2000)
Fitzgibbons Jr., R.J., Giobbie-Hurder, A., et al.: Watchful waiting vs repair of inguinal hernia in minimally symptomatic men: a randomized clinical trial. JAMA 295(3), 285–292 (2006)
Franz, M.G.: The biology of hernia formation. Surg. Clin. North. Am. 88(1), 1–15 (2008)
Goepel: Uber die verschliessung von bruchpforten durch einheilung geflocuhtener fertiger silberdrahtnetze. Verh. Deutsch. Ges. Chir. 9, 174–179 (1900)

Gonzalez, R., Ramshaw, B.J.: Comparison of tissue integration between polyester and polypropylene prostheses in the preperitoneal space. Am. Surg. 69(6), 471–476 (2003); discussion 476-7

Gonzalez, R., Rodeheaver, G.T., et al.: Resistance to adhesion formation: a comparative study of treated and untreated mesh products placed in the abdominal cavity. Hernia 8(3), 213–219 (2004)

Grevious, M.A., Cohen, M., et al.: The use of prosthetics in abdominal wall reconstruction. Clin. Plast. Surg. 33(2), v, 181–197 (2006)

Grevious, M.A., Cohen, M., et al.: Structural and functional anatomy of the abdominal wall. Clin. Plast. Surg. 33(2), v, 169–179 (2006)

Harrell, A.G., Novitsky, Y.W., et al.: Prospective evaluation of adhesion formation and shrinkage of intra-abdominal prosthetics in a rabbit model. Am. Surg. 72(9), 808–813 (2006); discussion 813-4

Heikkinen, T., Haukipuro, K., et al.: Total costs of laparoscopic and lichtenstein inguinal hernia repairs: a randomized prospective study. Surg. Laparosc. Endosc. 7(1), 1–5 (1997)

Heniford, B.T., Park, A., et al.: Laparoscopic repair of ventral hernias: nine years' experience with 850 consecutive hernias. Ann. Surg. 238(3), 391–399 (2003); discussion 399-400

Hiles, M.: Evidence-Based Medicine and Biologic Grafts: How Do They Stack Up for Hernia Repair? Fort Myers (2009)

Hiles, M., Ritchie, R.D., et al.: Are Biologic Grafts Effective for Hernia Repair? A Systematic Review of the Literature. Surg. Innov. (2009)

Hilling, D.E., Koppert, L.B., et al.: Laparoscopic correction of umbilical hernias using a transabdominal preperitoneal approach: results of a pilot study. Surg. Endosc. (2008)

Hjaltason, E.: Incarcerated hernia. Acta. Chir. Scand. 147(4), 263–267 (1981)

Jackson, O.J., Moglen, L.H.: Umbilical hernia. A retrospective study. Calif. Med. 113(4), 8–11 (1970)

Jansen, P.L., Mertens Pr, P., et al.: The biology of hernia formation. Surgery 136(1), 1–4 (2004)

Patino, J.F.: A history of the treatment of hernia. Lippincott, Philadelphia (1995)

Johansson, B., Hallerback, B., et al.: Laparoscopic mesh versus open preperitoneal mesh versus conventional technique for inguinal hernia repair: a randomized multicenter trial (SCUR Hernia Repair Study). Ann. Surg. 230(2), 225–231 (1999)

Junge, K., Klinge, U., et al.: Functional and morphologic properties of a modified mesh for inguinal hernia repair. World J. Surg. 26(12), 1472–1480 (2002)

Kaufman, Z., Engelberg, M., et al.: Fecal fistula: a late complication of Marlex mesh repair. Dis. Colon. Rectum. 24(7), 543–544 (1981)

Khoury, N.: A randomized prospective controlled trial of laparoscopic extraperitoneal hernia repair and mesh-plug hernioplasty: a study of 315 cases. J. Laparoendosc. Adv. Surg. Tech. A 8(6), 367–372 (1998)

Kirkpatrick, S., Schubert, T.: Umbilical hernia rupture in cirrhotics with ascites. Dig. Dis. Sci. 33(6), 762–765 (1988)

Klinge, U., Klosterhalfen, B., et al.: Impact of polymer pore size on the interface scar formation in a rat model. J. Surg. Res. 103(2), 208–214 (2002)

Klinge, U., Klosterhalfen, B., et al.: Modified mesh for hernia repair that is adapted to the physiology of the abdominal wall. Eur. J. Surg. 164(12), 951–960 (1998)

Klinge, U., Klosterhalfen, B., et al.: Shrinking of polypropylene mesh in vivo: an experimental study in dogs. Eur. J. Surg. 164(12), 965–969 (1998)

Klinge, U., Klosterhalfen, B., et al.: Foreign body reaction to meshes used for the repair of abdominal wall hernias. Eur. J. Surg. 165(7), 665–673 (1999)

Klinge, U., Prescher, A., et al.: Development and pathophysiology of abdominal wall defects. Chirurg 68(4), 293–303 (1997)

Klosterhalfen, B., Junge, K., et al.: The lightweight and large porous mesh concept for hernia repair. Expert Rev. Med. Devices 2(1), 103–117 (2005)

Klosterhalfen, B., Klinge, U., et al.: Morphologic correlation of functional abdominal wall mechanics after mesh implantation. Langenbecks Arch. Chir. 382(2), 87–94 (1997)

Klosterhalfen, B., Klinge, U., et al.: Functional and morphological evaluation of different polypropylene-mesh modifications for abdominal wall repair. Biomaterials 19(24), 2235–2246 (1998)

Koontz, A.R., Kimberly, R.C.: Further experimental work on prostheses for hernia repair. Surg. Gynecol. Obstet. 109(321) (1959)

Korenkov, M., Beckers, A., et al.: Biomechanical and morphological types of the linea alba and its possible role in the pathogenesis of midline incisional hernia. Eur. J. Surg. 167(12), 909–914 (2001)

Lang, B., Lau, H., et al.: Epigastric hernia and its etiology. Hernia 6(3), 148–150 (2002)

Lange, W.: On the inscriptiones tendinae in the human musculus rectus abdominis. Anat. Anz. 121(4), 372–380 (1967)

Larson, D.W., Farley, D.R.: Spigelian hernias: repair and outcome for 81 patients. World J. Surg. 26(10), 1277–1281 (2002)

Larson, G.M., Vandertoll, D.J.: Approaches to repair of ventral hernia and full-thickness losses of the abdominal wall. Surg. Clin. North. Am. 64(2), 335–349 (1984)

Lau, W.Y.: History of treatment of groin hernia. World J. Surg. 26(6), 748–759 (2002)

Leber, G.E., Garb, J.L., et al.: Long-term complications associated with prosthetic repair of incisional hernias. Arch. Surg. 133(4), 378–382 (1998)

Leveen, H.H., Barberio, J.R.: Tissue Reaction to Plastics Used in Surgery with Special Reference to Teflon. Ann. Surg. 129(1), 74–84 (1949)

Liang, H.C., Chang, Y., et al.: Effects of crosslinking degree of an acellular biological tissue on its tissue regeneration pattern. Biomaterials 25(17), 3541–3552 (2004)

Losanoff, J.E., Richman, B.W., et al.: Entero-colocutaneous fistula: a late consequence of polypropylene mesh abdominal wall repair: case report and review of the literature. Hernia 6(3), 144–147 (2002)

Luijendijk, R.W., Hop, W.C., et al.: A comparison of suture repair with mesh repair for incisional hernia. N. Engl. J. Med. 343(6), 392–398 (2000)

Lyons, A.S., Joseph Petrucelli, R.: Medicine: an Illustrated History, New York (1987)

Mark, A., Malangoni, M., Michael, J., Rosen, M.D.: Hernias. In: Courtney, M., Townsend Jr., M. (eds.) Townsend: Sabiston Textbook of Surgery. Saunders Elsevier, Philadelphia (2007)

Matthews, B.D., Pratt, B.L., et al.: Assessment of adhesion formation to intra-abdominal polypropylene mesh and polytetrafluoroethylene mesh. J. Surg. Res. 114(2), 126–132 (2003)

McCormack, K., Scott, N.W., et al.: Laparoscopic techniques versus open techniques for inguinal hernia repair. Cochrane Database Syst. Rev.(1), CD001785 (2003)

McCormack, K.W.B., Fraser, C., Vale, L., Perez, J., Grant, A.: Transabdominal pre-peritoneal (TAPP) versus totally extraperitoneal (TEP) laparoscopic techniques for inguinal hernia repair: a systematic review. Hernia 9(2), 109–114 (2005)

Memon, M.A., Cooper, N.J., et al.: Meta-analysis of randomized clinical trials comparing open and laparoscopic inguinal hernia repair. Br. J. Surg. 90(12), 1479–1492 (2003)

Muschaweck, U.: Umbilical and epigastric hernia repair. Surg. Clin. North. Am. 83(5), 1207–1221 (2003)

Nathan, J.D., Pappas, T.N.: Inguinal hernia: an old condition with new solutions. Ann. Surg. 238(suppl. 6), S148–S157 (2003)

Nguyen, N.T., Lee, S.L., et al.: Laparoscopic umbilical herniorrhaphy. J. Laparoendosc Adv. Surg. Tech. A 10(3), 151–153 (2000)
Read, R.C.: Milestones in the history of hernia surgery: prosthetic repair. Hernia 8(1), 8–14 (2004)
Riazz, A.A., Ismail, M., Barsam, A., Bunce, C.J.: Mesh erosion into the bladder: A late complication of incisional hernia repair: A case report and review of literature. Hernia 8(2), 158–159 (2004)
Ridings, P., Evans, D.S.: The transabdominal pre-peritoneal (TAPP) inguinal hernia repair: a trip along the learning curve. J. R. Coll. Surg. Edinb. 45(1), 29–32 (2000)
Salameh, J.R.: Primary and unusual abdominal wall hernias. Surg. Clin. North. Am. 88(1), viii, 45–60 (2008)
Santora, T.A., Roslyn, J.J.: Incisional hernia. Surg. Clin. North. Am. 73(3), 557–570 (1993)
Schmidbauer, S., Ladurner, R., et al.: Heavy-weight versus low-weight polypropylene meshes for open sublay mesh repair of incisional hernia. Eur. J. Med. Res. 10(6), 247–253 (2005)
Schumpelick, V., Klinge, U., et al.: Incisional abdominal hernia: the open mesh repair. Langenbecks Arch. Surg. 389(1), 1–5 (2004)
Shell, D.H.t., de la Torre, J., et al.: Open repair of ventral incisional hernias. Surg. Clin. North. Am. 88(1), viii, 61–83 (2008)
Usher, F.C., Ochsner, J., et al.: Use of marlex mesh in the repair of incisional hernias. Am. Surg. 24(12), 969–974 (1958)
Velasco, M., Hidalgo, G.-U.M., Hidalgo, M., Vega, V., Cohen, M., Carnero, F.J.: Current concepts on adult umbilical hernia (1999)
Wellwood, J., Sculpher, M.J., et al.: Randomised controlled trial of laparoscopic versus open mesh repair for inguinal hernia: outcome and cost. BMJ 317(7151), 103–110 (1998)
Welty, G., Klinge, U., et al.: Functional impairment and complaints following incisional hernia repair with different polypropylene meshes. Hernia 5(3), 142–147 (2001)
Wright, D.M., Kennedy, A., et al.: Early outcome after open versus extraperitoneal endoscopic tension-free hernioplasty: a randomized clinical trial. Surgery 119(5), 552–557 (1996)
Yahchouchy-Chouillard, E., Aura, T., et al.: Incisional hernias. I. Related risk factors. Dig. Surg. 20(1), 3–9 (2003)
Yavuz, N., Ipek, T., et al.: Laparoscopic repair of ventral and incisional hernias: our experience in 150 patients. J. Laparoendosc. Adv. Surg. Tech. A 15(6), 601–605 (2005)

Author Index

CPSIA information can be obtained at www.ICGtesting.com
Printed in the USA
LVOW10*1212190114

370038LV00010B/440/P

9 783642 005336